Lecture Notes in Computer Science 8287

Commenced Publication in 1973
Founding and Former Series Editors:
Gerhard Goos, Juris Hartmanis, and Jan van Leeuwen

T0236544

Peter Widmayer   Yinfeng Xu   Binhai Zhu (Eds.)

Combinatorial Optimization and Applications

7th International Conference, COCOA 2013
Chengdu, China, December 12-14, 2013
Proceedings

 Springer

Volume Editors

Peter Widmayer
ETH Zürich, Institut für Theoretische Informatik
Universitätstr. 6, 8092 Zürich, Switzerland
E-mail: widmayer@inf.ethz.ch

Yinfeng Xu
Sichuan University, School of Management
Chengdu, Sichuan 610000, China
E-mail: yfxu@scu.edu.cn

Binhai Zhu
Montana State University, Department of Computer Science
Bozeman, MT 59717, USA
E-mail: bhz@cs.montana.edu

ISSN 0302-9743   e-ISSN 1611-3349
ISBN 978-3-319-03779-0   e-ISBN 978-3-319-03780-6
DOI 10.1007/978-3-319-03780-6
Springer Cham Heidelberg New York Dordrecht London

Library of Congress Control Number: 2013954447

CR Subject Classification (1998): F.2, G.2.2, G.2, G.1.6, I.2.8, C.2

LNCS Sublibrary: SL 1 – Theoretical Computer Science and General Issues

Typesetting: Camera-ready by author, data conversion by Scientific Publishing Services, Chennai, India

Printed on acid-free paper

Springer is part of Springer Science+Business Media (www.springer.com)

Preface

The papers in this volume were presented at the 9th International Conference on Combinatorial Optimization and Applications (COCOA 2013), held during December 12–14, 2013, in Chengdu, China. It is expected that starting this year, the COCOA conferences will be held in winter instead of summer. The topics cover most areas in algorithms, combinatorial optimization, and their applications.

Submissions to the conference this year were conducted electronically. A total of 72 papers were submitted, of which 36 were accepted. The papers were evaluated by an international Program Committee overseen by the Program Committee co-chairs: Peter Widmayer, Yinfeng Xu, and Binhai Zhu. The Program Committee consists of Tatsuya Akutsu, Laurent Bulteau, Yongxi Cheng, Marek Chrobak, Shantanu Das, Bhaskar DasGupta, Yucheng Dong, Thomas Erlebach, Zhiping Fan, Mordecai Golin, Jiong Guo, Xiaodong Hu, Haitao Jiang, Iyad Kanj, Naoki Katoh, Adrian Kosowski, Elias Koutsoupias, Michael Langston, Guohui Lin, Xiaowen Liu, Bernard Mans, Matus Mihalak, Nicola Santoro, Jianxin Wang, Hamid Zarrabi-Zadeh, and Shanfeng Zhu. It is expected that most of the accepted papers will appear in a more complete form in scientific journals.

The submitted papers were from 25 countries/regions: Australia, Brazil, Canada, China, Denmark, France, Germany, Hong Kong, India, Iran, Japan, Korea, Lebanon, Malaysia, The Netherlands, Philippines, Poland, Singapore, Spain, Switzerland, Taiwan, Tunisia, Turkey, UK and USA. On average, each paper was evaluated by three Program Committee members, assisted in some cases by sub-reviewers. In addition to the selected papers, the conference also included two invited presentations by Franz Aurenhammer and Dimitrios Thilikos.

We thank all the people who made this meeting possible: the authors for submitting papers, the Program Committee members and external reviewers (listed in the proceedings) for their excellent work, and the two invited speakers. Finally, we thank NSF of China and Sichuan University for their support and the local organizers and colleagues for their assistance.

December 2013

Peter Widmayer
Yinfeng Xu
Binhai Zhu

Organization

Program Committee Co-chairs

Peter Widmayer   ETH Zurich, Switzerland
Yinfeng Xu   Sichuan University, China
Binhai Zhu   Montana State University, USA

Program Committee Members

Tatsuya Akutsu   Kyoto University, Japan
Laurent Bulteau   Université de Nantes, France
Yongxi Cheng   Xi'an Jiao Tong University, China
Marek Chrobak   University of California at Riverside, USA
Shantanu Das   Aix-Marseille Université, France
Bhaskar DasGupta   University of Illinois at Chicago, USA
Yucheng Dong   Sichuan University, China
Thomas Erlebach   University of Leicester, UK
Zhiping Fan   Northeasten University, China
Mordecai Golin   HKUST, Hong Kong
Jiong Guo   Universität des Saarlandes, Germany
Xiaodong Hu   Chinese Academy of Sciences, China
Haitao Jiang   Shandong University, China
Iyad Kanj   DePaul University, USA
Naoki Katoh   Kyoto University, Japan
Adrian Kosowski   Inria Bordeaux, France
Elias Koutsoupias   University of Athens, Greece
Michael Langston   University of Tennessee, USA
Guohui Lin   University of Alberta, Canada
Xiaowen Liu   Indiana University-Purdue University
     Indianapolis, USA
Bernard Mans   Macquarie University, Australia
Matus Mihalak   ETH Zurich, Switzerland
Nicola Santoro   Carleton University, Canada
Jianxin Wang   Central South University, China
Hamid Zarrabi-Zadeh   Sharif University of Technology, Iran
Shanfeng Zhu   Fudan University, China

Organizing Committee

Jiuping Xu   Sichuan University, China
Xin Gu   Sichuan University, China
Guanqun Ni   Sichuan University, China

Additional Reviewers

Sepehr Assadi   Mostafa Nouri Baygi   Michael Borokhovich
Yixin Cao   Arnaud Casteigts   Janos Csirik
Ehsan Emamjomeh-Zadeh   Guillaume Fertin   Florent Foucaud
Emanuele G. Fusco   Emmanuel Godard   Ronald Hagan
Yuya Higashikawa   Yasushi Kawase   Naoyuki Kamiyama
Christian Komusiewicz   Miroslaw Korzeniowski   Arnaud Labourel
Akaki Mamageishvili   Euripides Markou   Monaldo Mastrolilli
Luke Mathieson   Yoshio Okamoto   Yota Otachi
Linda Pagli   Dominik Pajak   Guillem Perarnau
Charles Phillips   Chung Keung Poon   Maurice Queyranne
Andre van Renssen   Gary Rogers   Rahmtin Rotabi
Kaveh Shahbaz   Farhad Shahmohammadi   Zuzanna Stamirowska
Weitian Tong   Yushi Uno   Przemyslaw Uznanski
Giovanni Viglietta   Kai Wang   Tony Wirth
Xingang Wen   Yongjie Yang

Invited Lectures

Theory and Applications of Bidimensionality*

Dimitrios M. Thilikos[1]

AlGCo project-team, CNRS, LIRMM
Department of Mathematics, National and Kapodistrian University of Athens

Bidimensionality Theory is a meta-algorithmic theory whose main ingredients are the *Grid Exclusion Theorem* of the Graph Minors series of Robertson and Seymour and the celebrated *Courcelle's Theorem*. The grid exclusion theorem states that if a graph excludes a bidimensional grid as a minor (a graph H is a *minor* of a graph G if H can be obtained by some subgraph of G by contracting edges) then its structure topologically resembles the structure of a tree (in technical terms, it has small treewidth). Intuitively, this result says that the absence of the bidimensional structure of a grid implies that a graph has the "mono-dimensional" structure of a tree. On the other side, Courcelle's theorem states that if a problem on graphs is expressible in Monadic Second Order Logic (MSOL), then it is possible to solve it in linear time when the treewidth of their input graphs is fixed. Intuitively, this theorem expresses the fact that the mono-dimensional structure of a tree (i.e., small treewidth) makes it possible to treat a graph as the input string of a finite-state tree automaton, where the finiteness of its states is guaranteed by the MSOL-expressibility of the problem. Combining these two results together, we derive that the *absence* of the bidimensional structure of a grid, enables the applicability of the "divide-and-conquer" technique for problems of certain descriptive complexity. It appears that for many graph theoretic problems the existence of a grid-minor (or other bidimensional strucures) on the input graph provides a certificate for an immediate negative (or positive) answer and, for the remaining instances, a dynamic programming approach on graphs of bounded treewidth may give an answer to the problem. This phenomenon reveals fruitful interleave between graph structure and logic in graph algorithms. Bidimensionality Theory aims at systematizing this idea and extending its applicability in diverse paradigms of algorithm design.

The notion of problem bidimensionality was proposed for the first time in [2]. Given some graph invariant $\mathbf{p}$, we denote by $\Pi_{\mathbf{p}}$ the problem of asking, for some pair (G, k), whether $\mathbf{p}(G) \leq k$ and we say that $\Pi_{\mathbf{p}}$ is *minor bidimensional* if

i) $\mathbf{p}$ is minor closed, i.e. if G_1 is a minor of G_2, then $\mathbf{p}(G_1) \leq \mathbf{p}(G_2)$.

ii) for every k, $\mathbf{p}(L_k) = \Omega(k^2)$ (here, L_k is the $(k \times k)$-grid).

* This work was co-financed by the European Union (European Social Fund - ESF) and Greek national funds through the Operational Program "Education and Lifelong Learning" of the National Strategic Reference Framework (NSRF) - Research Funding Program: "Thales. Investing in knowledge society through the European Social Fund.

Some of the main meta-algorithmic results of Bidimensionality Theory can be summarized as follows: Let $\Pi_{\mathbf{p}}$ be a minor bidimensional problem and let $\mathcal{G}$ be a class of graphs where

$$\forall_{G \in \mathcal{G}} \ treewidth(G) = O(\max\{k \mid L_k \text{ is a minor of } G\}). \tag{1}$$

Let $\Pi_{\mathbf{p}}^{\mathcal{G}}$ be the restriction of $\Pi_{\mathbf{p}}$ to the graphs in $\mathcal{G}$. Then the following hold:

1. If $\mathbf{p}(G)$ can be computed in $2^{O(treewidth(G))} \cdot n^{O(1)}$ steps, then there exists an algorithm that decides, given a graph G in $\mathcal{G}$ as input, whether $\mathbf{p}(G) \leq k$, in $2^{O(\sqrt{k})} \cdot n^{O(1)}$ steps (see [2]).
2. If $\mathbf{p}$ satisfies some separability property (see [3, 5] for the definition) and $\mathbf{p}(G) \leq k \iff \exists S \subseteq V(G) : |S| \leq k$ and $(G, S) \models \psi$ where ψ is a MSOL sentence, then the problem $\Pi_{\mathbf{p}}$ admits a *linear kernel*, i.e., there exists a polynomial algorithm reducing every instance (G, k) of $\Pi_{\mathbf{p}}$ to an equivalence instance (G', k') where $|V(G')| = O(k)$ and $k' \leq k$ (see [5]).
3. If $\mathbf{p}$ satisfies some separability property and is reducible (in the sense these notions are defined in [7]), then there is an EPTAS for computing $\mathbf{p}(G)$ on the graphs in $\mathcal{G}$ (see [3, 7]).

According to [4], every graph class excluding some fixed graph as a minor satisfies (1). Further extensions of the applicability of the above theory on geometric graphs have been given in [8] and [1].

All above results concern only minor-closed parameters. The counterpart of the above theory for contraction-closed parameters is based on the notion of *contraction bidimensionality*, uses slightly different versions of Conditions **i**, **ii**, and (1), and its algorithmic potential is investigated in [6]. Currently the combinatorial challenge of Bidimensionality Theory is to broaden its applicability by detecting graph classes where (1) holds or to suitable adapt/extend/modify Conditions **i**, **ii**, and (1) so that similar results can be derived for wider families of graph theoretic problems.

References

1. D.M. Thilikos, A. Grigoriev, A. Koutsonas. Bidimensionality of geometric intersection graphs. *CoRR*, arXiv:1308.6166, August 2013.
2. E.D. Demaine, F.V. Fomin, M. Hajiaghayi, and D.M. Thilikos. Subexponential parameterized algorithms on graphs of bounded genus and H-minor-free graphs. *Journal of the ACM*, 52(6):866–893, 2005.
3. E.D. Demaine and M. Hajiaghayi. Bidimensionality: new connections between FPT algorithms and PTASs. In *16th Annual ACM-SIAM Symposium on Discrete Algorithms (SODA 2005)*, pages 590–601, 2005.
4. E.D. Demaine and M. Hajiaghayi. Linearity of grid minors in treewidth with applications through bidimensionality. *Combinatorica*, 28(1):19–36, 2008.
5. F. V. Fomin, D. Lokshtanov, S. Saurabh, and D.M. Thilikos. Bidimensionality and kernels. In *21st Annual ACM-SIAM Symposium on Discrete Algorithms (SODA 2010)*, pages 503–510. ACM-SIAM, 2010.

6. F.V. Fomin, P.A. Golovach, and D.M. Thilikos. Contraction obstructions for treewidth. *J. Comb. Theory, Ser. B*, 101(5):302–314, 2011.
7. F.V. Fomin, D. Lokshtanov, V. Raman, and S. Saurabh. Bidimensionality and EPTAS. In *22st ACM–SIAM Symposium on Discrete Algorithms (SODA 2011)*, pages 748–759, 2011.
8. F. V. Fomin, D. Lokshtanov, and S. Saurabh. Bidimensionality and geometric graphs. In *Proceedings of the 23rd Annual ACM-SIAM Symposium on Discrete Algorithms, (SODA 2012)*, pages 1563–1575, 2012.

Recent Trends on Voronoi Diagrams

Franz Aurenhammer

Institute for Theoretical Computer Science,
University of Technology, Graz, Austria
auren@igi.tugraz.at

The Voronoi diagram is a versatile geometric graph whose usefulness as a geometric data structure is widely appreciated; see e.g. [1]. Given a set of n point sites in the Euclidean plane, their Voronoi diagram allots to each site the region of the plane at closest distance to it. These regions are bordered by edges that form a planar straight-line graph of size only $O(n)$, which can be computed in optimal time $O(n \log n)$ with various algorithmic techniques.

Voronoi diagrams have been a topic of ongoing interest, within and outside computer science. After reviewing some of their basic properties and methods for computing them, we present recent developments on this structure and its relatives. These include new construction methods, as well as generalizations which lead to scenarios quite different from the classical concept.

We discuss a novel divide & conquer approach to computing Voronoi diagrams, which actually delivers the medial axis of a planar shape, but directly applies to Voronoi diagrams for point sites, and also to sites of rather general shape. We then turn to the circle offset model (or growth model) of the Voronoi diagram, which captures the concept of weighting the sites, and leads to complex phenomena that recently have been dealt with using so-called abstract Voronoi diagrams. Non-circular offset models even lead to 'non-Voronoi' straight-line graphs, known as the straight skeleton, where we will mainly consider the three-dimensional case with its peculiarities concerning offsets of polytopes, and recent solutions. We conclude with some new results on visibility Voronoi diagrams, where distances to the sites are influenced by visibility constraints. Such constraints may stem from prespecified geometric objects, or may be implicit in the definition of the distance function. The latter case arises for quasi-Euclidean distances, which have a physical meaning in relativity theory.

Reference

1. F. Aurenhammer, R. Klein, and D.T. Lee. *Voronoi Diagrams and Delaunay Triangulations*. World Scientific, Singapore, 2013.

Table of Contents

Contributed Papers

Parameterized and Approximation Algorithms for Finding Two Disjoint Matchings

Zhi-Zhong Chen[1], Ying Fan[2], and Lusheng Wang[2]

[1] Division of Information System Design, Tokyo Denki University,
Hatoyama, Saitama 350-0394, Japan
zzchen@mail.dendai.ac.jp
[2] Department of Computer Science, City University of Hong Kong,
Tat Chee Avenue, Kowloon, Hong Kong SAR
yingying1988@gmail.com, lwang@cs.cityu.edu.hk

Abstract. We first present a fixed-parameter algorithm for the NP-hard problem of deciding if there are two matchings M_1 and M_2 in a given graph G such that $|M_1| + |M_2|$ is no less than a given number k. The algorithm runs in $O\left(m + k \cdot k! \cdot \left(2\sqrt{2}\right)^k \cdot n^2 \log n\right)$ time, where n (respectively, m) is the number of vertices (respectively, edges) in G. We then present a combinatorial approximation algorithm for the NP-hard problem of finding two disjoint matchings in a given edge-weighted graph G so that their total weight is maximized. The algorithm achieves an approximation ratio of roughly 0.76 and runs in $O\left(m + n^3\alpha(n)\right)$ time, where α is the inverse Ackermann function.

Keywords: Fixed-parameter algorithms, approximation algorithms, graph algorithms, matchings, NP-hardness.

1 Introduction

Throughout this paper, a graph means an undirected graph that may have parallel edges but no self-loops. A graph is *simple* if it has no parallel edges. A *matching* in a graph G is a set F of edges in G such that no two edges in F share an endpoint. A *maximum matching* in G is a matching in G whose cardinality is maximized over all matchings in G. Given a graph G, the *maximum matching problem* (MM for short) requires the computation of a maximum matching in G. MM is very fundamental in many areas and has been extensively studied in the literature.

In this paper, we consider a generalization of MM, called the *maximum two-matching problem* (MTM for short). Given a graph G, MTM requires the computation of two disjoint matchings in G whose total cardinality is maximized. Motivated by call admittance issues in satellite based telecommunication networks, Feige *et al.* [5] introduced MTM (among others) and showed its *APX*-hardness. They also observed that MTM is obviously a special case of the well-known *maximum coverage problem* (see [10]): We wish to cover the maximum number of edges of a given graph G with two sets each of which is a matching

P. Widmayer, Y. Xu, and B. Zhu (Eds.): COCOA 2013, LNCS 8287, pp. 1–12, 2013.

of G. Since this special case of the maximum coverage problem can be approximated by a greedy algorithm within a ratio of 0.75 [10], so can be MTM. They then gave a randomized approximation algorithm for MTM that achieves an expected ratio of $\frac{10}{13} \approx 0.769$. Their algorithm is based on an LP approach and random rounding. In particular, their LP has an exponential number of constraints and hence can only be solved by using the ellipsoid method together with a separation oracle. Hence, their algorithm is extremely slow although its running time is polynomial.

The *simple* case of MTM (SMTM for short) where the input graph is simple has been studied recently [5,12,3,2,11,1]. Feige et al. [5] gave a simple approximation algorithm for SMTM that achieves a ratio of 0.8. This ratio was then improved in a series of papers [12,3,2,11,1]. The best known ratio achieved by a polynomial-time approximation algorithm for SMTM is roughly 0.842 [1]. All known approximation algorithms for SMTM start by using Hartvigsen's polynomial-time algorithm [8] to compute a maximum-sized subgraph H of the input graph such that the degree of each vertex in H is at most 2 and there is no cycle of length 3 in H. Unfortunately, Hartvigsen's algorithm only works for simple graphs.

In this paper, we first consider the parameterized complexity of MTM. We show that MTM is fixed-parameter tractable by designing an algorithm that checks, in $O\left(m + k \cdot k! \cdot \left(2\sqrt{2}\right)^k \cdot n^2 \log n\right)$ time, if a given n-vertex m-edge graph G contains two disjoint matchings M_1 and M_2 such that $|M_1| + |M_2|$ is no less than a given number k. Our algorithm first reduces the problem for a given input (G, k) to the problem for (G', k) such that G' has a vertex set U with $|U| < k$ and $G' - U$ is edgeless. It then solves the problem for (G', k) with the help of Gabow's algorithm [7] for the maximum-weight degree-constrained subgraph problem.

We then consider the weighted version of MTM (MWTM for short), where each edge of the input graph G is given a nonnegative weight and the goal is to find two disjoint matchings whose total weight is maximized. MWTM is also a special case of the maximum coverage problem: We wish to cover the maximum-weight set of edges of a given graph G with two sets each of which is a matching of G. Since this special case of the maximum coverage problem can be approximated by a greedy algorithm within a ratio of 0.75 [10], so can be MWTM. However, all the ideas used in the known approximation algorithms [5,12,3,2,11,1] for MTM cannot be applied to MWTM because the algorithms call Hartvigsen's algorithm [8] which only works for nonweighted simple graphs.

We can observe that the algorithm of Feige *et al.* [5] for MTM can be slightly modified into a *randomized* approximation algorithm for MWTM that achieves an *expected* ratio of $\frac{10}{13} \approx 0.769$. However, as mentioned before, Feige *et al.*'s algorithm is extremely slow. So, in this paper, we present a completely new (*deterministic*) approximation algorithm for MWTM that achieves a ratio of roughly 0.76. Our new algorithm is combinatorial and runs in $O(m + n^3 \alpha(n))$ time, where α is the inverse Ackermann function. The algorithm is motivated by the approaches developed in [14,9,4] for the *maximum traveling salesman problem* which is the problem of finding a maximum-weight Hamiltonian cycle in a given edge-weighted complete graph.

Due to lack of space, some proofs are omitted.

2   Basic Definitions

Let G be a graph. We denote the vertex set of G by $V(G)$, and denote the edge set of G by $E(G)$. For a subset U of $V(G)$, $G - U$ denotes the graph obtained from G by removing the vertices in U (together with the edges incident to them). For a subset F of $E(G)$, $G - F$ denotes the graph obtained from G by removing the edges in F. The *degree* of a vertex v in G is the number of edges incident to v in G. Two edges of G are *adjacent* if they have at least one common endpoint.

A *cycle* in G is a connected subgraph of G in which each vertex is of degree 2. A *path* in G is either a single vertex of G or a connected subgraph of G in which exactly two vertices are of degree 1 and the others are of degree 2. The *length* of a cycle or path C is the number of edges in C and is denoted by $|C|$. A *k-cycle* is a cycle of length k. If the length of a cycle or path P is odd, then we say that P is *odd*; otherwise, we say that P is *even*. A *2-matching* of G is a subgraph H of G with $V(H) = V(G)$ in which the degree of each vertex is at most 2. Note that each connected component of a 2-matching is a path or cycle. A 2-matching $\mathcal{C}$ of G is *even* if each cycle in $\mathcal{C}$ is even. A *semi-path set* of G is a set F of edges in G such that each connected component of the graph $(V(G), F)$ is a path or a 2-cycle. A *matching* of G is a (possibly empty) set of pairwise nonadjacent edges of G. A *perfect matching* of G is a matching M of G such that each vertex of G is incident to an edge in M. An *independent set* of G is a set of vertices no two of which are adjacent in G.

3   The Parameterized Algorithm for MTM

Throughout this section, fix a graph G and a nonnegative integer k. We want to decide if G has two disjoint matchings M_1 and M_2 with $|M_1| + |M_2| \geq k$. In other words, we want to decide if G has an even 2-matching with at least k edges. To this end, we first perform the following five steps:

1. For each pair $\{u, v\}$ of vertices in G such that G has more than two edges between u and v, remove all but two edges between u and v from G. (*Comment:* This step removes redundant edges from G because a 2-matching of G uses at most two edges between each pair of vertices in G. After this step, G has $O(n^2)$ edges, where n is the number of vertices in G.)
2. Initialize $b = k$, $H = G$, and $\mathcal{C} = (V(G), \emptyset)$.
3. While $b > 0$ and H has at least one edge, perform the following two steps:
   (a) Add an arbitrary edge $\{u, v\}$ of H to $\mathcal{C}$, delete $\{u, v\}$ from H, and decrease b by 1.
   (b) Delete from H all edges e such that the graph obtained from $\mathcal{C}$ by adding e has a connected component that is not a path.

4. If $b = 0$, then output "yes" and halt.
5. Obtain a set I of vertices in $\mathcal{C}$ by initializing $I = \emptyset$ and then for each connected component P of $\mathcal{C}$, adding to I an arbitrary vertex of P whose degree in $\mathcal{C}$ is at most 1. (*Comment:* I contains all vertices of degree 0 in $\mathcal{C}$.)

Obviously, if our algorithm halts in Step 4, then $\mathcal{C}$ is an even 2-matching (indeed, a collection of vertex-disjoint paths) of G with k edges. So, for further discussion, we assume that our algorithm does not halt in Step 4.

Lemma 1. *I is an independent set of G.*

Lemma 2. *Let $U = V(G) - I$. Then, U contains at most $k - 1$ vertices.*

Our algorithm then constructs an edge-weighted graph by performing the following step:

6. Let $\mathcal{U}$ be the edge-weighted graph whose vertex set is U and whose edge set is constructed as follows.
   (a) For each edge e of G between two vertices of U, add e to $\mathcal{U}$ and assign a weight of 1 to e.
   (b) For each vertex $v \in I$ and for each (unordered) pair $\{u_1, u_2\}$ of distinct vertices in U such that both u_1 and u_2 are adjacent to v in G, add an edge between u_1 and u_2 to $\mathcal{U}$ and assign a weight of 2 to it.

Note that since $\mathcal{U}$ may have parallel edges, we need to assign distinct labels to the edges of $\mathcal{U}$ in order to distinguish them. So, each edge e of $\mathcal{U}$ has two endpoints, a weight, and a label. For convenience, we say that two even 2-matchings $\mathcal{C}_1$ and $\mathcal{C}_2$ of $\mathcal{U}$ are *the same* if ignoring the labels of edges of $\mathcal{C}_1$ and $\mathcal{C}_2$ yields the same graph. If P is a path or cycle in $\mathcal{U}$, then the *weight* of P is the total weight of edges in P. A 2-matching $\mathcal{C}$ in $\mathcal{U}$ is *even* if the weight of each cycle of $\mathcal{C}$ is even, and is *properly marked* if no vertex of degree at least 1 in $\mathcal{C}$ is marked but zero or more vertices of degree 0 in $\mathcal{C}$ are marked.

Lemma 3. *$\mathcal{U}$ has at most $(k - 1)! \cdot \left(2\sqrt{2}\right)^{k-1}$ distinct properly marked even 2-matchings.*

Proof. A simple way of enumerating all properly marked even 2-matchings in $\mathcal{U}$ is as follows.

First, we enumerate all partitions of U into cyclically ordered subsets. It is widely known that there are exactly $|U|!$ such partitions. So, there are at most $(k - 1)!$ such partitions for $|U| < k$.

Next, for each partition $\mathcal{P}$ of U into cyclically ordered subsets, we try all possible ways to transform $\mathcal{P}$ into a properly marked even 2-matching of $\mathcal{U}$. To see the details, let s be the number of singleton subsets in $\mathcal{P}$, and $S_1, \ldots, S_h$ be the nonsingleton subsets in $\mathcal{P}$. Consider an arbitrary $i \in \{1, \ldots, h\}$ and let $n_i = |S_i|$. Since S_i is cyclically ordered, we can view S_i as a cycle. Note that we can transform S_i into a path or a cycle of $\mathcal{U}$. To transform S_i into a cycle of $\mathcal{U}$, we have at most 2^{n_i} ways because S_i has n_i edges and we have at most two choices to handle each edge e with endpoints u_1 and u_2 in S_i as follows:

- If either $\mathcal{U}$ has no edge with endpoints u_1 and u_2, or $\mathcal{U}$ has only one edge with endpoints u_1 and u_2 but S_i is a 2-cycle, then there is no way to transform S_i into a cycle of $\mathcal{U}$.
- If all edges with endpoints u_1 and u_2 in $\mathcal{U}$ have the same weight, then the only choice is to let e have the same weight as the edges.
- If $\mathcal{U}$ has two edges f_1 and f_2 with endpoints u_1 and u_2 such that f_1 and f_2 have different weights (namely, 1 and 2), then we have two choices of assigning a weight (namely, 1 or 2) to e.

Note that even if we obtain a cycle C_i of $\mathcal{U}$ after handling each edge e of S_i as above, C_i may not be an even cycle and we just discard it if so.

To transform S_i into a path of $\mathcal{U}$, we first have n_i choices to break S_i into a path P_i, and then have at most 2^{n_i-1} ways to transform P_i into a path of $\mathcal{U}$ because P_i has $n_i - 1$ edges and we have at most two choices to handle each edge e of P_i as in the case of transforming S_i into a cycle of $\mathcal{U}$.

In total, there are at most $2^{n_i} + 2^{n_i-1}n_i$ ways to transform S_i into a path or an even cyle of $\mathcal{U}$. Thus, in total, there are at most $\prod_{i=1}^{h}\left(2^{n_i} + 2^{n_i-1}n_i\right) \leq 2^{k-1-s}\prod_{i=1}^{h}\left(1 + \frac{n_i}{2}\right)$ ways to transform $\mathcal{P}$ into an even 2-matching of $\mathcal{U}$, where the inequality holds because $\sum_{i=1}^{h} n_i \leq |U| - s < k - s$.

Recall that for each $i \in \{1, \ldots, h\}$, $n_i \geq 2$. Moreover, $\sum_{i=1}^{h} n_i \leq k-1-s$. We claim that these facts imply that $\prod_{i=1}^{h}\left(1 + \frac{n_i}{2}\right) \leq 2^{\frac{k-1-s}{2}}$ if $k - 1 - s$ is even, while $\prod_{i=1}^{h}\left(1 + \frac{n_i}{2}\right) \leq 2.5 \cdot 2^{\frac{k-4-s}{2}}$ if $k - 1 - s$ is odd. To see this claim, first note that for every even integer $m \geq 2$, $1 + \frac{m}{2} \leq \left(1 + \frac{2}{2}\right)^{\frac{m}{2}}$. Moreover, for every odd integer $m \geq 3$, $1 + \frac{m}{2} \leq \left(1 + \frac{2}{2}\right)^{\frac{m-3}{2}}\left(1 + \frac{3}{2}\right)$. Thus, under the conditions that $n_1 \geq 2, \ldots, n_h \geq 2$, and $\sum_{i=1}^{h} n_i \leq k - 1 - s$, the value of $\prod_{i=1}^{h}\left(1 + \frac{n_i}{2}\right)$ is maximized

- at $(n_1, n_2, \ldots, n_h) = (2, 2, \ldots, 2)$ if $k - 1 - s$ is even,
- at $(n_1, n_2, \ldots, n_h) = (3, 2, \ldots, 2)$ if $k - 1 - s$ is odd.

Therefore, the claim holds. By the claim, $\prod_{i=1}^{h}\left(1 + \frac{n_i}{2}\right) \leq 2^{\frac{k-1-s}{2}}$ no matter whether $k - 1 - s$ is even or odd. Hence, there at most $\left(2\sqrt{2}\right)^{k-1-s}$ ways to transform $\mathcal{P}$ into an even 2-matching of $\mathcal{U}$. Obviously, for each properly marked even 2-matching C transformed from $\mathcal{P}$, there are exactly 2^s ways to properly mark C. So, there are at most $\left(2\sqrt{2}\right)^{k-1-s} \cdot 2^s \leq \left(2\sqrt{2}\right)^{k-1}$ ways to transform $\mathcal{P}$ into a properly marked even 2-matching of $\mathcal{U}$. Now, since there are $(k - 1)!$ $\mathcal{P}$'s in total, the lemma holds.   ∎

Finally, our algorithm uses $\mathcal{U}$ to check if G has an even 2-matching with at least k edges by performing the following two steps:

7. For each properly marked even 2-matching C of $\mathcal{U}$, perform the following steps:

    (a) Construct an edge-weighted simple bipartite graph B_C as follows:

- The vertex set of B_C is $I \cup U_0 \cup U_1 \cup U_2$, where U_0 (respectively, U_1) consists of all $u \in U$ whose degree in C is 0 (respectively, 1) and $U_2 = \{u_e \mid e$ is an edge of weight 2 in $C\}$.
- For each edge e of weight 2 in C and for each vertex $v \in I$ such that v is adjacent to both endpoints of e in G, B_C has an edge of weight 0 between v and u_e.
- For each edge e of G such that one endpoint of e is in I and the other is a vertex of U_1 or an unmarked vertex of U_0, B_C has an edge of weight 1 between the endpoints of e.
- For each 2-cycle C of G such that one vertex of C is in I and the other is a marked vertex of U_0, B_C has an edge of weight 2 between the vertices of C.

(b) Compute a maximum-weight subgraph S_C of B_C such that (1) the degree of each vertex $v \in I \cup U_1$ in S_C is at most 1, (2) the degree of u_e in S_C is exactly 1 for each edge e of weight 2 in C, (3) the degree of each marked vertex $u \in U_0$ in S_C is exactly 1, and (4) the degree of each unmarked vertex $u \in U_0$ in S_C is at most 2.

(c) If S_C was found in Step 7b and the sum of the weights of C and S_C is at least k, then output "yes" and halt.

8. Output "no" and halt.

Lemma 4. *G has an even 2-matching with at least k edges if and only if our algorithm outputs "yes" in Step 7c for some properly marked even 2-matching C of $\mathcal{U}$.*

Theorem 1. *Given a nonnegative integer k and a graph G with n vertices and m edges, it takes $O\left(m + k \cdot k! \cdot \left(2\sqrt{2}\right)^k n^2 \log n\right)$ time to decide whether G has two disjoint matchings M_1 and M_2 such that $|M_1| + |M_2| \geq k$.*

4   The Approximation Algorithm for MWTM

Throughout this section, fix an instance (G, w) of MWTM, where G is an n-vertex m-edge graph and w is a function mapping each edge e of G to a nonnegative real number $w(e)$. After a simple $O(m + n^2)$-time preprocessing, we can assume that for every two vertices u and v of G, there are exactly two edges between u and v in G. To see that no generality is lost with this assumption, first observe that if there are three or more edges between two vertices u and v in G, then we can delete all but the heaviest two edges between u and v from G. On the other hand, if there is at most one edge between two vertices u and v in G, then we can add one or more edges of weight 0 between u and v so that G has exactly two edges between them. The *mate* of an edge e in G is the other edge in G that has the same endpoints as e. We may further assume that n is even, because if n is odd, we can add a new vertex and connect it to each of the other vertices with two edges of weight 0. For a subset F of $E(G)$, $w(F)$ denotes $\sum_{e \in F} w(e)$. The *weight* of a subgraph H of G is $w(H) = w(E(H))$.

Note that MWTM is equivalent to the problem of computing a maximum-weight even 2-matching of a given graph.

For a random event A, $\Pr[A]$ denotes the probability that A occurs. For two random events A and B, $\Pr[A \mid B]$ denotes the (conditional) probability that A occurs given the known occurrence of event B. For a random variable X, $\mathcal{E}[X]$ denotes the expected value of X.

In the remainder of this section, we first design a randomized approximation algorithm for MWTM and then derandomize it. Section 4.1 gives an outline of the randomized algorithm. Section 4.2 then describes the details that are missing in the outline. Section 4.3 estimates the time complexity and the expected approximation ratio achieved by the randomized algorithm. Finally, Section 4.4 derandomizes the algorithm.

4.1 Outline of the Algorithm

Our algorithm starts by computing a maximum-weight 2-matching $\mathcal{C}$ and a maximum-weight matching M of G. We may assume that M is a perfect matching of G because n is even. Our algorithm then uses $\mathcal{C}$ and M to perform a preprocessing as follows.

1. Construct a graph K as follows: Initially, K is the graph $(V(G), M)$. Next, for every 2-cycle C in $\mathcal{C}$, add the heavier edge of C to K. (*Comment:* Each connected component of K is a path, a 2-cycle, or an even cycle of length 4 or more.)
2. Modify $\mathcal{C}$ by performing the following step for every cycle C' of K with $|C'| \geq 4$:
   (a) Delete all 2-cycles C from $\mathcal{C}$ such that one edge of C appears in C'.
   (b) Add C' to $\mathcal{C}$.

Obviously, $\mathcal{C}$ remains to be a 2-matching of G after the preprocessing. So, the preprocessing does not increase $w(\mathcal{C})$ because $\mathcal{C}$ was originally a maximum-weight 2-matching of G. The preprocessing does not decrease $w(\mathcal{C})$ either, because otherwise we would be able to obtain a heavier matching of G than M by modifying it by deleting the edges added to $\mathcal{C}$ in the preprocessing while adding the edges deleted from $\mathcal{C}$ in the preprocessing.

We hereafter assume that our algorithm has done the preprocessing. If $\mathcal{C}$ is now even, then our algorithm just outputs $\mathcal{C}$ and stops. In the remainder of this paper, we assume that $\mathcal{C}$ is not an even 2-matching of G. Then, $\mathcal{C}$ has at least two connected components. Suppose that T is a maximum-weight even 2-matching of G. Let T_{int} denote the set of all edges $\{u, v\}$ of T such that some cycle C in $\mathcal{C}$ contains both u and v. Let T_{ext} denote the set of edges in T but not in T_{int}. Let $\beta = w(T_{\text{int}})/w(T)$.

Our algorithm then computes three even 2-matchings T_1, T_2, T_3 of G, outputs the heaviest one among them, and stops. T_1 is computed by modifying the odd cycles in $\mathcal{C}$ as follows. Fix a parameter $0 < \epsilon < 1$. For each odd cycle C in $\mathcal{C}$, if C has more than ϵ^{-1} edges, then remove the minimum-weight edge; otherwise, replace C by a maximum-weight even 2-matching of the graph obtained from G by deleting all vertices not in C. Then, C becomes an even 2-matching. Obviously, we have the following fact:

Fact 2. $w(T_1) \geq (1 - \epsilon)w(T_{\text{int}}) = (1 - \epsilon)\beta w(T)$.

When $w(T_{\text{ext}})$ is large, $w(T_{\text{int}})$ is small and $w(T_1)$ may be small, too. The two even 2-matchings T_2 and T_3 together are aimed at the case where $w(T_{\text{ext}})$ is large. Their computation is given in the next subsection.

4.2   Computation of T_2 and T_3

To compute T_2 and T_3, we first perform the following two steps:

1. Compute a maximum-weight matching M' in an auxiliary graph H, where $V(H) = V(G)$ and $E(H)$ consists of those $\{u, v\} \in E(G)$ such that u and v belong to different connected components of C.
2. Fix an arbitrary ordering $C_1, \ldots, C_r$ of the connected components of C such that the 2-cycles precede the others. (*Comment:* Since C is a maximum-weight 2-matching of G and the weight of each edge in G is nonnegative, the weight of the edge between the endpoints of each path P that is a connected component of C is 0. So, we can change P into a cycle by adding the edge between its endpoints without changing its weight. In the remainder of this paper, for ease of explanation, we assume that each connected component of C is a cycle.)

We then process the cycles $C_1, \ldots, C_r$ in turn. Roughly speaking, the processing can be sketched as follows:

3. For $i = 1, 2, \ldots, r$ (in this order), process C_i by performing the following:
   (a) Mark some suitable edges $\{u, v\} \in M'$ with $\{u, v\} \cap V(C_i) \neq \emptyset$.
   (b) Move some suitable edges of C_i to M while always maintain that M is a semi-path set of G.

To detail Substeps 3a and 3b, we need several definitions and lemmas. In the remainder of this paper, for each integer $i \in \{1, \ldots, r\}$, the phrase "at time i" means the time at which $C_1, \ldots, C_{i-1}$ have been processed and C_i is the next cycle to be processed. For each integer $i \in \{1, \ldots, r\}$, let M_i be the set M at time i. For convenience, let M_{r+1} and C_{r+1} be the values of M and C immediately after Step 3, respectively.

A set F of edges in G is *available at time i* if $F \subseteq E(C_i)$, $F \cap M_1 = \emptyset$, and $M_i \cup F$ is a semi-path set of G. Since $M_1 \subseteq M_i$ and M_1 is a perfect matching of G, each set available at time i is a matching in C_i. A *matching-pair* in C_i is an (unordered) pair $\{A, B\}$ such that both A and B are (possibly empty) matchings in C_i. An *available matching-pair at time i* is a matching-pair $\{A, B\}$ in C_i such that both A and B are available at time i. A matching-pair $\{A, B\}$ in C_i *covers* a vertex u of C_i if at least one edge in $A \cup B$ is incident to u.

If C_i is a 2-cycle, then we can obtain an available matching-pair $\{A_i, B_i\}$ at time i that covers both vertices of C_i, by simply letting A_i consist of one of the edges of C_i and letting B_i consist of the other. $\{A_i, B_i\}$ is available because every cycle C_j of C with $j < i$ is a 2-cycle and we have done the preprocessing. Thus, if C_i is a 2-cycle, the details of Substeps 3a and 3b are as follows (i.e., they are replaced by the following three substeps):

(a) Compute an available matching-pair $\{A_i, B_i\}$ at time i by letting A_i consist of one of the edges of C_i and letting B_i consist of the other.
(b) For each $v \in V(C_i)$, if some edge e of M' is incident to v, then mark e.
(c) Select one of A_i and B_i uniformly at random and move its edges from C to M.

If C_i is a cycle of length 3 or more, then it is easy to modify the proof of Lemma 1 in [9] to obtain a subroutine for computing an available matching-pair at time i that covers all vertices of C_i. Moreover, using the famous union-find data structure, we can implement this subroutine in $O\left(|C_i| \cdot \alpha(n)\right)$ time. The following lemma summarizes this result:

Lemma 5. *If C_i is a cycle of length 3 or more, then we can compute an available matching-pair $\{A_i, B_i\}$ at time i in $O\left(|C_i| \cdot \alpha(n)\right)$ time that covers all vertices of C_i.*

A *maximal available set at time i* is a set F available at time i such that for every $e \in E(C_i) - F$, $F \cup \{e\}$ is not available at time i. Now, we are ready to describe the details of Substeps 3a and 3b for those cycles C_i of C with $|C_i| \geq 3$. Indeed, they are replaced by the following four substeps:

(a) Compute an available matching-pair $\{A_i, B_i\}$ at time i in $O\left(|C_i| \cdot \alpha(n)\right)$ time that covers all vertices of C_i.
(b) Extend both A_i and B_i to maximal available sets at time i. (*Comment:* Using the famous union-find data structure, we can implement this substep in $O\left(|C_i| \cdot \alpha(n)\right)$ time by scanning the edges of C_i in an arbitrary order and checking if each of them can be added to A_i and/or B_i.)
(c) For each vertex $v \in V(C_i)$ such that both A_i and B_i have an edge incident to v, if some edge e of M' is incident to v, then mark e.
(d) Select one of A_i and B_i uniformly at random and move its edges from C to M.

We finish computing T_2 and T_3 by performing the following three steps:

4. Add to C those edges $\{u, v\} \in M'$ such that both u and v are of degree at most 1 in C. (*Comment:* Let M'_4 denote the set of edges in M' that are added to C at this step. For each cycle C in C, $|E(C) \cap M'_4| \geq 2$.)
5. For each odd cycle C in C, if $|E(C) \cap M'| = 2$ and exactly one edge in $E(C) \cap M'$ is marked, then delete one edge in $E(C) \cap M'$ from C at random in such a way that the marked edge is deleted with probability $2/3$; otherwise, select one edge in $E(C) \cap M'$ uniformly at random and delete it from C. (*Comment:* Let M'_5 denote the set of edges in M' that remain in C immediately after this step.)
6. Set T_2 and T_3 to be C and the graph $(V(G), M)$, respectively.

4.3   Analysis of the Approximation Ratio

Our algorithm is clearly correct. We next analyze its approximation ratio.

Lemma 6. *Let F be an available set at time i. Suppose that $e_1 = \{u_1, u_2\}$ and $e_2 = \{u_2, u_3\}$ are two adjacent edges in C_i such that F contains no edge incident to u_1, u_2, or u_3. Then, $F \cup \{e_1\}$ or $F \cup \{e_2\}$ is available at time i.*

Corollary 1. *Suppose that F is a maximal available set at time i. Then, $C_i - F$ is a collection of vertex-disjoint paths each of length 1, 2, or 3.*

A matching-pair $\{A, B\}$ in C_i *favors* a vertex u of C_i if A contains an edge $e_1 \in E(C_i)$ incident to u and B contains an edge $e_2 \in E(C_i)$ incident to u (possibly $e_1 = e_2$). An available set F at time i is *dangerous* for an (unordered) pair $\{e_1, e_2\}$ of edges in M' if $C_i - F$ contains a connected component that is a length-2 path one of whose endpoints is an endpoint of e_1 and the other is an endpoint of e_2.

Lemma 7. *Let $\{A, B\}$ be an arbitrary matching-pair in C_i that covers all vertices of C_i. If A (respectively, B) is dangerous for a pair $\{e_1, e_2\}$ of edges in M', then $\{A, B\}$ favors exactly one endpoint of the length-2 path in $C_i - A$ (respectively, $C_i - B$) between an endpoint of e_1 and an endpoint of e_2.*

A *maximal available matching-pair at time i* is an available matching-pair $\{A_1, A_2\}$ at time i such that both A_1 and A_2 are maximal available sets at time i.

Lemma 8. *Let $e = \{v_1, v_2\}$ be an edge in M'. Then, $\Pr[e \in M'_5] \geq \frac{1}{6}$.*

Recall T, T_{int}, T_{ext}, and β (they are defined in Section 4.1).

Lemma 9. *Let $\delta w(T)$ be the expected total weight of edges moved from $\mathcal{C}$ to M at Step 3. Then, $\mathcal{E}[w(T_3)] \geq (0.5 + \delta)w(T)$ and $\mathcal{E}[w(T_3)] \geq \left((1 - \delta) + \frac{1}{12}(1 - \beta)\right)w(T)$.*

Theorem 3. *For any fixed $\epsilon > 0$, there is an $O(m + n^3)$-time randomized approximation algorithm for MWTM achieving an expected approximation ratio of $\frac{19(1 - \epsilon)}{25 - 24\epsilon}$.*

Proof. We first estimate the running time of our algorithm. The computation of $\mathcal{C}$, M, and M' can be done in $O(n^3)$ time [7]. Step 3 can be done in $O(n \cdot \alpha(n))$ total time. The other steps take $O(n)$ time. Thus, the time complexity is $O(n^3)$.

We next estimate its approximation ratio. By Fact 2 and Lemma 9, we have the following three inequalities:

$$w(T_1) \geq (1 - \epsilon)\beta w(T), \tag{1}$$

$$\mathcal{E}[w(T_2)] \geq \left((1 - \delta) + \frac{1}{12}(1 - \beta)\right)w(T), \tag{2}$$

$$\mathcal{E}[w(T_3)] \geq (0.5 + \delta)w(T). \tag{3}$$

Adding Inequalities 2 and 3, we have

$$\mathcal{E}[w(T_2)] + \mathcal{E}[w(T_3)] \geq \left(1.5 + \frac{1}{12}(1 - \beta)\right) w(T). \tag{4}$$

Multiplying both sides of Inequality 4 by $12(1 - \epsilon)$ and adding the resulting inequality to Inequality 1, we get

$$w(T_1) + 12(1 - \epsilon)(\mathcal{E}[w(T_2)] + \mathcal{E}[w(T_3)]) \geq 19(1 - \epsilon)w(T). \tag{5}$$

By Inequality 5, we have

$$\mathcal{E}[\max\{w(T_1), w(T_2), w(T_3)\}] \geq \frac{19(1 - \epsilon)}{25 - 24\epsilon} \cdot w(T). \tag{6}$$

Therefore, the algorithm achieves an expected approximation ratio of $\frac{19(1-\epsilon)}{25-24\epsilon}$. ∎

4.4 Derandomization

The above randomized algorithm makes random choices only in Substep 3d and Step 5. To derandomize Step 5, we just modify it as follows:

5. For each odd cycle C in $\mathcal{C}$, delete one edge $e \in E(C) \cap M'$ from C such that the weight e is minimized over all edges in $E(C) \cap M'$.

When processing cycle C_i in Step 3, we need one random bit in Substep 3d. So, Step 3 needs r random bits in total. In the above analysis of the randomized algorithm, only the proof of Lemma 8 is based on the mutual independence between these random bits. Indeed, by carefully inspecting the proof, we can see that the proof is still valid even if the random bits are only pairwise independent. So, we can derandomize it via conventional approaches. Therefore, we have the following theorem:

Theorem 4. *For any fixed $\epsilon > 0$, there is an $O(m+n^3\alpha(n))$-time approximation algorithm for MWTM achieving a ratio of $\frac{19(1-\epsilon)}{25-24\epsilon}$.*

5   Open Problems

An obvious question is to ask if we can improve the running time of our parameterized algorithm for MTM. Preferably, we want a parameterized algorithm for MTM such that the exponent of the exponential factor in the time bound of the algorithm is linear in k. Another obvious question is to ask if we can design a combinatorial approximation algorithm for MWTM that achieves a better ratio than 0.769.

Feige *et al.* [5] have considered the problem of computing t disjoint matchings in a given graph G such that the total number of edges in the matchings is maximized, where t is a fixed positive integer. The special case of this problem where $t = 3$ and the input graph is *simple* has been considered by Kosowski [11] and Rizzi [13]. It seems interesting to design parameterized or approximation algorithms for this problem and its special cases.

References

1. Chen, Z.-Z., Konno, S., Matsushita, Y.: Approximating Maximum Edge 2-Coloring in Simple Graphs. Discrete Applied Mathematics 158, 1894–1901 (2010); A preliminary version appeared in Chen, B. (ed.) AAIM 2010. LNCS, vol. 6124, pp. 78–89. Springer, Heidelberg (2010)
2. Chen, Z.-Z., Tanahashi, R.: Approximating Maximum Edge 2-Coloring in Simple Graphs via Local Improvement. AAIM 2008 410, 4543–4553 (2009); A preliminary version appeared in Fleischer, R., Xu, J. (eds.) AAIM 2008. LNCS, vol. 5034, pp. 84–96. Springer, Heidelberg (2008)
3. Chen, Z.-Z., Tanahashi, R., Wang, L.: An Improved Approximation Algorithm for Maximum Edge 2-Coloring in Simple Graphs. Journal of Discrete Algorithms 6, 205–215 (2008); A preliminary version appeared in Kao, M.-Y., Li, X.-Y. (eds.) AAIM 2007. LNCS, vol. 4508, pp. 27–36. Springer, Heidelberg (2007)
4. Chen, Z.-Z., Wang, L.: An Improved Randomized Approximation Algorithm for Max TSP. Journal of Combinatorial Optimization 9, 401–432 (2005)
5. Feige, U., Ofek, E., Wieder, U.: Approximating Maximum Edge Coloring in Multigraphs. In: Jansen, K., Leonardi, S., Vazirani, V.V. (eds.) APPROX 2002. LNCS, vol. 2462, pp. 108–121. Springer, Heidelberg (2002)
6. Gabow, H.: Implementation of Algorithms for Maximum Matching on Nonbipartite Graphs. Ph.D. Thesis, Department of Computer Science, Stanford University, Stanford, California (1973)
7. Gabow, H.: An Efficient Reduction Technique for Degree-Constrained Subgraph and Bidirected Network Flow Problems. In: Proceedings of the 15th Annual ACM Symposium on Theory of Computing (STOC 1983), pp. 448–456. ACM (1983)
8. Hartvigsen, D.: Extensions of Matching Theory. Ph.D. Thesis, Carnegie-Mellon University (1984)
9. Hassin, R., Rubinstein, S.: Better Approximation for Max TSP. Information Processing Letters 75, 181–186 (2000)
10. Hochbaum, D.: Approximation Algorithms for NP-Hard Problems. PWS Publishing Company, Boston (1997)
11. Kosowski, A.: Approximating the Maximum 2- and 3-Edge-Colorable Subgraph Problems. Discrete Applied Mathematics 157, 3593–3600 (2009)
12. Kosowski, A., Malafiejski, M., Zylinski, P.: Packing $[1, \Delta]$-Factors in Graphs of Small Degree. Journal of Combinatorial Optimization 14, 63–86 (2007)
13. Rizzi, R.: Approximating the Maximum 3-Edge-Colorable Subgraph Problem. Discrete Mathematics 309, 4166–4170 (2009)
14. Serdyukov, A.I.: An Algorithm with an Estimate for the Traveling Salesman Problem of Maximum. Upravlyaemye Sistemy 25, 80–86 (1984) (in Russian)

Discretely Following a Curve

Tim Wylie

Department of Computer Science, The University of Alberta, Edmonton, AB, Canada T6G-2E8
twylie@cs.ualberta.ca

Abstract. Finding the similarity between paths is an important problem that comes up in many areas such as 3D modeling, GIS applications, ordering, and reachability. Given a set of points S, a polygonal curve P, and an $\varepsilon > 0$, the discrete set-chain matching problem is to find another polygonal curve Q such that the nodes of Q are points in S and $d_F(P, Q) \leq \varepsilon$. Here, d_F is the discrete Fréchet distance between the two polygonal curves. For the first time we study the set-chain matching problem based on the discrete Fréchet distance rather than the continuous Fréchet distance. We further extend the problem based on unique or non-unique nodes and on limiting the number of points used. We prove that three of the variations of the set-chain matching problem are **NP**-complete. For the version of the problem that we prove is polynomial, we give the optimal substructure and the recurrence for a dynamic programming solution.

1 Introduction

Matching geometric objects and finding paths through designated points are common problems in many areas of research such as pattern matching, computer vision, map routing, protein structure alignment, ordering, etc. Some of these path problems are fundamental, and are used to define complexity classes and completeness. A problem closely related to our study here is map matching where the goal is to find a path through an embedded graph that minimizes the distance from a given polygonal curve [4]. This has several useful applications, as mentioned by Alt et al., such as determining the path of a vehicle on a road network (graph) given noisy approximate GPS data (polygonal curve). For map matching, the distance measure used is the Fréchet distance.

The Fréchet distance was originally defined by Maurice Fréchet in 1906 as a measure of similarity between two parametric curves [9]. In the early 1990s, the Fréchet distance between polygonal curves was studied by Alt and Godau [5] who presented efficient algorithms and time bounds of $O(mn \log mn)$, where m, n are the number of vertices in the polygonal curves. Following in 1994 Eiter and Mannila [7] defined the discrete Fréchet distance as an approximate solution to the Fréchet distance based on polygonal curves where only the nodes are taken into consideration.

With the continuous Fréchet distance, the time complexity of map matching on a complete graph was further improved upon in [12] where a new problem was introduced-which we will call set-chain matching (it was unnamed in this work). Given a polygonal curve P, a set of points S, and a maximum distance $\varepsilon > 0$, the problem is to find another polygonal curve, Q, through the set of points such that the Fréchet distance between the new curve and the original are within an allowed distance, $d_F(P, Q) \leq \varepsilon$.

P. Widmayer, Y. Xu, and B. Zhu (Eds.): COCOA 2013, LNCS 8287, pp. 13–24, 2013.

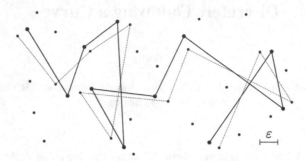

Fig. 1. An instance of the set-chain matching problem in 2D with one solution of $k \geq 11$

Beyond the original work, we investigate many variations. We look at the complexity of set-chain matching based on the discrete Fréchet distance, and although the original definition allowed points in the set to be reused in the path, we now consider both unique and non-unique points. We show that the unique point versions are **NP**-complete, and the non-unique point versions are **NP**-complete when restricting the size of the set of points used, but polynomial when limiting the size of the path. Figure 1 shows a simple instance of the set-chain matching problem, which is formally defined at the beginning of Section 3.

The variations of discrete set-chain matching have many applications. Suppose we have intermittent lossy GPS vehicle data where we can not guarantee the path of the vehicle between our data points. We can find the shortest (and arguably the most plausible) path of the vehicle based on the discrete Fréchet distance. If the points in our set represent signal towers (cellular, radio, etc.), which generally have a spherical range, then we can also consider several coverage problems. Assuming we know the path of a vehicle, what is the minimum number of towers needed to ensure that the signal is not lost. Simply knowing whether the path is covered is important, but optimizing it along multiple roads and areas is crucial. These types of problems are studied in many areas related to wireless sensor networks, graphics, scheduling, and ordering.

We first provide some background and related work in Section 2. We then cover the definitions and variations of the discrete set-chain matching problem in Section 3. Sections 4, 5, and 6 follow with the actual results of the problems. Finally, we conclude in Section 7 and give some future work related to this research.

2   Background

With respect to map matching, the problem of finding a path in a graph given a polygonal line was first posed by Alt et al. [4] as follows: Let $G = (V, E)$ be an undirected connected planar graph with a given straight-line embedding in $\mathbb{R}^2$ and a polygonal line P, find a path π in G which minimizes the Fréchet distance between P and π. They give an efficient algorithm which runs in $O(pq \log q)$ time and $O(pq)$ space where p is the number of line segments of P and q is the complexity of G, but it also allowed vertices and edges to be visited multiple times.

The recent work by Maheshwari et al. improved the running time for the case of a complete graph [12]. The original algorithm decides the map matching problem in

$O(pk^2 \log k)$ where k is the number of vertices in the graph, and the new algorithm solves it in $O(pk^2)$. Although they do not specify the name for the problem, we refer to it as set-chain matching to avoid confusion with other matching problems. Formally, the set-chain matching problem is defined as: Given a point set S and a polygonal curve P in $\mathbb{R}^d$ ($d \geq 2$), find a polygonal curve Q with its vertices chosen from S, which has a minimum Fréchet distance to P. They decide this problem in $O(pk^2)$, and also give an algorithm to find the minimal Fréchet distance in $O(pk^2 \log pk)$.

We originally noted the complexity of discrete set-chain matching with unique nodes, without the actual proof, in [18]. We not only prove it here, but we also show that the continuous version of the problem with unique points is **NP**-complete. This paper is a continuation of our earlier work, but the result was also independently proven by Accisano and Üngür [1] and Shahbaz proved it for non-unique points [15].

A variation of the discrete set-chain matching problem is also related to the discrete unit disk cover (DUDC) problem when limiting the number of points from S used. The DUDC problem is known to be **NP**-Hard, and is also difficult to approximate with the most recent results being an 18-approximation algorithm [6], a 15-approximation algorithm [8], and a $(9 + \varepsilon)$-approximation algorithm [2]. Nearly all of the constant factor approximations have been within the last decade. The problem does admit a PTAS [14], but this is infeasible for most instances of the problem. DUDC does not admit a Fully Polynomial Time Approximation Scheme (FPTAS) unless **P=NP**.

The discrete Fréchet distance was originally defined by Eiter and Mannila [7] in 1994, and was further expanded on theoretically by Mosig et. al. in 2005 [13]. Given two polygonal curves, we define the discrete Fréchet distance as follows. We use $d(a, b)$ to represent the euclidean distance between two points a and b, but it could be replaced with other distance measures depending on the application.

Definition 1. *The discrete Fréchet distance d_F between two polygonal curves $f : [0, m]$ $\rightarrow \mathbb{R}^k$ and $g : [0, n] \rightarrow \mathbb{R}^k$ is defined as:*

$$d_F(f,g) = \min_{\sigma:[1:m+n]\rightarrow[0:m],\beta:[1:m+n]\rightarrow[0:n]} \max_{s\in[1:m+n]} \left\{ d\Big(f(\sigma(s)), g(\beta(s))\Big) \right\}$$

where σ and β range over all discrete non-decreasing onto mappings of the form σ : $[1 : m + n] \rightarrow [0 : m], \beta : [1 : m + n] \rightarrow [0 : n]$.

The continuous Fréchet distance is typically explained as the relationship between a person and a dog connected by a leash walking along the two curves and trying to keep the leash as short as possible. However, for the discrete case, we only consider the nodes of these curves, and thus the man and dog must "hop" along the nodes of the curves. Consider the scenario in which a person walks along A and a dog along B. Intuitively, the definition of the paired walk is based on three cases:

1. $|B_i| > |A_i| = 1$: the person stays and the dog hops forward;
2. $|A_i| > |B_i| = 1$: the person hops forward and the dog stays;
3. $|A_i| = |B_i| = 1$: both the person and the dog hop forward.

By giving a dynamic programming solution for finding the discrete Fréchet distance between two polygonal curves, Eiter and Mannila proved:

Theorem 1. *The discrete Fréchet distance between two polygonal curves, with m and n vertices respectively, can be computed in $O(mn)$ time [7].*

Figure 2 shows the relationship between the discrete and continuous Fréchet distances. In Figure 2(a), we have two polygonal curves (or chains) $\langle a_1, a_2, a_3 \rangle$ and $\langle b_1, b_2 \rangle$, the continuous Fréchet distance between the two is the distance from a_2 to segment $\overline{b_1 b_2}$, i.e., $d(a_2, o)$. The discrete Fréchet distance is $d(a_2, b_2)$. The discrete Fréchet distance could be quite larger than the continuous distance. On the other hand, with enough sample points on the two curves, the resulting discrete Fréchet distance, i.e., $d(a_2, b)$ in Figure 2(b), closely approximates $d(a_2, o)$.

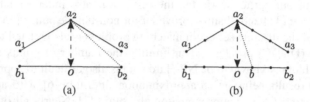

(a)   (b)

Fig. 2. The relationship between the discrete and continuous Fréchet distance where o is the continuous and the dotted line between nodes is the discrete. (a) shows a case where the curves have fewer nodes and a larger discrete Fréchet distance, while (b) is the same basic path with more nodes, and thus provides a better approximation of the Fréchet distance.

3   Discrete Set-Chain Matching

We begin with the formal definitions of the problem and the variations as well as some terminology. It is important to note that, as in the continuous version, we make no requirements that P or Q be planar. For discussion, we will refer to the number of nodes in a polygonal curve as the "size" of the curve and it will be denoted as $|A|$ for a polygonal curve A.

Definition 2 (The Discrete Set-Chain Matching Problem)
Instance: *Given a point set S, a polygonal curve P in $\mathbb{R}^d$ ($d \geq 2$), an integer $K \in \mathbb{Z}^+$, and an $\varepsilon > 0$.*
Problem: *Does there exist a polygonal curve Q with vertices chosen from S' where $S' \subseteq S$, such that $T \leq K$ and $d_F(P, Q) \leq \varepsilon$?*

T is defined in two ways. When limiting the number of nodes in the curve, $T = |Q|$, and if restricting the number of points used then $T = |S'|$. Figure 3 shows an example demonstrating the difference between minimizing $|Q|$ or $|S'|$. Here, minimizing $|Q|$ will always yield $|Q| = 3$ regardless of the points chosen. However, minimizing $|S'|$ will return $|S'| = 2$ and $|Q| = 3$, which is the only set of points that is minimal.

We look at three variations of discrete set-chain matching. They vary whether there is a uniqueness constraint on $s \in S$ being used as a node in Q (if points may be used more than once), and whether our goal is to limit the size of the curve Q or the set S'. We distinguish the problems as Unique/Non-unique(U/N) Set-Chain(S) Matching(M)

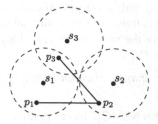

Fig. 3. The difference between minimizing $|Q|$ and $|S'|$. Minimizing $|S'|$ gives $Q = \langle s_1, s_2, s_1 \rangle$ where $|S'| = 2$ and $|Q| = 3$, but minimizing $|Q|$ will yield $|Q| = 3$ whether it uses the sequence $\langle s_1, s_2, s_1 \rangle$ or $\langle s_1, s_2, s_3 \rangle$

with a k Subset/Curve(S/C). The variants are thus NSMS-k, NSMC-k, and USM-k. When looking at unique nodes, limiting $|Q|$ is equivalent to limiting the set of points used, $|S'|$, since they can only be used once, so we do not separate the cases.

4   Set-Chain Matching with $T = |Q|$ (NSMC-k)

The original set-chain matching work dealt with the continuous version of NSMC-k. The discrete version is decidable with a straightforward dynamic programming solution. We first overview the recurrence relation and algorithm to solve the optimization version, and show that that NSMC-k exhibits an optimal substructure.

Figure 3 demonstrates that we must find at least one point $s_i \in S$ for every $p_j \in P$. The recurrence relation is shown in Equation 1. It assumes a 2D array, M, of size $|S| \times |P|$ where the columns represent the nodes in the polygonal curve P and the rows represent points in the set S. The initial condition assumes a column zero populated with 0's in every row. The recurrence can then be processed column by column until finished. The final optimal value will be $Opt = \min_{k=1}^{|S|}(M[k, |P|])$. This can be solved in $O(mn)$ time. A straightforward iterative algorithm that implements this method and solves the optimization version of the problem is easy to construct. The optimal result is then used to decide NSMC-k.

$$M[i,j] = \min \begin{cases} M[i, j-1], & \text{if } d(s_i, p_j) \leq \varepsilon, M[i, j-1] \neq \emptyset \\ \min_{k=1}^{|S|}(M[k, j-1]) + 1, & \text{if } d(s_i, p_j) \leq \varepsilon, M[i, j-1] = \emptyset \quad (1) \\ \emptyset, & \text{if } d(s_i, p_j) > \varepsilon \end{cases}$$

Theorem 2 (Optimal Substructure of NSMC-k). *Let $P = \langle p_1, ..., p_n \rangle$ be a polygonal chain, and $S = \{s_1, ..., s_m\}$ be a set of points such that there exists a $Q = \langle q_1, ..., q_k \rangle$ through a set $S' \subseteq S$ which is a minimum sequence such that $d_F(P, Q) \leq \varepsilon$.*
(1) If $d(p_{n-1}, q_k) \leq \varepsilon$ and $d(p_{n-1}, q_{k-1}) > \varepsilon$, then Q_k is an optimal solution for P_{n-1}.
(2) If $d(p_{n-1}, q_{k-1}) \leq \varepsilon$, then Q_{k-1} is an optimal solution for P_{n-1}.
(3) If $d(p_{n-1}, q_k) > \varepsilon$, then Q_{k-1} is an optimal solution for P_{n-1}.

Proof. (1) If $d(p_n, q_k) \leq \varepsilon$ and $d(p_{n-1}, q_k) \leq \varepsilon$, then the point q_k covers both points by an ε-ball. However, q_{k-1} does not cover p_{n-1}. Thus, Q_k is still the optimal solution.

(2) If $d(p_{n-1}, q_{k-1}) \leq \varepsilon$, then q_k only covers p_n. If $d(p_n, q_{k-1}) \leq \varepsilon$, then Q_{k-1} would be an optimal solution, but by definition Q was minimal so this can not be true. (3) If $d(p_{n-1}, q_k) > \varepsilon$, then we have the same argument with p_n only covered by q_k, and thus Q_{k-1} must be optimal for P_{n-1}. □

Theorem 3. *The discrete non-unique set-chain matching problem where $T = |Q|$ is polynomial, i.e., NSMC-k $\in$ **P**.*

Proof. Since we have shown that NSMC-k has an optimal substructure, given P, S, and K, we can find an optimal K' from a dynamic programming algorithm based on the recurrences (Equation 1). Then we decide NSMC-k by comparing whether $K \leq K'$. □

5 Set-Chain Matching with $T = |S'|$ (NSMS-k)

The discrete non-unique set-chain matching problem where we limit the number of points from S used as nodes in Q turns the problem into a coverage issue. This problem is equivalent to the discrete unit disk cover (DUDC) problem, which is known to be **NP**-Hard and is difficult to approximate.

Theorem 4. *The discrete non-unique set-chain matching (NSMS-k) problem where $T = |S'|$ is **NP**-complete.*

Proof. This can be shown via a straightforward reduction from the discrete unit disk cover (DUDC) problem which is **NP**-Hard [6]. Formally, we are given a set of points P and a set of disks $D = \{D_1, D_2, ..., D_N\}$ with centers $C = \{c_1, c_2, ..., c_N\}$ with all disks of radius r.

Now, let P' be a polygonal curve made of all points in P in any order. Let $S = C$ and $\varepsilon = r$. Now, $\exists$ a minimum-cardinality subset $D' \subseteq D$ with centers C' such that $\forall p \in P, \exists$ a $D_i \in D'$ that contains p if and only if $\exists$ a polygonal curve Q where the vertices are from points in $S' \subseteq S$ such that $|S'| = |D'|$ and $d_F(P', Q) \leq \varepsilon$.

We first prove the forward direction. Given an instance $I \subseteq D$ that is a minimum covering for all points in P. We construct P' by connecting all points in P in any order. Making a polygonal curve Q with the set of centers (C_I) of I is straightforward. We construct Q by finding the disk (D_i) that covers $p_1 \in P'$, and we set $q_1 = c_i$ where c_i is the center of disk D_i. Similarly, we walk through each $p_i \in P'$ and set the center of the disk $D_j \in I$ covering point p_i as $q_i = c_j$. Every ordered node in P' is now still within ε of a node in Q, thus $d_F(P', Q) \leq \varepsilon$, and the set of nodes used, $|S'|$, is equal to $|I|$.

In the other direction, if we have a polygonal curve $Q = \{q_1, q_2, \ldots, q_N\}$ such that the number of unique locations used for vertices is of minimum cardinality and $d_F(P', Q) \leq \varepsilon$. Suppose the set of unique locations S' that Q is made of is not a minimal disk cover of all the vertices of P' viewed as points in a set P. This implies there exists at least one q_i that is unnecessary for a covering by C, and there is a point p_j that can be covered by another c_k. Let C' be this smaller covering. Using the same construction as above we can build a P'' and Q'. This would mean $|C'| < |S'|$ which contradicts our assumption that S' in minimal. Thus, every node $p_i \in P'$ is within ε of at least one node $q_j \in Q$, and S' is a minimum cover.

Finally, we show the problem is in **NP**. Given an instance I we can check whether $d_F(P, I) \leq \varepsilon$ in $O(mn)$ time via Theorem 1. □

6   Unique Set-Chain Matching (USM-k)

We now address unique set-chain matching where any point from the set can be used at most once, and show that this problem is **NP**-complete via a reduction from planar 3-SAT [11]. Planar 3-SAT is any 3-SAT formula that can be drawn as a planar graph with vertices representing clauses and variables. This is a convenient form of 3-SAT for geometric reductions since a crossover gadget is unnecessary.

By standard convention, we first introduce several planar "gadgets" that we then arrange in our reduction. We will build up the gadgets in a piecewise manner, and then show how they are connected to form a single polygonal curve. Due to the length of this section, we cover the gadgets and then formally do the reduction with the assumption of their correctness.

Let φ be the 3-SAT formula represented by the input instance of planar 3-SAT with N variables and M clauses. Given an $\varepsilon > 0$, we construct a point set S and a polygonal curve P and let $K = |S_\varepsilon| = |S|$ requiring all points to be used. Here, $S_\varepsilon = \{s \in S | p \in P \text{ and } d(p, s) \leq \varepsilon\}$ and referred to as the set of reachable points. We show that φ is satisfiable if and only if with our construction there exists a polygonal curve Q with unique nodes from the set S such that $d_F(P, Q) \leq \varepsilon$, i.e. $|Q| = |S| \leq K$.

6.1   Choices and Chains

We first look at the main building block for our gadgets in this reduction, which is the choice gadget shown in Figure 4(a). There are two ways for a new curve to be constructed starting at a and using the points $\{a, b, c\}$ in order to "cover" the nodes of the curve $\langle x, y, z \rangle$. We label the curve $\langle a, b, c \rangle$ as **true**, and the curve $\langle a, c, b \rangle$ as **false**. This is because the second curve violates our ε constraint since $d(b, z) > \varepsilon$.

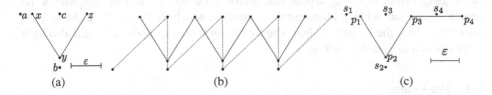

(a)   (b)   (c)

Fig. 4. (a) A choice gadget. (b) A chain with a **false** connection. (c) A variable gadget.

Choice gadgets are linked together to make a chain. Chain gadgets are important because they force a new curve to stay in a **true** or **false** orientation, and therefore transfer information. An example of a chain with a **false** curve is shown in Figure 4(b).

6.2   The Variable Gadget

The base of the variable gadget is shown in Figure 4(c). A **true** setting begins the new chain as $\langle s_1, s_2, s_3, s_6 \rangle$ while a **false** setting begins $\langle s_1, s_3, s_2, s_4, s_5 \rangle$. The different settings change whether s_4 is needed to keep $d_F(P, Q) \leq \varepsilon$. A **true** setting does not need the extra node while the **false** does. This free node is what is propagated to the clause

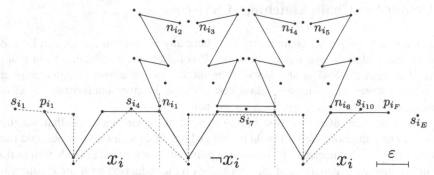

Fig. 5. Variable gadgets linked together for variable x_i where x_i is set to **false** (s_{i_4} is used) and thus $\neg x_i$ is **true** (s_{i_7} is free)

gadget. Figure 5 shows the full variable gadget. As is standard in many reductions, each variable is repeated some finite length while alternating between x and $\neg x$ based on what is needed in the equation.

Unfortunately, the variable gadget alone will not ensure that the new curve alternates between **true** and **false** configurations, which we need for a variable and its complement. Therefore, the variable gadget has a "switch" component, which makes the free point necessary at every other variable gadget, and thus alternates Q between **true** and **false** paths. It is important to note that these switch segments will not be connected to the variable gadgets within ε. Note in Figure 5 that the first and last instance of the variable gadget do not have the full switch component.

For our planar 3-SAT instance, there may be edges which need to connect from the top and the bottom of the variable gadget. Although an example is not given, this is possible with our variable gadget. Looking at Figure 5, imagine everything is rotated in the gadget from s_{i_7} to s_{i_E} around that vector. This flips the variable and half of the switch component without changing the reduction, which allow attaching chains onto the other side of the variable gadget. The following switch component would also have to be below and then flip back up.

6.3 The Clause

A clause gadget is straightforward. As shown in Figure 6(a), three chains meet within ε of each other ($c_{i_1}, c_{i_2}, c_{i_3}$), and there are only two points between them. Each chain is connected at the other end ($v_{i_1}, v_{i_2}, v_{i_3}$) to variable gadgets. The **true** or **false** setting from the variable is propagated up to the clause gadget and at least one of the chains must have the new curve in a **true** position. Only two of the chains can have a **false** setting or else one of the end nodes (C_{k_i}) in the clause gadget will not be within ε of any available point, which is equivalent to the clause being **false** in 3-SAT. Also note that in the clause gadget, if either point is not needed, they can be used by a **true** chain so that all points are used.

The chains from the clause gadgets are attached to the variable gadgets in the highlighted area of Figure 6(b). There is one point between the ends of the three chains. A segment is added from the clause endpoint v_{k_y} (for clause c_k where $1 \leq y \leq 3$) to the

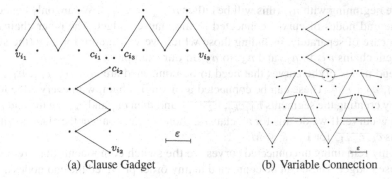

(a) Clause Gadget   (b) Variable Connection

Fig. 6. (a) The clause gadget. (b) The connection point between a variable gadget and a chain to the clause gadget.

opposite side of the switch component of the variable (or complement) desired, e.g., if x_1 is the third variable in the clause c_k and the connection point is $n_{1_i}(x_1)$ or $n_{1_j}(\neg x_1)$, then a segment is placed connecting the chain v_{k_3} to $n_{1_j}(\neg x_1)$.

6.4 Connecting the Gadgets

Although the polygonal curve P does not have to be planar, it must be a single continuous curve. Here, we will show that all the gadgets and segments can be connected to form P. The non-planarity allows us to focus on a single clause gadget to show one way in which everything can be connected. We have to be careful that we do not connect two nodes that would change the reduction such as connecting two end nodes at a clause– $c_{k_1}, c_{k_2}, c_{k_3}$ for clause C_k. For simplicity, we can connect all variables together and all the beginning and end switch points. Let $q_1 = p_{1_1}$ and then connect the variable gadgets by adding in the edge $\overline{p_{k_F} p_{k+1_1}}$ for all variables $1 \leq k \leq N - 1$, and the last variable node p_{N_F} connects to a vertex in C_1.

We show a simple example of three variables and a clause in Figure 7 without the connecting segments between gadgets. Let this be clause C_k, and the connected variables be x_1, x_2, x_3, at nodes n_{t_i} or n_{t_j} where $1 \leq t \leq 3$ and let n_{t_j} be the end node of

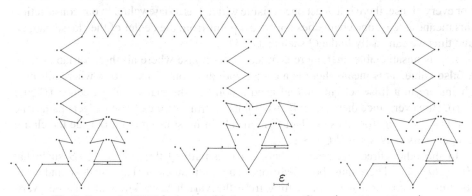

Fig. 7. Example USM-k clause with three variables $C_k = (\neg x_1 \cup x_2 \cup \neg x_3)$

the curve beginning with n_{t_i} (this will be either $n_{t_{i-1}}$ or $n_{t_{i+1}}$). We are only concerned about the end nodes of curves connected to the clause gadget. The other chains will be taken care of separately, including those which we will ignore for now (the switch component chains n_{1_3} to n_{1_4} and n_{3_3} to n_{3_4} in our example).

The end nodes of the curves that need to be connected are c_{k_1}, c_{k_2}, c_{k_3}, $n_{1_j}(n_{1_1})$, $n_{2_i}(n_{2_4})$, $n_{3_j}(n_{3_1})$. These can be connected as a single chain, with every edge longer than ε, by creating the segments $\overline{n_{1_j}c_{k_2}}$, $\overline{n_{2_i}n_{3_j}}$, and then c_{k_1} and c_{k_3} are the end nodes of the new curve. If we do this for all clauses, then we can connect the clauses with the segments $\overline{c_{k_3}c_{k+1_1}}$ for $1 \leq k \leq M - 1$.

The only remaining unconnected curves are the switch components that are not tied to a clause gadget. These can be connected in any order provided the end nodes are not within ε, and we do not introduce a loop. This is straightforward by connecting every other switch component curve (never creating the segments $\overline{n_{t_{i-1}}n_{t_i}}$ or $\overline{n_{t_i}n_{t_{i+1}}}$ for $1 \leq t \leq N$), and then connecting all the skipped curves.

6.5   The Reduction

Theorem 5. *The discrete unique set-chain matching (USM-k) problem is **NP**-complete.*

Proof. We are given a planar 3-SAT instance $G_\varphi = \{V, E\}$ with vertices $V = X \cup C$ such that the vertices represent variables $X = \{x_1, x_2, \ldots, x_N\}$ and clauses $C = \{C_1, C_2, \ldots, C_M\}$, and the edges $E = \{e_1, e_2, \ldots, e_Z\}$ connect variables to clauses with the degree of each $C_i \in C$ being three. Given the planar 3-SAT instance G_φ, we construct a polygonal curve P and a point set S using an $\varepsilon > 0$ based on the method described. This construction takes $O(|C| + |X| + |E|)$ for constructing P and S and is thus polynomial. The sizes of P and S are dependent on ε and the metric space. In general, for any edge $e_i \in E$ in the space, where $\|e_i\|$ is the length of the edge, there are $\lceil \|e_i\|/\varepsilon \rceil$ points in S and nodes of P used to transfer information along that edge.

We also refer to the 3-SAT equation φ derived from G_φ for the satisfiability of G_φ. The planar 3-SAT equation φ derived from G_φ is satisfiable if and only if there exists a polygonal curve Q with nodes from the set S such that $d_F(P, Q) \leq \varepsilon$ and each point represents a unique node in Q.

In the forward direction, we look at the value of φ. First, we assume φ is satisfiable. For every clause, there is at least one variable which has a **true** value. In our construction this means at least one chain does not need a point from the center of the clause gadget, and thus we can easily find a Q such that $d_F(P, Q) \leq \varepsilon$.

If φ is unsatisfiable, then there is at least one clause where all three variables have a **false** value. This means there is a clause gadget in our construction where all three chains are in a **false** setting, and all need a point in the clause gadget center (Figure 6(a)). However, since there are only two points within ε of the clause gadget chains (the points c_{i_1}, c_{i_2}, c_{i_3} for clause gadget C_i), one chain must use a point outside the clause gadget. This causes $d_F(P, Q) > \varepsilon$.

In the other direction, assume there exists a path Q through $S' \subset S$ such that $d_F(P, Q) \leq \varepsilon$. There must be at least one **true** chain at each clause gadget, and since the three chains propagate this setting from the variable, we know at least one variable (or complement) was **true**. Thus, for every variable attached to a clause, it has the correct **true** or **false** setting. Therefore, if $d_F(P, Q) \leq \varepsilon$, then the current assignment of each variable also satisfies φ.

If no path Q exists such that $d_F(P, Q) \leq \varepsilon$, then there is at least one clause gadget where all three chains had **false** settings and needed an extra point for Q within the clause gadget. Since the variable gadgets and switch components always have a path within ε, the problem must occur in a clause. Again, this only happens if all three chains have a **false** setting, and similarly to the previous example, these propagated along the chains from the attachments to the variable gadgets. Thus, there must also exist a clause in φ where all three variables are **false**.

Last, we know the problem is in **NP**. Given an instance I we can check whether $d_F(P, I) \leq \varepsilon$ in $O(mn)$ time via Theorem 1.   □

Our reduction is based on the discrete Fréchet distance, but our construction also ensures that any resulting path Q is within ε of P along the edges as well. Thus, our reduction can be adapted to prove that USM-k is also **NP**-complete for the continuous Fréchet distance. This result was also recently proven independently and with a unique reduction in [1]. Due to this result being known and for space concerns, we only supply the basic outline of the proof.

Corollary 1. *The unique set-chain matching (USM-k) problem based on the continuous Fréchet distance is **NP**-complete.*

Proof. This can be proven based on the polygonal curves P and Q being constructed of straight line segments. Given two line segments $a = \langle p_1, p_2 \rangle$ and $b = \langle p'_1, p'_2 \rangle$, it is straightforward to see that if $d(p_1, p'_1) \leq \varepsilon$ and $d(p_2, p'_2) \leq \varepsilon$, then under the continuous Fréchet distance $d_{\mathcal{F}}(a, b) \leq \varepsilon$.

Further, it is known that for any two polygonal curves, $d_{\mathcal{F}}(P, Q) \leq d_F(P, Q)$ [7]. Thus, if both P and Q are polygonal curves and the problem is **NP**-complete for the discrete Fréchet distance within ε, it will also hold for the continuous Fréchet distance within an $\varepsilon' \leq \varepsilon$ and an instance can be verified in $O(mn \log mn)$ [5].   □

7 Conclusion

In this paper we have outlined and extended the discrete set-chain matching problem and other variations based on restricting our selection to unique nodes, the number of nodes allowed in the curve, or the number of points to choose from. We proved that two variations are **NP**-complete, and the unique point variation is still **NP**-complete when based on the continuous Fréchet distance. We proved that the other variation is polynomial, and gave the recurrences for a dynamic programming implementation. We conclude with some open problems and further research directions for this work.

(1) What are the complexities based on maximizing the number of vertices in Q?

(2) We can also reverse the problem– if we are given a set size for Q, can we minimize the discrete Fréchet distance between P, Q, i.e., $d_F(P, Q) \leq \varepsilon$?

(3) What are the complexities with imprecise input? How difficult is it to find the minimum and maximum length Q while respecting the discrete Fréchet distance? This builds off computing the discrete Fréchet distance with imprecise input in general [3].

(4) What are the approximation bounds for the optimization versions? We know NSMS-k is equivalent to DUDC which generally only admits high approximations.

References

1. Accisano, P., Üngör, A.: Hardness results on curve/point set matching with Fréchet distance. In: Proc. of the 29th European Workshop on Computational Geometry, EuroCG 2013, pp. 51–54 (March 2013)
2. Acharyya, R., Manjanna, B., Das, G.K.: Unit disk cover problem. CoRR, abs/1209.2951 (2012)
3. Ahn, H.-K., Knauer, C., Scherfenberg, M., Schlipf, L., Vigneron, A.: Computing the discrete Fréchet distance with imprecise input. In: Cheong, O., Chwa, K.-Y., Park, K. (eds.) ISAAC 2010, Part II. LNCS, vol. 6507, pp. 422–433. Springer, Heidelberg (2010)
4. Alt, H., Efrat, A., Rote, G., Wenk, C.: Matching planar maps. J. Algorithms 49(2), 262–283 (2003)
5. Alt, H., Godau, M.: Computing the Fréchet distance between two polygonal curves. International Journal of Computational Geometry and Applications 5, 75–91 (1995)
6. Das, G.K., Fraser, R., Lòpez-Ortiz, A., Nickerson, B.G.: On the discrete unit disk cover problem. In: Katoh, N., Kumar, A. (eds.) WALCOM 2011. LNCS, vol. 6552, pp. 146–157. Springer, Heidelberg (2011)
7. Eiter, T., Mannila, H.: Computing discrete Fréchet distance. Technical Report CD-TR 94/64, Information Systems Department, Technical University of Vienna (1994)
8. Fraser, R., Lòpez-Ortiz, A.: The within-strip discrete unit disk cover problem. In: Proc. of the 24th Canadian Conf. on Computational Geometry, CCCG 2012, pp. 53–58 (2012)
9. Fréchet, M.: Sur quelques points du calcul fonctionnel. Rendiconti del Circolo Matematico di Palermo (1884 - 1940) 22(1), 1–72 (1906)
10. Jiang, M.: Map Labeling with Circles. PhD thesis, Montana State University (2005)
11. Lichtenstein, D.: Planar Formulae and Their Uses. SIAM Journal on Computing 11(2), 329–343 (1982)
12. Maheshwari, A., Sack, J.-R., Shahbaz, K., Zarrabi-Zadeh, H.: Staying close to a curve. In: Proc. of the 23rd Canadian Conf. on Computational Geometry, CCCG 2011, August 10-12 (2011)
13. Mosig, A., Clausen, M.: Approximately matching polygonal curves with respect to the Fréchet distance. Computational Geometry: Theory and Applications 30(2), 113–127 (2005)
14. Mustafa, N.H., Ray, S.: Improved results on geometric hitting set problems. Discrete and Computational Geometry 44(4), 883–895 (2010)
15. Shahbaz, K.: Applied Similarity Problems Using Fréchet Distance. PhD thesis, Carleton University (2013)
16. Wolff, A.: A simple proof for the NP-hardness of edge labeling. Technical Report W-SPNPH-00, Institut für Mathematik und Informatik, Universität Greifswald (2000)
17. Wylie, T.: The Discrete Fréchet Distance with Applications. PhD thesis, Montana State University (2013)
18. Wylie, T., Zhu, B.: Discretely following a curve (short abstract). In: Computational Geometry: Young Researchers Forum, CG:YRF 2012, pp. 33–34 (2012)

NF-Based Algorithms for Online Bin Packing with Buffer and Item Size Limitation

Feifeng Zheng[1], Li Luo[2], and E. Zhang[3]

[1] Glorious Sun School of Business and Management,
Donghua University, Shanghai, 200051, P.R. China
[2] Business School of Sichuan University, Chengdu, 610064, P.R. China
[3] School of Information Management and Engineering,
Shanghai University of Finance and Economics, Shanghai, 200433, P.R. China
ffzheng@dhu.edu.cn

Abstract. This paper studies a variation of online bin packing where there is a capacitated buffer to temporarily store items during packing, and item size is within interval $(\alpha, 1/2]$ for some $0 \leq \alpha < 1/2$. The problem is motivated by surgery scheduling such that we regard the planned uniform available time interval in each day as a unit size bin and surgeries as items to be packed. We investigate the asymptotic performance of algorithm NF (Next Fit) and NF-based online algorithms. The classical NF algorithm without use of the buffer is proved to be asymptotic $\frac{2}{1+\alpha}$-competitive. We mainly propose an NF-based algorithm which uses the buffer and has an asymptotic competitive ratio of 13/9 for any constant buffer size not less than one. We also prove a lower bound of 4/3. Experimental results further reveal that the proposed algorithm using buffer is of effective practical performance.

Keywords: Bin packing, Online algorithm, Asymptotic competitive ratio.

1 Introduction

This paper studies the following scenario of online bin packing problem. A set of items, which are with lower and upper bounds of size, are released over list and to be packed into unit size bins. Released items are required to be temporarily stored in a capacitated buffer before packing into bins. The problem is motivated in the area of *OR (Operating Room)* scheduling. In each day there is a uniform time interval available for an OR to process surgical operations. Each surgical request with specific planned operation time is temporarily stored in a request waiting pool after its arrival. In each day, a surgery scheduler selects in the pool a subset of requests to be processed in the OR on the next day, satisfying that the total planned operation time of selected requests cannot exceed the length of available time of the OR in the day. To improve its running efficiency, an OR is generally used to process surgeries from a single department, such as urological surgeries from department of urology. These surgeries are of similar

P. Widmayer, Y. Xu, and B. Zhu (Eds.): COCOA 2013, LNCS 8287, pp. 25–36, 2013.

planned operation time, and it is reasonable to assume lower and upper bounds of operation time for surgical requests. We may regard the uniform available time interval in a day as the unit size space of a bin, the surgical requests as items, and the request waiting pool as a buffer. Focusing on the task to assign efficiently an available combination of waiting requests to each day and disregarding other uncertainties, we describe the above OR scheduling problem as online bin packing with buffer and item size limitation.

Online bin packing problem was first studied by Ullman [1] and has been extensively investigated in recent decades (see Galambos and Woeginger [2] and Seiden [3]). In one-dimension online bin packing problem, each item with size at most one is released over list such that it has to be packed irrecoverably into unit size open bins on its arrival without knowledge of any future items. Johnson [4] proved that NF(Next Fit) algorithm has a worst-case ratio of 2. NF keeps exactly one bin open at any time during packing, closing a current open bin to open the next one provided that a released item cannot be packed into the current open bin. Simchi-Levi [5] proved that FF (First Fit) and BF (Best Fit) algorithms have a worst-case ratio of no more than 7/4. Zhang et al. [6] presented a 7/4-competitive online algorithm that runs in linear time and keeps at most four bins open at any time.

Some authors studied long term performance of online algorithms which is measured by parameter ACR (*asymptotic competitive ratio*). We define ACR as follows. For any item input instance σ, let $n_{\mathcal{A}}(\sigma), n^*(\sigma)$ be the number of bins used by an online algorithm $\mathcal{A}$ and by an optimal offline algorithm OPT respectively. The algorithm $\mathcal{A}$ is defined as

$$R_{\mathcal{A}}^{\infty} = \lim_{u \to \infty} \sup_{\sigma} \left\{ \frac{n_{\mathcal{A}}(\sigma)}{n^*(\sigma)} \middle| n^*(\sigma) = u \right\}.$$

We also say $\mathcal{A}$ is asymptotic $R_{\mathcal{A}}^{\infty}$-competitive. For online bin packing algorithms, Johnson et al. [7] defined an ANY FIT class such that a new bin is never opened unless one released item cannot be packed into any of current open bins, and showed that any algorithm in the ANY FIT class cannot have an ACR smaller than 1.7. Note that NF, FF and BF are all in the ANY FIT class. Yao[8] presented a revised FF algorithm that is not in the ANY FIT class and has an ACR of 5/3. Ramanan et al. [9] provided a linear time online algorithm Modified Harmonic (MH) with ACR strictly less than 1.615. Seiden [3] developed an online algorithm named Harmonic++ which has an ACR of 1.58889. The best known lower bound 1.5403 of ACR is due to Balogh et al. [10].

Since NF keeps at most one bin open at any time and runs in $O(u)$ time and $O(1)$ space for processing u items, it is probably the simplest and definitely one of the fastest online algorithms. This implies that NF is an easiest algorithm to operate in practice. Thus, we focus on the performance of NF-based algorithms with the use of buffer for the problem considered in this paper.

1.1   Related Work

Galambos [11] studied an online bin packing problem with unit size *buffer-bins*, which are used to temporarily store released items before they are packed into open bins, and proposed an online algorithm using two buffer-bins. This idea was further developed by Galambos and Woeginger [12] for an algorithm that uses three buffer-bins and has an ACR of $\prod_\infty \approx 1.69103$. They showed that any bounded-space online algorithm with repack has an ACR not less than $\prod_\infty$.

Another related variation of online bin packing is allowed to repackage a finite number of already packed items among open bins (see Gambosi et al. [13], and Balogh and Galambos [14]). Gambosi et al. [13] presented two algorithms with ACR of 3/2 and 4/3 respectively. The first algorithm is allowed to move some small items only one time for each, while the second one is allowed to move each item more than one time. Balogh et al. [15] studied the case where at most k items can be repackaged at any time during packing. They proved that as k goes to infinity, both upper and lower bounds of ACR approach 3/2.

Some authors studied various scenarios on the constraint of item size. Gutin et al. [16] considered a special case where there are exactly two possible item sizes and both item sizes are bounded from above by $1/k$ for some natural number k. They proposed an optimal 4/3-competitive online algorithm. Epstein and Levin [17] gave a further study on specific k and presented an online algorithm with optimal competitive ratio of at most $\frac{(k+1)^2}{k^2+k+1}$. Coffman et al. [18] investigated the asymptotic performance of FF algorithm in a dynamic bin packing model where items are released over time and are of size at most $1/k$ where k is a natural number. Han et al. [19] reinvestigated FF algorithm in the case with item size $1/i$ where i is a natural number not less than k.

In this paper we study an online bin packing scenario with a capacitated buffer and item size limitation. The rest of the paper is organized as follows. Section 2 describes the problem formally and gives some results on the asymptotic performance of NF algorithm. Sections 3 presents an NF-based online algorithm as well as its asymptotic competitive analysis. In Section 4 we present experimental results on practical performance of the two online algorithms. Finally Section 5 concludes this paper.

2   Basic Description and Analysis of NF Algorithm

2.1   Problem Description

There are sufficient unit size bins to pack a set of items $\{J_1, J_2, \ldots\}$. Items are released over list such that the decision on packing each item into some open bin has to be made on its arrival. There is an auxiliary buffer S with capacity $|S| \geq 1$, and each released item is temporarily stored in the buffer before it is moved into one of current open bins. The decision on packing each released item into an open bin is thus delayed to some extent due to the use of the buffer. It is assumed that the size s_j of any item J_j satisfies $s_j \in (\alpha, 1/2]$ where $0 \leq \alpha < 1/2$ due to the observation that for some specific departments, the planned operation time of any surgery cannot be arbitrarily small and is generally less than half of the length of available time interval in a day. We also assume that there is at most one bin open for loading items at any time, that is, the NF rule is adopted during bin packing. Items can be exchanged between an open bin and the buffer during packing. The objective is to minimize the total number of used bins for packing all the released items.

We observe that if $1/3 \leq \alpha < 1/2$, then each item is of size within $(1/3, 1/2]$. This is a trivial case since any algorithm may produce an optimal solution by packing exactly two items into each open bin. Thus, it is assumed that $0 \leq \alpha < 1/3$ in the remaining of this paper. For any item input instance, let $(B_1, B_2, \ldots, B_n)$ be the bins used by an online algorithm and indexed according to their opening order. Let l_i be the final load of each bin B_i $(1 \leq i \leq n)$. By the assumption that each item is of size within $(\alpha, 1/2]$, the following lemma is straightforward.

Lemma 1. *Given that item size is within $(\alpha, 1/2]$ where $0 \leq \alpha < 1/3$. For any online algorithm, each bin B_i $(1 \leq i < n)$ contains at least two items, and its load $l_i > 1/2$.*

2.2 Algorithm NF and Its ACR Analysis

Johnson [4] proved that NF has a worst-case ratio of 2 when item size is within $(0, 1]$. We observe that the algorithm has an ACR of 2, even if item size is within $(0, 1/2]$. This can be verified by a constructed instance, in which there are u copies of two items with sizes $\{1/2, \epsilon\}$ where $0 < \epsilon < 1/u$ and u is an arbitrarily large even number. NF spends u bins while an optimal offline algorithm OPT spends exactly $u/2 + 1$ bins, implying an ACR of 2. Below we restate the algorithm, and then prove its ACR in the case with a tighter item size limitation $(\alpha, 1/2]$.

> **Algorithm NF:**
> On the release of the first item J_1, open a bin B_1 to pack the item. On the release of any item J_j $(j = 2, 3, \ldots)$, pack the item into the current open bin B_i $(i \geq 1)$ if its current load is at most $1 - s_j$ before packing J_j; otherwise close B_i and open the next bin B_{i+1} to pack J_j.

Note that NF does not make use of the buffer S. The following theorem shows the asymptotic competitiveness of NF with item size limitation.

Theorem 1. *For the online bin packing problem with buffer S and item size in $(\alpha, 1/2]$ where $0 \leq \alpha < 1/3$, NF has an ACR of $\frac{2}{1+\alpha}$.*

Proof. For an arbitrary item input instance σ, assume that NF spends totally n bins to pack all the items. By Lemma 1, there are at least two items in bin B_i for $1 \leq i < n$. Let $J_{i,1}, J_{i,2}$ be the first and second packed items in B_i, and their sizes satisfy $\alpha < s_{i,j} \leq 1/2$ $(j = 1, 2)$.

By the description of algorithm NF, we conclude that for $1 < i \leq n$, the first packed item in B_i, i.e., $J_{i,1}$, can not be packed into the preceding bin B_{i-1}. This implies for $1 \leq i < n$ that

$$l_i + s_{i+1,1} > 1.$$

The load of B_i satisfies $l_i \geq s_{i,1} + s_{i,2} \geq s_{i,1} + \alpha$ for $1 \leq i < n$. Combining $l_n \geq s_{n,1} \geq \alpha$ with $l_1 > 1/2$ by Lemma 1, we have

$$2 \sum_{i=1}^{n} l_i \geq \sum_{i=1}^{n} l_i + l_1 + \sum_{i=2}^{n-1}(s_{i,1} + \alpha) + s_{n,1}$$

$$= l_n + l_1 + \sum_{i=1}^{n-1}(l_i + s_{i+1,1}) + (n-2)\alpha$$

$$> \alpha + 1/2 + (n-1) + (n-2)\alpha$$

$$= (1+\alpha)n + 1/2 - \alpha.$$

Thus $\sum_{i=1}^{n} l_i > (1+\alpha)n/2 + 1/4 - \alpha/2$, which is the least number of bins spent by an optimal algorithm OPT for instance σ. The ratio between the number of bins used by NF and that by OPT is equal to $\frac{n}{(1+\alpha)n/2+1/4-\alpha/2} \to \frac{2}{1+\alpha}$ as $n \to \infty$. The theorem follows. □

Below we further provide a lower bound of NF's ACR.

Theorem 2. *For the online bin packing problem with buffer S and item size in $(\alpha, 1/2]$ where $0 \leq \alpha < 1/3$, NF has an ACR of at least $\frac{2}{1+3\alpha/2}$.*

Proof. It suffices to construct an item input instance σ to make NF be at best asymptotic $\frac{2}{1+3\alpha/2}$-competitive. The instance σ consists of u copies of the following four items $\{1/2, \alpha, (1-\alpha)/2, \alpha\}$ where $u > 0$ is an arbitrarily large natural number. NF spends two bins for packing the four items in each copy, spending $2u$ bins in total. OPT spends $u/2$ bins for packing the u items with size $1/2$, and $u/2$ bins to pack the u items with size $(1-\alpha)/2$ together with $u/2$ items with size α. The rest $2u - u/2 = 3u/2$ items with size α consumes totally $\lceil 3u\alpha/2 \rceil < 3u\alpha/2 + 1$ bins. Thus OPT spends at most $u/2 + u/2 + 3u\alpha/2 + 1 = u + 3u\alpha/2 + 1$ bins. It implies an ACR of at least $2u/(u + 3u\alpha/2 + 1) \to 2/(1 + 3\alpha/2)$. □

By Theorems 1 and 2, both upper and lower bounds of NF's ACR approaches 2 as $\alpha \to 0$. As $\alpha \to 1/3$, the upper bound and lower bound approach 3/2 and 4/3 respectively.

3 An NF-Based Algorithm and Its ACR Analysis

In this section, we present an NF-based algorithm called NFB (NF with Buffer) that makes use of buffer S. The basic idea of the algorithm is that each bin is loaded with the largest items in the buffer and it is closed given that its load as well as any packed item in the bin is sufficiently large; otherwise it is repackaged with relatively small items in the buffer. On the release of each item J_j $(j \geq 1)$, let B_i $(i \geq 1)$ be the current open bin, and τ_i the set of items in B_i, S and $\{J_j\}$. The number of items in τ_i increases over the release of items during B_i is open.

We first consider a special case where $|S| = 1$, and prove the ACR of algorithm NFB. The result is then extended to the general case where S is of any constant capacity, i.e., $|S| \geq 1$. The algorithm is formally described as follows.

Algorithm NFB:

Step 0. Open the first bin B_1 on the release of the first item J_1 which is packed into the buffer temporarily.

Step 1. Repeatedly pack each released item into the buffer S. If there are no more items to be released, go to Step 4. Once a released item J_j cannot be packed into the buffer, move the largest item, denoted by J, in S and $\{J_j\}$ into the current open bin B_i $(i \geq 1)$ and store the rest items in S. Repeat this step until the J cannot be packed into B_i on the release of some J_j.

Step 2. Exchange the items between the open bin B_i and the buffer S. Let S_i be the set of items in S currently, and $\delta_i = \{J_{i,1}, J_{i,2}, \ldots, J_{i,n_i}\}$ the set of items in both S and B_i. The items in δ_i are indexed in non-increasing order of item size such that $s_{i,j+1} \leq s_{i,j}$ for $1 \leq j \leq n_i - 1$. Consider three cases.

- **Case 1.** There are exactly two items $J_{i,1}$ and $J_{i,2}$ in B_i, and $s_{i,2} > 1/3$. Further pack the items in S into B_i as many as possible in non-increasing order of item size.

- **Case 2.** There are exactly two items $J_{i,1}$ and $J_{i,2}$ in B_i and $s_{i,2} \leq 1/3$. In this case B_i is completely re-packed by the items in $\delta_i/\{J_{i,1}\}$ in non-increasing order of item size.

- **Case 3.** There are at least three items $J_{i,1}, J_{i,2}, J_{i,3}$ in B_i. Further pack the items in S into B_i in non-increasing order of item size.

Step 3. Close B_i and open the next bin B_{i+1}. Go back to Step 1.

Step 4. Pack the rest items in S into the current open bin in arbitrary order and close the bin, and finally pack all the rest items in S, if any, into another open bin.

Assume that NFB uses totally n bins for any item input instance. By Step 1 of the algorithm, if an item J is packed into B_i in the step, then J is of size not less than any other items in S at the time, which implies that the items in B_i contains the largest items in δ_i at the beginning of Step 2. Moreover, at each time packing J into B_i in Step 1, the total size of J and the rest items in S is strictly larger than $|S| = 1$; otherwise J is temporarily stored in the buffer. We conclude that the total size of items in S_i and the smallest item in B_i is strictly larger than one at the beginning of Step 2.

Lemma 2. *If B_i ($1 \leq i \leq n - 2$) is closed by Case 1 in Step 2, then $l_i > 2/3$; otherwise if it is closed by either Case 2 or Case 3 in Step 2, then $l_i > 3/4$.*

Proof. First, B_i contains at least the largest two items in δ_i at the beginning of Step 2 by Lemma 1 and Step 1 of the algorithm. If B_i is closed by Case 1 in Step 2, then B_i contains at least the largest two items $J_{i,1}$ and $J_{i,2}$ with $l_i > s_{i,1} + s_{i,2} > 1/3 + 1/3 = 2/3$.

If B_i is closed by Case 2 in Step 2, we already have by previous analysis that the total size of items in $\delta_i/\{J_{i,1}\}$ is strictly larger than one, i.e., $\sum_{j=2}^{n_i} s_{i,j} > 1$, at the beginning of Step 2. Since $s_{i,j} \leq s_{i,2} \leq 1/3$ for $3 \leq j \leq n_i$, we conclude $n_i \geq 5$, and B_i contains at least the largest three items $J_{i,2}, J_{i,3}, J_{i,4}$ in $\delta_i/\{J_{i,1}\}$ after repack. Assume that $J_{i,2}, J_{i,3}, \ldots, J_{i,k}$ are repackaged into B_i while $J_{i,k+1}$ cannot be repackaged into the bin where $k \geq 4$. We claim that $s_{i,2} + \ldots + s_{i,k} > (k-1)/k$ since otherwise $s_{i,k+1} \leq s_{i,k} \leq 1/k$, implying that $J_{i,k+1}$ can be further repackaged into B_i. A contradiction. Since $k \geq 4$, $(k - 1)/k \geq 3/4$.

For the case where B_i is closed by Case 3 in Step 2, the analysis is similar to the previous case. Assume that $J_{i,1}, J_{i,2}, \ldots, J_{i,k}$ are packed into B_i while $J_{i,k+1}$ cannot be further packed into B_i where $k \geq 3$. We claim that $s_{i,1} + \ldots + s_{i,k} > k/(k + 1)$ since otherwise $s_{i,k+1} \leq s_{i,k} \leq 1/(k + 1)$, implying that $J_{i,k+1}$ can be further packed into B_i. A contradiction. Since $k \geq 3$, $k/(k + 1) \geq 3/4$.

The lemma follows. □

According to the above lemma, we observe that each bin B_i ($1 \leq i \leq n - 2$) has a final load $l_i > 2/3$. For $1 \leq i \leq n - 2$, B_i is defined as a *light* bin if $l_i \in (2/3, 3/4]$; otherwise it is a *heavy* bin if $l_i \in (3/4, 1]$. We further define an item with size strictly larger than $1/3$ as a *large* item; otherwise it is a *small* item if its size is at most $1/3$. By Lemma 2, a light bin is due to Case 1 in Step 2 of the algorithm, and it contains exactly two large items by the description of Case 1 in Step 2.

Theorem 3. *For the online bin packing problem with buffer S and item size in $(\alpha, 1/2]$ where $0 \leq \alpha < 1/3$, if the buffer size $|S| = 1$, NFB has an ACR of $13/9$.*

Proof. Given an arbitrary item input instance σ. Assume that NFB produces a packing solution $\Gamma = \{B_1, B_2, \ldots, B_n\}$ such that except the last two bins B_{n-1}, B_n there are x light bins and y heavy bins. Thus $x + y = n - 2$. We already have that in each light bin, there are exactly two large items. So there are totally $2x$ large items in the x light bins. Let X be the set of the $2x$ large items, and Y the set of all the items in the y heavy bins. Note that there may be some small items in the x light bins.

We observe that any three large items in X are of a total size strictly larger than 1 and cannot be packed into one bin by any algorithms. Hence, an optimal algorithm OPT either assigns two large items in X to one bin, resulting in a rest room less than $1/3$ in the bin for other items, or assigns only one large item in X to one bin. Assume that for instance σ, OPT produces an optimal packing solution Γ^*, in which there are u bins each of which contains two large items in set X, v bins each of which contains only one large item in X, and w bins that contain no large items in X. So, $2u + v = 2x$.

Now we need to bound from above the number of heavy bins by NFB, i.e., the value of y, in the worst case. First, each item in the w bins in Γ^* is from either set Y or one of the x light bins. At most $\lceil \frac{w}{3/4} \rceil < \frac{4w}{3} + 1$ bins among the y heavy bins are used to pack the items in the w bins. Now consider each of the u bins in Γ^* that contains two large items in X. Since there is a room less than $1/3$ in each bin for OPT to pack small items from set Y, we conclude that the u bins contain small items with a total size less than $u/3$. NFB spends at most $\lceil \frac{u/3}{3/4} \rceil < \frac{4u}{9} + 1$ bins among the y heavy bins to pack the small items. With similar reasoning, for each of the v bins in Γ^*, there is a room less than $1 - 1/3 = 2/3$ for OPT to pack other items from set Y. Hence, at most $\lceil \frac{2v/3}{3/4} \rceil < \frac{8v}{9} + 1$ bins among the y heavy bins are used to pack the items from Y in the v bins. It is concluded that $y < (\frac{4w}{3} + 1) + (\frac{4u}{9} + 1) + (\frac{8v}{9} + 1) = \frac{4u+8v+12w}{9} + 3$.

OPT spends totally $u+v+w$ bins, and $u+v+w \to \infty$ in terms of asymptotic measure. NFB spends totally $n = x + y + 2$ bins. Together with $u + v/2 = x$, the ratio between the number of bins spent by NFB and by OPT is bounded as follows.

$$\frac{n}{u+v+w} < \frac{(u+v/2) + (4u + 8v + 12w)/9 + 3 + 2}{u+v+w}$$
$$= \frac{13u/9 + 25v/18 + 4w/3 + 5}{u+v+w}$$
$$\leq \frac{13}{9}, \quad as \ u+v+w \to \infty.$$

The theorem follows.   □

We observe that for the general case where $|S| \geq 1$, Lemma 2 still holds for each bin B_i ($1 \leq i < n - \lceil |S| \rceil - 1$). With similar analysis as in the proof of Theorem 3, we have the following corollary.

Corollary 1. *For the online bin packing problem with buffer S and item size in $(\alpha, 1/2]$ where $0 \leq \alpha < 1/3$, if the buffer size $|S| \geq 1$, NFB has an ACR of 13/9.*

The following theorem shows that any online algorithm cannot have an ACR less than $4/3$.

Theorem 4. *For the online bin packing problem with buffer S and item size in $(\alpha, 1/2]$ where $0 \leq \alpha < 1/3$ and the buffer size $|S| \geq 1$, any online algorithm cannot have an ACR less than 4/3.*

Proof. We first consider the case where $|S| = 1$. It suffices to construct an item input instance σ with $3u$ items in total such that the ratio between the number of bins used by any online algorithm $\mathcal{A}$ and that of OPT approaches $4/3$ asymptotically. In σ, the first $2u$ items are with uniform size $1/3 + \epsilon$ and the rest u items are with uniform size $1/3 - 2\epsilon$, where $\epsilon > 0$ is arbitrarily small and satisfies $1/3 - 2\epsilon \geq \alpha$. Since any three of the first $2u$ items are of a total size strictly larger than one, $\mathcal{A}$ spends $(u - 1)$ bins for packing the first $2(u - 1)$ items with a load of $2/3 + 2\epsilon$ in each bin. For the rest $u + 2$ items, $\mathcal{A}$ packs exactly three of them into each bin. The total number of bins used by $\mathcal{A}$ is equal to $\lceil (u-1) + (u+2)/3 \rceil \geq 4u/3 - 1/3$. For an optimal offline algorithm OPT, it consumes totally u bins to pack all the $3u$ items, each of which contains two items with size $1/3 + \epsilon$ and one item with size $1/3 - 2\epsilon$. The ratio between the number of bins used by $\mathcal{A}$ and that by OPT is at least $\frac{4u/3-1/3}{u}$, which approaches $4/3$ as $u \to \infty$.

For the other case where $|S| > 1$, the same instance is contructed. Similarly $\mathcal{A}$ spends $(u - \lfloor |S| \rfloor /2)$ bins for packing the first $2u - \lfloor |S| \rfloor$ items with a load of $2/3 + 2\epsilon$ in each bin. For the rest $u + \lfloor |S| \rfloor$ items, $\mathcal{A}$ packs exactly three of them into each bin. The total number of bins used by $\mathcal{A}$ is equal to $\lceil (u - \lfloor |S| \rfloor /2) + (u + \lfloor |S| \rfloor)/3 \rceil \geq 4u/3 - \lfloor |S| \rfloor /6$. OPT spends u bins to pack all the items. The ratio approaches $4/3$ again as $u \to \infty$.

The theorem follows.   □

By Theorems 3 and 4, we observe that provided that $|S| \geq 1$, the size of the buffer does not affect the ACR of online algorithms. That is, a larger size of the buffer cannot improve the asymptotic performance of algorithm NFB. In other words, a smallest size of one is enough for the buffer S during bin packing in algorithm NFB.

4   Computational Tests

In this section, we present experimental results of algorithms NF and NFB, to get an impression of their practical performance. We set the buffer size $|S| = 1$, and assume that item size follows the uniform distribution in $(\alpha, 1/2]$ where $0 \leq \alpha < 1/3$.

Let (n_{item}, n_{repe}) represent an experiment setting where for any given lower bound α $(0 \leq \alpha < 1/3)$ of item size, we produce n_{repe} instances each of which contains n_{item} randomly produced items with size in $(\alpha, 1/2]$. For each instance, we calculate the ratio between the number of bins used by algorithm NF (or NFB) and that of an optimal algorithm OPT. The average performance of algorithm NF (or NFB) is then measured by an average of ratios over the n_{repe} instances. For each setting combination (n_{item}, n_{repe}), we test the performance of the two online algorithms for various α.

Since it is well-known that the classical offline bin packing problem is NP-hard and it is almost impossible to obtain exact solutions for instances with large input scale. In the experiment, the number of bins used by OPT in each instance is replaced by its lower bound via rounding-up the total size of the n_{item} items. Thus, as α is relatively large, say $\alpha > 1/4$, the error between the theoretical optimal number of bins and the experimental value by the above rounding-up method is not negligible, and it results in inexact experimental ratios. In especial, we has previously mentioned that any online algorithms perform optimally provided that $\alpha \geq 1/3$, while experimental results show that both algorithms have an experimental average ratio of $6/5$ when $\alpha = 1/3$. Below we set $\alpha = 0, 0.0125, 0.025, \ldots, 0.2375, 0.25$, and test the variation of the average performance of both online algorithms for different α.

We first test the performance stability of algorithms NF and NFB in randomly generated instances with any given α. We test four setting combinations $(n_{item}, n_{repe}) = (5000, 20), (5000, 50), (10000, 20), (10000, 50)$ as well as some other combinations, and observe that when n_{item} is large, say $n_{item} \geq 5000$, the ratio between the number of bins used by online algorithm NF (or NFB) and that of bins by OPT varies little for any specific value of $\alpha \in [0, 1/4]$. Figure 1 is an illustration of the experimental ratios of algorithms NF and NFB in 20 instances where $\alpha = 1/10$ and $n_{item} = 10000$. The ratio for algorithm NF varies around 1.2, and it varies around 1.03 for NFB. All these ratios are much smaller than their respective theoretical ratios. By Theorems 1 and 3, the ACRs of NF and NFB are equal to 20/11 and 13/9 respectively when $\alpha = 1/10$. Figure 1 also reveals that algorithm NFB has much better practical performance than the classical algorithm NF with the help of buffer. When $\alpha = 1/10$, for instance, the use of buffer makes experimental ratios improved by around $0.17(= 1.2 - 1.03)$.

We next investigate the average performance of algorithms NF and NFB. Again experiment results show that both algorithms perform stably for any α. Figure 2 illustrates the average ratios of the online algorithms. Each point in the figure represents an average ratio over 20 randomly generated instances with $(n_{item}, n_{repe}) = (10000, 20)$. As the value of α increases from zero to 0.25, the ratios of both algorithms rise gradually. One reasonable explanation is the computational error of the optimal number of bins used by OPT rises as α increases, as mentioned previously. For any $\alpha \in [0, 1/4]$, the average ratio of NFB is much smaller than that of NF, i.e., a difference of at least 0.14. We conclude by the above experimental results that the use of buffer makes NF-based algorithms perform much better in asymptotic competitiveness.

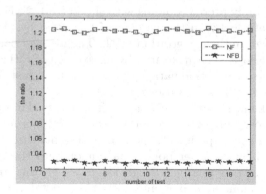

Fig. 1. A comparison of experimental ratios between algorithms NF and NFB in 20 instances ($\alpha = 1/10$, $n_{item} = 10000$)

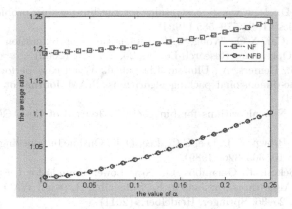

Fig. 2. A comparison of average ratios between algorithms NF and NFB $((n_{item}, n_{repe}) = (10000, 20))$

5 Conclusion

This paper investigates an scenario of online bin packing problem with a capacitated buffer and lower and upper bounds of item size. We mainly propose an NF-based online algorithm with an ACR of 13/9, and prove a lower bound of 4/3 as well, leaving a gap of 1/9 between the upper and lower bounds of ACR in the considered problem. Both theoretical and experimental results show that using buffer really helps in improving asymptotic competitive performance of NF-based algorithms. One interesting related problem is how much the buffer helps in asymptotic competitive performance of other packing rules in the ANY FIT class, where it is allowed to keep a finite number of bins other than a single bin open at any time.

Acknowledgements. This work was partially supported by the National Natural Science Foundation of China under Grants 71172189, 71071123 and 61221063, Program for New Century Excellent Talents in University (NCET-12-0824), and the Fundamental Research Funds for the Central Universities.

References

1. Ullman, J.D.: The performance of a memory allocation algorithm. Tech. Rep. 100, Princeton University, Princeton (1971)
2. Galambos, G., Woeginger, G.J.: On-line Bin Packing-A Restricted Survey. Mathematical Methods of Operations Research 42, 25–45 (1995)
3. Seiden, S.S.: On the online bin packing problem. Journal of the ACM 49(5), 640–671 (2002)
4. Johnson, D.S.: Fast algorithms for bin packing. Journal of Computer and System Sciences 8, 272–314 (1974)

5. Simchi-Levi, D.: New worst-case results for the bin-packing problem. Naval Research Logistics 41(4), 579–585 (1994)
6. Zhang, G.C., Cai, X.Q., Wong, C.K.: Linear time-approximation algorithms for bin packing. Operations Research Letters 26, 217–222 (2000)
7. Johnson, D.S., Demers, A., Ullman, J.D., et al.: Worst-case performance bounds for simple one-dimensional packing algorithms. SIAM Journal on Computing 3, 256–278 (1974)
8. Yao, A.C.C.: New algorithms for bin packing. Journal of the ACM 27, 207–227 (1980)
9. Ramanan, P., Brown, D.J., Lee, C.C., Lee, D.T.: On-line bin packing in linear time. J. Algorithms 10, 305–326 (1989)
10. Balogh, J., Békési, J., Galambos, G.: New Lower Bounds for Certain Classes of Bin Packing Algorithms. In: Jansen, K., Solis-Oba, R. (eds.) WAOA 2010. LNCS, vol. 6534, pp. 25–36. Springer, Heidelberg (2011)
11. Galambos, G.: A new heuristic for the classical bin packing problem. Tech. Rept 82, Institut für Mathematik, Universität Augsburg (1985)
12. Galambos, G., Woeginger, G.J.: Repacking helps in bounded space on-line bin packing. Computing 49, 329–338 (1993)
13. Gambosi, G., Postiglione, A., Talamo, M.: Algorithms for the relaxed online bin-packing model. SIAM Journal on Computing 30, 1532–1551 (2000)
14. Balogh, J., Galambos, G.: Algorithms for the on-line bin packing problem with repacking. Alkalmazott Matematikai Lapok 24, 117–130 (2007)
15. Balogh, J., Békési, J., Galambos, G., et al.: On-line bin packing with restricted repacking. Journal of Combinatorial Optimization (published online, 2013)
16. Gutin, G., Jensen, T., Yeo, A.: Optimal on-line bin packing with two item sizes. Algorithmic Operations Research 1(2), 72–78 (2006)
17. Epstein, L., Levin, A.: More on online bin packing with two item sizes. Discrete Optimization 5, 705–713 (2008)
18. Coffman, E.G., János, J., Rónyai, C.L., et al.: Dynamic bin packing. SIAM Journal of Computing 12(2), 227–258 (1983)
19. Han, X., Peng, C., Ye, D., et al.: Dynamic bin packing with unit fraction items revisited. Information Processing Letters 110, 1049–1054 (2010)

A Comparative Study of Multi-objective Evolutionary Algorithms for the Bi-objective 2-Dimensional Vector Packing Problem

Nadia Dahmani[1], Saoussen Krichen[1], François Clautiaux[2], and El-Ghazali Talbi[2]

[1] Institut Supérieur de Gestion de Tunis, LARODEC, 41 Avenue de la Liberté, Cité Bouchoucha, 2000 Le Bardo, Tunisie
[2] Université de Lille 1, LIFL CNRS UMR 8022, INRIA Lille-Nord Europe, Parc de la Haute Borne, 59655 Villeneuve d'Ascq, France

Abstract. This paper presents a comparative study of multi-objective evolutionary algorithms on the bi-objective 2-dimensional vector packing problem. Three state-of-the-art methods which prove their efficiency for a large variety of multi-objective optimization problems were designed to approximate the whole Pareto set of the problem. Computational experiments are performed on well-known benchmark test instances. The proposed algorithms are extensively compared to each other using different performance metrics.

Keywords: 2-dimensional vector packing problem, Multi-objective optimization, Evolutionary algorithms.

1 Introduction

The bi-objective version of the 2-dimensional vector packing problem (Mo2-DBPP) considers a set N of items $i = \{1, \ldots, N\}$ with two sizes in two independent dimensions c_i (weight) and h_i (height), and an unlimited number of identical bins with two sizes C and H. The 2-dimensional vector packing problem (2-DVPP) was always addressed in a single-objective way where the goal is to pack the items into a minimum number of bins without violating the capacity constraint in each dimension ([11,3]). However, from a practical point of view it is often necessary to simultaneously satisfy multiple objectives, and the problem investigated in this paper highlights what can typically be found in the industry. Indeed, as pointed out in [5], a large number of packing problems can be formulated as multi-objective optimization problems. In our work, we relax the height constraint, which becomes an objective to minimize. The main objective is to find a trade-off between the number of used bins and the maximum height of a bin.

The Mo2-DBPP is an $\mathcal{NP}$-hard problem since it is a generalization of the 2-DVPP. Moreover, large-scale problem instances generally cannot be solved exactly. Therefore, approximation search methods are considered to generate the set of potentially efficient solutions.

P. Widmayer, Y. Xu, and B. Zhu (Eds.): COCOA 2013, LNCS 8287, pp. 37–48, 2013.

A surge of research activities are observed in recent years on multi-objective evolutionary metaheuristics. These algorithms generate in a single run a set of solutions that approximate the whole or a part of the Pareto set. The main advantages of such approaches is that they do not need the transformation of the multi-objective problem into a single objective one.

Three grounded multi-objective evolutionary algorithms, namely the Indicator Based Evolutionary Algorithm (IBEA) [12], the Non-dominated Sorting Genetic Algorithm II (NSGA-II) [6] and the Strength Pareto Evolutionary Algorithm 2 (SPEA2) [13] are designed to tackle the bi-objective problem. A comparative study of these search methods for the Mo2-DBPP is investigated based on benchmark test instances [3] and their respective behavior is discussed.

The reminder of the paper is organized as follows. Section 2 is devoted to the Mo2-DBPP by providing a formal definition and a mathematical formulation of the addressed problem. Section 3 deals with the proposed multi-objective evolutionary algorithms. A general presentation of these methods and their application to the Mo2-DBPP is stated. In section 4, we report our experiments. A conclusion and perspectives are drawn in the last section.

2   The Bi-objective 2-Dimensional Vector Packing Problem

The Mo2-DBPP can be described as follows. Let $\{1, \ldots, N\}$ be a set of items. Each item i has a weight c_i and a height h_i. Let also $\{1, \ldots, \bar{M}\}$ be a set of bins, where $\bar{M}$ is an upper bound on the number of bins that can be used (obviously $\bar{M} \leq N$). Each bin j has a weight capacity C. The two conflicting objectives that have to be concurrently minimized are the number of used bins and the maximum height of a bin.

We state in (1)-(8) a mathematical formulation of the problem based on integer linear programming. The decision variables are defined as follows.

$$x_{ij} = \begin{cases} 1 & \text{if item } i \text{ is placed in bin } j \\ 0 & \text{otherwise} \end{cases}$$

$$y_j = \begin{cases} 1 & \text{if bin } j \text{ is used} \\ 0 & \text{otherwise} \end{cases}$$

H: is a integer variable that expresses the maximum height loaded into one bin.

$$\min \ f_1(s) = \sum_{j=1}^{\bar{M}} y_j \tag{1}$$

$$\min \ f_2(s) = H \tag{2}$$

$$\text{s.t} \ \sum_{j=1}^{\bar{M}} x_{ij} = 1, \ i = 1, \ldots, N \tag{3}$$

$$\sum_{i=1}^{N} c_i x_{ij} \leq C y_j, \ j = 1, \ldots, \bar{M} \tag{4}$$

$$\sum_{i=1}^{N} h_i x_{ij} - H \leq 0, \ j = 1, \ldots, \bar{M} \tag{5}$$

$$x_{ij} \in \{0, 1\}, \ i = 1, \ldots, N, \ j = 1, \ldots, \bar{M} \tag{6}$$

$$y_j \in \{0, 1\}, \ j = 1, \ldots, \bar{M} \tag{7}$$

$$H \in \mathbb{N} \tag{8}$$

where $s = (x_{ij}, y_j)$ $i = \{1, 2, \ldots, N\}$ and $j = \{1, \ldots, \bar{M}\}$.

Objectives

- The first objective (1) seeks to minimize the number of used bins.
- The second objective (2) tries to minimize the maximum height H of a bin.

Constraints

- The partition constraints (3) ensure that each item i is assigned to exactly one bin j.
- Inequalities (4) express the maximum weight capacity for each bin.
- Inequalities (5) assure that each item size combination into a single bin j must not exceed the maximum height H.

As the Mo2-DBPP is a bi-objective problem, a definition of some basic concepts related to multi-objective optimization is needed.

A solution $s \in S$ is evaluated using a vector $(f_1(s), f_2(s))$. A decision vector s dominates another vector s' if $f_k(s) \leq f_k(s') \ \forall k \in \{1, 2\}$ and $\exists l | f_l(s) < f_l(s')$ for $l \neq k$. Consequently, solving the Mo2-DBPP problem seeks to identify all efficient solutions or the Pareto set.

Due to the complexity of the considered problem and the potentially exponential number of efficient solutions, it is usually impossible to generate the entire Pareto set. Therefore, an alternative is to identify a good approximation of it, using multi-objective evolutionary algorithms.

3   Multi-objective Evolutionary Algorithms for the Mo2-DBPP

3.1   Multi-objective Evolutionary Algorithms

The multi-objective evolutionary algorithms designed to tackle the Mo2-DBPP are small variations of three state-of-the-art search method, namely IBEA [12], NSGA-II [6], and SPEA2 [13]. Small modifications were carried out to save the whole set of non-dominated solutions found during the search process.

IBEA. Presented by Zitzler and Künzli [12], the main idea of the *Indicator Based Evolutionary Algorithm* (IBEA) is to perform of a pairwise comparison of solutions contained in a population, using a binary quality indicator for the fitness assignment scheme. A fitness value $F(u)$ is assigned to each individual u to measure the "loss in quality" if u was removed from the current population P, (*i.e.*, $F(u) = \sum_{u' \in P \setminus \{u\}} (-e^{-I(u',u)/k})$), where $k > 0$ is a user-defined scaling factor. Different indicators can be used for such a purpose and the binary additive ϵ-indicator (Eps_2) is used in this work. As defined in [12], the $Eps_2(u, u')$ gives the minimum value by which a solution $u \in U$ has to or can be translated in the objective space to weakly dominate another solution $u' \in U$. The selection scheme is a binary tournament between randomly chosen individuals. The replacement strategy consists of deleting, one-by-one, the worst individuals and updating the fitness values of the remaining solutions each time there is a deletion. This process is iterated until the reach of the required population size. An archive is used to store the non-dominated solutions in order to prevent their loss during the stochastic search process.

NSGA-II. Introduced by Deb et al. [6], the *Non-dominated Sorting Genetic Algorithm II* (NSGA-II) sorts the solutions contained in the population into many classes at each generation. Individuals from the first front belong to the first efficient set. Individuals from the second front belong to the second best efficient set, etc. Two values are then computed for every solutions of the population. The first one consists of the *rank* of the corresponding solution. This value represents the quality of the solution in terms of convergence. The second one, deals with the *crowding distance* which is a density estimation of solutions surrounding a particular point of the objective space. This value represents the quality of the solution in terms of diversity. A dominant solution is characterized by a best rank, or in the case of a tie, by the best crowding distance. The selection strategy is a deterministic tournament between two random solutions. Only the best individuals survive at the replacement step with respect to the population size. Likewise, an external population is added to the steady-state NSGA-II in order to store every potentially efficient solution found during the search.

SPEA2. As an improved version of the *Strength Pareto Evolutionary Algorithm* (SPEA) , SPEA2 was elaborated by Zitzler et al. [13]. The algorithm adapts a fine-grained fitness assignment strategy that incorporates density information. During the selection process, SPEA2 manages a fixed size archive to generate the offspring solutions. When the number of non-dominated solutions exceeds the archive size, a bounded archive mechanism based on fitness and diversity informations, is performed. Contrariwise, if the number of non-dominated solution is less than the predefined archive size, it is filled up by dominated individuals. At each iteration, a strength value $S(x)$ is assigned to each individual x from the current population and the archive. $S(x)$ is equal to the number of solutions which are dominated by x according to the Pareto dominance properties. Then, the fitness value $F(x)$ of each solution is computed by adding all $S(x)$ values of dominated solutions by x. To preserve the diversity of solutions, a strategy based on k^{th} nearest neighbor, is incorporated. The selection scheme is elitist and consists of a binary tournament with replacement applied only on the archive.

3.2   Application to the Mo2-DBPP

Solution Encoding. A classical way for solving packing problems with metaheuristics is to use an indirect encoding based on a permutation $\sigma = (\sigma(1), \ldots, \sigma(N))$ of N items. The major asset of these indirect approaches is that there is no need for a penalty function as only feasible solutions are generated. All the problem-specific knowledge is handled by the decoder, which is generally a greedy heuristic.

For single objective problems, the main goal is to minimize the number of bins. Hence, a single decoder devoted to this criterion is sufficient. However, in the multi-objective framework, using a only one decoder may introduce a bias towards certain regions of the objective space. Thus, two well-known heuristics from the literature: The *first-fit* heuristic and the *least-loaded* heuristic (both described below) are adapted so as to get a balanced distribution of the generated potentially efficient solutions.

The following two additional genes are added to the chromosome and used as parameters by the decoder. Figure 1 illustrates our chromosome encoding.

- A binary variable $x \in \{0, 1\}$ that informs about the procedure used for decoding the chromosome.
- An integer variable lb that represents a classical lower bound on the number of bins proposed in [7].

Fig. 1. Chromosome encoding

Solution Evaluation. The evaluation process consists on decoding a solution in order to evaluate its merit of fitness. In order to obtain a balanced distribution of the generated solutions, we adapt the following two decoding procedures.

1. **The First fit algorithm** (FF) [4] is well known approximation algorithm that aims to minimize the number of bins. Thus, it maximize the height of a bin. FF considers the items in the predefined order of the permutation and starts from the first item and the first bin. Each item i is iteratively packed into the partially filled bin with the smallest index and which has sufficient residual capacity in terms of the second dimension (*i.e.*, the height) and respecting the weight capacity constraints. If no such a bin is available, then item i is packed into a new bin. The process is stopped when there no more items to pack.

2. **The Least loaded algorithm** (LL) is another greedy heuristic that was proposed in the server consolidation context [1]. The aim of this heuristic is to balance the load between the bins. Consequently, it deals with the minimization of the maximum height. However, the minimization of the number of bins is not guaranteed.

    The algorithm initializes a list of m empty bins that corresponds to a lower bound on the number of used bins and starts the iterative process by considering the items in the order given by the permutation. At each iteration, LL tries to pack the incoming item i into one of the current open bins by choosing the least loaded one in terms of the height while satisfying the weight capacity constraints. If LL fails to pack a selected item into the current open bins, then a new bin is opened. The process is stopped when there are no more items to pack.

    Since this heuristic needs a theoretical lower bound on the number of used bins as an input, we simply use the basic continuous lower bound. This parameter is embedded into the solution encoding (as described above) and modified during the search process.

3.3 Evolutionary Algorithms Features

Crossover Operator. We use the *Two points crossover* [8]. A pair of crossing points is randomly selected. The generated offspring preserves the items outside the selected two points from the first parent chromosome. The remaining items are inserted from the second parents while respecting the order of their appearance. For each gene added as a parameter (*i.e.*, x and lb), if it is the same for the two parents, then this information is kept for the generated offspring. If no, the offspring has 50% of chance to inherit the corresponding information from one of the parents. Figure 2 illustrates the manner in which this operator performs.

Mutation Operators. We adapted the following mutation operators:

- Shift mutation: A pair of randomly choosing components are shifted from the permutation part of the chromosome. The lb parameter is changed by removing a random number from the old values in the chromosome.

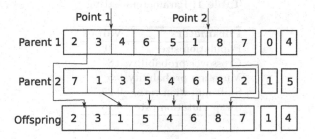

Fig. 2. Two points crossover

- Swap mutation: A pair of randomly choosing components are swapped from the the permutation part of the chromosome. The lb is changed by adding a random number to the old values in the chromosome.

For both operators, the x parameter is randomly changed in $[0, 1]$ in order to decide about the used decoding strategy (*i.e.*, FF or LL). From preliminary experiments, we noted that a good performance of the above described mutation operators is achieved by adding or removing to the lb parameter a random number in $[0, \frac{N}{5}]$.

4 Experiments

The presented multi-objective evolutionary algorithms stated in the foregoing section were implemented in C++ using the ParadisEO framework [10]. All algorithms share the same base components for a fair comparison. Computational runs were performed on an Intel Core 2 Duo 6600 (2.40 GHz) machine, with 2 GB RAM running Linux.

4.1 Experimental Design

Benchmarks. Experiments were conducted based on the benchmarks of Caprara and Toth [3] for 2-DVPP. About 24 instances are randomly selected by considering the problem classes in $\{1, \dots, 6\}$ according to their difficulty and the problem sizes related to the values $N \in \{25, 50, 100, 200\}$.

Parameters Setting. The parameters for the proposed multi-objective evolutionary algorithms are adjusted after some preliminary experiments. Indeed, we did not tune but set the parameters in order to obtain a relatively good performance on all instances. Table 1 presents the parameters setting used by all search methods. The initial population is randomly generated. The stopping criterion is related to the computational time. We arbitrarily set the amount

Table 1. Parameters setting

Parameters	Values
Population size	200
Crossover probability	0.8
Mutation probability	0.5
Shift mutation rate	0.25
Swap mutation rate	0.25

of runtime according to the size of the instance under consideration. For each value $N \in \{25, 50, 100, 200\}$, the runtime is equal to $\{60, 120, 240, 360\}$ seconds respectively for a single simulation run per instance and per algorithm.

Furthermore, specific parameters exist for some algorithms: SPEA2 requires an internal, fixed-size archive which is set to 100 individuals. Moreover, the scaling factor k of IBEA is set to 0.05 following [12].

Performance Assessment. For each search method, a set of 20 runs *per* instance were performed with different initial populations. In order to assess the quality of the approximated Pareto front obtained for every test instance, we follow the protocol given in [9]. First, we compute a reference set F_N^* of non-dominated solutions extracted from the union of all outputs. Then, we define $f^{max} = (f_1^{max}, f_2^{max})$, where f_1^{max} and f_2^{max} denote the upper bounds of both objective functions for all fronts approximations.

To evaluate the quality of an output set X regarding F_N^*, we use two different multi-objective performance indicators that inform about the convergence and the diversity of the generated fronts approximations. The unary hyper-volume metric [14,15] (Hyp) computes the portion of the objective space that is weakly dominated by F_N^* and not by X. We also consider the unary additive ϵ-indicator (Eps) proposed in [15]. Contrary to Eps_2 used in the IBEA algorithm, this indicator is used to compare non-dominated approximations sets, and not solutions. Eps gives the minimum value by which an approximation X has to be translated in the objective space to weakly dominate the reference set F_N^*. Note that f^{max} is considered as the reference point for both indicators. For each test instance, we obtain 20 Hyp and 20 Eps measures, corresponding to the 20 runs, *per* algorithm.

As suggested by Knowles et al. [9], once all these values are computed, we perform a statistical analysis for a pairwise comparison of algorithms. As the collected samples here can be considered as matched samples, the most appropriate statistical test is the Wilcoxon signed rank test. Indeed, for a given run, both the initial population and the random seed are identical for all algorithms. Thus, the final indicator values can be taken as pairs. Consequently, for each test instance, and according to the $p - value$ and to the metric under consideration, this statistical test indicates if the sample of approximation sets obtained by a given search method is significantly better than the ones of another search method, or if there is no significant difference between both. All the performance assessment procedures were conducted using the performance assessment tool suite provided in PISA[1]

4.2   Computational Results

Table 2 compares IBEA, NSGA-II and SPEA2 algorithms with respect to Hyp and Eps metrics respectively. Our experimental protocol gave rise to 120 scenarios for both indicators. For each class under consideration, we plotted the average metrics values (multiplied by 10^{-3}) for the corresponding algorithm. A lower average for both indicators signifies a "better" approximation set.

Column $\# \succ$ reports the number of times for which the current algorithm significantly dominates the other algorithms. This was achieved by performing the Wilcoxon statistical test and considering a $p - value$ equal to 5%.

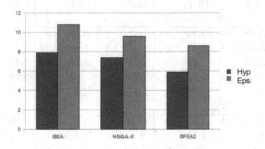

Fig. 3. Overall averages of Hyp and Eps indicator metrics

According to the experimental design used in this study, it can be noted that IBEA was better than NSGA-II and SPEA2 for small size instances ($N = 25$) of Class 1, 2 and 3 where each bin contains on average about 4, 2 and 2 items respectively. However, this does not hold for small size instances of Class 4, 5 and 6 where more items per bin are packed. Indeed, both algorithms (NSGA-II and SPEA2) concurrently outperform IBEA for these classes.

For medium size instances ($N = 50$ and $N = 100$), there was no significant difference between the considered algorithms. Their corresponding approximations sets were generally equivalent for both indicator metrics.

However, SPEA2 is clearly useful for large size instances ($N = 200$) and performs significantly better than IBEA and NSGA-II algorithms. Furthermore, the difference between the algorithm can be large for both metrics. For instance, for Class 5, $N = 200$, SPEA2 leads to values 1.5 and 3.88, whereas $IBEA$ leads to 32.27 and 39.39.

---

[1] The package is available at: http://www.tik.ee.ethz.ch/pisa/assessment.html [2]

Table 2. Algorithms comparison according with respect to Hyp and Eps metrics

Class	N	IBEA				NSGA − II				SPEA2			
		Hyp	$\#\succ$	Eps	$\#\succ$	Hyp	$\#\succ$	Eps	$\#\succ$	Hyp	$\#\succ$	Eps	$\#\succ$
	25	2.04	2	4.25	1	1.57	1	4.58	1	8.05	0	9.17	0
	50	6.48	0	9.25	0	6.83	0	9	0	5.92	1	8.97	0
1	100	4.23	0	6.91	0	4.71	0	7.32	0	4.16	0	6.4	0
	200	5.94	0	8.46	0	5.88	1	9.5	0	4.96	1	7.24	1
Average		5.55		7.22		4.75		7.6		5.77		7.95	
Total			2		1		2		1		2		1
	25	8.4	1	9.75	2	9.81	0	9.81	0	10.57	0	11.91	0
	50	8.38	0	11.97	0	9.3	0	11.91	0	8.61	0	11.96	0
2	100	9.96	0	8.46	0	10.11	0	8.95	4	9.27	0	9.1	0
	200	9.25	0	10.62	0	8.23	1	8.46	1	6.55	2	8.09	1
Average		9		10.2		9.36		9.78		8.75		10.27	
Total			1		2		1		5		2		1
	25	4.71	2	12.17	1	8.32	0	16.53	0	6.21	1	13.03	1
	50	8.26	1	16.16	1	11.48	0	17.21	0	9.33	1	16.49	1
3	100	11.86	0	14.17	0	10.93	0	13.52	0	11.48	0	13.06	0
	200	7.97	0	11.15	0	7.88	0	9.43	2	7.97	0	10.69	0
Average		8.2		13.41		9.65		14.17		8.75		13.32	
Total			3		2		0		2		2		2
	25	5.28	1	7.43	0	1.31	2	3.83	2	5.77	0	7.43	0
	50	3.62	0	6.41	0	3.37	1	5.83	2	3.22	0	6.12	0
4	100	3.01	2	5.61	0	3.65	0	5.03	1	3.13	1	4.7	2
	200	4.3	0	6.46	0	2.36	0	4.8	1	2.78	2	4.89	2
Average		4.05		6.48		2.67		4.87		3.73		5.79	
Total			3		0		3		6		3		4
	25	1.19	0	3.97	0	1.09	0	3.61	1	0.99	1	3.84	0
	50	1.72	0	4.15	0	3.4	0	5.52	1	1.5	1	3.8	0
5	100	24.5	0	18.04	0	10.82	0	15.05	1	1.7	2	3.94	1
	200	32.27	0	39.39	0	24.9	1	17.36	0	1.5	1	3.88	1
Average		14.92		16.39		10.05		10.39		1.42		3.87	
Total			0		0		1		3		5		2
	25	5.15	0	9.94	0	7.07	1	10.84	1	5.91	2	9.26	2
	50	7.72	1	10.23	0	6.98	0	9.71	0	6.09	1	9.11	1
6	100	9.89	0	12.62	0	9.76	1	11.57	0	8.64	2	12.35	0
	200	7.99	0	11.6	0	7.56	0	10.59	1	6.67	2	10.69	0
Average		5.69		11.1		7.84		10.68		6.83		10.35	
Total			1		0		2		2		7		3
Overall average		7.9		10.8		7.39		9.58		5.88		8.59	
Overall total			10		5		9		19		21		13

Finally, we can compare the overall performance of the considered algorithms. In Figure 3, we plotted the overall averages for both indicator metrics (Hyp and Eps respectively). Figure 4 presents the total occurrence number for which the current algorithm significantly dominates the other algorithms with respect to Hyp and Eps indicators and using the Wilcoxon statistical test with a $p-value$

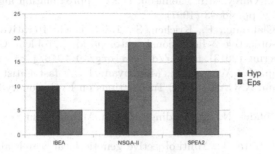

Fig. 4. Total number of occurrence where the current algorithm significantly dominates the other algorithms in terms of both indicators Hyp and Eps using the Wilcoxon statistical test with a $p - value = 5\%$

equal to 5%. SPEA2 outperforms IBEA and NSGA-II in overall Hyp and Eps averages (Figure 3). According to Hyp metric, it leads to 5.88 against 7.9 and 7.39 for both algorithms respectively. Moreover, it was 21 times better against 10 times for IBEA and 9 times for NSGA-II (see Figure 4).

5   Conclusion

In this paper, three multi-objective evolutionary algorithms were proposed for the Mo2-DBPP in order to provide good approximations for the whole Pareto set. Experiments were conducted using various test instances inspired from the literature. A comparative study for the search methods is stated based on different multi-objective performance indicators. Computational results show that although IBEA was effective for solving some small size instances and SPEA2 and NSGA-II were generally equivalent for solving medium size instances, SPEA2 was significantly better than IBEA and NSGA-II when solving large size instances. Furthermore, SPEA2 clearly outperforms both algorithms on overall average performance.

As a future work, it could be interesting to investigate and compare the performance of SPEA2 with the iterative approaches presented in [5] in solving the Mo2-DBPP.

References

1. Ajiro, Y., Tanaka, A.: Improving packing algorithms for server consolidation. In: International CMG Conference, pp. 399–406. Computer Measurement Group (2007)
2. Bleuler, S., Laumanns, M., Thiele, L., Zitzler, E.: Pisa: A platform and programming language independent interface for search algorithm. In: Fonseca, C.M., Fleming, P.J., Zitzler, E., Deb, K., Thiele, L. (eds.) EMO 2003. LNCS, vol. 2632, pp. 494–508. Springer, Heidelberg (2003)
3. Caprara, A., Toth, P.: Lower bounds and algorithms for the 2-dimensional vector packing problem. Discrete Applied Mathematics 111(3), 231–262 (2001)

4. Coffman, J.E.G., Garey, M.R., Johnson, D.S.: Approximation algorithms for bin packing: a survey, pp. 46–93 (1997)
5. Dahmani, N., Clautiaux, F., Krichen, S., Talbi, E.-G.: Iterative approaches for solving a multi-objective 2-dimensional vector packing problem. Computers & Industrial Engineering (2013), http://dx.doi.org/10.1016/j.cie.2013.05.016
6. Deb, K., Pratap, A., Agarwal, S., Meyarivan, T.: A fast elitist multi-objective genetic algorithm: NSGA-II. IEEE Transactions on Evolutionary Computation 6, 182–197 (2000)
7. Eilon, S., Christofides, N.: The loading problem. Management Science 17, 259–267 (1971)
8. Ishibuchi, H., Murata, T.: Multi-objective genetic local search algorithm and its application to flowshop scheduling. IEEE Transactions on Systems, Man and Cybernetics 28(3), 392–403 (1998)
9. Knowles, J., Thiele, L., Zitzler, E.: A Tutorial on the Performance Assessment of Stochastic Multiobjective Optimizers. TIK Report 214, Computer Engineering and Networks Laboratory (TIK), ETH Zurich (2006)
10. Liefooghe, A., Basseur, M., Jourdan, L., Talbi, E.-G.: ParadisEO-MOEO: A framework for evolutionary multi-objective optimization. In: Obayashi, S., Deb, K., Poloni, C., Hiroyasu, T., Murata, T. (eds.) EMO 2007. LNCS, vol. 4403, pp. 386–400. Springer, Heidelberg (2007)
11. Spieksma, F.C.R.: A branch-and-bound algorithm for the two-dimensional vector packing problem. Computers & Operations Research 21, 19–25 (1994)
12. Zitzler, E., Künzli, S.: Indicator-based selection in multiobjective search. In: Yao, X., Burke, E.K., Lozano, J.A., Smith, J., Merelo-Guervós, J.J., Bullinaria, J.A., Rowe, J.E., Tiňo, P., Kabán, A., Schwefel, H.-P. (eds.) PPSN 2004. LNCS, vol. 3242, pp. 832–842. Springer, Heidelberg (2004)
13. Zitzler, E., Laumanns, M., Thiele, L.: SPEA2: Improving the strength pareto evolutionary algorithm for multiobjective optimization. In: Giannakoglou, K.C., Tsahalis, D.T., Périaux, J., Papailiou, K.D., Fogarty, T. (eds.) Evolutionary Methods for Design Optimization and Control with Applications to Industrial Problems, Athens, Greece, pp. 95–100. International Center for Numerical Methods in Engineering (2001)
14. Zitzler, E., Thiele, L.: Multiobjective evolutionary algorithms: a comparative case study and the strength pareto approach. IEEE Transactions on Evolutionary Computation 3(4), 257–271 (1999)
15. Zitzler, E., Thiele, L., Laumanns, M., Fonseca, C.M., da Fonseca, V.G.: Performance assessment of multiobjective optimizers: An analysis and review. IEEE Transactions on Evolutionary Computation 7, 117–132 (2003)

Approximation Algorithms for the Maximum Multiple RNA Interaction Problem

Weitian Tong[1], Randy Goebel[1], Tian Liu[2], and Guohui Lin[1,*]

[1] Department of Computing Science, University of Alberta
Edmonton, Alberta T6G 2E8, Canada
{weitian,rgoebel,guohui}@ualberta.ca
[2] Key Laboratory of High Confidence Software Technologies, Ministry of Education
Institute of Software, School of Electronic Engineering and Computer Science
Peking University, Beijing 100871, China
lt@pku.edu.cn

Abstract. RNA interactions are fundamental in many cellular processes, which can involve two or more RNA molecules. Multiple RNA interactions are also believed to be much more complex than pairwise interactions. Recently, multiple RNA interaction prediction has been formulated as a maximization problem. Here we extensively examine this optimization problem under several biologically meaningful interaction models. We present a polynomial time algorithm for the problem when the order of interacting RNAs is known and pseudoknot interactions are allowed; for the general problem without an assumed RNA order, we prove the NP-hardness for both variants (allowing and disallowing pseudoknot interactions), and present a constant ratio approximation algorithm for each of them.

Keywords: RNA interaction, maximum weight *b*-matching, acyclic 2-matching, approximation algorithm, worst case performance ratio.

1 Introduction

RNA interaction is one of the fundamental mechanisms underlying many cellular processes, in particular genome regulatory code processes, such as mRNA translation, editing, gene silencing, and synthetic self-assembling RNA design. In the literature, pairwise RNA interaction prediction has been independently formulated as a computational problem [15,2,11]. While these variants are all motivated by specific biological considerations, the general formulation is usually NP-hard and many special scenarios have been extensively studied [13,4,5,16,8,9].

In more complex instances, biologists have found that multiple small nucleolar RNAs (snoRNAs) interact with ribosomal RNAs (rRNAs) in guiding the methylation of the rRNAs [12], and multiple small nuclear RNAs (snRNA) interact with an mRNA in the splicing of introns [17]. Multiple RNA interactions are believed much more complex than pairwise RNA interactions, where only

---

* Corresponding author.

P. Widmayer, Y. Xu, and B. Zhu (Eds.): COCOA 2013, LNCS 8287, pp. 49–59, 2013.
© Springer International Publishing Switzerland 2013

two RNA molecules are involved. In fact, even if we have a perfect computational framework for pairwise RNA interactions, it might still be difficult to deal with multiple RNA interactions since, for a given pool of RNA molecules, it is non-trivial to predict their interaction order without sufficient prior biological knowledge.

Motivated by biological goals, Ahmed *et al.* have developed a system for multiple RNA interaction prediction, denoted as MRIP [1]. Here we provide basic definitions to formally introduce the MRIP problem. An RNA molecule is a sequence of nucleotides (A, C, G, and U). A basepair in the RNA is presented as (i, j), where $i < j$, indicating that the i-th nucleotide and the j-th nucleotide form a canonical pairing (*i.e.*, the two nucleotides are either A and U or C and G). The molecule folds into a *structure* which is described as a set of basepairs. In general, every nucleotide can participate in at most one basepair, and if not, it is a *free* base (or nucleotide). The set of basepairs is *nested* (a.k.a. secondary structure), if for any two basepairs (i_1, j_1) and (i_2, j_2) with $i_1 < i_2$, either $j_1 < i_2$ or $j_2 < j_1$; otherwise the set is *crossing* (a.k.a. tertiary structure) containing *pseudoknots*. An interaction between two RNAs is a basepair which consists of one free base from each RNA. In the sequel, we use interaction and basepair interchangeably.

In the MRIP problem, we are given a pool of RNA sequences denoted as $\mathcal{R} = \{R_1, R_2, \ldots, R_m\}$. Without loss of generality, we assume m is even and these RNA sequences have the same length n. We use $R_{i\ell}$ to denote the ℓ-th base of R_i. Following the formulation by Ahmed *et al.* [1], the possible interactions between every pair of RNAs are assumed known. In fact, these possible interactions can be predicted using existing pairwise RNA interaction predictors [13,4,5,16,8,9]. For a possible interaction $(R_{i_1\ell_1}, R_{i_2\ell_2})$, its weight $w(R_{i_1\ell_1}, R_{i_2\ell_2})$ can be set using a probabilistic model or using an energy model or simply at 1 to indicate its contribution to the structure stability. The problem goal is to find out the order of RNAs in which they interact, that the first RNA interacts with the second RNA, which in turn interacts with the third RNA, and so on, and how every two consecutive RNAs interact, so as to maximize the total weight of the interactions (to achieve the most structure stability). Throughout this paper, we consider the uni-weight case, that is to maximize the total number of interactions. Two interactions $(R_{i_1\ell_1}, R_{i_2\ell_2})$ and $(R_{i_1k_1}, R_{i_2k_2})$ are pseudoknot-like if $\ell_1 < k_1$ but $\ell_2 > k_2$. The MRIP problem can allow or disallow pseudoknot-like interactions, depending on application details similar to RNA structure prediction.

For a very special case of MRIP (the *Pegs and Rubber Bands* problem in [1]), where the order of interacting RNAs is assumed and pseudoknot-like interactions are disallowed, Ahmed *at al.* proved its NP-hardness and presented a polynomial-time approximation scheme [1]. Given that predicting the interaction order is nontrivial, they also proposed a heuristic for the more general case with unknown interacting order but still disallowing pseudoknot-like interactions.

In this paper, we first show that the MRIP allowing pseudoknot-like interactions and with an assumed RNA interaction order can be solved in polynomial time.

Secondly, notice that the interactions are basepairs and thus follow the Watson-Crick basepairing rule. For four RNAs $R_{i_1}, R_{i_2}, R_{i_3}, R_{i_4}$, when there are possible interactions $(R_{i_1\ell_1}, R_{i_2\ell_2}), (R_{i_2\ell_2}, R_{i_3\ell_3}), (R_{i_3\ell_3}, R_{i_4\ell_4})$ (for example, if they are basepairs (A, U), (U, A), (A, U), respectively), then it is naturally to assume another possible interaction $(R_{i_1\ell_1}, R_{i_4\ell_4})$ between RNAs R_{i_1} and R_{i_4}. If the given interactions satisfy the above property then the MRIP problem is said to have the "*transitivity*" property. We show that the MRIP problem without an assumed RNA interaction order, either allowing or disallowing pseudoknot-like interactions, is NP-hard. In this case we present a constant ratio approximation algorithm for each variant.

2   Algorithmic and Hardness Results

2.1   MRIP with a Known RNA Interaction Order

In this subsection, we consider the MRIP problem with a known RNA interaction order, and we assume the order is $(R_1, R_2, \ldots, R_m)$. When disallowing pseudoknot-like interactions, Ahmed *et al.* [1] showed that the problem is NP-hard via a reduction from the *Longest Common Subsequence* problem.

Theorem 1. [1] *The MRIP problem disallowing pseudoknot-like interactions is NP-hard.*

When allowing pseudoknot-like interactions, we first construct a graph $H = (U, F)$ where every vertex $u_{i\ell}$ corresponds to nucleotide $R_{i\ell}$ and two vertices are connected by an edge if there is a given possible interaction between them. Clearly, one can see that a matching M of graph H gives a feasible solution to the MRIP problem allowing pseudoknot-like interactions, and vice versa. Therefore, the *MRIP* problem allowing pseudoknot-like interactions can be solved in polynomial time.

2.2   The General MRIP

By general MRIP, we mean the MRIP problem in which no RNA interaction order is assumed. Instead, the possible interactions are given for every pair of RNAs and the problem goal is to find an interaction order achieving the maximum number of interactions.

Theorem 2. *The general* MRIP *problem, either allowing or disallowing pseudoknot-like interactions, is NP-hard.*

Proof. Given a 0-1 matrix $A_{m \times n}$, two consecutive 1's in a column of the matrix is said to form a *bandpass*. When counting the total number of bandpasses in the matrix, no two bandpasses in the same column are allowed to share any common 1. The *Bandpass* problem is to find a row permutation for the input matrix to achieve the maximum total number of bandpasses. Lin proved that the Bandpass problem is NP-hard via a reduction from the *Hamiltonian Path* problem [10].

Let the i-th RNA be the i-th row of matrix A, so there is a possible interaction between $R_{i_1 \ell_1}$ and $R_{i_2 \ell_2}$ if and only if both positions have a 1. Though such constructed RNAs and interactions are not necessarily biologically meaningful, this reduction shows the general MRIP problem is NP-hard. Furthermore, no two possible interactions between a pair of RNAs are *crossing* each other, and thus there are no pseudoknot-like interactions. Hence, the general MRIP problem, either allowing or disallowing pseudoknot-like interactions, is NP-hard.   □

Given an instance I of a maximization problem Π, let $C^*(I)$ ($C(I)$, respectively) denote the value of the optimal solution (the value of the solution produced by an algorithm, respectively). The performance ratio of the algorithm on I is $\frac{C(I)}{C^*(I)}$. The algorithm is a ρ-approximation if $\inf_I \frac{C(I)}{C^*(I)} \geq \rho$, that is, it guarantees, on any instance, a solution of value at least a fraction ρ of the optimum.

Using the possible interactions between the pair of RNAs R_i and R_j, we construct a bipartite graph $BG(i, j) = (V_i \cup V_j, E(i, j))$, where the vertex subset V_i (V_j, respectively) corresponds to the set of nucleotides in R_i (R_j, respectively) and the edge set $E(i, j)$ corresponds to the set of given possible interactions between R_i and R_j. That is, if $(R_{i\ell_1}, R_{j\ell_2})$ is a possible interaction, then there is an edge between $R_{i\ell_1}$ and $R_{j\ell_j}$ in $BG(i, j)$. One clearly sees that, when allowing pseudoknot-like interactions, the size of the maximum matching in $BG(i, j)$ is exactly the maximum total number of interactions between RNAs R_i and R_j; when pseudoknot-like interactions are not allowed, the maximum total number of interactions between RNAs R_i and R_j can be computed by a dynamic programming algorithm similar to one for computing the longest common subsequence between two given sequences. Either way, this maximum number of interactions is set as the weight between RNAs R_i and R_j, denoted as $w(R_i, R_j)$.

We next construct an edge-weighted complete graph G, in which a vertex corresponds to an RNA and the weight between two vertices (RNAs) R_i and R_j is $w(R_i, R_j)$ computed above. Since the optimal solution to the MRIP problem, either allowing or disallowing pseudoknot-like interactions, can be decomposed into two matchings by including alternate edges in the solution, the maximum weight matching M^* of graph G has a weight that is at least half of the total number of interactions in the optimal solution. It follows that this maximum weight matching-based algorithm, described in Fig. 1, is a 0.5-approximation to the MRIP problem.

Theorem 3. APPROX I *is a 0.5-approximation algorithm for the general MRIP problem, either allowing or disallowing pseudoknot-like interactions.*

Proof. When allowing pseudoknot-like interactions, $w(R_i, R_j)$ can be computed by a maximum matching algorithm in $O(n^3)$ time, where n is the (common) length of the given RNAs.

When disallowing pseudoknot-like interactions, $w(R_i, R_j)$ can be computed by a dynamic programming algorithm in $O(n^2)$ time.

Input: m RNAs R_i, $i = 1, 2, \ldots, m$;

Output: a permutation π of $[m]$ and interactions between RNAs $R_{\pi(i)}$ and $R_{\pi(i+1)}$, for $i = 1, 2, \ldots, m - 1$

1. for each RNA pair R_i and R_j,
    1.1. construct bipartite graph $BG(i, j)$;
    1.2. compute $w(R_i, R_j)$;
2. construct edge-weighted complete graph G;
3. compute the maximum weight matching M^* of G;
4. stack RNA pairs in M^* arbitrarily to form a permutation π;
5. output π and the interactions in $w(R_{\pi(i)}, R_{\pi(i+1)})$.

Fig. 1. A high-level description of APPROX I

It follows that the time for constructing graph G is $O(m^2 n^3)$. Graph G contains m vertices, and its maximum weight matching M^* can be computed in $O(m^3)$ time. Subsequent construction of the solution permutation π takes linear time.

Therefore, APPROX I is an $O(\max\{m^3, m^2 n^3\})$-time 0.5-approximation algorithm for the MRIP problem allowing pseudoknot-like interactions. For the MRIP problem disallowing pseudoknot-like interactions, its worst-case performance ratio remains 0.5, but its running time is $O(\max\{m^3, m^2 n^2\})$. □

3 Better Approximations for General MRIP with Transitivity

In the previous section, we proved the NP-hardness for the general MRIP problem, and presented a 0.5-approximation algorithm. One can imagine that this performance ratio of 0.5 must be tight, if the given possible interactions are arbitrary. Now we consider a biologically meaningful special case where the given possible interactions are *transitive*, that is, for any four RNAs $R_{i_1}, R_{i_2}, R_{i_3}, R_{i_4}$, when there are possible interactions $(R_{i_1 \ell_1}, R_{i_2 \ell_2}), (R_{i_2 \ell_2}, R_{i_3 \ell_3}), (R_{i_3 \ell_3}, R_{i_4 \ell_4})$ (for example, they are basepairs (A, U), (U, A), (A, U), respectively), then $(R_{i_1 \ell_1}, R_{i_4 \ell_4})$ is also a possible interaction between RNAs R_{i_1} and R_{i_4}. We call it the general MRIP problem with transitivity. Note that in the proof of NP-hardness in Theorem 2, the constructed instance of the MRIP problem satisfies the transitivity property. Therefore the general MRIP problem with transitivity, either allowing or disallowing pseudoknot-like interactions, is also NP-hard. We next show that we can explore the transitivity property to design approximation algorithms with performance ratios better than 0.5.

3.1 A 0.5328-Approximation for Disallowing Pseudoknots

The improved approximation algorithm for the general MRIP with transitivity and disallowing pseudoknot-like interactions is denoted as APPROX II, and its high-level description in provided in Fig. 2.

Input: m RNAs R_i, $i = 1, 2, \ldots, m$, with transitivity;
Output: a permutation π of $[m]$ and interactions between RNAs $R_{\pi(i)}$ and $R_{\pi(i+1)}$,
      for $i = 1, 2, \ldots, m - 1$

1. for each RNA pair R_i and R_j,
   1.1. construct bipartite graph $BG(i, j)$;
   1.2. compute $w(R_i, R_j)$ disallowing pseudoknot-like interactions;
2. construct edge-weighted complete graph G using edge weight function w;
3. compute the maximum weight matching M^* of G;
   3.1. delete nucleotides involved in the interactions of M^*;
   3.2. reconstruct bipartite graph $BG(i, j)$;
   3.3. compute $w'(R_i, R_j)$ disallowing pseudoknot-like interactions;
4. construct edge-weighted complete graph G' using edge weight function w';
   4.1. compute the maximum weight 4-matching $\mathcal{C}$ of G';
   4.2. compute an approximate acyclic 2-matching $\mathcal{P}$ of G';
   4.3. compute a matching M out of $\mathcal{C}$ and $\mathcal{P}$ to extend M^*;
5. stack RNA paths in $G[M^* \cup M]$ arbitrarily to form a permutation π;
6. output π and the interactions in $w(R_{\pi(i)}, R_{\pi(i+1)}) + w'(R_{\pi(i)}, R_{\pi(i+1)})$.

Fig. 2. A high-level description of APPROX II

Note that to compute the maximum number of interactions between two RNAs R_i and R_j (Step 1.2) while disallowing pseudoknot-like interactions, we can use the same dynamic programming algorithm used in APPROX I, which runs in $O(n^2)$-time. In Step 4.2, the best approximation algorithm for the Maximum-TSP (with a performance ratio of $\frac{7}{9}$ [14]) is used to compute an acyclic 2-matching. In Step 4.3, to compute a matching M to extend M^*, the union of the edge sets of M and M^*, i.e. $G[M \cup M^*]$, is an acyclic 2-matching (sub-tour is an alternative terminology often used in the literature). Basically algorithm APPROX II adds to the maximum weight matching M^* of graph G a subset of edges that contains a proven fraction of interactions.

Let I denote the set of interactions in the optimal solution. Let J be the set of interactions extracted from the weights of the edges in the maximum weight matching M^* of graph G. Note that neither I or J contains crossing interactions. Similarly as in the MRIP problem with a known RNA interaction order (Section 2), we construct another graph $H = (U, F)$ for the instance where every vertex $u_{i\ell}$ corresponds to nucleotide $R_{i\ell}$ and two vertices are connected by an edge if there is a given possible interaction between them. With respect to graph H, both I and J are non-crossing matchings. Therefore, the subgraph of H induced by the interactions of I and J, $H[I \cup J]$, is a 2-matching of graph H, denoted by T. Using this 2-matching T, we partition I into 4 subsets of interactions, $I = I_1 \cup I_2 \cup I_3 \cup I_4$, and at the same time partition J into 4 subsets of interactions, $J = J_1 \cup J_2 \cup J_3 \cup J_4$.

Since T is a 2-matching, there are only alternating paths and cycles in T. First we consider paths. For a path of length 1, say $P = \langle u_1, u_2 \rangle$, if its only edge/interaction is in $I \cap J$, then the edge belongs to I_1 and belongs to J_1 too;

if the edge is in $I - J$, then the edge belongs to I_4; if the edge is in $J - I$, then the edge belongs to J_4. For a path of length 3, say $P = \langle u_1, u_2, u_3, u_4 \rangle$, if $(u_1, u_2), (u_3, u_4) \in I$, then they belong to I_2 and edge (u_2, u_3) belongs to J_2. For a path other than the above cases, the edges of I all belong to I_3 and the edges of J all belong to J_3. And for each cycle, the edges of I all belong to I_3 and the edges of J all belong to J_3.

Lemma 1. *Let $|X_i|$ denote the size of, that is the number of interactions in, set X_i, for $X = I, J$ and $i = 1, 2, 3, 4$. We have $|J_1| = |I_1|$, $|J_2| = \frac{1}{2}|I_2|$, and $|J_3| \geq \frac{2}{3}|I_3|$.*

Proof. By the definition of I_1, J_1, I_2, J_2, we can easily see $|J_1| = |I_1|$ and $|J_2| = \frac{1}{2}|I_2|$. For I_3 and J_3, from each path or cycle, the number of edges assigned to J_3 is either greater than or equal to the number of edges assigned to I_3, or 1 less; but in the latter case the length of the path must be at least 5. Therefore, the worst case happens when two and three edges are assigned to J_3 and I_3 respectively, which implies $|J_3| \geq \frac{2}{3}|I_3|$.   □

Corollary 1. *We have*

$$|I| = |I_1| + |I_2| + |I_3| + |I_4|, \tag{1}$$

$$w(M^*) = |J_1| + |J_2| + |J_3| + |J_4|, \tag{2}$$

$$w(M^*) \geq |I_1| + \frac{1}{2}|I_2| + \frac{2}{3}|I_3|, \tag{3}$$

$$w(M^*) \geq \frac{1}{2}|I| = \frac{1}{2}\left(|I_1| + |I_2| + |I_3| + |I_4|\right). \tag{4}$$

Proof. The first two equations are straightforward, following the description of partitioning process and that $w(M^*) = |J|$. The last two inequalities follow from Lemma 1 and Theorem 3, respectively.   □

After deleting the bases involved in the interactions of the maximum weight matching M^*, graph G' is constructed the same as graph G except using weight function w'. For a path of length 3, $P = \langle u_1, u_2, u_3, u_4 \rangle$, such that (u_1, u_2), $(u_3, u_4) \in I_2$, the transitivity property ensures that there is a possible interaction between u_1 and u_4. Clearly, this interaction is left in graph G', and such an interaction is called an *induced* interaction. Let G'_s be the subgraph of G' that contains exactly those edges each of which is contributed by at least one induced interaction.

Lemma 2. *G'_s is a 4-matching in G', and its weight $w'(G'_s) \geq \frac{1}{2}|I_2|$.*

Proof. To prove the first part, we only need to prove that every RNA can have induced interactions with at most 4 other RNAs. By the definition of I_2, there is an induced interaction (u_1, u_4) if and only if there is an alternating length-3 path

$P = \langle u_1, u_2, u_3, u_4 \rangle$, such that $(u_1, u_2), (u_3, u_4) \in I$ and $(u_2, u_3) \in J$. Suppose $u_k \in R_{i_k}$, for $k = 1, 2, 3, 4$. It follows that R_{i_1}, R_{i_2} (R_{i_3}, R_{i_4}, respectively) are adjacent in the optimal permutation and R_{i_2}, R_{i_3} are matched in M^*. Since each RNA can be adjacent to at most two other RNAs in the optimal solution, R_{i_1} and every RNA can have induced interactions with at most 4 other RNAs.

The second part of the lemma follows directly from the definition of an induced interaction, which corresponds to a distinct pair of interactions of I_2.   □

It is known that in $O(n^{2.5})$ time, a 4-matching can be decomposed into two 2-matchings [6,7], and a 2-matching can be further decomposed for our purpose in the next few lemmas.

Lemma 3. [3,18] *Let C be a 2-matching of graph G such that $M^* \cap C = \emptyset$. Then, we can partition the edge set of C into 4 matchings X_0, X_1, X_2, X_3 each of which extends M^*. Moreover, the partitioning takes $O(n\alpha(n))$ time, where $\alpha(n)$ is the inverse Ackerman function.*   □

The maximum weight 4-matching C of graph G' can be decomposed into two 2-matchings C_1 and C_2. By Lemma 3, C_1 can be partitioned into 4 matchings X_0, X_1, X_2, X_3 and C_2 can be partitioned into 4 matchings Y_0, Y_1, Y_2, Y_3, each of which extends M^*.

Lemma 4. [18] *Let C be a 4-matching of graph G such that $M^* \cap C = \emptyset$. Then, we can partition the edge set of C into 8 matchings such that each of them extends M^* and the maximum weight among them is at least $\frac{2}{15}w'(C)$. Moreover, the partitioning takes $O(n^{2.5})$ time.*   □

Lemma 5. *The maximum weight acyclic 2-matching $\mathcal{D}$ of graph G' has weight $w'(\mathcal{D}) \geq |I_4|$.*

Proof. Note that graph G' contains all interactions of I_4 because the only bases deleted are those involved in the interactions of M^*. The subgraph of graph G' containing exactly the edges contributed by at least one interaction of I_4 is a subgraph of the optimal solution, and thus it is an acyclic 2-matching in graph G'. Therefore,

$$w'(\mathcal{D}) \geq |I_4|.$$

This proves the lemma.   □

Lemma 6. [3,18] *Let $\mathcal{P}$ be an acyclic 2-matching of G such that $M^* \cap \mathcal{P} = \emptyset$. Then, we can partition the edge set of $\mathcal{P}$ into three matchings Y_0, Y_1, Y_2 each of which extends M^*. Moreover, the partitioning takes $O(n\alpha(n))$ time.*   □

Lemma 7. [14] *The Max-TSP admits an $O(n^3)$-time $\frac{7}{9}$-approximation algorithm, where n is the number of vertices in the graph.*   □

Corollary 2. *The weight of the second matching M to extend M^* has weight $w'(M) \geq \max\{\frac{1}{15}|I_2|, \frac{7}{27}|I_4|\}$.*

Proof. Using Lemmas 2 and 4, we have

$$w'(M) \geq \frac{2}{15}w'(\mathcal{C}) \geq \frac{1}{15}|I_2|.$$

Using Lemmas 5–7, we have

$$w'(M) \geq \frac{1}{3}w'(\mathcal{P}) \geq \frac{7}{27}w'(\mathcal{D}) \geq \frac{7}{27}|I_4|.$$

The corollary holds.   □

Theorem 4. *Algorithm* APPROX II *is a 0.5328-approximation for the general MRIP problem with transitivity and disallowing pseudoknot-like interactions.*

Proof. Combining Corollaries 1 and 2, we have for any real $x, y \in [0, 1]$,

$$\begin{aligned}
w(\pi) &= w(M^*) + w'(M) \\
&\geq x(|I_1| + \frac{1}{2}|I_2| + \frac{2}{3}|I_3|) + (1-x)\frac{1}{2}(|I_1| + |I_2| + |I_3| + |I_4|) \\
&\quad + y\frac{1}{15}|I_2| + (1-y)\frac{7}{27}|I_4| \\
&= \frac{1+x}{2}|I_1| + \frac{15+2y}{30}|I_2| + \frac{3+x}{6}|I_3| + \frac{41-27x-14y}{54}|I_4| \\
&\geq \frac{255}{426}|I_1| + \frac{227}{426}|I_2| + \frac{227}{426}|I_3| + \frac{227}{426}|I_4| \\
&\geq \frac{227}{426}|I|,
\end{aligned}$$

where the second last inequality holds by setting $x = \frac{14}{71}$ and $y = \frac{35}{71}$.   □

3.2   A 0.5333-Approximation for Allowing Pseudoknots

The improved approximation algorithm for the general MRIP with transitivity and allowing pseudoknot-like interactions is denoted as APPROX III, and its high-level description is provided in Fig. 3.

APPROX III is very similar to APPROX II, and only differs at two places. First, since the problem allows pseudoknot-like interactions, we run a maximum weight bipartite matching algorithm to compute those edge weights, in Steps 1.2 and 3.3. Second, computing a matching M to extend M^* is now based only on the maximum weight 4-matching $\mathcal{C}$, of which the weight can be better estimated because pseudoknot-like interactions are allowed.

The analysis of the algorithm is similar to that of the previous section. We do exactly the same interaction partitioning for the optimal solution and the maximum weight matching M^*. One can easily verify Lemma 1, Corollary 1, and Lemma 2. The following lemma is the key to the improvement, which provides an improved lower bound on the weight of the maximum weight 4-matching.

Input: m RNAs R_i, $i = 1, 2, \ldots, m$, with transitivity;
Output: a permutation π of $[m]$ and interactions between RNAs $R_{\pi(i)}$ and $R_{\pi(i+1)}$,
    for $i = 1, 2, \ldots, m - 1$

---

1. for each RNA pair R_i and R_j,
    1.1. construct bipartite graph $BG(i, j)$;
    1.2. compute $w(R_i, R_j)$ allowing pseudoknot-like interactions;
2. construct edge-weighted complete graph G using edge weight function w;
3. compute the maximum weight matching M^* of G;
    3.1. delete nucleotides involved in the interactions of M^*;
    3.2. reconstruct bipartite graph $BG(i, j)$;
    3.3. compute $w'(R_i, R_j)$ allowing pseudoknot-like interactions;
4. construct edge-weighted complete graph G' using edge weight function w';
    4.1. compute the maximum weight 4-matching $\mathcal{C}$ of G';
    4.2. compute a matching M out of $\mathcal{C}$ to extend M^*;
5. stack RNA paths in $G[M^* \cup M]$ arbitrarily to form a permutation π;
6. output π and the interactions in $w(R_{\pi(i)}, R_{\pi(i+1)}) + w'(R_{\pi(i)}, R_{\pi(i+1)})$.

Fig. 3. A high-level description of APPROX III

Lemma 8. *The weight of the maximum weight 4-matching $\mathcal{C}$ of graph G' is*

$$w'(\mathcal{C}) \geq \max\{\frac{1}{2}|I_2|, \frac{1}{4}|I_2| + |I_4|\}. \tag{5}$$

Proof. The first component straightly follows from Lemma 2 since G'_s is a 4-matching in graph G'. Note also that graph G' contains all the edges of the optimal solution, each of which is contributed by at least one interaction of I_4. This remaining optimal solution, denoted as $\mathcal{P}$, is an acyclic 2-matching in G', and has weight $w'(\mathcal{P}) \geq |I_4|$.

Since G'_s is a 4-matching, it can be decomposed into two 2-matchings denoted as $\mathcal{D}_1$ and $\mathcal{D}_2$. One clearly see that both $\mathcal{P} \cup \mathcal{D}_1$ and $\mathcal{P} \cup \mathcal{D}_2$ are 4-matchings in graph G'. The interactions of I_4 counted towards $\mathcal{P}$ are not counted towards G'_s. Therefore, we have

$$\begin{aligned}
w'((\mathcal{C})) &\geq \max\{w'(\mathcal{P} \cup \mathcal{D}_1), w'(\mathcal{P} \cup \mathcal{D}_2)\} \\
&\geq \frac{1}{2}(w'(\mathcal{D}_1) + w'(\mathcal{D}_2)) + w'(\mathcal{P}) \\
&= \frac{1}{2}w'(G'_s) + |I_4| \\
&\geq \frac{1}{4}|I_2| + |I_4|.
\end{aligned}$$

This proves the lemma. □

Theorem 5. *Algorithm APPROX III is a 0.5333-approximation for the general MRIP problem with transitivity and allowing pseudoknot-like interactions.*

Proof. The estimation of the performance ratio of 0.5333 is very similar to that of ratio 0.5328 in Theorem 4, and is omitted from here. □

Acknowledgement. Weitian Tong, Randy Goebel, and Guohui Lin are supported in part by NSERC, AITF and iCORE. Weitian Tong, Tian Liu, and Guohui Lin thank the Open Fund of Top Key Discipline of Computer Software and Theory in Zhejiang Provincial Colleges at the Zhejiang Normal University for sponsoring a workshop where this work was started.

References

1. Ahmed, S.A., Mneimneh, S., Greenbaum, N.L.: A combinatorial approach for multiple RNA interaction: Formulations, approximations, and heuristics. In: Du, D.-Z., Zhang, G. (eds.) COCOON 2013. LNCS, vol. 7936, pp. 421–433. Springer, Heidelberg (2013)
2. Alkan, C., Karakoç, E., Nadeau, J.H., Sahinalp, S.C., Zhang, K.: RNA-RNA interaction prediction and antisense RNA target search. Journal of Computational Biology 13, 267–282 (2006)
3. Chen, Z.-Z., Wang, L.: An improved approximation algorithm for the Bandpass-2 problem. In: Lin, G. (ed.) COCOA 2012. LNCS, vol. 7402, pp. 188–199. Springer, Heidelberg (2012)
4. Chitsaz, H., Backofen, R., Sahinalp, S.C.: biRNA: Fast RNA-RNA binding sites prediction. In: Salzberg, S.L., Warnow, T. (eds.) WABI 2009. LNCS, vol. 5724, pp. 25–36. Springer, Heidelberg (2009)
5. Chitsaz, H., Salari, R., Sahinalp, S.C., Backofen, R.: A partition function algorithm for interacting nucleic acid strands. Bioinformatics 25, 365–373 (2009)
6. Diestel, R.: Graph Theory. Graduate Texts in Mathematics. Springer (2005)
7. Harary, F.: Graph Theory. Addison-Wesley (1969)
8. Huang, F.W.D., Qin, J., Reidys, C.M., Stadler, P.F.: Partition function and base pairing probabilities for RNA-RNA interaction prediction. Bioinformatics 25, 2646–2654 (2009)
9. Li, A.X., Marz, M., Qin, J., Reidys, C.M.: RNA-RNA interaction prediction based on multiple sequence alignments. Bioinformatics 27, 456–463 (2011)
10. Lin, G.: On the Bandpass problem. Journal of Combinatorial Optimization 22, 71–77 (2011)
11. Saad, M.: On the approximation of optimal structures for RNA-RNA interaction. IEEE/ACM Transactions on Computational Biology and Bioinformatics 6, 682–688 (2009)
12. Meyer, I.M.: Predicting novel RNA-RNA interactions. Current Opinion in Structural Biology 18, 387–393 (2008)
13. Mückstein, U., Tafer, H., Hackermüller, J., Bernhart, S.H., Stadler, P.F., Hofacker, I.L.: Thermodynamics of RNA-RNA binding. Bioinformatics 22, 1177–1182 (2006)
14. Paluch, K., Mucha, M., Mądry, A.: A 7/9- approximation algorithm for the maximum traveling salesman problem. In: Dinur, I., Jansen, K., Naor, J., Rolim, J. (eds.) APPROX 2009. LNCS, vol. 5687, pp. 298–311. Springer, Heidelberg (2009)
15. Pervouchine, D.D.: Iris: intermolecular RNA interaction search. Genome Informatics 15, 92–101 (2004)
16. Salari, R., Backofen, R., Sahinalp, S.C.: Fast prediction of RNA-RNA interaction. Algorithms for Molecular Biology 5, 5 (2010)
17. Sun, J.S., Manley, J.L.: A novel U2-U6 snRNA structure is necessary for mammalian mRNA splicing. Genes & Development 9, 843–854 (1995)
18. Tong, W., Chen, Z.-Z., Wang, L., Xu, Y., Xu, J., Goebel, R., Lin, G.: An approximation algorithm for the Bandpass-2 problem. ArXiv e-print 1307.7089 (July 2013)

On the Clustered Steiner Tree Problem

Bang Ye Wu

National Chung Cheng University, ChiaYi, Taiwan 621, R.O.C.
bangye@cs.ccu.edu.tw

Abstract. We investigate the *Clustered Steiner tree* problem on metric graphs, which is a variant of Steiner minimum tree problem. The required vertices are partitioned into clusters, and in a feasible clustered Steiner tree, the subtrees spanning two different clusters must be disjoint. In this paper, we show that the problem remains NP-hard even if the topologies of all clusters and the inter-cluster tree are given. We propose a $(\rho+2)$-approximation algorithm for the general case, in which ρ is the approximation ratio for the Steiner tree problem. When the topologies for all clusters are given, we show a $(\rho+1)$-approximation algorithm. We also discuss the Steiner ratio for this problem. We show the ratio is lower and upper bounded by three and four, respectively.

Keywords: approximation algorithm, Steiner tree, NP-hard, graph algorithm.

1 Introduction

For a simple undirected graph $G = (V, E, c)$ and a *required vertex* set $R \subseteq V$, a *Steiner tree* is a connected and acyclic subgraph of G that spans all the vertices in R. Due to the large number of applications, Steiner tree problems are extensively studied. The Steiner Minimum Tree (SMT) problem is a classical and well-known NP-hard problem which involves finding a Steiner tree with minimum total edge cost [10, 15]. Numerous variants of the SMT problem have been studied, for example, the versions on the Euclidean metric [8] and the rectilinear metric [9], the Steiner forest problem [1], the group Steiner tree problem [11], the terminal Steiner tree problem [4, 6, 17–19], and the internal-selected Steiner tree problem [13, 14, 16]. The best approximation ratio ρ on general metrics achieved in polynomial time is an important parameters for many graph problems. From the first non-trivial result $11/6$ [23], it has been improved several times [3, 20]. The current best approximation ratio is 1.39 [3].

In this paper, we consider another variant of SMT, the *Clustered Steiner tree* (CLUSTEINER) problem. In addition to a metric graph $G = (V, E, c)$ and required vertex set R, we are also given a partition $\mathcal{R} = \{R_1, R_2, \ldots, R_k\}$ of R. A Steiner tree T is a clustered Steiner tree for $\mathcal{R}$ if all the vertices in the same cluster (R_i) are *clustered together* in T. More formally, the local tree of a cluster R_i in T is the minimal subtree of T spanning R_i. A Steiner tree T is a clustered Steiner tree if the local trees are mutually disjoint. That is, T can be cut into k subtrees

P. Widmayer, Y. Xu, and B. Zhu (Eds.): COCOA 2013, LNCS 8287, pp. 60–71, 2013.

by removing $k-1$ edges such that each subtree is a Steiner tree for one cluster R_i. An equivalent definition is that for $s_i, t_i \in R_i$ and $s_j, t_j \in R_j$ the unique $s_i t_i$-path and $s_j t_j$-path in T are disjoint for all $i \neq j$. If there is only one cluster or each required vertex is itself a cluster, the problem degenerates to the original Steiner minimum tree problem. The contribution of this paper is as follows.

- When $R = V$, or equivalently no Steiner vertex can be used, the problem can be simply solved in polynomial time.
- The Steiner ratio for CLUSTEINER is lower and upper bounded by three and four, respectively, in which the Steiner ratio is defined as the largest possible ratio of the minimal cost without using any Steiner vertex to the optimal cost.
- The problem remains NP-hard even if the topologies of all clusters and the inter-cluster tree are given.
- The problem can be $(\rho+2)$-approximated, in which ρ is the best approximation ratio for the Steiner tree problem. The ratio can be improved to $(\rho+1)$ if the topologies for all clusters are given.

A possible application of this variant is as follows. When designing transportation or computer networks, the links are usually divided into two levels: inter-clustered or intra-clustered, possibly with different costs, qualities, and capacities. Also, after the network is built, the communications between nodes in the same cluster should be routed locally rather than globally for the sake of capacity consideration or the simpleness of routing protocols. Another application is for the case that the local topologies for all clusters are given. In this case the task is to design the inter-cluster topology, as well as the possible insertion of local Steiner vertices without violating their topologies. A similar consideration was also studied for the *traveling salesperson problem* (TSP), named *clustered TSP* problem [2, 12]. For this problem, the goal is to find a minimum cost Hamiltonian path such that the vertices of each cluster are visited consecutively.

The rest of the paper is organized as follows. In Section 2, we give some notation and definitions. In Section 3, we discuss some properties and the Steiner ratio. The NP-hardness and the approximation algorithms are shown in Sections 4 and 5 respectively. Finally some remarks are given in Section 6.

2   Notation and Definitions

For a graph $G = (V, E, c)$, V and E are the vertex and edge sets, respectively, and c is the edge cost. In this paper we consider only undirected graphs. An edge between vertices u and v is denoted by (u, v), and its cost is denoted by $c(u, v)$. For a subgraph T of G, $c(T)$ denotes the total cost of all edges of T. For a graph G, $V(G)$ and $E(G)$ denote the vertex and the edge sets, respectively. For a vertex subset U, the subgraph of G induced by U is denoted by $G[U]$. By $\mathrm{smt}(G, R)$, we denote a Steiner minimum tree with instance (G, R) and also its cost. We use $\mathrm{mst}(R)$ to denote a minimum spanning tree (MST), and also its cost, of $G[R]$. A path with end vertices s and t is called an st-path. An undirected graph $G = (V, E, c)$ is a *metric graph* if it is a complete graph with nonnegative edge costs satisfying the triangle inequality. For a set S, a collection $\mathcal{S}$ of subsets of S is a *partition* of S if the subsets are mutually disjoint and their union is exactly S.

Definition 1. *For a tree T spanning R, i.e., $R \subseteq V(T)$, the* local tree *of R on T is the minimal subtree of T spanning all vertices in R.*

By definition, any leaf of a local tree must be in R.

Definition 2. *Let $\mathcal{R} = \{R_i | 1 \leq i \leq k\}$ be a partition of R. A Steiner tree T for R is a* clustered Steiner tree *for $\mathcal{R}$ if the local trees of all $R_i \in \mathcal{R}$ are mutually disjoint, i.e., there exists a cut set $C \subset E(T)$ with $|C| = k - 1$ such that each component of $T - C$ is a Steiner tree T_i for R_i, $1 \leq i \leq k$.*

The problem is formally defined as follows.

PROBLEM: Clustered Steiner Tree problem (CLUSTEINER)
INSTANCE: A metric graph $G = (V, E, c)$, required vertices $R \subseteq V$, and a partition $\mathcal{R} = \{R_1, R_2, \ldots, R_k\}$ of R.
GOAL: Find a minimum-cost clustered Steiner tree for $\mathcal{R}$.

A vertex not in R is a *Steiner vertex*. In the remainder of this paper, we assume that $(G, \mathcal{R})$ is the instance of the problem, in which $G = (V, E, c)$ and $\mathcal{R} = \{R_1, R_2, \ldots, R_k\}$ is a partition of R. We also use $n = |V|$ and $m = |E|$.

An Eulerian path/cycle is a path/cycle traveling all the edges exactly once. A graph is Eulerian if there exists an Eulerian cycle (or Eulerian tour). A connected undirected graph is Eulerian if and only if all the degree of vertices are even. There exists an Eulerian path if and only if there are exactly two vertices of odd degree. A Hamiltonian path/cycle is a path/cycle traveling all the vertices exactly once.

For a graph H, *contraction* of $(u, v) \in E(H)$ replaces u, v with a new vertex w. For any other vertex s, the edge cost is set to $c(s, w) = \min\{c(s, u), c(s, v)\}$. For a subgraph S, contracting S in H means contracting all the edges $E(S)$ in H, and the resulting graph is denoted by H/S. For convenience, we also use H/S to denote $H/H[S]$ when S is a vertex subset. Let $G/\mathcal{R}$ denote the graph resulted from contracting all R_i for all $R_i \in \mathcal{R}$.

For a graph T and $(u, r), (r, v) \in E(T)$, "taking a shortcut between u, v" means we replace edges (u, r) and (r, v) with (u, v). Similarly, for a uv-path, taking a shortcut between u, v replaces the path with edge (u, v).

Definition 3. *For a local tree T_i of R_i, the* topology *of T_i is the tree obtained by repeatedly taking shortcuts between the two neighbors of Steiner vertices with degree two and removing such Steiner vertices in T_i until there is no such vertex.*

Figure 1 depicts an example of the topology of a local tree. For a clustered Steiner tree T, contracting all the local trees in T results in a tree, denoted by $T/\mathcal{R}$, called the *inter-cluster tree* of T. Since a Steiner vertex with degree two in an inter-cluster tree is meaningless, the topology of an inter-cluster tree is itself.

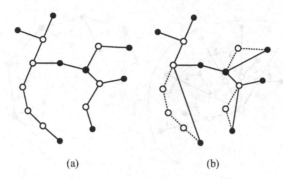

(a)   (b)

Fig. 1. Topology of a local tree. White vertices are Steiner vertices and black ones are required vertices. (a) The original local tree. (b) The topology (solid lines). Note that the degree-2 Steiner vertices are not vertices of the topology.

3   Steiner Ratio of CLuStEINER

Possibly the simplest way to approximate the SMT is by MST. The *Steiner ratio* (for the classic SMT problem) is the largest possible ratio between the cost of an MST and the cost of an SMT. The inequality (1) is well-known, see for example [22], which shows the Steiner ratio is two for general metric spaces.

$$\text{mst}(R) \leq 2\text{smt}(G, R). \tag{1}$$

The inequality can be simply shown as follows. Let $T = \text{smt}(G, R)$. By doubling $E(T)$, we can obtain an Eulerian multigraph and therefore an Eulerian tour Y with $c(Y) = 2c(T) = 2\text{smt}(G, R)$. Traveling along the Eulerian cycle and taking shortcuts between consecutive unvisited required vertices, we can obtain a Hamiltonian path of $G[R]$ with cost at most $c(Y)$ by the triangle inequality. Since MST is the cheapest way to connect R, we have that $\text{mst}(R) \leq c(Y)$ and the inequality follows.

When $R = V$, i.e., the minimum clustered spanning tree problem, the problem is equivalent to the case in which no Steiner vertex is allowed. We now show a simple algorithm for this variant. Since no Steiner vertex is allowed, the local tree of R_i in the optimal tree is an $\text{mst}(R_i)$ for each i. Similarly the inter-cluster topology is an MST of $G/\mathcal{R}$. The next result is simple, in which an MST can be solved in $O(n \log n + m)$ time [5].

Proposition 1. *The minimum clustered spanning tree problem can be solved with the same asymptotic time complexity as the MST problem.*

However, we found that the Steiner ratio 2 does not hold for CLuStEINER. Figure 2 gives a simple example. The left tree (a) is a minimum clustered Steiner tree with cost $p(3 + \epsilon)$, where $p = |R_1|$. If no Steiner vertex can be used, the right tree (b) is the best and the cost is $6(p - 1) + p(2 + \epsilon) \approx 8p$. The ratio is about $8/3 > 2$. Note that an MST consists of a path connecting required vertices

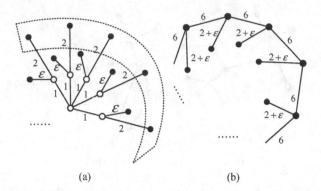

(a)   (b)

Fig. 2. An example for Steiner ratio larger than 2 for CLUSTEINER. Black vertices are required vertices and white ones are Steiner vertices. The required vertices circled by dotted line are in one cluster R_1, and all the other clusters are singletons. ϵ is the smallest possible edge cost, i.e., zero if edge costs are defined to be nonnegative. For any vertices u, v, $c(u, v)$ is the same of the cost of the uv-path in the tree on the left. (a) A clustered Steiner tree. (b) A feasible solution (clustered Steiner tree) without any Steiner vertex.

not in R_1 and linking vertices in R_1 to the path individually. However, it is not feasible for CLUSTEINER since R_1 is not clustered together, i.e., the local tree contains other required vertices.

Let the Steiner ratio of CLUSTEINER be defined by the ratio of the minimum cost without any Steiner vertex to the optimal cost. The above example shows that the Steiner ratio is at least $8/3$. Figure 3 shows an even worse example. The optimal tree (a) has cost $q(p(2 + \epsilon) + 1) \approx 2pq + 2q$. The right tree (b) is the best possible without Steiner vertex, and its cost is $(q - 1)(4p + 2) + qp(2 + \epsilon)$. The ratio is asymptotically three when pq is large. By this example, we have the lower bound in Lemma 1.

Lemma 1. *The Steiner ratio of* CLUSTEINER *is at least three.*

4   NP-Hardness for Fixed Topologies

At the first glance of CLUSTEINER, the hardness of CLUSTEINER seems from determining the best local and inter-cluster topologies. In this section we shall show that NP-hardness remains even when the topologies are given.

A *caterpillar* is a tree of which all the internal vertices form a path. Let s, t be two leaves adjacent to the two endpoints of the path, respectively. We call the tree an *st-caterpillar*.

PROBLEM: STEINER CATERPILLAR
INSTANCE: A metric graph $G = (V, E)$, required vertices $R \subset V$, and two vertices $x, y \in R$.

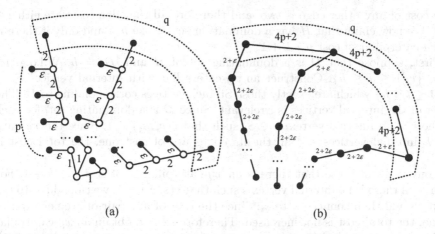

(a)   (b)

Fig. 3. An example with Steiner ratio three. The setting is similar to Figure 2. R_1 consists of the q required vertices circled by dotted line. As indicated, each path has p internal Steiner vertices. (a) The optimal solution. (b) The best one without any Steiner vertex.

GOAL: Find a minimum-cost xy-caterpillar spanning R such that all internal vertices of the xy-path are not in R.

The $(1, 2)$-STEINER CATERPILLAR is the special version in which all the edge costs are either one or two. Note that a complete graph with all edge costs of one or two is a metric graph. We show the NP-hardness of STEINER CATERPILLAR by reducing the following well-known NP-complete problem to it.

PROBLEM: DOMINATING SET
INSTANCE: A simple undirected graph H and an integer h.
QUESTION: Is there a dominating set with size h, i.e., a set $S \subseteq V(H)$ with $|S| \leq h$ such that for all $u \notin S$ there exists $v \in S$ for which $(u, v) \in E$?

Lemma 2. $(1, 2)$-STEINER CATERPILLAR *is NP-hard.*

Proof. We reduce DOMINATING SET to $(1, 2)$-STEINER CATERPILLAR, and the result follows from the NP-completeness of DOMINATING SET [10]. Let (H, h) be an instance of DOMINATING SET. We construct an instance $(G, \mathcal{R})$ of STEINER CATERPILLAR as follows. Let $V(H) = \{v_i | 1 \leq i \leq p\}$. For each $v_i \in V(H)$, we create a Steiner vertex s_i. Let $S = \{s_i | 1 \leq i \leq p\}$ and $R = V(H) \cup \{x, y\}$, in which $x, y \notin V(H)$ are two added vertices. Then, $V(G) = R \cup S$ and the edge costs are as follows.

$$\begin{cases} c(s_i, s_j) = 1 & \text{for } 1 \leq i < j \leq p \\ c(s_i, x) = c(s_i, y) = 1 & \text{for } 1 \leq i \leq p \\ c(s_i, v_j) = 1 & \text{for } (v_i, v_j) \in E(H) \text{ or } i = j \end{cases}$$

The cost of any other edge is two, and therefore all the edge costs are either 1 or 2. We now claim that H has a dominating set of size h if and only if there is an xy-caterpillar of cost $p + h + 1$.

First, suppose that D is a dominating set of H and $|D| = h$. W.l.o.g. let $D = \{v_i | 1 \leq i \leq h\}$. Construct an xy-caterpillar with internal vertices $S' = \{s_i | 1 \leq i < h\}$ which are exactly those Steiner vertices corresponding to D. The order of the internal vertices is irrelevant. Since D is a dominating set, for each v_i there is an internal vertex $s_j \in S'$ such that $c(v_i, s_j) = 1$. Since the xy-path has h internal vertices and all the its edges are of cost one, the total cost is $p + h + 1$.

Conversely, suppose that there is an xy-caterpillar T of cost $p + h + 1$. For each v_i, if there is no internal vertex s such that $c(s, v_i) = 1$, we can add s_i to the xy-path and then connect v_i to s_i. Since the cost of any pair of Steiner vertices is one, the total cost is not increased. Therefore we can obtain an xy-caterpillar of the same cost such that all leaves except for x, y are connected to the xy-path with cost one. Since the total cost is $p + h + 1$, the number of internal vertices is h, and its corresponding vertex subset in $V(H)$ is a dominating set of H. □

Theorem 1. CLUSTEINER *is NP-hard even when all the local topologies and the inter-cluster topology are given.*

Proof. By Lemma 2, it is sufficient to show that $(1, 2)$-STEINER CATERPILLAR is a special case. For an instance $(G = (V, E, c), R, x, y)$ of STEINER CATERPILLAR, we transform it into an instance $(G' = (V, E, c'), \mathcal{R})$ of CLUSTEINER. Let $\mathcal{R} = \{R_i | 1 \leq i \leq k\}$, in which $k = |R| - 1$, $R_1 = \{x, y\}$ and $|R_i| = 1$ for $i \geq 2$. Since every cluster contains no more than two vertices, the local topologies are trivial. The inter-cluster topology is a star with center R_1. Let $c'(x, y) = 2L$. Let $c'(x, v) = c(x, v) + L$ and $c'(y, v) = c(y, v) + L$ for all $v \neq x, y$, in which $L = 5$. All the other edge costs remain the same. We assume that $k \geq 3$ since it can be trivially solved in polynomial time if there are only constant number of required vertices.

First we show that G' is also a metric graph. It is sufficient to show the triangle inequality holds for any three vertices involving x or y. For any vertex $v \notin \{x, y\}$, $c'(x, y) + c'(x, v) > c'(y, v)$ since $c'(x, y)$ is the largest edge cost, and $c'(x, v) + c'(v, y) = c(x, v) + c(v, y) + 2L > c'(x, y) = 2L$. For $u, v \notin \{x, y\}$, $c'(x, u) + c'(u, v) = L + c(x, u) + c(u, v) \geq L + c(x, v) = c'(x, v)$.

Let T be a minimum clustered Steiner tree. Since the inter-cluster topology is a star with center R_1, if there is no Steiner vertex on T, the cost is larger than $(k + 1)L$ since an edge connecting any vertex to x or y has cost more than L. But adding any Steiner vertex to subdivide (x, y) and connecting all the required vertices to it reduces the cost to at most $2L + (k + 1)2 < (k + 1)L$ since $L > 4$ and $k \geq 3$. Recall that $c'(u, v) \leq 2$ for vertices $u, v \notin \{x, y\}$. Since there exists at least one Steiner vertex in T, we claim that no required vertex is connected to x or y. Otherwise, re-connecting this vertex to any Steiner vertex reduces the total cost. We conclude that x and y are leaves in T.

Therefore the optimal solution of the CLUSTEINER problem is the same as the one of the STEINER CATERPILLAR except for the additional cost $2L$. □

5   Approximation Algorithms

For a clustered Steiner tree T, let $\alpha(T)$ denote the total cost of all its local trees and $\beta(T) = c(T) - \alpha(T)$ the cost of its inter-cluster topology, i.e., $\beta(T) = c(T/\mathcal{R})$. By Algorithm 1, we shall show that any clustered Steiner tree T can be transformed into a clustered Steiner tree T' of which the local trees have no Steiner vertex. Figure 4 illustrates an example.

Algorithm 1

Input: a clustered Steiner tree T.
Output: a clustered Steiner tree T'.

 1: $T' \leftarrow T$;
 2: **for all** local tree T_i of T' **do**
 3:     construct a multigraph H_i by doubling the edges of T_i;
 4:     construct an Eulerian tour Y on H_i;
 5:     pick any required vertex r in T_i;
 6:     **while** existing non-visited required vertex in Y **do**   ▷ traveling along Y
 7:         let r' be the next non-visited required vertex in Y;
 8:         let s be the previous vertex of r' in Y;
 9:         replace (s, r') with (r, r') in T_i;   ▷ no change if $s = r$
10:         $r \leftarrow r'$;
11:     **end while**
12: **end for**
13: output T'.

In fact, we replace each local tree with a Hamiltonian path and cut some edges to break cycles. Since the Hamiltonian path consists of shortcuts of the cycle Y, the next result follows from the triangle inequality.

Claim. Each local tree T_i is replaced with a Hamiltonian path of R_i with cost at most $2c(T_i)$.

After the transformation, each local tree is the added Hamiltonian path, and cutting the edges makes its topology a part of the inter-cluster topology. Since no other edge is added, the increment is at most the original cost of the local trees.

Claim. $\beta(T') \leq \beta(T) + \alpha(T) = c(T)$.

The next lemma comes from the above two claims.

Lemma 3. *There exists a clustered Steiner tree T' with no Steiner vertex in its local trees and $\beta(T') \leq \beta(T^*) + \alpha(T^*) = c(T^*)$, in which T^* is a minimum clustered Steiner tree.*

Now we show a proposed approximation algorithm for CLUSTEINER in Algorithm 2.

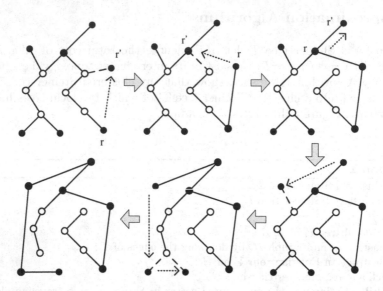

Fig. 4. An example illustrates Algorithm 1. For each step, r and r' are the tail and head of the arrow, respectively. The dashed edge is the one to be removed.

Algorithm 2. Approximating the minimum clustered Steiner tree

---

Input: an instance $(G, \mathcal{R})$ of the problem.

Output: a clustered Steiner tree T^a.

1: construct $T_i^a \leftarrow \mathrm{mst}(G[R_i])$ for each i;
2: construct $G/\mathcal{R}$ and let $R' = \{r_i'|1 \leq i \leq k\}$, in which r_i' is the vertex resulted from the contraction of R_i;
3: construct a ρ-approximation T_0^a of $\mathrm{smt}(G/\mathcal{R}, R')$;
4: replace r_i' with T_i^a to obtain a clustered Steiner tree T^a;
5: output T^a.

Theorem 2. CLUSTEINER *can be* $(2 + \rho)$-*approximated in* $O(n \log n + f(m, n))$ *time, in which* ρ *and* $f(m, n)$ *are the approximation ratio and the time complexity of an approximation algorithm for Steiner minimum tree on a graph with* m *edges and* n *vertices, respectively.*

Proof. Let T^* be a minimum cluster Steiner tree. Let T^a be the tree constructed by Algorithm 2 and T' the tree satisfying Lemma 3. We have that $\alpha(T^a) \leq 2\alpha(T^*)$. Since T' has no Steiner vertex in its local tree, $\beta(T') \geq \mathrm{smt}(G/\mathcal{R}, R')$, and therefore $\beta(T^a) \leq \rho\beta(T')$. Since $\beta(T') \leq \alpha(T^*) + \beta(T^*)$ by Lemma 3,

$$c(T^a) = \alpha(T^a) + \beta(T^a) \leq 2\alpha(T^*) + \rho(\alpha(T^*) + \beta(T^*))$$
$$\leq (2 + \rho)\alpha(T^*) + \rho\beta(T^*) \leq (2 + \rho)c(T^*).$$

□

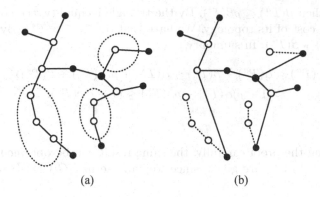

(a)   (b)

Fig. 5. Transformation from T_i to its topology Y_i. (a) A local tree in which the white vertices are Steiner vertices. The paths circled by dotted line will be cut into the inter-cluster tree. (b) After the transformation. The dotted lines are now a part of the inter-cluster tree.

In Algorithm 2, if we use $\mathrm{mst}(R')$ instead of the Steiner tree T_0^a, we obtain the best clustered Steiner tree without any Steiner vertex. Let Y be the tree. Since $\mathrm{mst}(R') \leq 2\beta(T')$ by (1), we have that $\beta(Y) \leq 2\beta(T') \leq 2(\alpha(T^*) + \beta(T^*))$, and therefore $c(Y) \leq 4\alpha(T^*) + 2\beta(T^*)$.

Corollary 1. *The Steiner ratio for* CLUSTEINER *is at most four.*

Next we consider the case that the topologies of local trees are given. For simplicity, we call the topology of a local tree the "local topology". Let Y_i be the local topology for R_i. Clearly $R_i \subseteq V(Y_i)$. Let $S_i = V(Y_i) - R_i$. The vertices in S_i are the Steiner vertices with degree at least three in the local tree. We may assume that $S_i \cap S_j = \emptyset$ for $i \neq j$. Otherwise there is no solution.

We can modify Algorithm 1 such that the shortcuts are taken between vertices in $V(Y_i)$ but not only R_i. The local tree T_i of the optimal tree T^* is now transformed to Y_i. Figure 5 illustrates an example. The proof of the next lemma is similar to Lemma 3 and is omitted.

Lemma 4. *Let* T^* *be a minimum cluster Steiner tree. Suppose that the local topology* Y_i *is given for each* i. *There exists a clustered Steiner tree* T' *with* $\beta(T') \leq \beta(T^*) + \alpha(T^*) = c(T^*)$ *such that the vertex set of each local tree is exactly* $V(Y_i)$.

Theorem 3. *When the local topologies are given, the problem* CLUSTEINER *can be* $(1 + \rho)$-*approximated in* $O(n \log n + f(m, n))$ *time.*

Proof. Let $S' = V - R - \bigcup_i S_i$ be the possible Steiner vertices in the inter-cluster tree. The approximation algorithm is similar to Algorithm 2 except that we use Y_i as the local tree T_i^a and construct the ρ-approximation of $\mathrm{smt}(G[R' \cup S'], R')$ as the inter-cluster tree, in which, as defined in Algorithm 2, R' consists of the vertices resulted from the contraction of all R_i. Therefore $\beta(T') \geq \mathrm{smt}(G[R' \cup$

S'], R'), and then $\beta(T^a) \leq \rho\beta(T')$. By the triangle inequality, the cost of a tree is at least the cost of its topology. We have that $\alpha(T^a) \leq \alpha(T^*)$. By Lemma 4, $\beta(T') \leq \alpha(T^*) + \beta(T^*)$. In summary,

$$c(T^a) = \alpha(T^a) + \beta(T^a) \leq \alpha(T^*) + \rho(\alpha(T^*) + \beta(T^*))$$
$$\leq (1 + \rho)\alpha(T^*) + \rho\beta(T^*) \leq (1 + \rho)c(T^*).$$

$\square$

By observing the proof carefully, the same result can be obtained as long as all $V(Y_i)$ instead of Y_i are given, since we can use $\mathrm{mst}(G[V(Y_i)])$ as the local tree for each i.

Corollary 2. *When the Steiner vertices in every local topology are given, the problem* CLUSTEINER *can be* $(1 + \rho)$-*approximated in* $O(n \log n + f(m, n))$ *time.*

6   Conclusion

In this paper, we show the Steiner ratio for CLUSTEINER is lower and upper bounded by three and four, respectively. It is interesting to improve the gap of the two bounds. Another interesting open problem is the approximability of CLUSTEINER. Both improving the ratio $(2 + \rho)$ and showing the inapproximability are interesting. We propose another variant of the problem studied in this paper. In applications to network design, there may be two cost functions. An edge (u, v) in a local tree has cost $c(u, v)$ and costs $c'(u, v)$ if it is in the inter-cluster tree. Usually $c'(u, v) > c(u, v)$. Now the problem is to design a clustered Steiner tree with minimum total cost.

Acknowledgment. This work was supported in part by NSC 100-2221-E-194-036-MY3 and NSC 101-2221-E-194-025-MY3 from the National Science Council, Taiwan, R.O.C.

References

1. Agrawal, A., Klein, P., Ravi, R.: When trees collide: An approximation algorithm for the generalized Steiner problem in networks. SIAM Journal on Computing 24(3), 445–456 (1995)
2. Bao, X., Liu, Z.: An improved approximation algorithm for the clustered traveling salesman problem. Information Processing Letters 112, 908–910 (2012)
3. Byrka, J., Grandoni, F., Rothvoß, T., Sanitá, L.: An improved LP-based approximation for Steiner tree. In: Proc. 42nd ACM Symposium on Theory of Computing, pp. 583–592 (2010)
4. Chen, Y.H., Lu, C.L., Tang, C.Y.: On the full and bottleneck full Steiner tree problems. In: Warnow, T.J., Zhu, B. (eds.) COCOON 2003. LNCS, vol. 2697, pp. 122–129. Springer, Heidelberg (2003)

5. Cormen, T.H., Leiserson, C.E., Rivest, R.L., Stein, C.: Introduction to Algorithms. MIT Press and McGraw-Hill (2001)
6. Drake, D.E., Hougardy, S.: On approximation algorithms for the terminal Steiner tree problem. Information Processing Letters 89(1), 15–18 (2004)
7. Fuchs, B.: A note on the terminal Steiner tree problem. Information Processing Letters 87, 219–220 (2003)
8. Garey, M.R., Graham, R., Johnson, D.: The complexity of computing Steiner minimal trees. SIAM Journal on Applied Mathematics 32, 835–859 (1977)
9. Garey, M.R., Johnson, D.: The rectilinear Steiner problem is NP-complete. SIAM Journal on Applied Mathematics 32, 826–834 (1977)
10. Garey, M.R., Johnson, D.S.: Computers and Intractability: A Guide to The Theory of NP-Completeness. Freeman, NewYork (1979)
11. Garg, N., Konjevod, G., Ravi, R.: A polylogarithmic approximation algorithm for the group Steiner problem. In: Proc. 9th ACM-SIAM Symposium on Discrete Algorithms, pp. 253–259 (1998)
12. Guttmann-Beck, N., Hassin, R., Khuller, S., Raghavachari, B.: Approximation algorithms with bounded performance guarantees for the clustered traveling salesman problem. Algorithmica 28, 422–437 (2000)
13. Hsieh, S.Y., Yang, S.C.: Approximating the selected-internal Steiner tree. Theoretical Computer Science 381(1-3), 288–291 (2007)
14. Huang, C.W., Lee, C.W., Gao, H.M., Hsieh, S.Y.: The internal Steiner tree problem: Hardness and approximations. Journal of Complexity 29, 27–43 (2013)
15. Karp, R.: Reducibility among combinatorial problems. In: Miller, R.E., Thatcher, J.W. (eds.) Complexity of Computer Computations, pp. 85–103. Plenum Press, New York (1972)
16. Li, X., Zou, F., Huang, Y., Kim, D., Wu, W.: A better constant-factor approximation for selected-internal Steiner minimum tree. Algorithmica 56, 333–341 (2010)
17. Lin, G.H., Xue, G.L.: On the terminal Steiner tree problem. Information Processing Letters 84(2), 103–107 (2002)
18. Lu, C.L., Tang, C.Y., Lee, R.C.T.: The full Steiner tree problem. Theoretical Computer Science 306(1-3), 55–67 (2003)
19. Martinez, F.V., de Pina, J.C., Soares, J.: Algorithms for terminal Steiner trees. Theoretical Computer Science 389, 133–142 (2007)
20. Robins, G., Zelikovsky, A.: Tighter bounds for graph Steiner tree approximation. SIAM Journal on Discrete Mathematics 19(1), 122–134 (2005)
21. Sebő, A.: Eight-fifth approximation for the path TSP. In: Goemans, M., Correa, J. (eds.) IPCO 2013. LNCS, vol. 7801, pp. 362–374. Springer, Heidelberg (2013)
22. Wu, B.Y., Chao, K.M.: Spanning Trees and Optimization Problems. Chapman & Hall (2004)
23. Zelikovsky, A.: An 11/6-approximation algorithm for the network Steiner problem. Algorithmica 9, 463–470 (1993)

Integrated Job Scheduling with Parallel-Batch Processing and Batch Deliveries

Xin Feng[1,3,4] and Feifeng Zheng[2]

[1] School of Management, Xi'an Jiaotong University, Xi'an, Shaanxi, 710049, China
[2] Glorious Sun School of Business and Management,
Donghua University, Shanghai, 200051, P.R. China
[3] State Key Lab for Manufacturing Systems Engineering,
Xi'an, Shaanxi, 710049, China
[4] Ministry of Education Key Lab for Process Control and Efficiency Engineering,
Xi'an, Shaanxi, 710049, China
fengxin.xjtu@stu.xjtu.edu.cn

Abstract. This paper studies an integrated scheduling problem which consists of production and distribution stages. Jobs are processed on a parallel-batch machine in the production stage, and then transported to customers in the distribution stage. The aim is to find an integrated schedule in the two stages with the objective to optimize both total delivery time and total transportation cost. We focus on two models with either unbounded or bounded batch processing capacity in the production stage. In both models we assume sufficient vehicles each with infinite transportation capacity. For the unbounded model, we derive an $O\left(g(n/g)^{2g}\right)$ time dynamic programming algorithm where $g \geq 1$ and n are the number of customers and jobs respectively. For the bounded model, we focus on a special case with m different processing times of jobs, and present an $O\left(n^3 2^{3m}\right)$ time algorithm.

Keywords: Scheduling, Parallel-batch processing, Dynamic programming, Batch delivery.

1 Introduction

For the manufacturing enterprise, a well-known JIT (just-in-time) production mode has been widely applied. JIT production mode means the operation at every stage of supply chain is determined by the demand or the order. In such a production mode, one manufacturer is generally required to respond to customer demands in a rapid speed and with the least cost as well. To make a satisfactory and cost saving performance, it is critical to coordinate processing schedule and delivery schedule of jobs. This problem has been extensively studied in the area of operation research in the last decades. Hall and Potts[8] firstly introduced the word "supply chain scheduling" and pointed out that its essential issue is collaborative scheduling decision. As they have introduced, jobs are processed by a manufacturer and then delivered in batches to customers. They considered a trade-off between customer service level and total transportation cost. Chen and Vairaktarakis[5] extended this problem by considering transportation time

P. Widmayer, Y. Xu, and B. Zhu (Eds.): COCOA 2013, LNCS 8287, pp. 72–83, 2013.

in transportation cost. Their objective is to minimize $\alpha S(D_1, \cdots, D_n) + (1-\alpha)T$, where D_j is the delivery time of job J_j ($1 \leq j \leq n$) when it is transported to its customer, function $S(D_1, \cdots, D_n)$ measures customer service level, and T denotes total transportation cost. The constant parameter α represents the relative preference between the customer service level and the total transportation cost. They presented either heuristics or efficient exact algorithms for various cases. For the different objective function, Pundoor and Chen [14] also studied this integrated production and distribution problem aiming at optimize the trade-off between maximum delivery tardiness and total transportation cost. Cheng et al.[6] studied the objective of minimizing the sum of batch delivery cost and job earliness penalties. Yang [17] considered a related problem under a special situation where jobs can only be delivered in batches at fixed delivery dates. In all the above studies, it is assumed that a manufacturer processes at most one job at any time. We refer the reader to Chen [4] for a comprehensive review on integrated production and distribution schedule.

In semi-conductor and steel manufacturing industries or some chemical processing, some tasks or jobs are processed simultaneously in batches. This special case is refined as the parallel-batch scheduling model. The parallel-batch machine processes multiple jobs simultaneously in a batch in its capacity. The processing time of a batch is equal to the longest processing time of the job within this batch as Webster and Baker [16] introduced. Motivated by the above application, we focus on the parallel-batch scheduling model in this paper. There are generally two kinds of models in literature considering either bounded or unbounded batch processing capacity. The bounded model arises in integrated circuit manufacturing industry where the batch processing capacity is bounded(see Lee et al. [10]), while the unbounded model arises in some real scenarios where jobs or compositions with relative small sizes need to be handled in kilns that are sufficiently large (see Brucker et al. [1]).

Many researches studied the parallel-batch scheduling model in production stage without considering distribution. Brucker et al.[1] studied single batch machine scheduling problems. They proposed two $O(n \log n)$ and $O(n^{B(B-1)})$ time dynamic programming algorithms for the unbounded model and the bounded model to minimize total completion time, where B is the batch processing capacity of the machine. Especially, for a special case with m different processing times in the bounded model, they gave an $O(B^2 m^2 2^m)$ time algorithm. For the bounded model, Poon and Yu [12] gave an improved $O(n^{6B})$ time algorithm for minimizing total completion time for the case where B is sufficiently large, i.e., $B \geq 16$. None of the above literature has considered the distribution stage. We refer the reader to Potts and Kovalyov[13] for a comprehensive review on batch scheduling.

Very few studies have cast light on integrated scheduling where jobs are handled in batches in both production and distribution stages. Tang and Gong[15] studied a related problem where jobs are transported via vehicles from a holding area to a parallel-batch machine for processing. Each vehicle transports a single job at a time and the machine is of limited batch capacity. They proved it

Table 1. A comparison between our main results and previous results

Objective Function	Processing Batch Capacity	Distribution Stage	Customers number g	Time Complexity
$\sum D_j + T$	$B = 1$	(∞, ∞)	$g \geq 1$	$O\left(n^{g+1}\right)$, in [8]
$\alpha \sum D_j + (1-\alpha)T$	$B \geq n$	(∞, ∞)	$g \geq 1$	$O\left(g(n/g)^{2g}\right)$ in this paper
$\sum C_j$	$B < n*$	null	$g = 1$	$O\left(B^2 m^2 2^m\right)$, in [1]
$\alpha \sum D_j + (1-\alpha)T$	$B = 1$	(∞, c)	$g = 1$	$O(n \log n)$, in [5]
$\alpha \sum D_j + (1-\alpha)T$	$B < n*$	(∞, ∞)	$g = 1$	$O\left(n^3 2^{3m}\right)$ in this paper

*jobs with m different processing times.

NP-hard to minimize the sum of total completion time and total processing cost. Li et al.[11] studied a model of unbounded parallel-batch machine with family jobs delivered by capacitated vehicles, aiming at minimizing the completion time of the last delivery batch. They provided a heuristic algorithm with a worst-case performance ratio of $3/2$. Gong and Tang[7] studied an integrated scheduling with one parallel-batch machine and one vehicle, both of which are of bounded capacity. They provided a polynomial-time algorithm for the objective of minimizing the makespan, and proved it NP-hard for the objective of minimizing total weighted completion time. In these researches, few consideration of the transportation cost has been taken.

In this paper, we study an integrated scheduling problem where there is a single parallel-batch machine in the production stage and sufficient vehicles each with infinite transportation capacity in the distribution stage. We consider two models with either unbounded or bounded batch processing capacity in the production stage. The aim is to find an optimal integrated schedule with the consideration of both total delivery time and total distribution cost.

We sum up our main results and give a comparison with previous results in Table 1, where the parameters in the table are defined in the next section.

The rest of this paper is organized as follows. In Section 2, we describe the problem under consideration and present some basic properties of an optimal schedule. In Sections 3 and 4, we present exact algorithms for the unbounded and bounded models respectively. Finally Section 5 concludes this paper.

2   Description and Basic Properties

The problem is formally described as follows. There are n jobs, which are requested by $g \geq 1$ customers at time 0, to be processed in the production stage and then transported to customers. Each job J_j is of a non-negative processing time p_j. In the production stage, there is a single parallel-batch machine with a batch capacity of B to process jobs in the production stage. If $B \geq n$, then we say it is in unbounded batch processing model, otherwise it is in bounded model if $B < n$. Assume there is a processing batch denoted as B_i. Then the processing time of B_i is $p(B_i) = \max_{J_j \in B_i} \{p_j\}$. For any $J_j \in B_i$, the completion time of its processing is $C_j = C(B_i)$. Let $|B_i|$ be the number of jobs in batch B_i.

In distribution stage, each job after processing is transported from the production stage to the customer who request for it. We assume that jobs are transported in the *direct* delivery mode introduced by Chen[4] such that each vehicle only delivers jobs from the same customer at a time. Moreover, there are sufficient vehicles each with infinite transportation capacity. Let t_i be the transportation time from the production center to customer i, $i = 1, 2, \ldots, g$. It consumes T_i transportation cost in a round trip between the production center and customer i. Assume there are totally q_i delivery batches for customer i in the distribution stage. Then the total transportation cost $T = \sum_{i=1,\ldots,g} q_i T_i$. Let D_j be the delivery time of job J_j ($j = 1, 2, \ldots, n$) when the job is transported its own customer i. The total delivery time is equal to $\sum_{i=1,\ldots,n} D_j$.

Similar to Chen and Vairaktarakis[5], we use $\sum_{i=1,\ldots,n} D_j$, T and α to represent customer service level, total transportation cost and relative preference on them respectively. The aim is to find an integrated processing and delivery schedule to minimize the sum of total delivery time and total transportation cost with the relative preference α for the parallel-batch scheduling model.

We mainly study two models with either unbounded or bounded batch processing capacity of the single batch machine, and adopt the five-field notation $\alpha |\beta| \pi |\delta| \gamma$ introduced by Chen[4] to represent the models as follows.

M1: $1|B - batch, B \geq n|V(\infty, \infty), direct|g|\alpha \sum_{j=1,\cdots,n} D_j + (1 - \alpha)T$, which represents the unbounded batch processing model with g customers.

M2: $1|B - batch, B < n|V(\infty, \infty), direct|1|\alpha \sum_{j=1,\cdots,n} D_j + (1 - \alpha)T$, which represents the bounded batch processing model with a single customer.

By the objective formula, it is a regular function which is non-decreasing in the completion times of jobs. Thus in an optimal production and distribution schedule, there is no idle time between any two consecutive processing batches. Besides, in an optimal integrated processing and delivery schedule of problems M1 and M2, all the jobs from the same customer in one processing batch are transported in the same delivery batch due to the infinite transportation capacity and each delivery batch starts immediately on the completion time of processing of the last job within this delivery batch.

3   The Unbounded Batch Processing Model M1

In this section, we investigate the unbounded model where the single machine processes up to B ($\geq n$) jobs simultaneously and jobs are requested by $g \geq 1$ customers. Assume that there are n_k jobs $J_1^k, J_2^k, \ldots, J_{n_k}^k$ requested by customer k ($k = 1, 2, \ldots, g$), and $\sum_{k=1,\cdots,g} n_k = n$. By observation, we can get two straightforward lemma as follows, the proof of lemma 1 is omitted.

Lemma 1. *For problem M1, there exists an optimal integrated processing and delivery schedule such that all the n_k jobs for customer k ($k = 1, 2, \ldots, g$) are processed in SPT (shortest processing time) order, i.e., $p_j^k \leq p_{j+1}^k$ for $1 \leq j < n_k - 1$.*

Lemma 2. *For problem M1, there exists an optimal integrated processing and delivery schedule such that in any delivery batch all the jobs for customer k are processed in a single processing batch.*

Proof. Assume otherwise in an optimal processing and delivery schedule σ, there exists a delivery batch ζ_u^k for customer k such that ζ_u^k consists of jobs from different processing batches $B_{u_1}, \ldots, B_{u_{j-1}}, B_{u_j}$, then processing all the jobs of customer k in ζ_u^k within a single batch B_{u_j} due to the unbounded processing capacity results in the same delivery schedule and objective value is reduced. The lemma follows.   □

By Lemma 1, we rearrange jobs $J_1^k, J_2^k, \ldots, J_{n_k}^k$ by SPT order for each customer k such that $p_j^k \leq p_{j+1}^k$. We observe that each processing batch may contain jobs from different customers. We adopt g-tuple $(j_1, \ldots, j_k, \ldots, j_g)$ where $1 \leq j_k \leq n_k$ to represent a state such that for customer k, the shortest j_k jobs has been assigned to previous processing batches and there are $n_k - j_k$ jobs left unscheduled for processing. If the next processing batch B_u from this state contains b_k jobs of customer k, i.e., $J_{j_k+1}^k, \ldots, J_{j_k+b_k}^k$, then its processing time

$$p(B_u) = \max \left\{ p\left(J_{j_k+b_k}^k\right) \mid b_k \neq 0, k = 1, \ldots, g \right\}.$$ According to lemma 2, all the jobs in B_u are transported to customers immediately on the completion of the processing batch.

We are now ready to present a backward dynamic programming algorithm DP1, in which the state is added to the beginning, to solve problem M1.

Algorithm DP1

Initialization: Arrange all the n_k jobs for customer k in SPT order, i.e., $p_j^k \leq p_{j+1}^k$ for $1 \leq j \leq n_k$. Set $F(n_1, \ldots, n_k, \ldots, n_g) = 0$.
Recursion: For $k = 1, 2, \ldots, g$ and $j_k = n_k, \ldots, 0$,

$$F(j_1, \ldots, j_k, \ldots, j_g) = \min_{b_k=0,\ldots,n_k-j_k, k=1,\ldots,g} \{F(j_1 + b_1, \ldots, j_k + b_k, \ldots, j_g + b_g)$$
$$+ (1 - \alpha) \sum_{b_k \neq 0, k=1,\ldots,g} T_k + \alpha \sum_{k=1}^g b_k \left(\max \left\{p\left(J_{j_k+b_k}^k\right) + t_k \mid b_k \neq 0, k = 1, \ldots, g\right\}\right)$$
$$+ \alpha \sum_{k=1}^g (n_k - j_k - b_k) \left(\max \left\{p\left(J_{j_k+b_k}^k\right) \mid b_k \neq 0, k = 1, \ldots, g\right\}\right)\}$$

Output: The optimal objective value $F(0, \ldots, 0)$.

In this backward dynamic programming algorithm, the state $(j_1, \ldots, j_g)$ comes from $(j_1 + b_1, \ldots, j_g + b_g)$ in the recursion implies adding a new processing batch B_u containing jobs set $\left\{J_{j_k+b_k}^k \mid b_k = 1, \ldots, n_k - j_k, k = 1, \ldots, g\right\}$, after the shortest j_k jobs for customer k has been scheduled. In the formula of $F(j_1, \ldots, j_k, \ldots, j_g)$, the item $I_1 = (1 - \alpha) \sum_{b_k \neq 0, k=1,\ldots,g} T_k$ represents the total transportation time for delivering all the jobs in B_u to customers. The item $I_2 = \alpha \sum_{k=1}^g b_k \left(\max \left\{p\left(J_{j_k+b_k}^k\right) + t_k \mid b_k \neq 0, k = 1, \ldots, g\right\}\right)$ counts the total delivery time of jobs in B_u. Since the arrangement of batch B_u makes the rest $n_k - (j_k + b_k)$ jobs for customer k deferred for $p(B_u)$ units of time for processing, the item $I_3 = \alpha \sum_{k=1}^g (n_k - j_k - b_k) \left(\max \left\{p\left(J_{j_k+b_k}^k\right) \mid b_k \neq 0, k = 1, \ldots, g\right\}\right)$

counts the increment of the total processing completion time of all the rest unscheduled jobs due to the deferment of the processing batch B_u.

Theorem 1. *Algorithm DP1 solves problem M1 in* $O\left(g(n/g)^{2g}\right)$ *time.*

Proof. First, in initialization, sorting jobs in SPT order takes $O\left(\sum n_k \log n_k\right) = O\left(n \log n\right)$ time. In the recursion, there are totally $\prod_{k=1}^{g}(n_k + 1)$ recurrences. In each recurrence, there are at most $\prod_{k=1}^{g}(n_k + 1)$ possible combinations of jobs in batch B_u, and it takes $O\left(g\right)$ time to calculate the values of items I_1, I_2 and I_3 for any given B_u. Thus it consumes $O\left(g \cdot \prod_{k=1}^{g}(n_k + 1)\right)$ time to calculate the value of $F\left(j_1, \ldots, j_k, \ldots, j_g\right)$ in each recurrence. By $\sum_{k=1,\cdots,g} n_k = n$, $\prod_{k=1}^{g}(n_k + 1) \le ((n + g)/g)^g$, algorithm DP1 solves the problem M1 in $O\left(g((n + g)/g)^{2g}\right) = O\left(g(n/g)^{2g}\right)$ time.   □

Theorem 1 means that when the number of customers g is fixed, Algorithm DP1 solves M1 in polynomial time. When g is arbitrary, according to the existing literatures (see Hall and Potts[8]), the complexity of this problem remains unsolved even the batch processing capacity is degenerated to $B = 1$.

4   The Bounded Batch Processing Model M2

In this section we consider the bounded model where the processing capacity B of the batch machine is strictly less than n, the number of jobs. Brucker et al.[1] investigated one related problem within the production stage, and provided an algorithms running in $O\left(n^{B(B-1)}\right)$ time. They pointed out that when B is arbitrary, the complexity of this problem remains unsolved. So they focused on the case where there is a single customer that requests m different processing times of jobs, and presented an algorithm running in $O\left(2^m m^2 B^2\right)$ time. This case is reasonable in some manufacturing industries where a manufacturer produces m types of products with different processing times, which is called multi-variety and small batch production for m is large or low-variety and large batch production for m is small. Chandru et al.[2,3] and Hochbaum and Landy[9] also proposed $O\left(m^3 B^{m+1}\right)$ and $O\left(m^2 3^m\right)$ algorithms respectively for this kind of problem with the objective of minimizing total completion time. We extend the above research with the further consideration of distribution stage. We assume that it consumes transportation time t and transportation cost T_0 to deliver any job from the production center to the customer.

Let $\bar{p}_1 < \bar{p}_2 < \cdots < \bar{p}_m$ be the m different processing times of the jobs, and n_j be the number of jobs with processing time of $\bar{p}_j$, $(j = 1, 2, \ldots, m)$. $n = \sum_{j=1}^{m} n_j$. For notational convenience, a job with processing time $\bar{p}_j$ is called a j-job, and a processing batch B_k with processing time $p(B_k) = \bar{p}_j$ is called a j-batch. If $|B_k| = B$, then it is called a full processing batch, otherwise it is non-full. If B_k is full and contains j-jobs only, it is called a pure j-batch, otherwise B_k is a non-pure batch if it contains non-uniform length jobs or it is non-full.

Completed jobs are transported to the customer in FCFS (first come first serve) rule in the distribution stage because waiting is meaningless unless it can be transported together with latter jobs, which means jobs in any processing batch are transported to the customer not later than its following processing batches. Moreover, since all the jobs in one processing batch are assigned to the same delivery batch, if we assign all the jobs in processing batch B_k to some delivery batch ζ_i, we simply say to assign the batch B_k to ζ_i. We next present some lemmas on the properties of an optimal schedule of problem M2.

Lemma 3. *In an optimal integrated processing and delivery schedule of problem M2, if there exist two processing batches B_i and B_j satisfying $|B_i| \leq |B_j|$ and $p(B_i) \geq p(B_j)$, then B_i cannot be processed and delivered earlier than B_j.*

Proof. Assume otherwise in an optimal integrated processing and delivery schedule σ, two different processing batches B_i and B_j with $|B_i| \leq |B_j|$ and $p(B_i) \geq p(B_j)$ are assigned to delivery batches ζ_1 and ζ_2 respectively, and ζ_1 starts earlier than ζ_2. As previously mentioned, jobs are transported in FCFS rule and thus B_i is processed earlier than B_j. Both ζ_1 and ζ_2 may contain several processing batches. Let n_1 and n_2 represent the number of other jobs delivered in ζ_1 and ζ_2 excluding B_i and B_j respectively. We use p_1 and p_2 to denote the total processing time among these n_1 and n_2 jobs. τ is the start time of the first processing batch delivered in ζ_1. Since the delivery batches between ζ_1 and ζ_2 have no influence in this proof, we ignore them for conciseness. Then the objective value is

$$F(\sigma) = F_0 + \alpha(n_1 + |B_i|)[\tau + p(B_i) + p_1 + t] + (1 - \alpha)T_0$$
$$+ \alpha(n_2 + |B_j|)[\tau + p(B_i) + p_1 + p(B_j) + p_2 + t] + (1 - \alpha)T_0$$

(F_0 is the objective value generated by the other delivery batches except ζ_1 and ζ_2).

Now we produce another schedule σ' from σ such that the schedule of all processing batches in σ' is the same as in σ except that the processing and the delivery of B_i and B_j are exchanged in the two schedules. Since the completion time of jobs delivered in the delivery batches except ζ_1 and ζ_2 is kept the same, F_0 is kept unchanged. Then the objective value is

$$F(\sigma') = F_0 + \alpha(n_1 + |B_j|)[\tau + p(B_j) + p_1 + t] + (1 - \alpha)T_0$$
$$+ \alpha(n_2 + |B_i|)[\tau + p(B_j) + p_1 + p(B_i) + p_2 + t] + (1 - \alpha)T_0$$

Then $F(\sigma) - F(\sigma') = \alpha n_1[p(B_i) - p(B_j)] + \alpha[|B_j|p(B_i) - |B_i|p(B_j)] + \alpha p_2(|B_j| - |B_i|)$. Since $|B_i| \leq |B_j|$ and $p(B_i) \geq p(B_j)$, i.e., $p(B_i) - p(B_j) \geq 0$, $|B_j|p(B_i) - |B_i|p(B_j) \geq 0$, $|B_j| - |B_i| \geq 0$, it can be concluded that $F(\sigma) \geq F(\sigma')$. A contradiction to the assumption that σ is an optimal schedule in this case. This establishes the lemma. □

Lemma 3 implies that all the full processing batches each with B jobs are processed in SPT order, and a non-full j-batch can only be scheduled after any full j-batch since the number of jobs in a non-full batch is less than B.

Lemma 4. *There exists an optimal integrated processing and delivery schedule of problem M2 such that all the non-full processing batches are processed in SPT order.*

Proof. Assume in an optimal integrated processing and delivery schedule σ, there are two non-full batches B_u and B_{u+k} with $|B_u|, |B_{u+k}| < B$. B_u is scheduled for processing before B_{u+k} while $p(B_{u+k}) < p(B_u)$. We may reassign any $\kappa = \min\{B - |B_u|, |B_{u+k}|\}$ jobs in B_{u+k} to B_u without changing the completion time of B_u's processing. The reassigned κ jobs' completion time is shifted to an earlier time so that B_u becomes a full processing batch or all the jobs in B_{u+k} are removed. Such reassignment can be applied to any two non-full batches in σ until all the non-full batches are in SPT order. The lemma follows.   □

We use $b_j = \lfloor n_j/B \rfloor$ to represent the maximum number of pure j-batches in the processing schedule.

Lemma 5. *In an optimal integrated processing and delivery schedule of problem M2, there exist b_j pure j-batches in the production stage for $j = 1, 2, \ldots, m$.*

Proof. If $b_j = 0$, the Lemma is straightforward. Below we focus on the case $b_j \geq 1$. Assume otherwise that there are at most $b_j - 1$ pure j-batches in an optimal integrated processing and delivery schedule σ ($1 \leq j \leq m$). By the definition of b_j, we claim that at least $n_j - B \cdot (b_j - 1) \geq B$ j-jobs are assigned to two or more non-pure batches in σ.

Consider any two such processing batches B_u and B_w that contain j-jobs in σ. Assume without loss of generality that $p(B_u) \leq p(B_w)$. In this case, swap the j-jobs in B_w and the non-j-jobs in B_u. Then the completion time of B_u kept unchange while that of B_w is reduced or unchange. The reassignment will not make the jobs' total completion time larger without changing the number of jobs contained in them. We can repeat such interchanging between two new batches B_u' and B_w' which contain j-jobs and are not pure j-batches. The above repetition stops provided that there are b_j pure j-batches for $1 \leq j \leq m$ and obtain a new schedule σ'. Since the processing times of batches, which are selected for interchanging jobs, are either reduced or unchanged during interchanging and the delivery arrangement keeps unchanged since the number of jobs contained in each processing batch is kept the same, we claim that the new schedule σ' cannot be worse than σ. The lemma follows.   □

Lemma 6. *In an optimal integrated processing and delivery schedule of problem M2, there exists at most one non-pure j-batch.*

Proof. Assume otherwise there exist two non-pure j-batches $B_u^{(j)}$ and $B_w^{(j)}$ in an optimal schedule σ where $B_u^{(j)}$ precedes $B_w^{(j)}$. By Lemma 5 there are $b_j = \lfloor n_j/B \rfloor$ pure j-batches in σ and thus the total number of j-jobs in $B_u^{(j)}$ and $B_w^{(j)}$ is at most $n_j - B \cdot b_j < B$. By interchanging each j-job in $B_u^{(j)}$ with a non-j-job in $B_w^{(j)}$, we obtain an alternative optimal processing schedule with one non-pure j-batch. Since the delivery schedule keeps unchanged, the lemma follows.   □

In the production stage, according to Lemmas 5 and 6, there are exactly $b_j = \lfloor n_j/B \rfloor$ pure j-batches and at most one non-pure j-batch for j-jobs in an optimal schedule. Hence, there is either zero or exactly one full non-pure j-batch for each j ($j = 1, 2, \ldots, m$). We use the **Batch Filling Procedure** introduced by Brucker et al.[1] to form full non-pure batches, and represent a given configuration of these batches by a set of indices $\chi \subset \{1, 2, \cdots, m\}$. We have $j \in \chi$ if and only if there exists one full non-pure j-batch in the configuration; otherwise $j \notin \chi$, there is no full non-pure j-batch. There are 2^m possible combinations of set χ, and the Batch Filling Procedure runs in $O(m)$ time for any given set χ (please refer to Brucker et al.[1] for details).

For any given set χ, all the full processing batches have been determined above. We next determine non-full processing batches for the remaining unassigned jobs. Assume there are totally n_χ jobs left to be assigned to some non-full processing batches for a given set χ. By Lemma 5, $n_\chi \leq m(B-1)$. We re-index the jobs as $J'_1, J'_2, \cdots, J'_{n_\chi}$ in SPT order. Let s_j $(0 \leq s_j \leq B-1)$ be the number of j-jobs among the n_χ unassigned jobs. To determine how to interleave the jobs contained in non-full batches with the full batches, all the s_j j-jobs can only be processed after all full j-batches but before all the s_{j+1} $(j+1)$-jobs due to Lemmas 3 and 4. Since there are m distinct jobs among the n_χ jobs and m different lengths among all the full processing batches, the number of permutation and combinations for the n_χ jobs and all the full batches is a **Catalan number**
$C_m = \binom{2m}{m} - \binom{2m}{m+1} \rightarrow \frac{4^m}{m^{3/2}\sqrt{\pi}}$ for a given set χ. Each permutation and combination specifies one processing sequence of all the n jobs.

For each of the C_m processing sequences above, we observe that some distinct jobs $J'_u, J'_{u+1}, \cdots, J'_w$ with $p(J'_u) < p(J'_w)$ may be arranged consecutively after the last full m-batch or between two full batches, i.e., between a full j-batch and a full $(j+1)$-batch for some $1 \leq j \leq m$. The consecutively arranged jobs $J'_u, \cdots, J'_w$ may form various combinations of non-full processing batches. The following lemma is useful for forming non-full processing batches.

Lemma 7. *In an integrated optimal processing and delivery schedule of problem M2, all the j-jobs among the n_χ jobs are assigned to the same processing batch.*

Proof. Assume otherwise in an optimal schedule σ, there exist two non-full processing batches B_u and B_w that both contain j-jobs, and B_u is scheduled before B_w. Since there are at most $B-1$ j-jobs $(1 \leq j \leq m)$ in all the non-full processing batches, we can always move the j-jobs in B_w to B_u provided that B_u is not full, and then exchange all the rest j-jobs, if any, in B_w with the same number of non-j-jobs in B_u given that B_u already contains $B-1$ jobs. The objective function will not be increased. The lemma follows.   □

Lemma 7 implies that during the assignment of these n_χ jobs, all j-jobs to be assigned can be treated as one group G_j since they are assigned to one batch. Below we assign the n_χ jobs into at most m groups by their processing time, i.e., $\{G_1, G_2, \cdots, G_m\}$. Group G_j has s_j j-jobs and it does not exist if $s_j = 0$. For one of C_m processing sequences, we first divide $\{G_1, G_2, \cdots, G_m\}$ into k subsets $\{G_1, G_2, \cdots, G_{u_1}\}, \{G_{u_1+1}, \cdots, G_{u_2}\}, \ldots, \{G_{u_{k-1}+1}, \cdots, G_m\}$ such that there exist some full processing batches between G_{u_i} and G_{u_i+1} for $i = 1, 2, \ldots, k-1$ in the sequence. Then jobs in different subsets cannot be assigned to the same non-full batch. Especially $k = 1$ implies that all the m groups are scheduled in SPT order after all the full processing batches. Since the $u_i - u_{i-1}$ groups of jobs ($i = 1, 2, \ldots, k$; $u_0 = 0$; $u_k = m$) in the ith subset may be jointed into one or more non-full processing batches, we use a new set of indices $\gamma \subset \{1, 2, \ldots, m\}$ to represent all the possible configurations of non-full batch combination for groups $G_1, G_2, \cdots, G_m$. We claim that for each processing sequence, there are at most 2^m combinations of γ. If $i \in \gamma$, all the i-jobs in group G_i is assigned to a non-full batch different from that of G_{i+1} ($1 \leq i \leq m$); otherwise G_i is assigned to the same non-full batch as G_{i+1} provided that both groups are in the same subset. We introduce new indicator variables d_i ($1 \leq i \leq m$) such that $d_i = 0$ if $i \in \gamma$ and $d_i = 1$ otherwise. That is, G_i and G_{i+1} are in the same non-full processing batch if $d_i = 1$. For any given set γ, we can produce a unique feasible processing schedule, if exists, by the following Non-full Batch Filling Procedure.

Non-full Batch Filling Procedure

Input: Any given set γ of non-full processing batch indices and the values of d_i ($1 \leq i \leq m$) for γ.

Step 1: Divide $\{G_1, G_2, \cdots, G_m\}$ into k subsets $\{G_1, G_2, \cdots, G_{u_1}\}$, $\{G_{u_1+1}, \cdots, G_{u_2}\}, \ldots, \{G_{u_{k-1}+1}, \cdots, G_m\}$, where $u_0 = 0, u_k = m$ and the value of k is specified by the given set γ. Set $w = 1$ and $h = w - 1$.

Step 2: Find the smallest index $i \in \gamma$ and $i > u_h$ in set $\{G_{u_h+1}, G_{u_h+2}, \cdots, G_{u_w}\}$. If there is no such index, then

Case 1: if $\sum_{j=u_h+1,\ldots,u_w} d_j s_j < B$, then groups $G_{u_h+1}, G_{u_h+2}, \cdots, G_{u_w}$ form a non-full processing batch. Update $w = w + 1$ and $h = w - 1$. If $w = k + 1$, terminate the procedure with a feasible batch processing schedule; otherwise if $w \leq k$, repeat this step.

Case 2: if $\sum_{j=u_h+1,\ldots,u_w} d_j s_j \geq B$, then terminate the procedure for the given set γ because it contradicts the requirement that the related jobs form a non-full batch.

Step 3: If such i exists and $\sum_{j=u_h+1,\ldots,i} d_j s_j \geq B$, then terminate the procedure with the same reasoning as in Case 2 of Step 1; otherwise if such i exists and $\sum_{j=u_h+1,\ldots,i} d_j s_j < B$, groups $G_{u_h+1}, G_{u_h+2}, \cdots, G_{u_i}$ form a non-full processing batch.

Step 4: Set $u_h = i$ and go back to step 2.

The above procedure produces one unique processing schedule $\tilde{\sigma}(\gamma)$ of all the n jobs in $O(l) \leq O(m)$ time for a given γ. The processing schedule $\tilde{\sigma}(\gamma)$ consists of $\sum_{j=1,\cdots,m} b_j = \sum_{j=1,\cdots,m} \lfloor n_j/B \rfloor \leq n/B$ pure batches, at most m full non-pure batches and no more than m non-full processing batches. Thus there are at most $N_\gamma < \frac{n}{B} + 2m$ processing batches, denoted as $\tilde{B}_1, \tilde{B}_2, \ldots, \tilde{B}_{N_\gamma}$, in $\tilde{\sigma}(\gamma)$. Let $p(\tilde{B}_j)$ be the processing time of the processing batch $\tilde{B}_j$ ($j = 1, 2, \ldots N_\gamma$) and $\left|\tilde{B}_j\right|$ be the number of jobs in the batch. The completion time of $\tilde{B}_j$ is equal to $\tilde{C}_j = \sum_{i=1,\ldots,j} p(\tilde{B}_j)$; $j = 1, 2, \ldots, N_\gamma$ since there is no idle time between any two consecutive processing batches.

Now we are ready to produce an optimal delivery schedule for each processing schedule $\tilde{\sigma}(\gamma)$ by the following forward dynamic programming algorithm DP2.

Algorithm DP2
Input: A feasible processing schedule $\tilde{\sigma}(\gamma)$ given by Batch Filling Procedure and Non-full Batch Filling Procedure.
Initialization: Set $F(0) = 0$.
Recursion: For $j = 1, 2, \ldots, N_\gamma$,

$$F_\gamma(j) = \min_{i=1,\ldots,j} \left\{ F(j-i) + \alpha(\tilde{C}_j + t)\left(\sum_{u=j-i+1,\cdots,j} \left|\tilde{B}_u\right|\right) + (1-\alpha)T_0 \right\}$$

Output: An optimal objective value $F(N_\gamma)$ for schedule $\tilde{\sigma}(\gamma)$.

Then the optimal solution for this problem is $F(opt) = \min_\gamma \{F_\gamma(N_\gamma)\}$. The optimal schedule is $\tilde{\sigma}(\gamma^*)$ for the corresponding γ^* in the optimal solution.

In the recursion of algorithm DP2, there are totally N_γ recurrences, in each of which it consumes $O(N_\gamma)$ time to calculate the value of item $\sum_{u=j-i+1,\cdots,j} \left|\tilde{B}_u\right|$ and then $O(N_\gamma^2)$ time to calculate the value of $F(j)$. Hence algorithm DP2 runs in $O(N_\gamma^3)$ time.

Theorem 2. *Algorithm DP2 solves problem M2 in $O(n^3 2^{3m})$ time.*

Proof. For all the n jobs, there are at most 2^m combinations of set χ, each of which produces at most C_m processing sequences via the Batch Filling Procedure running in $O(m)$ time. In each processing sequence, there are at most 2^m combinations of set γ. Thus there are totally $2^m \cdot 2^m \cdot C_m = 2^{2m}C_m$ combinations of γ. Each γ specified a unique processing schedule $\tilde{\sigma}(\gamma)$ via the Non-full Batch Filling Procedure which runs in $O(n_\chi) = O(m)$ time, and an optimal delivery schedule for $\tilde{\sigma}(\gamma)$ via algorithm DP2 which runs in $O\left(N_\gamma^3\right) = O((\frac{n}{B} + 2m)^3) < O(n^3)$ time. Together with $C_m \to \frac{4^m}{m^{3/2}\sqrt{\pi}}$, model M2 can be solved in $O(m2^m + 2^m C_m(m + (n + 2m)^3)) = O((n + 2m)^3 \frac{2^{3m}}{m^{3/2}}) = O(n^3 2^{3m})$ time. The theorem is established. □

Theorem 2 means that when the number of job types m is fixed, Algorithm DP2 solves M2 in polynomial time. When m is arbitrary, according to the existing literatures (see Brucker et al.[1]), the complexity of this problem remains unsolved even without considering the distribution stage.

5   Conclusions and Remarks

In this paper we investigate an integrated scheduling problem considering both production and distribution operations, in which jobs are processed on a parallel-batch machine and then delivered to customers via sufficient vehicles. The objective function embodies both customer service level and total transportation cost. We mainly present dynamic programming algorithms respectively for both models with unbounded or bounded batch processing capacity. Some other objective functions would be a direction for further research. Another further work is to design other algorithms with smaller time complexities.

Acknowledgements. This work was partially supported by the National Natural Science Foundation of China under Grants 71172189, 71071123 and 61221063, Program for Changjiang Scholars and Innovative Research Team in University (No.IRT1173), New Century Excellent Talents in University (NCET-12-0824), and the Fundamental Research Funds for the Central Universities.

References

1. Brucker, P., Gladky, A., Hoogeveen, H., Kovalyov, M.Y., Potts, C.N., Tautenhahn, T., Van de Velde, S.L.: Scheduling a batching machine. Journal of Scheduling 1, 31–54 (1998)
2. Chandru, V., Lee, C.Y., Uzsoy, R.: Minimizing total completion time on a batch processing machines. International Journal of Production Research 31(9), 2097–2122 (1993a)
3. Chandru, V., Lee, C.Y., Uzsoy, R.: Minimizing total completion time on a batch processing machines. Operations Research Letters 13(2), 61–65 (1993b)
4. Chen, Z.L.: Integrated production and outbound distribution scheduling: review and extensions. Operation Research 58(1), 130–148 (2010)
5. Chen, Z.L., Vairaktarakis, G.L.: Integrated scheduling of production and distribution operations. Management Science 51(4), 614–628 (2005)
6. Cheng, T.C.E., Gordon, V.S., Kovalyov, M.Y.: Single machine scheduling with batch deliveries. European Journal of Operational Research 94(2), 277–283 (1996)
7. Gong, H., Tang, L.: A scheduling problem on a single batching machine with batch deliveries. In: Proceedings of the 30th Chinese Control Conference, July 22-24 (2011)
8. Hall, N.G., Potts, C.N.: Supply chain scheduling: batching and delivery. Operations Research 51(4), 566–584 (2003)
9. Hochbaum, D., Landy, D.: Scheduling semiconductor burn-in operations to minimize total flowtime. Operation Research 45(6), 874–885 (1997)
10. Lee, C.Y., Uzsoy, R., Martin-Vega, L.A.: Efficient algorithms for scheduling semiconductor burn-in operations. Operation Research 40(4), 764–775 (1992)
11. Li, S.S., Yuan, J.J., Fan, B.Q.: Unbounded parallel-batch scheduling with family jobs and delivery coordination. Information Processing Letters 111(12), 575–582 (2011)
12. Poon, C.K., Yu, W.: On minimizing total completion time in batch machine scheduling. International Journal of Foundations of Computer Science 15(4), 593–607 (2004)
13. Potts, C.N., Kovalyov, M.Y.: Scheduling with batching: a review. European Journal of Operational Research 120(2), 228–249 (2000)
14. Pundoor, G., Chen, Z.L.: Scheduling a production-distribution system to optimize the tradeoff between delivery tardiness and distribution cost. Naval Research Logistics 52(6), 571–589 (2005)
15. Tang, L., Gong, H.: The coordination of transportation and batching scheduling. Applied Mathematical Modelling 33(10), 3854–3862 (2009)
16. Webster, S., Baker, K.R.: Scheduling groups of jobs on a single machine. Operation Research 43(4), 692–703 (1995)
17. Yang, X.: Scheduling with generalized batch delivery dates and earliness penalties. IIE Transactions 32(8), 735–741 (2000)

The Fractional Strong Metric
Dimension of Graphs

Cong X. Kang and Eunjeong Yi

Texas A&M University at Galveston, Galveston, TX 77553, USA
{kangc,yie}@tamug.edu

Abstract. For any two vertices x and y of a graph G, let $S\{x, y\}$ denote
the set of vertices z such that either x lies on a $y - z$ geodesic or y lies
on a $x - z$ geodesic. For a function g defined on $V(G)$ and $U \subseteq V(G)$,
let $g(U) = \sum_{x \in U} g(x)$. A function $g : V(G) \to [0, 1]$ is a *strong resolving
function* of G if $g(S\{x, y\}) \geq 1$, for every pair of distinct vertices x, y of
G. The *fractional strong metric dimension, $sdim_f(G)$*, of a graph G is
$\min\{g(V(G)) : g$ is a strong resolving function of $G\}$. For any connected
graph G of order $n \geq 2$, we prove the sharp bounds $1 \leq sdim_f(G) \leq \frac{n}{2}$.
Indeed, we show that $sdim_f(G) = 1$ if and only if G is a path. If G
contains a cut-vertex, then $sdim_f(G) \leq \frac{n-1}{2}$ is the sharp bound. We determine $sdim_f(G)$ when G is a tree, a cycle, a wheel, a complete k-partite
graph, or the Petersen graph. For any tree T, we prove the sharp inequality $sdim_f(T + e) \geq sdim_f(T)$ and show that $sdim_f(G + e) - sdim_f(G)$
can be arbitrarily large. Lastly, we furnish a Nordhaus-Gaddum-type result: Let G and $\overline{G}$ (the complement of G) both be connected graphs
of order $n \geq 4$; it is readily seen that $sdim_f(G) + sdim_f(\overline{G}) = 2$ if
and only if $n = 4$; further, we characterize unicyclic graphs G attaining
$sdim_f(G) + sdim_f(\overline{G}) = n$.

Keywords: (strong) metric dimension, fractional metric dimension, fractional strong metric dimension, Nordhaus-Gaddum-type result, tree, unicyclic graph, cut-vertex.

1 Introduction

Let $G = (V(G), E(G))$ be a finite, simple, undirected, connected graph of order
$|V(G)| \geq 2$ and size $|E(G)|$. For a vertex $v \in V(G)$, the *open neighborhood of*
v is the set $N(v) = \{u \in V(G) \mid uv \in E(G)\}$. The *degree* $\deg_G(v)$ of a vertex
$v \in V(G)$ is $|N(v)|$; a *leaf* (or an *end-vertex*) is a vertex of degree one, and
a *major vertex* is a vertex of degree at least three. The distance between two
vertices $u, v \in V(G)$, denoted by $d_G(u, v)$, is the length of the shortest path
between u and v; we omit G when ambiguity is not a concern. The diameter,
$diam(G)$, of a graph G is given by $\max\{d(u, v) \mid u, v \in V(G)\}$. A leaf u is
called *a terminal vertex of a major vertex* v if $d(u, v) < d(u, w)$ for every other
major vertex w. The *terminal degree* $ter(v)$ of a major vertex v is the number
of terminal vertices of v. A major vertex v is an *exterior major vertex* if it has

P. Widmayer, Y. Xu, and B. Zhu (Eds.): COCOA 2013, LNCS 8287, pp. 84–95, 2013.

positive terminal degree. Let $\sigma(G)$ denote the number of leaves of G, and let $ex_1(G)$ denote the number of exterior major vertices of terminal degree exactly one in G. The complement $\overline{G}$ of a graph G is the graph whose vertex set is $V(G)$ and $uv \in E(\overline{G})$ if and only if $uv \notin E(G)$ for $u, v \in V(G)$. We denote by P_n the path on n vertices.

For an ordered set $S = \{w_1, w_2, \ldots, w_k\} \subseteq V(G)$ and a vertex $v \in V(G)$, the k-vector $r(v|S) = (d(v, w_1), d(v, w_2), \ldots, d(v, w_k))$ is called the *metric representation of v with respect to S*. The set S is a *resolving set* for G if $r(x|S) = r(y|S)$ implies that $x = y$ for all pairs x, y of vertices of G. The *metric dimension* of G, denoted by $dim(G)$, is the minimum cardinality amongst all resolving sets of G. Metric dimension was introduced independently by Slater [13] and by Harary and Melter [6]. Some applications of metric dimension can be found in robot navigation [8], sonar [13], combinatorial optimization [12], and pharmaceutical chemistry [2]. It was noted in [5] that determining the metric dimension of a graph is an NP-hard problem. Currie and Oellermann [3] defined fractional metric dimension by relaxing a condition in the integer programming problem formulated for metric dimension. A formulation of fractional metric dimension as a linear programming problem is found in [4] and recounted in [1]. For discussions on the fractionalization of other graph parameters, see [11]. Arumugam and Mathew obtained some bounds for fractional metric dimension and computed the parameter for certain classes of graphs in [1]. For a function g defined on $V(G)$, let $g(U) = \sum_{s \in U} g(s)$ for $U \subseteq V(G)$. For $R\{x, y\} = \{z \in V(G) \mid d(x, z) \neq d(y, z)\}$, a real-valued function $g : V(G) \to [0, 1]$ is a *resolving function* of G if $g(R\{x, y\}) \geq 1$ for any two distinct vertices $x, y \in V(G)$. The *fractional metric dimension* $dim_f(G)$ of a graph G is $\min\{g(V(G)) : g$ is a resolving function of $G\}$. Notice that $dim_f(G)$ reduces to $dim(G)$, the well-known metric dimension, if the codomain of resolving functions is restricted to $\{0, 1\}$.

A vertex $z \in V(G)$ *strongly resolves* a pair of vertices $x, y \in V(G)$ if there exists a shortest $y - z$ path containing x or a shortest $x - z$ path containing y. A set of vertices $S \subseteq V(G)$ *strongly resolves* G if every pair of distinct vertices of G is strongly resolved by some vertex in S; then, S is called a *strong resolving set* of G. The *strong metric dimension* of G, denoted by $sdim(G)$, is the minimum cardinality over all strong resolving sets of G; a strong resolving set of minimum cardinality is also called a *strong basis*. Sebö and Tannier [12] introduced the strong metric dimension. In [10], Oellermann and Peters-Fransen showed that determining the strong metric dimension of a graph is an NP-hard problem. We say that $u \in V(G)$ is *maximally distant* from $v \in V(G)$ if for every $w \in N(u)$, $d(w, v) \leq d(u, v)$. If u is maximally distant from v and v is maximally distant from u, then we say that u and v are *mutually maximally distant*, or u MMD v for short. It was shown in [10] that if x MMD y in G, then any strong resolving set of G must contain either x or y.

Sebö and Tannier [12] observed that if S is a strong resolving set, then the vectors $\{r(v|S) \mid v \in V(G)\}$ uniquely determine the graph G, whereas if S is a resolving set, then the vectors $\{r(v|S) \mid v \in V(G)\}$ may not uniquely determine G. It's easy to construct two non-isomorphic graphs on a common set of vertices, a common minimum resolving set, and a common collection of metric vectors (see [12] for an example). For the other assertion, let us identify $\mathcal{C} = \{r(v|S) : v \in V(G)$ and S is a strong basis of $G\}$ with $V(G)$. Then $x, y \in \mathcal{C}$ *are adjacent if and only if* $\|x - y\|_\infty = 1$, where $\|x\|_\infty = \max_{1 \le i \le k} |x_i|$ is the ℓ_∞-norm on a vector space and $k = sdim(G)$. To see why, first observe that $xy \in E(G)$ implies $\|x - y\|_\infty = 1$. On the other hand, assume $\|x - y\|_\infty = 1$ and let δ_i be the unique element in S having 0 at the i-th coordinate which strongly resolves x from y. Since it is assumed that $|x_i - y_i| \le 1$ and δ_i strongly resolves x from y, we must have $|x_i - y_i| = 1$. Since, up to a transposition of x and y, there is a $x - \delta_i$ geodesic which is an extension by length 1 of a $y - \delta_i$ geodesic, $xy \in E(G)$.

We define fractional strong metric dimension as follows. Let $S\{x, y\}$ denote the set of vertices z such that x lies on $y - z$ geodesic or y lies on $x - z$ geodesic. A real valued function $g : V(G) \to [0, 1]$ is a strong resolving function of G if $g(S\{x, y\}) \ge 1$ for any two distinct vertices $x, y \in V(G)$. The *fractional strong metric dimension* of G, denoted by $sdim_f(G)$, is $\min\{g(V(G)) : g$ is a strong resolving function of $G\}$. Notice that $sdim_f(G)$ reduces to $sdim(G)$ if the codomain of strong resolving functions is restricted to $\{0, 1\}$. The problem of finding the fractional strong metric dimension of a graph can be formulated as a linear programming problem in exactly the same manner as detailed in [4] and recounted in [1], as long as the "resolving graph $R(G)$ of G" is replaced by the "strongly resolving graph $S(G)$ of G": Given a connected graph G of order n, let V be the vertex set of G and W be the set of all $\binom{n}{2}$ pairs of vertices of G. Then $S(G)$ is the bipartite graph with partite sets V and W, such that $x \in V$ is joined to $\{u, v\} \in W$ if and only if x strongly resolves u and v in G.

We first fill a hole in the literature on fractional metric dimension by proving $dim_f(G) = 1$ if and only if G is a path (Theorem 4); with which we readily show $sdim_f(G) = 1$ if and only G is a path (Corollary 1). Then, for any connected graph G of order $n \ge 2$, we show that $1 \le sdim_f(G) \le \frac{n}{2}$; if G contains a cut-vertex, then we prove the sharp bound $sdim_f(G) \le \frac{n-1}{2}$. We determine $sdim_f(G)$ when G is a tree, a cycle, a wheel, a complete k-partite graph, or the Petersen graph. For any tree T, we prove the sharp inequality $sdim_f(T + e) \ge sdim_f(T)$, and we give an example showing that $sdim_f(G + e) - sdim_f(G)$ can be arbitrarily large for some edge e. We obtain a Nordhaus-Gaddum-type result (see [9]) on fractional strong metric dimension: for connected graphs G and $\overline{G}$ of order $n \ge 4$, we show that $2 \le sdim_f(G) + sdim_f(\overline{G}) \le n$. It is readily seen that $sdim_f(G) + sdim_f(\overline{G}) = 2$ if and only if $n = 4$, and there is no tree T satisfying $sdim_f(T) + sdim_f(\overline{T}) = n$. We characterize unicyclic graphs G attaining $sdim_f(G) + sdim_f(\overline{G}) = n$.

2   Some Results on $dim_f(G)$, as a Preliminary to $sdim_f(G)$

Theorem 1. *[1] Let G be a connected graph of order n. Then $dim_f(G) \le \frac{n}{2}$. Further $dim_f(G) = \frac{n}{2}$ if and only if there exists a bijection $\alpha : V(G) \to V(G)$ such that $\alpha(v) \ne v$ and $|R\{v, \alpha(v)\}| = 2$ for all $v \in V(G)$.*

The following proposition extends part (iv) of Cor. 2.7 of [1] to allowing for $a_i = 1$.

Proposition 1. *[14] For $k \geq 2$, let $K_{a_1,a_2,\ldots,a_k}$ be a complete k-partite graph of order $n = \sum_{i=1}^{k} a_i$. Then*

$$dim_f(K_{a_1,a_2,\ldots,a_k}) = \begin{cases} \frac{n-1}{2} & \text{if } a_i = 1 \text{ for exactly one } i \in \{1,2,\ldots,k\} \\ \frac{n}{2} & \text{otherwise .} \end{cases} \quad (1)$$

Theorem 2. *[1]*

(a) *For the Petersen graph $\mathcal{P}$, $dim_f(\mathcal{P}) = \frac{5}{3}$.*
(b) *For the cycle C_n,*

$$dim_f(C_n) = \begin{cases} \frac{n}{n-1} & \text{if } n \text{ is odd,} \\ \frac{n}{n-2} & \text{if } n \text{ is even.} \end{cases} \quad (2)$$

(c) *For the wheel W_n, $n \geq 5$,*

$$dim_f(W_n) = \begin{cases} 2 & \text{if } n = 5 \\ \frac{3}{2} & \text{if } n = 6 \\ \frac{n-1}{4} & \text{if } n \geq 7. \end{cases} \quad (3)$$

(d) *For the grid graph $G = P_s \square P_t$ $(s, t \geq 2)$, $dim_f(G) = 2 = dim(G)$.*

Theorem 3. *[14] For any tree T, $dim_f(T) = \frac{1}{2}[\sigma(T) - ex_1(T)]$.*

Theorem 4. *For any graph G of order n, $dim_f(G) = 1$ if and only if $G = P_n$.*

Proof. ($\Longleftarrow$) Since $1 \leq dim_f(P_n) \leq dim(P_n) = 1$, $dim_f(P_n) = 1$.

($\Longrightarrow$) Let $G \neq P_n$ for any n. We will show that $dim_f(G) > 1$. Let $\mathcal{N} := \bigcap_{u,v \in V(G)} R\{u, v\}$, where the intersection is taken over all pairs of distinct vertices of G. Observe that

$$\mathcal{N} \neq \emptyset \text{ if and only if } G = P_n \text{ for some } n.$$

Suppose, for the sake of contradiction, $dim_f(G) = 1$ and $G \neq P_n$ for any n. Let $u, v, w, z \in V(G)$ such that $u \neq v$ and $w \neq z$. Let $A = R\{u, v\}$, $B = R\{w, z\}$, $A' = A - B$, $B' = B - A$, and let $g : V(G) \rightarrow [0, 1]$ be a resolving function with $g(V(G)) = 1$. By definition, $g(A') + g(A \cap B) \geq 1$ and $g(B') + g(A \cap B) \geq 1$. By the assumption that $dim_f(G) = 1$, $g(A') + g(A \cap B) + g(B') = 1$. So, $g(B') = 0 = g(A')$. Since u, v, w, z are arbitrary, g is zero except on $\mathcal{N}$. Since $G \neq P_n$, $\mathcal{N} = \emptyset$. Thus $g(V(G)) = 0$, a contradiction. $\square$

Theorem 5. *Let B_m be a bouquet of m circles with a cut-vertex (i.e., the vertex sum of m cycles at one common vertex), where $m \geq 2$. Then $dim_f(B_m) = m$.*

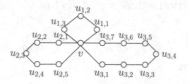

Fig. 1. A bouquet of three circles B_3, and its labeling

Proof. Let B_m be a bouquet of $m \geq 2$ circles C^1, C^2, ..., C^m, with the cut-vertex v (see Fig. 1). For each cycle C^i with $|V(C^i)| = 1 + k_i$, let the vertices of C^i be labeled cyclically, $v = u_{i,0}, u_{i,1}, \ldots, u_{i,k_i}$, where $i \in \{1, 2, \ldots, m\}$.

We first show that $dim_f(B_m) \geq m$. For each $i \in \{1, 2, \ldots, m\}$, $R\{u_{i,1}, u_{i,k_i}\} \cap [V(B_m) - (V(C^i) - \{v\})] = \emptyset$. So, for each $i \in \{1, 2, \ldots, m\}$, $g(V(C^i) - \{v\}) \geq 1$ for any resolving function $g : V(B_m) \rightarrow [0, 1]$; thus $dim_f(B_m) \geq m$.

Next, we show that $dim_f(B_m) \leq m$. For $w \in V(B_m)$, let $h : V(B_m) \rightarrow [0, 1]$ be a function defined by

$$h(w) = \begin{cases} 1 & \text{if } w = u_{i,\frac{k_i}{2}} \text{ and } C^i \text{ is an odd cycle} \\ \frac{1}{2} & \text{if } w \in \{u_{j,\lceil\frac{k_j}{2}\rceil-1}, u_{j,\lceil\frac{k_j}{2}\rceil+1}\} \text{ and } C^j \text{ is an even cycle} \\ 0 & \text{otherwise .} \end{cases} \quad (4)$$

Notice that $h(V(C^i) - \{v\}) = 1$ for each $i \in \{1, 2, \ldots, m\}$. We will show that h is a resolving function of B_m. Let $x, y \in V(B_m)$ and $x \neq y$. First, suppose that $x, y \in V(C^i)$ for some $i \in \{1, 2, \ldots, n\}$. If $d(x, v) \neq d(y, v)$, then $R\{x, y\} \supseteq V(B_m) - [V(C^i) - \{v\}]$; if $d(x, v) = d(y, v)$, then $R\{x, y\} \supseteq \{u_{i,\frac{k_i}{2}}\}$ when C^i is an odd cycle, and $R\{x, y\} \supseteq \{u_{i,\lceil\frac{k_i}{2}\rceil-1}, u_{i,\lceil\frac{k_i}{2}\rceil+1}\}$ when C^i is an even cycle. So, $h(R\{x, y\}) \geq 1$ in each case. Second, suppose that $x \in V(C^i)$ and $y \in V(C^j)$ for two distinct cycles C^i and C^j. If C^i or C^j, say C^i, is an odd cycle, then $R\{x, y\} \supseteq \{u_{i,\frac{k_i}{2}}\}$. If both C^i and C^j are even cycles, we may assume that x lies on $v - u_{i,\lceil\frac{k_i}{2}\rceil-1}$ geodesic and y lies on $v - u_{j,\lceil\frac{k_j}{2}\rceil-1}$ geodesic, by relabeling if necessary. Then $R\{x, y\} \supseteq \{u_{i,\lceil\frac{k_i}{2}\rceil-1}, u_{j,\lceil\frac{k_j}{2}\rceil-1}\}$. In each case, $h(R\{x, y\}) \geq 1$. So h is a resolving function of B_n with $h(V(B_m)) = m$; thus $dim_f(B_m) \leq m$. Therefore, $dim_f(B_m) = m$. $\square$

3   Basic Results on the Fractional Strong Metric Dimension of Graphs

Observation 6. *Let G be a connected graph. Then*

(a) $sdim_f(G) \leq sdim(G)$,
(b) $dim_f(G) \leq sdim_f(G)$,
(c) If x MMD y, then $S\{x, y\} = \{x, y\}$ and hence $g(x) + g(y) \geq 1$ for any strong resolving function g of G.

Corollary 1. *For any graph G of order n, $sdim_f(G) = 1$ if and only if $G = P_n$.*

Proof. Since an end-vertex belongs to every strongly resolving set, $sdim_f(P_n) = 1$. For the other direction, apply Theorem 4 and Observation 6(b). □

Theorem 7. *Let G be a connected graph of order $n \geq 2$. Then $1 \leq sdim_f(G) \leq \frac{n}{2}$. Further, $sdim_f(G) = \frac{n}{2}$ if and only if there exists a bijection α on $V(G)$ such that $\alpha(v) \neq v$ and $S\{v, \alpha(v)\} = \{v, \alpha(v)\}$ for every $v \in V(G)$.*

Proof. The proof (with $R\{x, y\}$ replaced by $S\{x, y\}$) to Theorem 2.6 found in [1] applies here. For sharpness of the lower bound, note $sdim_f(P_n) = 1$ since an end-vertex strongly resolves any pair of vertices of P_n (also see Corollary 1). For sharpness of the upper bound, note $sdim_f(K_n) = \frac{n}{2}$, where K_n is the complete graph on n vertices (see Theorem 8). □

Proposition 2. *Let G be a connected graph of order $n \geq 3$ with a cut-vertex. Then $sdim_f(G) \leq \frac{n-1}{2}$, and the bound is sharp.*

Proof. Let u be a cut-vertex of a connected graph G of order $n \geq 3$. For $w \in V(G)$, let $g : V(G) \to [0, 1]$ be a function defined by

$$g(w) = \begin{cases} 0 & \text{if } w = u \\ \frac{1}{2} & \text{otherwise.} \end{cases} \tag{5}$$

We will show that g is a strong resolving function of G; thus $sdim_f(G) \leq \frac{n-1}{2}$. Let $x, y \in V(G)$ with $x \neq y$. If $u \notin \{x, y\}$, then $S\{x, y\} \supseteq \{x, y\}$; thus $g(S\{x, y\}) \geq g(x) + g(y) = 1$. So, let $u \in \{x, y\}$, say $u = x$. Let G_1 and G_2 be two components of $G - v$. Without loss of generality, we may assume that $y \in V(G_1)$. Then there exists a vertex, say $z \in V(G_2)$, such that u lies on $y - z$ geodesic, since $d_G(y, z) = d_G(y, u) + d_G(u, z)$. So, $S\{u, y\} \supseteq \{y, z\}$; thus $g(S\{u, y\}) \geq 1$.

For the sharpness of the bound, let $G = B_m$ be a bouquet of m 3-cycles, where $m \geq 2$; notice that $|V(G)| = 2m + 1$. Since G contains a cut-vertex, $sdim_f(G) \leq \frac{|V(G)|-1}{2}$ by the present proposition. By Theorem 5 and Observation 6(b), $sdim_f(G) \geq m = \frac{|V(G)|-1}{2}$. So, $sdim_f(G) = \frac{|V(G)|-1}{2}$. □

Theorem 8. *For $k \geq 2$, let $K_{a_1, a_2, \ldots, a_k}$ be a complete k-partite graph of order $n = \sum_{i=1}^{k} a_i$. Then*

$$sdim_f(K_{a_1, a_2, \ldots, a_k}) = \begin{cases} \frac{n-1}{2} & \text{if } a_i = 1 \text{ for exactly one } i \in \{1, 2, \ldots, k\} \\ \frac{n}{2} & \text{otherwise .} \end{cases} \tag{6}$$

Proof. Let $G = K_{a_1, a_2, \ldots, a_k}$ be a complete k-partite graph of order $n = \sum_{i=1}^{k} a_i$, where $k \geq 2$; let $V_1, V_2, \ldots, V_k$ be the partition of $V(G)$ with $|V_i| = a_i$ $(1 \leq i \leq k)$. Let $g : V(G) \to [0, 1]$ be any strong resolving function of G. We consider two cases.

Case 1: $a_i \geq 2$ *for each* $i \in \{1, 2, \ldots, k\}$. In this case, $diam(G) = 2$. Since any two vertices in V_i are mutually maximally distant in G, we have $(a_i - 1) \cdot g(V_i) \geq \binom{a_i}{2} \iff g(V_i) \geq \frac{a_i}{2}$ for each $i \in \{1, 2, \ldots, k\}$. By summing over the k inequalities, we have $g(V(G)) = \sum_{i=1}^{k} g(V_i) \geq \frac{1}{2} \sum_{i=1}^{k} a_i = \frac{n}{2}$; thus $sdim_f(G) \geq \frac{n}{2}$. By Theorem 7, $sdim_f(G) = \frac{n}{2}$.

Case 2: $a_i = 1$ *for some* $i \in \{1, 2, \ldots, k\}$. Let s be the number of partite sets consisting of only one vertex. Without loss of generality, assume that $|V_i| = 1$ for $i \in \{1, 2, \ldots, s\}$. First, let $s = 1$. Then $g(V_j) \geq \frac{a_j}{2}$ for each $j \in \{2, 3, \ldots, k\}$; thus $g(V(G)) \geq \frac{1}{2} \sum_{j=2}^{k} a_j = \frac{n-1}{2}$, i.e., $sdim_f(G) \geq \frac{n-1}{2}$. If we let $h : V(G) \to [0, 1]$ be a function defined by

$$h(v) = \begin{cases} 0 & \text{if } v \in V_1 \\ \frac{1}{2} & \text{otherwise,} \end{cases} \tag{7}$$

then g is a strong resolving function of G; thus $sdim_f(G) \leq \frac{n-1}{2}$. So, $sdim_f(G) = \frac{n-1}{2}$ when $s = 1$. Second, let $1 < s < n$. If we let $W = \cup_{i=1}^{s} V_i$, then x MMD y for any pair $x, y \in W$; thus $g(W) = g(\cup_{i=1}^{s} V_i) \geq \frac{1}{2} \sum_{i=1}^{s} a_i$. Further, $g(V_j) \geq \frac{a_j}{2}$ for each $j \in \{s+1, s+2, \ldots, k\}$. So, $g(V(G)) = g(W) + \sum_{j=s+1}^{k} g(V_j) \geq \frac{1}{2} \sum_{i=1}^{k} a_i + \frac{1}{2} \sum_{j=s+1}^{k} a_j = \frac{n}{2}$, i.e., $sdim_f(G) \geq \frac{n}{2}$. By Theorem 7, $sdim_f(G) = \frac{n}{2}$. Third, let $s = n$. Then $diam(G) = 1$ and x MMD y for any pair $x, y \in V(G)$. So, $g(V(G)) \geq \frac{n}{2}$, i.e., $sdim_f(G) \geq \frac{n}{2}$. By Theorem 7, $sdim_f(G) = \frac{n}{2}$.   □

Theorem 9. *For the Petersen graph* $\mathcal{P}$, $sdim_f(\mathcal{P}) = 5$.

Proof. Let $V(\mathcal{P}) = \{u_i : 1 \leq i \leq 10\}$ and $g : V(\mathcal{P}) \to [0, 1]$ be any strong resolving function. Notice that $\mathcal{P}$ is vertex transitive (see [7]), $diam(\mathcal{P}) = 2$, and each $u_i \in V(\mathcal{P})$ has 6 non-neighbors. Since, for each u_i and a non-neighbor u_{i_j} of u_i, we have u_i MMD u_{i_j} in G, we must have $g(u_i) + g(u_{i_j}) \geq 1$ for each $j = 1 \ldots 6$. Then, we have

$$6g(u_i) + \sum_{j=1}^{6} g(u_{i_j}) \geq 6 \Rightarrow \sum_{i=1}^{10} \left[6g(u_i) + \sum_{j=1}^{6} g(u_{i_j}) \right] \geq 60$$

$$\Rightarrow 6g(V(\mathcal{P})) + \sum_{i=1}^{10} \sum_{j=1}^{6} g(u_{i_j}) \geq 60$$

Now, in the double summation, due to vertex transitivity, the number of times a vertex of $\mathcal{P}$ appears in the multiset $\{u_{i_j}\}(i = 1 \ldots 10$ and $j = 1 \ldots 6)$ is constant. Thus, $\sum_{i=1}^{10} \sum_{j=1}^{6} g(u_{i_j}) = 6 \sum_{i=1}^{10} g(u_i) = 6g(V(\mathcal{P}))$. Thus, the last inequality reduces to $12g(V(\mathcal{P})) \geq 60$, and whence $g(V(\mathcal{P})) \geq 5$. By Theorem 7, $sdim_f(\mathcal{P}) = 5$.   □

Theorem 10. *For the cycle* C_n *on* $n \geq 3$ *vertices,* $sdim_f(C_n) = \frac{n}{2}$.

Proof. Let $V(C_n)$ be labeled cyclically by $I = \{0, 1, \ldots n - 1\}$. By Theorem 7, it suffices to find a permutation α on I such that α is fix-point free and $|S\{v, \alpha(v)\}| = 2$ for each $v \in I = V(G)$. Let d be the diameter of C_n, and put $\alpha(v) = v + d \pmod{n}$. This α clearly is a fix-point free permutation on $V(C_n)$. Further, by Observation 6(c), $|S\{v, \alpha(v)\}| = 2$. $\qquad\square$

Theorem 11. *For the wheel W_n on $n \geq 4$ vertices, we have*

$$sdim_f(W_n) = \begin{cases} 2 & \text{if } n = 4 \\ \frac{1}{2}(n-1) & \text{if } n \geq 5. \end{cases} \tag{8}$$

Proof. Let $G = W_n$ for $n \geq 4$. If $n = 4$, then $G \cong K_4$ and $sdim_f(G) = 2$ by Theorem 8.

So, we consider for $n \geq 5$. Let g be any strong resolving function on $V(W_n)$. Let the vertices of the outer cycle be labeled cyclically by $I = \{0, 1, \ldots n - 2\}$. Let $d = \lfloor \frac{n}{2} \rfloor$ and define a permutation α on I such that $\alpha(v) = v + d \pmod{n-1}$. Notice that v MMD $\alpha(v)$ on the graph W_n. Hence, $|S\{v, \alpha(v)\}| = 2$ and $g(\{v, \alpha(v)\}) \geq 1$ for each $v \in I$. Then $\sum_{v \in I} g(\{v, \alpha(v)\}) \geq n - 1$ and, since each $v \in I$ appears twice in the sum, we have $2g(I) \geq n - 1$. Therefore, we obtain $g(V(W_n)) \geq g(I) \geq \frac{n-1}{2}$.

Next, let $h : V(W_n) \to [0, 1]$ be defined by $h(v) = 0$ for $v \notin I$ and $h(v) = \frac{1}{2}$ for each $v \in I$. Then h is a strong resolving function of G with $h(V(W_n)) = \frac{1}{2}(n-1)$. Therefore, $sdim_f(G) = \frac{1}{2}(n-1)$ for $n \geq 5$. $\qquad\square$

Proposition 3. *For $s, t \geq 2$, $sdim_f(P_s \square P_t) = 2$, where $P_s \square P_t$ is the Cartesian product of two paths P_s and P_t.*

Proof. First, realize $G = P_s \square P_t$ as the rectangular (and integral) lattice in the first quadrant of the xy-plane such that the base and left side of G lie on the x-axis and y-axis, respectively. Label the corners of G cyclically as $u_1 = (0, 0)$, $u_2 = (s - 1, 0)$, u_3, and u_4. Then, we have u_1 MMD u_3 and u_2 MMD u_4 in G. So, $g(u_1) + g(u_3) \geq 1$ and $g(u_2) + g(u_4) \geq 1$ for any strong resolving function g on G. Thus $sdim_f(G) \geq 2$.

Now, define a function h on $V(G)$ by $h(v) = 1$ for $v \in B = \{u_1, u_2\}$ and $h(v) = 0$ for $v \notin B$. We contend that h is a strong resolving function of G; i.e., given $v_1 = (a, b)$ and $v_2 = (c, d)$, an arbitrary pair of distinct vertices of G, there exists $u \in B$ such that there is a geodesic passing through v_1, v_2, and u (up to a transposition of v_1 and v_2). One can readily check that u_1 (u_2, respectively) strongly resolves v_1 and v_2 if $a \leq c$ and $b \leq d$ ($a \leq c$ and $b \geq d$, respectively); note that, up to relabeling, these are the only two distinct cases. Therefore, we conclude $sdim_f(G) = 2$. $\qquad\square$

Remark 1. We note that $sdim_f$ is not a monotone parameter with respect to subgraph inclusion in any sense.

First, consider subgraphs on the same set of vertices. Let $G = P_m \square P_2$ ($m \geq 2$), then C_{2m} is a subgraph of G. We have $sdim_f(G) = 2$, whereas $sdim_f(C_{2m}) = m$. On the other hand, we have $sdim_f(P_m) = 1$, whereas $sdim_f(K_m) = \frac{m}{2}$.

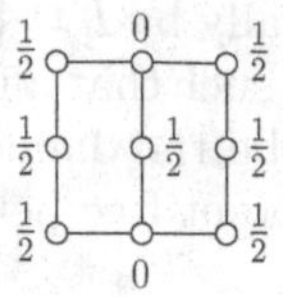

Fig. 2. A graph G and an induced subgraph C_8 of G with $sdim_f(C_8) > sdim_f(G)$

Second, consider induced subgraphs. In Fig. 2, the function g, with $g(v)$ indicated next to each vertex $v \in V(G)$, is a minimum strong resolving function of G. Thus $sdim_f(G) = \frac{7}{2}$. Notice C_8 is an *induced subgraph* of G, and $sdim_f(C_8) = 4 > sdim_f(G)$.

4 The Fractional Strong Metric Dimension of Trees and Unicyclic Graphs

Proposition 4. *For any connected graph G, $sdim_f(G) \geq \frac{1}{2}\sigma(G)$.*

Proof. Let $\ell_1, \ell_2, \ldots, \ell_\sigma$ be leaves of G, where $\sigma = \sigma(G)$. Let $g : V(G) \to [0,1]$ be any strong resolving function. Since ℓ_i MMD ℓ_j, $i \neq j$, in G, $g(\ell_i) + g(\ell_j) \geq 1$. By summing over $\binom{\sigma}{2}$ inequalities, we have $(\sigma - 1)\sum_{i=1}^{\sigma} g(\ell_i) \geq \binom{\sigma}{2}$, i.e., $\sum_{i=1}^{\sigma} g(\ell_i) \geq \frac{\sigma}{2}$. Thus $sdim_f(G) \geq \frac{\sigma}{2}$. □

Corollary 2. *For any tree T, $sdim_f(T) = \frac{1}{2}\sigma(T)$.*

Proof. By Proposition 4, $sdim_f(T) \geq \frac{1}{2}\sigma(T)$. Let $L = \{v \in V(G) : \deg(v) = 1\}$, and let $g : V(T) \to [0,1]$ be a function defined by $g(v) = \frac{1}{2}$ if $v \in L$ and $g(v) = 0$ if $v \notin L$. Since any two vertices, say x and y, in T lie in a path containing two leaves of T, $|S\{x,y\} \cap L| \geq 2$; thus $g(S\{x,y\}) \geq 1$. So, g is a strong resolving function of T; thus $sdim_f(T) \leq \frac{1}{2}\sigma(T)$. □

Since $\sigma(T + e) \geq \sigma(T) - 2$ for $e \in E(\overline{T})$, by Proposition 4, $sdim_f(T + e) \geq sdim_f(T) - 1$. In fact, we will show that $sdim_f(T + e) \geq sdim_f(T)$.

Theorem 12. *For a tree T, $sdim_f(T + e) \geq sdim_f(T)$, where $e \in E(\overline{T})$.*

Proof. Let $g : V(T + e) \to [0,1]$ be any strong resolving function of $T + e$. Notice that $\sigma(T) - 2 \leq \sigma(T + e) \leq \sigma(T)$ for $e \in E(\overline{T})$. If $\sigma(T + e) = \sigma(T)$, then $sdim_f(T + e) \geq \frac{1}{2}\sigma(T + e) = \frac{1}{2}\sigma(T) = sdim_f(T)$ by Proposition 4 and Corollary 2.

So, suppose that $\sigma(T + e) = \sigma(T) - 1$. Then there exists a leaf ℓ in T such that $\deg_{T+e}(\ell) = 2$. Let $\mathcal{C}$ be the unique cycle of $T + e$; we may view the unicyclic graph $T + e$ as $\mathcal{C}$ together with subtrees rooted at vertices on $\mathcal{C}$. Since $\ell \in V(\mathcal{C})$, there exists a vertex $u \in V(\mathcal{C})$ with $d_{\mathcal{C}}(\ell, u) = diam(\mathcal{C})$. Let T_u be the subtree of $T + e$ rooted at u. Let U be the leaves of T_u if $T_u \neq \{u\}$, and let $U = \{u\}$ if $T_u = \{u\}$. Then, $g(x) + g(y) \geq 1$ for distinct vertices $x, y \in U \cup \{\ell\}$, since x MMD y in $T + e$. By summing over the $\binom{|U|+1}{2}$ inequalities, we have $|U|(g(U) + g(\ell)) \geq \binom{|U|+1}{2}$, i.e., $g(U) + g(\ell) \geq \frac{1}{2}(|U| + 1)$. Then, $g(V(T + e)) = g(V(T_u) \cup \{\ell\}) + g(V(T + e) - V(T_u) - \{\ell\}) \geq g(U) + g(\ell) + g(V(T + e) - V(T_u) - \{\ell\}) \geq \frac{1}{2}(|U| + 1) + \frac{1}{2}(\sigma(T + e) - |U|) = \frac{1}{2}(\sigma(T + e) + 1) = \frac{1}{2}\sigma(T)$, since the number of leaves of $T + e$ not on the subtree T_u equals $\sigma(T + e) - |U|$ or $\sigma(T + e)$ (when $U = \{u\}$). Therefore, $sdim_f(T + e) \geq sdim_f(T)$ when $\sigma(T + e) = \sigma(T) - 1$.

Suppose now $\sigma(T + e) = \sigma(T) - 2$, and let ℓ_1 and ℓ_2 be the two (adjacent) vertices on $\mathcal{C}$ which are leaves of T. Suppose there are distinct vertices u_1 and u_2 such that $diam(\mathcal{C}) = d(\ell_i, u_i)$ for $i = 1, 2$, then we apply the above argument to the three subset partition $V(T+e) = (U_1 \cup \{\ell_1\}) \cup (U_2 \cup \{\ell_2\}) \cup (\text{the rest})$, where U_i ($i \in \{1, 2\}$) is the leaves of T_{u_i} if $T_{u_i} \neq \{u_i\}$, and $U_i = \{u_i\}$ if $T_{u_i} = \{u_i\}$. The above argument, modified mutatis mutantis, yields the desired result. If there is only one vertex u on $\mathcal{C}$ such that $diam(\mathcal{C}) = d(\ell_i, u)$ for $i = 1, 2$, then write $g(V(T + e)) = g(U \cup \{\ell_1\}) + g(U \cup \{\ell_2\}) - g(U) + g(V(T + e) - U - \{\ell_1, \ell_2\})$. The desired result again follows from a similar argument. □

Remark 2. For the sharpness of the bound of Theorem 12, let T be a tree with k exterior major vertices $u_1, u_2, \ldots, u_k$ and $ter(u_i) = 2$ for each $i \in \{1, 2, \ldots, k\}$ such that $|V(T)| = 3k$ and $u_i u_{i+1} \in E(T)$ for $1 \leq i \leq k - 1$. If we let $e = u_1 u_k$, then $sdim_f(T+e) = k = sdim_f(T)$. By Corollary 2, $sdim_f(T) = k$. We will show that $sdim_f(T+e) = k$. Since $\sigma(T+e) = 2k$, $\sigma(T+e) \geq k$ by Proposition 4. If we let $g : V(T + e) \to [0, 1]$ be a function defined by $g(v) = \frac{1}{2}$ if $v \notin \{u_i \mid 1 \leq i \leq k\}$ and $g(v) = 0$ if $v \in \{u_i \mid 1 \leq i \leq k\}$, then g is a strong resolving function of $T + e$; thus $sdim_f(T + e) \leq k$.

Remark 3. There exists a graph G such that $sdim_f(G + e) - sdim_f(G)$ is arbitrary large. If $G = P_{2k}$ and $G + e \cong C_{2k}$, then $sdim_f(G) = 1$ by Corollary 1 and $sdim_f(G + e) = k$ by Theorem 10.

Remark 4. Let T be a tree. Then

(a) $sdim_f(T) = sdim(T)$ if and only if T is a path.
(b) $dim_f(T) = sdim_f(T)$ if and only if each exterior major vertex u of T satisfies $ter(u) \geq 2$.

5 Nordhaus-Gaddum-Type Result on Fractional Strong Metric Dimension

Let both G and $\overline{G}$ be connected graphs of order $n \geq 4$. First, we show that $2 \leq sdim_f(G) + sdim_f(\overline{G}) \leq n$. It is readily seen that $sdim_f(G) + sdim_f(\overline{G}) = 2$ if and only if $n = 4$. By Theorem 7, $sdim_f(G) + sdim_f(\overline{G}) = n$ is equivalent to $sdim_f(G) = \frac{n}{2} = sdim_f(\overline{G})$. It is readily seen that $sdim_f(T) + sdim_f(\overline{T}) < n$ for any tree T, since $sdim_f(T) = \frac{n}{2}$ if and only if $T = P_2$ by Corollary 2. We characterize unicyclic graphs G attaining $sdim_f(G) + sdim_f(\overline{G}) = n$, i.e., $sdim_f(G) + sdim_f(\overline{G}) = n$ if and only if $G = C_n$ for $n \geq 5$.

Theorem 13. *Let G and $\overline{G}$ be connected graphs of order $n \geq 4$. Then*

$$2 \leq sdim_f(G) + sdim_f(\overline{G}) \leq n.$$

Moreover, $sdim_f(G) + sdim_f(\overline{G}) = 2$ if and only if $n = 4$.

Proof. Let G and $\overline{G}$ be connected graphs of order $n \geq 4$. The bounds follow from Theorem 7. If $G = \overline{G} = P_4$ (the only possible decomposition when $n = 4$), then $sdim_f(G) = sdim_f(\overline{G}) = 1$, achieving the lower bound of the present theorem. If $G = \overline{G} = C_5$, then $sdim_f(G) + sdim_f(\overline{G})$ achieves the upper bound of the present theorem. Next, we show that $sdim_f(G) + sdim_f(\overline{G}) = 2$ if and only if $n = 4$. Notice that

$$
\begin{aligned}
sdim_f(G) + sdim_f(\overline{G}) = 2 &\Longleftrightarrow sdim_f(G) = sdim_f(\overline{G}) = 1 \\
&\Longleftrightarrow G \cong \overline{G} \cong P_n \text{ by Corollary 1} \Longleftrightarrow |E(G)| = |E(\overline{G})| = n - 1.
\end{aligned}
\tag{9}
$$

Now, $|E(K_n)| = \frac{n(n-1)}{2} = 2(n-1)$, which implies that $n = 4$. □

Next, we characterize unicyclic graphs G for which $sdim_f(G) + sdim_f(\overline{G}) = |V(G)|$.

Theorem 14. *Let G and $\overline{G}$ be connected graphs of order $n \geq 5$. Also, let G be a unicyclic graph. Then $sdim_f(G) + sdim_f(\overline{G}) = n$ if and only if $G = C_n$.*

Proof. Let G be a unicyclic graph of order $n \geq 5$, and let $\mathcal{C}$ be the unique cycle of G such that the vertices of $\mathcal{C}$ are labeled cyclically, say $u_1, u_2, \ldots, u_k$, where $k \geq 3$.

($\Longleftarrow$) Let $G = \mathcal{C} = C_n$ for $n \geq 5$. Then $sdim_f(G) = \frac{n}{2}$ by Theorem 10. Next, we consider $\overline{G}$; notice that $diam(\overline{G}) = 2$. Let $g : V(\overline{G}) \to [0,1]$ be any strong resolving function of $\overline{G}$. Since u_1 MMD u_2 and u_1 MMD u_n in $\overline{G}$, $g(u_1) + g(u_2) \geq 1$ and $g(u_1) + g(u_n) \geq 1$. Since $\overline{G}$ is vertex-transitive, by summing over n inequalities, we have $2 \sum_{i=1}^{n} g(u_i) \geq n$. So, $\sum_{i=1}^{n} g(u_i) \geq \frac{n}{2}$, and hence $sdim_f(\overline{G}) \geq \frac{n}{2}$. By Theorem 7, $sdim_f(\overline{G}) = \frac{n}{2}$.

($\Longrightarrow$) Note that, in order for both G and $\overline{G}$ to be connected and of order n, $sdim_f(G) + sdim_f(\overline{G}) = n$ is equivalent to $sdim_f(G) = \frac{n}{2} = sdim_f(\overline{G})$. If there exists a vertex $u \in V(\mathcal{C})$ with $\deg_G(u) \geq 3$, then u is a cut-vertex and $sdim_f(G) < \frac{n}{2}$ by Proposition 2. Thus, $sdim_f(G) = \frac{n}{2}$ implies that each vertex of $\mathcal{C}$ is of degree two in G, that is, $G = \mathcal{C} = C_n$. □

We conclude this paper with some open problems.

Question 1. Can we characterize graphs G such that $sdim_f(G) = sdim(G)$?

Question 2. Can we characterize graphs G such that $dim_f(G) = sdim_f(G)$?

Question 3. Can we characterize graphs G attaining $sdim_f(G) = \frac{|V(G)|}{2}$?

Acknowledgement. The authors wish to thank the anonymous referees for some helpful comments and suggestions which improved the presentation of the paper.

References

1. Arumugam, S., Mathew, V.: The fractional metric dimension of graphs. Discrete Math. 312, 1584–1590 (2012)
2. Chartrand, G., Eroh, E., Johnson, M.A., Oellermann, O.R.: Resolvability in graphs and the metric dimension of a graph. Discrete Appl. Math. 105, 99–113 (2000)
3. Currie, J., Oellermann, O.R.: The metric dimension and metric independence of a graph. J. Combin. Math. Combin. Comput. 39, 157–167 (2001)
4. Fehr, M., Gosselin, S., Oellermann, O.R.: The metric dimension of Cayley digraphs. Discrete Math. 306, 31–41 (2006)
5. Garey, M.R., Johnson, D.S.: Computers and intractability: A guide to the theory of NP-completeness. Freeman, New York (1979)
6. Harary, F., Melter, R.A.: On the metric dimension of a graph. Ars Combin. 2, 191–195 (1976)
7. Holton, D.A., Sheehan, J.: The Petersen graph. Cambridge University Press (1993)
8. Khuller, S., Raghavachari, B., Rosenfeld, A.: Landmarks in graphs. Discrete Appl. Math. 70, 217–229 (1996)
9. Nordhaus, E.A., Gaddum, J.W.: On complementary graphs. Amer. Math. Monthly 63, 175–177 (1956)
10. Oellermann, O.R., Peters-Fransen, J.: The strong metric dimension of graphs and digraphs. Discrete Appl. Math. 155, 356–364 (2007)
11. Scheinerman, E.R., Ullman, D.H.: Fractal graph theory: A rational approach to the theory of graphs. John Wiley & Sons, New York (1997)
12. Sebö, A., Tannier, E.: On metric generators of graphs. Math. Oper. Res. 29, 383–393 (2004)
13. Slater, P.J.: Leaves of trees. Congress. Numer. 14, 549–559 (1975)
14. Yi, E.: The fractional metric dimension of permutation graphs (submitted)

Online Scheduling on Two Parallel Machines with Release Times and Delivery Times[*]

Peihai Liu and Xiwen Lu

Department of Mathematics, East China University of Science and Technology,
Shanghai, People's Republic of China, 200237
{pliu,xwlu}@ecust.edu.cn

Abstract. We consider an online scheduling problem where jobs arrive over time. A set of independent jobs has to be scheduled on two parallel machines, where preemption is not allowed and the number of jobs is unknown in advance. The characteristics of each job, i.e., processing time and delivery time, become known at its release time. Each job is delivered to the destination independently and immediately at its completion time on the machines. The objective is to minimize the time by which all jobs have been delivered. We present an online algorithm which has a competitive ratio of $(1 + \sqrt{5})/2 \approx 1.618$.

Keywords: Scheduling, Delivery times, Parallel machines, Online algorithm.

1 Introduction

In this paper, we investigate an online scheduling on two parallel machines with release times and delivery times. There are n jobs, say $\{J_1, J_2, \ldots, J_n\}$, which arrive over time. Each job J_j has a release time r_j, a processing time p_j, and a delivery time q_j. The characteristics of each job become known at its arrival time and the number of jobs is unknown in advance. Preemption is not allowed. There are sufficiently many vehicles. Once the processing of a job is completed, it is delivered to the destination by a vehicle immediately. The objective is to minimize the time by which all jobs have been delivered. Let L_j be the time by which job J_j is delivered in a schedule and $L_{max} = \max L_j$. Then this problem can be denoted by $P2|r_j, q_j, online|L_{max}$.

Due to their theoretical and practical interests, scheduling problems with release times and delivery times have been the subject of numerous papers. The classical one machine problem $1|r_j, q_j|L_{max}$ is known to be strongly NP-hard [3]. Potts [14] presented a heuristic algorithm with the worst-case performance ratio $3/2$ for the problem. Several polynomial-time approximation schemes for the same problem are provided by Hall and Shmoys [7,8] and Mastrolilli [12]. Hoogeveen and Vestjens [10] studied the online version of this problem.

---

[*] This work was supported by the National Nature Science Foundation of China (11101147,11371137).

P. Widmayer, Y. Xu, and B. Zhu (Eds.): COCOA 2013, LNCS 8287, pp. 96–105, 2013.

They provided an online algorithm with a competitive ratio of $\frac{\sqrt{5}+1}{2}$ and showed that it is best possible. If all jobs have small delivery times, i.e., for each job $J_j, q_j \leq p_j$, Tian et al. [16] gave a best possible online algorithm with a competitive ratio of $\sqrt{2}$. Liu et al. [11] considered another online restricted version. Stougie and Vestjens [15] derive a randomized lower bound of 1.302 for the single machine problem.

$P|r_j, q_j|L_{max}$ is strongly NP-hard since it is a generalization of the classical one-machine scheduling problem with heads and tails. When jobs have identical delivery times, it becomes the classical identical parallel machine problem denoted by $P|r_j|C_{max}$. A great deal of work has been done on this problem. Here we only mention some online results. Chen and Vestjens [2] showed that LPT which starts the job with the largest processing time whenever a machine becomes available is $3/2$ competitive for all m. In the smae paper, they also show a lower bound of $(5 - \sqrt{5})/2 \approx 1.38197$ for $m = 2$ and a general lower bound of 1.34729. Noga and Seiden [13] presented an online algorithm called Sleepy which matches the lower bound when $m = 2$. Some randomized lower bounds are given in [15] and [13]. When jobs have identical release times and general delivery times, Woeginger [18] gave a heuristic with a worst-case performance guarantee of $2 - 2/(m + 1)$. When jobs have both general release times and general delivery times, Mastrolilli [12] presented an efficient polynomial-time approximation schemes. Hall and Shomys [7] observed that List Scheduling is 2-competitive even in the presence of precedence constrains for problem $P|r_j, q_j, online|L_{max}$. Vestjens [17] showed that no online algorithm can be better than $3/2$-competitive. Some more researches can be referred to [1,5,4,9]. Stougie and Vestjens [15] derive a randomized lower bound of 1.265 for any $m \geq 2$.

This paper is organized as follows: in the next section, we introduce some notions and lower bounds for the problem. In Section 3, we present an online algorithm and show that it has a competitive ratio of $\frac{\sqrt{5}+1}{2} \approx 1.618$. This leaves a gap from 1.5 to 1.618.

2   Preliminaries

Given a job set S, let $p(S)$ denote the total processing time of jobs in S. Denote by $S_j(\sigma), C_j(\sigma)$ and $L_j(\sigma)$ the starting time, completion time and the time by which job J_j is delivered to the destination in schedule σ, respectively. We use σ and π to denote the schedule produced by an online algorithm and an offline optimal schedule, respectively. Let $L_{max}(\sigma), L_{max}(\pi)$ be the objective value of σ and π, respectively.

Now, we will introduce some lower bounds. An obvious lower bound is

$$LB_0(S) = \max_{j \in S}(r_j + p_j + q_j) \tag{1}$$

Another simple lower bound given in Reference [1] is

$$LB_1(S) = \min_{j \in S} r_j + \frac{1}{2} \sum_{j \in S} p_j + \min_{j \in S} q_j \qquad (2)$$

This bound is based on the fact that all release times and delivery times are assumed to be equal to $\min_{j \in S} r_j$ and $\min_{j \in S} q_j$.

The following lower bound, proposed in Reference [1], tries to take into account release times and delivery times in a more effective way:

$$LB_2(S) = \frac{1}{2}(\bar{r}_1 + \bar{r}_2 + \sum_{j \in S} p_j + \bar{q}_1 + \bar{q}_2) \qquad (3)$$

where $\bar{r}_i$ and $\bar{q}_i (i = 1, 2)$ denote the ith smallest release time and delivery time in S, respectively.

3   An Online Algorithm

In this section, we will present an online algorithm and show that it has a competitive ratio of $1 + \alpha$, where $\alpha = (\sqrt{5} - 1)/2$.

We say that a job is available if it has been released, but not yet scheduled. A machine that is not processing a job is called idle. Our algorithm, which we call MLDT, is quite simple. If both machines are processing jobs or there are no available jobs, then we have no choice but to wait. So, assume that there is at least one idle machine and at least one available job. If both machines are idle then start the available job with largest delivery time. If at time t, one machine is idle and the other machine is running job J_j then start the available job with largest delivery time if and only if $t \geq \alpha p_j(t)$, where $p_j(t)$ is the remaining processing time of job J_j at time t. This is equivalent to $t \geq \alpha^2 C_j(\sigma)$, as $\alpha^2 + \alpha = 1$.

We use $U(t)$ to denote the set of released and unscheduled jobs at time t.

Algorithm $MLDT$

Step 0. Set t=0;
Step 1. If $U(t) = \emptyset$, goto Step 5.
Step 2. If both machines are idle at time t and $U(t) \neq \emptyset$, do the following: select a job with the largest delivery time in $U(t)$ and schedule it on one of the idle machines.
Step 3. If only one machine is idle at time t and the busy machine is running a job J_j at time t, do the following:
  • If $t \geq \alpha^2 C_j(\sigma)$, select a job with the largest delivery time in $U(t)$ and schedule it on the idle machine.
  • If $t < \alpha^2 C_j(\sigma)$, reset t be $t^* = \alpha^2 C_j(\sigma)$. Then goto Step 1.
Step 4. If both machines are busy at time t, reset t to be the first time instant after t at which at least one machine is idle. Then goto Step 1.

Step 5. Waiting a new job arrives and reset t to be the release time of such a job. Then goto Step 1.

Note that if at time t, one machine is idle and the other machine is running job J_j, then the machine can be idle for two reasons. If one machine is idle because $t < \alpha p_j(t)$, we say the machine is sleeping. If one machine is idle because there are no available jobs we say that the machine is waiting. In this situation, $t \geq \alpha p_j(t)$ (or $t \geq \alpha^2 C_j(\sigma)$). If one machine is waiting we call the next job released tardy, or a tardy job.

Let J_l denote the first job in σ that assumes the value $L_{max}(\sigma)$. To analyze the algorithm, we first give a structure property that the schedule σ satisfies.

Lemma 1. *Without loss of generality, we may assume that, at any time before $S_l(\sigma)$ in σ, at least one machine is not idle.*

Proof. First, we show that, if there is a common idle period in σ before time $S_l(\sigma)$, during which both machines are idle, then the schedule must start with the common idle period. Suppose this is not the case, i.e. at least one job has finished before the common idle period. Note that, according to the MLDT algorithm, jobs that are scheduled after the common idle period must be released after this period. If we remove all the jobs that finish before this idle period, then the objective value of the schedule created by the MLDT algorithm does not change, whereas the corresponding optimal value does not increase. Hence the ratio $L(\sigma)/L_{max}(\pi)$ for the new instance does not decrease. Therefore, we can assume that schedule σ starts with a common idle period, if it exists.

Suppose that the first starting job is released at time r_1, which is the end of this idle period. We add a new job J_0 with $r_0 = 0, p_0 = r_1, q_0 = 0$. This new job will start at 0 and be completed at r_1 in σ. Therefore, it does not change the objective value of schedule created by the MLDT algorithm. Since all jobs except J_0 are released at or after r_1, there exists an optimal schedule in which J_0 will start at 0 and be completed at r_1. Hence, J_0 does not change the objective value of the optimal schedule.

Thus, we may assume that, at any time before $S_l(\sigma)$ in σ, at least one machine is not idle in σ.

This completes the proof. $\square$

Without loss of generality, we index jobs with their nondecreasing starting times in σ, i.e., $S_1(\sigma) \leq S_2(\sigma) \leq \ldots \leq S_n(\sigma)$. Let j_x be the last job in σ before J_l with a delivery time smaller than q_l, i.e., $q_x < q_l$ and $S_x(\sigma) < S_l(\sigma)$. If such J_x does not exist, let J_x be the first starting job J_1 which starts at time 0. Let t_b be the minimum time such that both machines are busy processing jobs in the interval $[t_b, S_l(\sigma)]$.

We now prove that Algorithm MLDT has a competitive ratio no grater than $1 + \alpha$. The proof consists of two situations: (1) $t_b > S_x(\sigma)$ and (2) $t_b \leq S_x(\sigma)$.

Before the analysis of the two situations, we first give some basic properties for σ and π as follows.

Proposition 1. *For any two jobs J_i, J_j in σ, if $S_i(\sigma) < S_j(\sigma)$ and $q_i < q_j$, then $r_j > S_i(\sigma)$.*

Proposition 2. *Suppose that J_i is the running job when J_j starts in σ, i.e., $S_i(\sigma) \le S_j(\sigma) \le C_i(\sigma)$. Then $S_j(\sigma) \ge \alpha^2 C_i(\sigma)$ and $S_j(\sigma) \ge \alpha p_i'$, where p_i' is the remaining processing time at time $S_j(\sigma)$.*

Proposition 3. *Suppose at time t, one machine is idle and the other machine is running job J_j. If $t > \alpha p_j(t)$, then all jobs which start no earlier than t are released tardy, i.e., $r_i \ge t$ for each J_i with $S_i(\sigma) \ge t$.*

Proposition 4. $L_{max}(\pi) \ge S_x(\sigma) + p_l + q_l$.

Lemma 2. *If $t_b > S_x(\sigma)$, then $L_{max}(\sigma) \le (1+\alpha)L_{max}(\pi)$.*

Proof. $t_b > S_x(\sigma)$ means that immediately before t_b one machine is idle and one machine is busy. Let J_r be the running job immediately before t_b, i.e., $C_r(\sigma) \ge t_b, S_r(\sigma) < t_b$. Define $A = \{J_j | t_b \le S_j(\sigma) < S_l(\sigma)\}$, $B = \{J_j | S_x(\sigma) < S_j(\sigma) < S_l(\sigma)\} \cup \{J_l\}$. Thus, $q_j \ge q_l$ and $r_j \ge S_x(\sigma)$ for each $j \in B$. Let $p_r' = C_r(\sigma) - t_b$.

We claim that the optimal schedule has the following lower bound

$$L_{max}(\pi) \ge \frac{1}{2}(t_b + p_r' + p(A) + p_l + q_l) \tag{4}$$

In fact, we can check two cases. If $S_r(\sigma) \le S_x(\sigma)$, then $p_r \ge C_r(\sigma) - S_x(\sigma) = p_r' + t_b - S_x(\sigma)$. On basic of the jobs in $\{J_r, J_l\} \cup A$ and LB_2, we derive the following lower bound on $L_{max}(\pi)$,

$$L_{max}(\pi) \ge \frac{1}{2}(0 + S_x(\sigma) + p_r + p(A) + p_l + q_l + 0)$$

$$= \frac{1}{2}(t_b + p_r' + p(A) + p_l + q_l)$$

If $S_r(\sigma) > S_x(\sigma)$, then $J_r \in B$ and $q_r \ge q_l$. Let t' be the minimum time instant such that during $[t', t_b]$ at least one machine is running jobs in B. Thus $p(B) \ge t_b - t' + p_r' + p(A) + p_l$. By Lemma 1, there exists a running job immediately before t', say $J_y(C_y \ge t')$. Then, we know that $S_y(\sigma) \le S_x(\sigma)$. Otherwise $S_y(\sigma) > S_x(\sigma)$. Thus, $J_y \in B$ which contradicts to the definition of t'. So, $p_y \ge t' - S_x(\sigma)$. On basic of the jobs in $\{J_y\} \cup B$ and LB_2, we deduce the following lower bound,

$$L_{max}(\pi) \ge \frac{1}{2}(0 + S_x(\sigma) + p_y + p(B) + q_l + 0)$$

$$= \frac{1}{2}(t_b + p_r' + p(A) + p_l + q_l)$$

From the above discussion follows the inequality (4). By the definition of t_b and p_r', we can derive that

$$L_{max}(\sigma) \le t_b + \frac{1}{2}(p_r' + p(A)) + p_l + q_l \tag{5}$$

From (4) and (5), we deduce that

$$L_{max}(\sigma) - L_{max}(\pi) \leq \frac{1}{2}(t_b + p_l + q_l) \tag{6}$$

According to MLDT algorithm, we know that $t_b \geq \alpha p_r'$. If $t_b > \alpha p_r'$, then at time t_b, the machine is waiting and $r_j \geq t_b$ for each J_j with $S_j(\sigma) \geq t_b$. Thus, $r_l \geq t_b$ and $L_{max}(\pi) \geq t_b + p_l + q_l$. Therefore, $L_{max}(\sigma) - L_{max}(\pi) \leq \alpha L_{max}(\pi)$.

If $t_b = \alpha p_r'$, we discuss two cases according to the relationship between $p(A)$ and p_r'.

Case 1 $p(A) \geq p_r'$.

If $p_l + q_l \geq (3 + 2\alpha)t_b$, then

$$L_{max}(\pi) \geq p_l + q_l \geq \frac{1+\alpha}{2}(p_l + q_l) + \frac{1-\alpha}{2}(p_l + q_l)$$

$$\geq \frac{1+\alpha}{2}(t_b + p_l + q_l)$$

If $p_l + q_l < (3 + 2\alpha)t_b$, then $p_r' + p(A) \geq 2(1 + \alpha)t_b \geq \alpha(t_b + p_l + q_l)$. Thus,

$$L_{max}(\pi) \geq \frac{1}{2}(t_b + p_r' + p(A) + p_l + q_l)$$

$$\geq \frac{1+\alpha}{2}(t_b + p_l + q_l)$$

Therefore, $L_{max}(\sigma) - L_{max}(\pi) \leq \alpha L_{max}(\pi)$.

Case 2 $p(A) < p_r'$, then all the jobs of A are processed on the same machine in σ. Thus, $S_l(\sigma) = t_b + p(A)$ and $S_l(\sigma) \leq t_b + p_r' \leq (1 + \alpha)p_r'$ since $t_b \leq \alpha p_r'$.

From (4) we have

$$L_{max}(\pi) \geq \frac{1}{2}(t_b + p_r' + p(A) + p_l + q_l)$$

$$= \frac{1}{2}(S_l(\sigma) + p_r' + p_l + q_l) \tag{7}$$

If $p_l + q_l \leq S_l(\sigma) + p_r'$, then by (7) and $L_{max}(\sigma) = S_l(\sigma) + p_l + q_l$ we obtain that

$$\frac{L_{max}(\sigma) - L_{max}(\pi)}{L_{max}(\pi)} \leq \frac{S_l(\sigma) + p_l + q_l - p_r'}{S_l(\sigma) + p_l + q_l + p_r'}$$

$$\leq \frac{S_l(\sigma)}{S_l(\sigma) + p_r'}$$

Recall that $p_r' \geq \alpha S_l(\sigma)$. Therefore, $L_{max}(\sigma) - L_{max}(\pi) \leq \alpha L_{max}(\pi)$.

If $p_l + q_l > S_l(\sigma) + p_r'$, then $p_l + q_l \geq (1 + \alpha)S_l(\sigma)$, as $S_l(\sigma) \leq (1 + \alpha)p_r'$. Since $L_{max}(\pi) \geq p_l + q_l$ and $L_{max}(\sigma) = S_l(\sigma) + p_l + q_l$, we have $L_{max}(\sigma) - L_{max}(\pi) \leq S_l(\sigma) \leq \alpha L_{max}(\pi)$.

This completes the proof.   □

Lemma 3. *For any job J_j, define $B = \{J_i | S_i(\sigma) < S_j(\sigma)\}$. If there is a job J_h such that $r_h \leq \alpha^2 S_j(\sigma)$ and $S_h(\sigma) \geq S_j(\sigma)$, then $p(B) \geq (1+\alpha)S_j(\sigma)$.*

Proof. Let t_f be the minimum time such that there is no idle machines during interval $[t_f, S_j(\sigma)]$. Thus there exists an idle interval on one of the machines immediately before t_f.

If $t_f \leq \alpha^2 S_j(\sigma)$, then it is clear that $p(B) \geq 2S_j(\sigma) - t_f \geq (1+\alpha)S_j(\sigma)$.

If $t_f > \alpha^2 S_j(\sigma)$, let J_k be the running job immediately before t_f. We claim that $t_f \leq \alpha^2 C_k(\sigma)$ by Step 3 of Algorithm MLDT and the fact $r_h \leq \alpha^2 S_j(\sigma) < t_f, S_h(\sigma) \geq S_j(\sigma)$. Otherwise, J_h will start before t_f which contradicts the fact provided above. From $t_f > \alpha^2 S_j(\sigma)$ and $t_f \leq \alpha^2 C_k(\sigma)$, we can obtain that $C_k(\sigma) > S_j(\sigma)$ and $C_k(\sigma) > (2+\alpha)t_f \geq \alpha S_j(\sigma) + t_f$. Thus, $p(B) \geq S_j(\sigma) + C_k(\sigma) - t_f \geq (1+\alpha)S_j(\sigma)$.

This completes the proof. $\qquad\qquad\square$

Lemma 4. *If $t_b \leq S_x(\sigma)$, then $L_{max}(\sigma) \leq (1+\alpha)L_{max}(\pi)$.*

Proof. Since in the interval $[t_b, S_l(\sigma)]$ both machines are busy processing jobs and $t_b \leq S_x(\sigma) < S_l(\sigma)$, there exists a job other than J_x, say J_y, such that $S_y(\sigma) \leq S_x(\sigma)$ and $C_y(\sigma) > S_x(\sigma)$. Without loss of generality, suppose J_y is assigned before J_x by the algorithm, i.e., $S_y(\sigma) \leq S_x(\sigma)$. Denote by p'_y the remaining processing time of J_y at time $S_x(\sigma)$. By the algorithm, $S_x(\sigma) \geq \alpha p'_y$. Define $A = \{J_j | S_x(\sigma) < S_j(\sigma) < S_l(\sigma)\}$, $Q = A \cup \{J_l\}$. Then $q_j \geq q_l$ and $r_j \geq S_x(\sigma)$ for each $j \in Q$. Thus we have

$$L_{max}(\sigma) \leq S_x(\sigma) + \frac{1}{2}(p'_y + p_x + p(A)) + p_l + q_l \qquad (8)$$

Now we consider four cases depending on the assignment of J_x and J_y in the optimal schedule π.

Case 1: $S_x(\pi) \geq S_x(\sigma)$ and $S_y(\pi) \geq S_y(\sigma)$.

In this case, there are at least $p_{y'} + p_x + p(A) + p_l$ time unit of jobs are processed no earlier than $S_x(\sigma)$ since $r_j \geq S_x(\sigma)$ for each $j \in Q$. Hence,

$$L_{max}(\pi) \geq S_x(\sigma) + \frac{1}{2}(p'_y + p_x + p(A) + p_l) \qquad (9)$$

On basic of the jobs in $\{J_x, J_l\} \cup A$, similar to LB_2, we have

$$L_{max}(\pi) \geq S_x(\sigma) + \frac{1}{2}(p_x + p(A) + p_l + q_l) \qquad (10)$$

Recall that $S_x(\sigma) \geq \alpha p'_y$ and $r_l \geq S_x(\sigma)$. We obtain

$$L_{max}(\pi) \geq S_x(\sigma) + p_l + q_l \geq \alpha p'_y + p_l + q_l \qquad (11)$$

Thus, from $\alpha^3(9) + 2\alpha^2(10) + \alpha(11)$, we deduce

$$(1+\alpha)L_{max}(\pi) \geq S_x(\sigma) + \frac{1}{2}(p'_y + p_x + p(A)) + p_l + q_l$$

which follows that $L_{max}(\sigma) \le (1+\alpha)L_{max}(\pi)$.

Case 2: $S_x(\pi) \ge S_x(\sigma)$ and $S_y(\pi) < S_y(\sigma)$.

Case 2.1: $C_y(\pi) \le \max_{j \in Q} C_j(\pi)$. Similar to LB_2, we have

$$L_{max}(\pi) \ge \frac{1}{2}(p_y + S_x(\sigma) + p_x + p(A) + p_l + q_l) \tag{12}$$

From (8) and (12), we derive that

$$L_{max}(\sigma) - L_{max}(\pi) \le \frac{1}{2}(S_x(\sigma) + p_l + q_l) \le \alpha L_{max}(\pi)$$

Case 2.2: $C_y(\pi) > \max_{j \in Q} C_j(\pi)$. Thus all jobs in Q are processed on the same machine in the optimal schedule π since $r_j \ge S_y(\sigma)$ for each $j \in Q$. Hence,

$$L_{max}(\pi) \ge S_x(\sigma) + p(A) + p_l + q_l \tag{13}$$

It is clear that $L_{max}(\sigma) \le S_x(\sigma) + \min\{p_x, p'_y\} + p(A) + p_l + q_l$. Combine (13) and the above inequality, we deduce that $L_{max}(\sigma) - L_{max}(\pi) \le \min\{p_x, p'_y\}$. Recall that $S_x(\pi) \ge S_x(\sigma)$ and $S_x(\sigma) \ge \alpha p'_y$. Then $L_{max}(\pi) \ge S_x(\sigma) + p_x \ge (1+\alpha)\min\{p_x, p'_y\}$ and $L_{max}(\sigma) - L_{max}(\pi) \le \alpha L_{max}(\pi)$.

Case 3: $S_x(\pi) < S_x(\sigma)$ and $S_y(\pi) \ge S_y(\sigma)$.

Similar to Case 2, we can deduce that $L_{max}(\sigma) - L_{max}(\pi) \le \alpha L_{max}(\pi)$.

Case 4: $S_x(\pi) < S_x(\sigma)$ and $S_y(\pi) < S_y(\sigma)$.

If $C_y(\pi) > \max_{j \in Q} C_j(\pi)$ or $C_x(\pi) > \max_{j \in Q} C_j(\pi)$, similar to Case 2.2, we can deduce that $L_{max}(\sigma) - L_{max}(\pi) \le \alpha L_{max}(\pi)$. Thus we can only check situation where $C_y(\pi) \le \max_{j \in Q} C_j(\pi)$ and $C_x(\pi) \le \max_{j \in Q} C_j(\pi)$.

Case 4.1: $\min\{r_x, r_y\} \ge \alpha^2 S_y(\sigma)$ or $q_l \ge 2\alpha^2 S_y(\sigma)$. Then

$$\begin{aligned} L_{max}(\pi) &\ge \frac{1}{2}(r_y + p_y + r_x + p_x + p(A) + p_l) + q_l \\ &\ge \frac{1}{2}(p_y + p_x + p(A) + p_l + q_l) + \frac{1}{2}(r_y + r_x + q_l) \\ &\ge \frac{1}{2}(p_y + p_x + p(A) + p_l + q_l) + \alpha^2 S_y(\sigma) \end{aligned} \tag{14}$$

Thus, from the inequality (8), (14) and $S_y(\sigma) \le S_x(\sigma)$, we derive that

$$\begin{aligned} L_{max}(\sigma) - L_{max}(\pi) &\le \frac{1}{2}(\alpha^3 S_y(\sigma) + S_x(\sigma) + p_l + q_l) \\ &\le \alpha(S_x(\sigma) + p_l + q_l) \\ &\le \alpha L_{max}(\pi) \end{aligned}$$

Case 4.2: $\min\{r_y, r_x\} < \alpha^2 S_y(\sigma)$ and $q_l < 2\alpha^2 S_y(\sigma)$. Then by Lemma 3, we have that $p(B) \ge (1+\alpha)S_y(\sigma)$, where $B = \{J_j | S_j(\sigma) < S_y(\sigma)\}$. Thus,

$$\begin{aligned} L_{max}(\pi) &\ge \frac{1}{2}(p(B) + p_y + p_x + p(A) + p_l) \\ &\ge \frac{1+\alpha}{2}S_y(\sigma) + \frac{1}{2}(p_y + p_x + p(A) + p_l) \end{aligned} \tag{15}$$

Noting that

$$L_{max}(\sigma) \leq \frac{1}{2}(S_x(\sigma) + S_y(\sigma) + p_y + p_x + p(A)) + p_l + q_l \qquad (16)$$

Thus, from (15) and (16), we obtain that

$$
\begin{aligned}
L_{max}(\sigma) - L_{max}(\pi) &\leq \frac{1}{2}(S_x(\sigma) + p_l) + q_l - \frac{\alpha}{2}S_y(\sigma) \\
&\leq \alpha(S_x(\sigma) + p_l + q_l) \\
&\leq \alpha L_{max}(\pi)
\end{aligned}
$$

This completes the proof.   □

Theorem 1. *Algorithm MLDT has a competitive ratio no greater than $\frac{\sqrt{5}+1}{2}$ and the bound is tight.*

Proof. From Lemma 2 and 4, we know that Algorithm MLDT has a competitive ratio not greater than $1 + \alpha = \frac{\sqrt{5}+1}{2}$. Now we construct an instance to show that the bound is tight as follows: There are three jobs J_1, J_2, J_3 which are released at time 0. The processing times and delivery times of these jobs are given by $p_1 = 1, p_2 = \alpha, p_3 = 1 + \alpha, q_1 = 2\varepsilon, q_2 = \varepsilon, q_3 = 0$.

In the schedule σ generated by the online algorithm MLDT, J_1, J_2 start at times 0, α^2 respectively and J_3 cannot start until the completion of J_1, J_2. However, there exists an optimal schedule π in which J_1, J_2 are assigned on the same machine and start at times 0 and 1 respectively. And J_3 starts at 0 on the other machine.

Thus $L_{max}(\sigma) = 2 + \alpha$ and $L_{max}(\pi) = 1 + \alpha + \varepsilon$. As ε tends to zero, the ratio $L_{max}(\sigma)/L_{max}(\pi)$ tends to $1 + \alpha$.   □

Acknowledgements. The authors would thank the anonymous referees for many constructive comments and hints which have improved this paper.

References

1. Carlier, J.: Scheduling jobs with release dates and tails on identical machines to minimize the makespan. European Journal of Operational Research 29, 298–306 (1987)
2. Chen, B., Vestjens, A.P.A.: Scheduling on identical machines: How good is LPT in an online setting? Operations Research Letters 21(4), 165–169 (1997)
3. Garey, M.R., Johnson, D.S.: Strong NP-completeness results: Motivation, examples and implications. Journal of the Association of Computer Machinery 25, 499–508 (1978)
4. Gharbi, A., Haouari, M.: Minimizing makespan on parallel machines subject to release dates and delivery times. Journal of Scheduling 5, 329–355 (2002)
5. Gharbi, A., Haouari, M.: An approximate decomposition algorithm for scheduling on parallel machines with heads and tails. Computers & Operations Research 34, 868–883 (2007)

6. Grabowski, J., Nowicki, E., Zdrzalka, S.: A block approach for single-machine scheduling with release dates and due dates. European Journal of Operational Research 26, 278–285 (1986)
7. Hall, L., Shmoys, D.: Approximation algorithms for constrained scheduling problems. In: Proceedings of the 30th IEEE Symposium on Foundations of Computer Science, pp. 134–139. IEEE Computer Society Press, New York (1989)
8. Hall, L., Shmoys, D.: Jacksons rule for single-machine scheduling: Making a good heuristic better. Mathematics of Operations Research 17, 22–35 (1992)
9. Haouari, M., Gharbi, A.: Lower Bounds for Scheduling on Identical Parallel Machines with Heads and Tails. Annals of Operations Research 129, 187–204 (2004)
10. Hoogeveen, J.A., Vestjean, A.P.A.: A best possible deterministic online algorithm for minimizing maximum delivery times on a single machine. SIAM Journal on Discrete Mathematics 13, 56–63 (2000)
11. Liu, M., Chu, C., Xu, Y., Zheng, F.: An optimal online algorithm for single machine scheduling with bounded delivery times. European Journal of Operational Research 201(3), 693–700 (2010)
12. Mastrolilli, M.: Efficient approximation schemes for scheduling problems with release dates and delivery times. Journal of Scheduling 6, 521–531 (2003)
13. Noga, J., Seiden, S.S.: An optimal online algorithm for scheduling two machines with release times. Theoretical Computer Science 268(1), 133–143 (2001)
14. Potts, C.N.: Analysis of a heuristic for one machine sequencing with release dates and delivery times. Operations Research 28, 1436–1441 (1980)
15. Stougie, L., Vestjens, A.P.A.: Randomized on-line scheduling: How low can't you go? Operations Research Letters 30(2), 89–96 (2002)
16. Tian, J., Fu, R., Yuan, J.: A best on-line algorithm for single machine scheduling with small delivery times. Theoretical Computer Science 393, 287–293 (2008)
17. Vestjens, A.P.A.: Online Machine Scheduling. Ph.D. Thesis, Department of Mathematics and Computing Science, Eindhoven University of Technology, Eindhoven, The Netherlands (1997)
18. Woeginger, G.J.: Heuristics for parallel machine scheduling with delivery times. Acta Informatica 31(6), 503–512 (1994)

Parallel Machine Scheduling with a Single Server: Loading and Unloading*

Jueliang Hu, Qinghui Zhang, Jianming Dong, and Yiwei Jiang**

Department of Mathematics, Zhejiang Sci-Tech University,
Hangzhou 310018, China
ywjiang@zstu.edu.cn

Abstract. This paper addresses the non-preemptive scheduling on two parallel identical machines sharing a single server in charge of loading and unloading jobs. Each job has to be loaded by the server before being processed on one of the machines and unloaded immediately by the server after its processing. The goal is to minimize the makespan. This paper considers classical algorithms *LS* and *LPT* for the problem where the loading and unloading times both are equal to one time unit. We first analyze the structure of list scheduling for the problem and then show that the tight worst case ratios of *LS* and *LPT* are 12/7 and 4/3, respectively. Finally, we provide a proof of computational complexity for the problem.

Keywords: scheduling, server, algorithm, worst case ratio, makespan.

1 Introduction

In this paper, we consider parallel machine scheduling problem with a single server, which is used to load/unload the job on the machine before/after its processing. The problem can be described as follows. We are given a sequence $\mathcal{J} = \{J_1, J_2, \ldots, J_n\}$ of independent jobs, which must be scheduled on one of two identical machines M_1 and M_2. Job J_j is associated with a loading time s_j, a processing time p_j and a unloading time t_j. All of which are positive integers. One server exists for loading and unloading jobs. Namely, prior to processing, each job J_j is loaded by the server onto one machine and the loading time is s_j. Once the loading is finished, the machine starts to process the job immediately and the server is free to load/unload other job again. Similarly, after finishing processing, the job will be unloaded immediately by the server from the machine and the unloading time is t_j. The server cannot load/unload job onto/from a machine when then machine is processing a job. Each machine processes one job at a time and job preemption is not allowed. The objective is to minimize the makespan, i.e., the completion time of the latest job. Using the three parameter notation [8,11], our problem can be denoted by $P2, S1|s_j, t_j|C_{\max}$.

---

* Supported by the National Natural Science Foundation of China (11001242, 11071220) and Zhejiang Province Natural Science Foundation of China (LY13A010015).
** Corresponding author.

P. Widmayer, Y. Xu, and B. Zhu (Eds.): COCOA 2013, LNCS 8287, pp. 106–116, 2013.
© Springer International Publishing Switzerland 2013

There have been numerous literatures of parallel machine scheduling with servers. For the problem with unit setup times on two machines $P2, S1|s_j = 1|C_{\max}$, Kravchenko and Werner [11] showed that it is binary NP-hard and proposed a pseudo-polynomial algorithm. Hall et al. [8] solved the problem $P2, S1|p_j = 1|C_{\max}$ in polynomial time and showed that the equal setup times problem $P2, S1|s_j = s|C_{\max}$ is strongly NP-hard. Brucker et al.[4] showed that the equal processing times problem $P2, S1|p_j = p|C_{\max}$ is NP-hard. Abdekhodaee and Wirth [1] and Abdekhodaee et al. [2,3] studied the computational complexity of some special and general cases and provided some effective heuristics for them. Besides the objective of minimizing makespan, Hall et al.[8] considered the objectives of minimizing the maximum lateness of any job and the total completion time of all jobs. They provided the computational complexity of the problems and polynomial or pseudo-polynomial algorithms. Wang and Cheng [15] proposed an approximation algorithm for minimizing the total weighted completion time.

For the online version of the scheduling on two machines with a single server, Zhang and Andrew [20] considered LS algorithm for three special cases where jobs arrive over list. Su [14] studied the online LPT algorithm for the online version where jobs arrive over time. Jiang et.al [10] considered the preemptive version of the scheduling on two identical machines with a single server and presented an algorithm with a tight worst case ratio of $4/3$, which can produce optimal schedules for two special cases: equal processing times and equal setup times. In addition, Ou et al. [12], Werner and Kravchenko [16] studied the problem with multiple servers. In Cheng and Kovalyov[6], Brucker et al. [5], Iravani and Teo [9], Su and Lee [13], the authors studied the flowshop problem with a single server.

All the above variants of the scheduling with servers only considered the loading operation (setup) and ignored the unloading operation. However, Xie et.al [17] considered the problem with both loading and unloading operations, where the unloading operation can be delayed after the job processing is finished. They showed that the tight worst case ratio of LPT is $\frac{3}{2}$ $\frac{1}{2m}$ for the special case where $s_j = t_j = 1$ and p_j is an arbitrary positive real number. Unfortunately, the result is uncorrect. In fact, the worst case ratio of LPT is at least $3/2$ for $m = 2$, while their result showed that the tight ratio is $5/4$ when $m = 2$. For the case where the jobs is partitioned previously, Xie et.al [18] proposed an algorithm with tight bound of 2. Yip et al. [19] considered both loading and unloading operations in a flow-shop problem with multiple servers.

In this paper, we consider the problem with both loading and unloading, where no delay is allowed between loading and processing, as well as between processing and unloading. Hence, the model in [17], where the unloading is allowed to be delayed, can be regarded as one relaxation of our problem.

In this section, we provided a proof of computational complexity of our problem. Note that problem $P2, S1|s_j = 1|C_{\max}$ is shown to binary NP-complete [8], it is easy to show that our problem $P2, S1|s_j = t_j = 1|C_{\max}$ is also binary NP-complete by the similar argument. For our problem, we first discuss the structure and properties of list scheduling. Then we show that the tight worst case ratios of algorithms LS and LPT are 12/7 and 4/3, respectively.

The rest of the paper is organized as follows. Section 2 introduces the preliminaries. In section 3, we analyze the structure of list scheduling and show the worst case ratio of algorithm LS. In Section 4, we consider the performance of algorithm LPT. Finally, Section 5 presents some concluding remarks.

2   Preliminaries

In this section, we mainly provide the lower bound of the optimal makespan in the offline version and introduce several definitions.

Before giving the lower bound of the optimal makespan, we first give a couple of notations as follows. Let e_j denote the execution time of J_j, which represents the period of time from the beginning of loading to the end of unloading. That is, $e_j = s_j + p_j + t_j$. Clearly, $e_j \geq 3$ since we assume that s_j, p_j and t_j are positive integers. Let $E = \sum_{i=1}^{n} e_i$ be the total execution time of all jobs and $e_{\max}$ be the largest execution time among all jobs. The completion time of a job is defined as the time by which it has been unloaded. Denote by C^A and C^* the makespan produced by algorithm A and the optimal makespan, respectively.

Now we present the following conclusion about the optimal makespan.

Lemma 1. *For the problem $P2, S1|s_j = t_j = 1|C_{\max}$, the makespan of the optimal scheduling satisfies*

$$C^* \geq \max\{e_{\max}, \frac{E+2}{2}\}.$$

Proof. It is clear that $C^* \geq e_{\max}$. On the other hand, when the server loads the first job onto one of the machines, the other machine must be idle. Similarly, a machine must be idle when the server unloads the last completion job. Thus, during the time interval from zero to the makespan in the optimal scheduling, there are at least two idle time units . Together with the total execution time E, we can conclude that the makespan of the optimal scheduling is at least $\frac{E+2}{2}$ and thus the result holds.

We next introduce some definitions for simplifying expression in the subsequent sections.

Definition 1. *A time unit is called single loading time unit if one machine is busy loading a job and the other is idle in this time unit. Denote by SST the start time of the single loading time units.*

Definition 2. *Two adjacent time units are called double unloading time units if both of the machines occupy one time unit for unloading a job, respectively . Denote by EDT the end time of the double unloading time unit.*

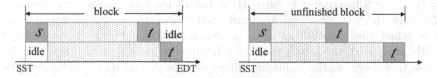

Fig. 1. Definitions of SST, EDT, block and unfinished block

Definition 3. *A block in a schedule is defined as a period of time from an SST to the earliest EDT after the SST. If there is no EDT from the SST until all jobs have been scheduled, this period of time is called unfinished block.*

Remark. (a) As shown in Figure 1, the difference between the completion times of two machines is one time unit in a block and at least two time units in a unfinished block. (b) There is at least two jobs in each block.

3   List Scheduling

Graham [7] proposed List Scheduling heuristic (LS) which assigns the current job to the first available machine at the earliest possible time. We apply LS to our problem and it can be described in detail as follows.

Algorithm 1. List Scheduling (LS)

[1.] For each job $J_i, 1 \leq i \leq n$, let l_1 and l_2 be the current completion times of machine M_1 and M_2, respectively.

[2.] (Job assignment) If $l_1 \leq l_2$, assign it to machine M_1, otherwise, assign it to M_2.

[3.] (Job execution) The loading for the job starts at the earliest time such that the server is free to perform both loading and unloading for the job.

3.1   Structure of List Scheduling

We provide some properties of a schedule produced by algorithm LS.

Proposition 1. *In any block $\mathcal{B}$, there must be two or alternatively three idle time units.*

Proof. It is clear that there is one idle time unit at each end of the block as shown in Figure 1. We only need to show that there is at most one idle time unit in the middle of the block. In fact, before the formation of the block (that is, it is currently a unfinished block.) no new idle time is introduced except the idle time at the beginning of the block. For any job $J_k \in \mathcal{B}$ except the first job in the block $\mathcal{B}$, let l_i, $i = 1, 2$, be the completion time of machine M_i before scheduling J_k. Without loss of generality, we assume $l_1 \leq l_2$. Then we must have

$\Delta = l_2 - l_1 \geq 2$ because it is currently a unfinished block. Moreover, according to LS rule, the last job J_i on M_2 must start before time l_1 as shown in Figure 2. The scheduling possibilities of J_k is demonstrated in Figure 2. According to LS rule, if $e_k \neq \Delta$, the job J_k must be scheduled at time l_1, which is the earliest feasible time because the unloading times of jobs J_i and J_k do not overlap(see Figure 2 (a) and (b)). In this case, no new idle time is introduced. On the other hand, if $e_k = \Delta$, J_k can not be scheduled at time l_1 due to the overlapping of the unloading times of J_i and J_k. Therefore, algorithm LS must schedule it at time $l_1 + 1$ and an idle time unit is consequently introduced. At the same time, however, a block is formed after scheduling this job (see Figure 2 (c)). Hence, there is at most one idle time unit in the middle of the block and the proof is complete.

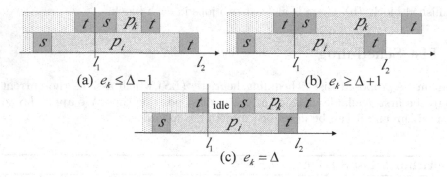

(a) $e_k \leq \Delta - 1$

(b) $e_k \geq \Delta + 1$

(c) $e_k = \Delta$

Fig. 2. Scheduling possibilities of job J_k

Proposition 2. *In an unfinished block, there is only one idle time unit before the smaller completion time of the two machines.*

Proof. By the proof of Proposition 1, we can obtain that there is no more idle time before formation of a block, except one idle time unit introduced at the beginning of the unfinished block.

Proposition 3. *The schedule σ produced by LS is a unfinished block or consists of a number of successive blocks and followed by at most one unfinished block.*

Proof. By LS rule, we can conclude that from time zero, i.e., the first SST, a block must be formed unless all jobs have been scheduled, which implies that the whole schedule σ is a unfinished block. On the other hand, once a block is formed, algorithm LS must schedule the next job at the end of the block (EDT), which can also be regarded as an SST for the coming block or unfinished block. Hence, we can obtain the structure of schedule σ as shown in Figure 3.

For convenience, denote by $\sigma = \mathcal{B}'$ and $\sigma = (\mathcal{B}_1, \mathcal{B}_2, \cdots, \mathcal{B}_k, \mathcal{B}')$ two structures of schedule σ mentioned in Proposition 3.

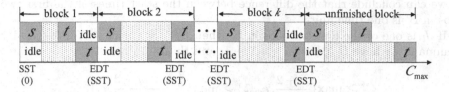

Fig. 3. Structure of list scheduling

3.2   Worst Case Ratio

In this subsection, we show that the tight worst case of algorithm LS is $\frac{12}{7}$. Before this, we first give a lemma about the blocks produced in the schedule σ. Let $l(\mathcal{B})$ and $e(\mathcal{B})$ be the time length of a block $\mathcal{B}$ and the total execution time of all jobs in $\mathcal{B}$, respectively. We can obtain the following result.

Lemma 2. $e(\mathcal{B}) \geq \frac{7}{5}l(\mathcal{B})$.

Proof. Since there are at least two jobs in any block and each job has a execution time no less than 3, we have $l(\mathcal{B}) \geq 4$. Furthermore, it is not hard to obtain that $l(\mathcal{B}) = 4$ if and only if the block $\mathcal{B}$ exactly consists of two same jobs with execution time of 3. In that case, we have $e(\mathcal{B}) = 6$ and thus $e(\mathcal{B}) = \frac{3}{2}l(\mathcal{B}) > \frac{7}{5}l(\mathcal{B})$.

If $l(\mathcal{B}) \geq 5$, then we have $e(\mathcal{B}) \geq 2l(\mathcal{B}) - 3$ because there are at most three idle time units by Proposition 1. It follows that $\frac{e(\mathcal{B})}{l(\mathcal{B})} \geq 2 - \frac{3}{l(\mathcal{B})} \geq \frac{7}{5}$.

Theorem 1. *Algorithm LS has a tight worst case ratio of $\frac{12}{7}$.*

Proof. Let J_l be the last completed job and let T be the start time of J_l, then we have $C^{LS} = T + e_l$. We distinguish two cases according to the structure of the schedule σ by Proposition 3.

Case 1. $\sigma = \mathcal{B}'$ is a single unfinished block. Note that there is only one idle time unit in σ by Proposition 2, then the total execution time of all jobs is at least $2T + e_l - 1$ and thus $C^* \geq \max\{\frac{2T+e_l+1}{2}, e_l\}$ due to Lemma 1. Hence,

$$\frac{C^{LS}}{C^*} \leq \frac{T + e_l}{\max\{\frac{2T+e_l+1}{2}, e_l\}} \leq \begin{cases} \frac{2T+2e_l}{2T+e_l+1}, & \text{for } e_l \leq 2T \\ \frac{T+e_l}{e_l}, & \text{for } e_l > 2T \end{cases} \leq \frac{3}{2} < \frac{12}{7}. \quad (1)$$

Case 2. $\sigma = (\mathcal{B}_1, \mathcal{B}_2, \cdots, \mathcal{B}_k, \mathcal{B}')$. Let $X = \sum_{i=1}^{k} l(\mathcal{B}_i)$ be the total length of all blocks and let $Y = \sum_{i=1}^{k} e(\mathcal{B}_i)$ be the total execution time of all jobs in the k blocks. Then we have $Y \geq \frac{7}{5}X$ by Lemma 2.

If $\mathcal{B}' = \emptyset$, then $C^{LS} = X$ and $E = Y \geq \frac{7}{5}X$. Thus, $C^* \geq \frac{E+2}{2} > \frac{7}{10}X = \frac{7}{10}C^{LS}$, i.e., $\frac{C^{LS}}{C^*} < \frac{10}{7} < \frac{12}{7}$.

We now focus on $\mathcal{B}' \neq \emptyset$. Note that $\mathcal{B}'$ is scheduled after all blocks, the last completed job J_l must belong to $\mathcal{B}'$. From the definition of $\mathcal{B}'$ as shown in Figure

1, we can conclude that the difference between the start times of the first two jobs in $\mathcal{B}'$ is one time unit.

If J_l is one of the first two jobs in $\mathcal{B}'$, then $C^{LS} = T + e_l \leq X + 1 + e_l$. By Lemma 2, we have $E \geq Y + e_l \geq \frac{7}{5}X + e_l$ and thus

$$C^* \geq \max\{\frac{E+2}{2}, e_{\max}\} \geq \max\{\frac{\frac{7}{5}X + e_l + 2}{2}, e_l\}.$$

Hence,

$$\frac{C^{LS}}{C^*} \leq \frac{X + e_l + 1}{\max\{\frac{7}{10}X + e_l/2 + 1, e_l\}} < \frac{12}{7},$$

the last inequality can be obtained similar to (1).

If J_l is not one of the first two jobs in $\mathcal{B}'$, then there is at least one job J_x in $\mathcal{B}'$ which is scheduled before J_l, that is, $\Delta = T - X \geq e_x \geq 3$. Let $e(\Delta)$ be the total execution time of jobs processed on two machines during the period of time between X and T. Note that there is only one idle time unit during the period of time from X to T, then $e(\Delta) = 2\Delta - 1$. Thus $e(\Delta) > \frac{7}{5}\Delta$ due to $\Delta \geq 3$. In this case, we have $C^{LS} = T + e_l = X + \Delta + e_l$ and $E \geq \frac{7}{5}X + e(\Delta) + e_l \geq \frac{7}{5}(X + \Delta) + e_l$. It implies that

$$\frac{C^{LS}}{C^*} \leq \frac{X + \Delta + e_l}{\max\{\frac{7}{10}(X + \Delta) + e_l/2 + 1, e_l\}} < \frac{12}{7}.$$

Tight Example. We now give an instance $\mathcal{I}$ to show that the worst case ratio of algorithm LS is tight. Consider the instance $\mathcal{I}$ including $2n+1$ jobs: $e_{2i-1} = 4$ and $e_{2i} = 3$ for all $1 \leq i \leq n$, and $e_{2n+1} = 7n + 2$. It is easy to obtain that each pair of two jobs J_{2i-1} and J_{2i} for any $1 \leq i \leq n$ forms a block by LS. Then σ consists of n same blocks with total length of $5n$ and followed by the last job J_{2n+1}. Thus, we have $C^{LS} = 12n + 2$. In optimal schedule, the last job is scheduled on one machine at time zero, and the remainder $2n$ jobs with total execution time of $7n$ are scheduled on another machine from time 1 to $7n + 1$, implying $C^* = 7n + 2$. Hence,

$$\frac{C^{LS}}{C^*} = \frac{12n + 2}{7n + 2} \rightarrow \frac{12}{7}$$

as n tends to infinity.

4　Algorithm LPT

In this section, we apply LPT algorithm to our problem and show that its tight worst case ratio is $4/3$. The LPT algorithm can be described as follows.

As a matter of fact, the schedule σ produced by LPT can also be regarded as a list scheduling but the jobs are listed in a non-increasing order of processing time. Therefore, Propositions 1, 2 and 3 still apply in this section. Let J_l be the last completed job and its start time is T. We provide the worst case ratio of LPT in two cases according to Proposition 3.

Algorithm 2. Largest Processing Time (LPT)

[1.] Sort all the jobs such that $e_1 \geq e_2 \geq \cdots \geq e_n$.
[2.] For any job J_i, $1 \leq i \leq n$, schedule it by algorithm LS.

Lemma 3. If $\sigma = \mathcal{B}'$, we have $\frac{C^{LPT}}{C^*} \leq \frac{4}{3}$.

Proof. If J_l is one of the first two jobs in the schedule, then we have $C^{LPT} \leq e_l + 1$. Obviously, $C^* \geq e_l$ and thus $\frac{C^{LPT}}{C^*} \leq \frac{e_l+1}{e_l} \leq \frac{4}{3}$ due to $e_l \geq 3$. On the other hand, if J_l is not one of the first two jobs, then at least one job J_x is scheduled on the same machine before J_l. It implies that $T \geq e_x \geq e_l$. Note that there is only one idle time unit before time T in the unfinished block, we obtain $E \geq 2T - 1 + e_l$ and thus $C^* \geq \frac{2T + e_l + 1}{2}$ due to Lemma 1. Hence,

$$\frac{C^{LPT}}{C^*} \leq \frac{T + e_l}{(2T + e_l + 1)/2} \leq \frac{4}{3},$$

the last inequality holds because of $T \geq e_l$.

Next, similar to Case 2 in the proof of Theorem 1, we assume that the schedule σ consists of k ($k \geq 1$) blocks $\mathcal{B}_1, \mathcal{B}_2, \cdots, \mathcal{B}_k$ and a unfinished block $\mathcal{B}'$. Denote $X = \sum_{i=1}^{k} l(\mathcal{B}_i)$ and $Y = \sum_{i=1}^{k} e(\mathcal{B}_i)$. Let q be the execution time of the smallest job in all k blocks, then we can draw a conclusion as follows.

Lemma 4. $Y \geq \frac{2q+3}{q+3} X - \frac{3}{q+3}$.

Proof. Since q is the execution time of the smallest job and there are at least two jobs in each block $\mathcal{B}_i$, $1 \leq i \leq k$, we have $l(\mathcal{B}_i) \geq q + 1$.

(a) If $l(\mathcal{B}_i) = q + 1$, then the block $\mathcal{B}_i$ exactly consists of two same jobs with execution time of q, that is $e(\mathcal{B}_i) = 2q$. It implies that

$$e(\mathcal{B}_i) = \frac{2q}{q+1} l(\mathcal{B}_i) \geq \frac{2q+3}{q+3} l(\mathcal{B}_i),$$

where the last inequality holds because $q \geq 3$.

(b) If $l(\mathcal{B}_i) = q + 2$, then the block $\mathcal{B}_i$ must consist of two jobs with execution times of q and $q + 1$, respectively. In this block, there are three idle time units and thus $e(\mathcal{B}_i) = 2q + 1$.

(c) If $l(\mathcal{B}_i) \geq q + 3$, we have $e(\mathcal{B}_i) \geq 2l(\mathcal{B}_i) - 3$ because there are at most three idle time units in the block by Proposition 1. It follows that

$$e(\mathcal{B}_i) \geq (2 - \frac{3}{l(\mathcal{B}_i)}) l(\mathcal{B}_i) \geq (2 - \frac{3}{q+3}) l(\mathcal{B}_i) = \frac{2q+3}{q+3} l(\mathcal{B}_i).$$

By algorithm LPT, we can conclude that there is at most one block whose length is $q+2$. If no block's length is $q+2$, by the above arguments (a) and (c), we have $e(\mathcal{B}_i) \geq \frac{2q+3}{q+3} l(\mathcal{B}_i)$ for all $1 \leq i \leq k$ and thus

$$Y \geq \frac{2q+3}{q+3} X > \frac{2q+3}{q+3} X - \frac{3}{q+3}.$$

If there is one block $\mathcal{B}_j$ such that $l(\mathcal{B}_j) = q + 2$, then we have $e(\mathcal{B}_i) \geq \frac{2q+3}{q+3}l(\mathcal{B}_i)$ for any $i \neq j$. Thus, by the above arguments, we obtain

$$Y = \sum_{i \neq j} e(\mathcal{B}_i) + e(\mathcal{B}_j) \geq \frac{2q+3}{q+3} \sum_{i \neq j} l(\mathcal{B}_i) + 2q + 1$$

$$= \frac{2q+3}{q+3}(X - (q+2)) + 2q + 1 = \frac{2q+3}{q+3}X - \frac{3}{q+3}.$$

Hence, the proof is complete.

Lemma 5. *If* $\sigma = (\mathcal{B}_1, \mathcal{B}_2, \cdots, \mathcal{B}_k, \mathcal{B}')$, *we have* $\frac{C^{LPT}}{C^*} \leq \frac{4}{3}$.

Proof. If $\mathcal{B}' = \emptyset$, we have $C^{LPT} = X$ and $E = Y$. It follows that $C^* \geq \frac{Y+2}{2}$ by Lemma 1. Combining Lemma 4 and $q \geq 3$, we obtain that

$$\frac{C^{LPT}}{C^*} \leq \frac{2X}{Y+2} \leq \frac{2X}{\frac{2q+3}{q+3}X - \frac{3}{q+3} + 2} \leq \frac{2q+6}{2q+3} \leq \frac{4}{3}.$$

We next consider the case where $\mathcal{B}' \neq \emptyset$, which implies $J_l \in \mathcal{B}'$. By the *LPT* rule, the execution time of J_l is not greater than any one of jobs in $\bigcup_{1 \leq i \leq k} \mathcal{B}_i$, that is,

$$e_l \leq q. \tag{2}$$

Similar to Case 2 in the proof of Theorem 1, the lemma can be proved according to the position of the job J_l.

Case 1. J_l is the first job in $\mathcal{B}'$. It follows that $C^{LPT} = X + e_l$ and $E \geq Y + e_l$. We first consider the case where there is only one block $\mathcal{B}_1$ in the schedule, that is, $X = l(\mathcal{B}_1)$ and $Y = e(\mathcal{B}_1)$. If $X = q + 1$, we must have $Y = 2q$, that is, $C^{LPT} = q + 1 + e_l$ and $E \geq Y + e_l = 2q + e_l$. By (2), we have

$$\frac{C^{LPT}}{C^*} \leq \frac{q + e_l + 1}{\frac{2q+e_l+2}{2}} = \frac{2q + 2e_l + 2}{2q + e_l + 2} \leq \frac{4q+2}{3q+2} < \frac{4}{3}.$$

If $X \geq q + 2$, we have $Y \geq 2X - 3$ by Proposition 1 and thus $E \geq Y + e_l \geq 2X - 3 + e_l$. By (2) and $X \geq q + 2$, we obtain that

$$\frac{C^{LPT}}{C^*} \leq \frac{X + e_l}{\frac{2X-3+e_l+2}{2}} = \frac{2X + 2e_l}{2X + e_l - 1} \leq \frac{2X + 2q}{2X + q - 1} \leq \frac{2(q+2) + 2q}{2(q+2) + q - 1} = \frac{4}{3}.$$

Now we focus on the case where there is at least two blocks. In that case, we have $X \geq 2(q+1)$. It is easy to get that $C^* \geq \frac{\frac{2q+3}{q+3}X - \frac{3}{q+3} + e_l + 2}{2}$ according to Lemmas 1 and 4. By (2), we can obtain

$$\frac{C^{LPT}}{C^*} \leq \frac{2X + 2e_l}{\frac{2q+3}{q+3}X - \frac{3}{q+3} + e_l + 2} \leq \frac{2X + 2q}{\frac{2q+3}{q+3}X - \frac{3}{q+3} + q + 2}$$

$$\leq \frac{4(q+1) + 2q}{\frac{2q+3}{q+3}2(q+1) - \frac{3}{q+3} + q + 2} = \frac{6q^2 + 22q + 12}{5q^2 + 15q + 9} \leq \frac{4}{3},$$

where the last inequality holds because of $q \geq 3$.

Case 2. If J_l is the second job in $\mathcal{B}'$, then we have $C^{LPT} = X + e_l + 1$ by the definition of the unfinished block as shown in Figure 1. Note that the execution time of the first job in $\mathcal{B}'$ is not less than that of J_l, we can obtain that $E \geq Y + 2e_l$ and thus $C^* \geq \frac{\frac{2q+3}{q+3}X - \frac{3}{q+3} + 2e_l + 2}{2}$ due to Lemmas 1 and 4. It is easy to obtain that $\frac{2q+3}{q+3} \geq \frac{3}{2}$ due to $q \geq 3$ and $q \geq e_l \geq 3$ by (2). Thus

$$\frac{C^{LPT}}{C^*} \leq \frac{2X + 2e_l + 2}{\frac{2q+3}{q+3}X - \frac{3}{q+3} + 2e_l + 2} < \frac{2X + 2e_l + 2}{\frac{2q+3}{q+3}X + 2e_l}$$

$$\leq \frac{2X + 8}{\frac{2q+3}{q+3}X + 6} \leq \frac{2X + 8}{\frac{3}{2}X + 6} = \frac{4}{3}.$$

Case 3. If J_l is not one of the first two jobs in $\mathcal{B}'$, then at least one job J_x is scheduled before J_l on the same machine, that is, $\Delta = T - X \geq e_x \geq e_l$ and thus $C^{LPT} = T + e_l = X + \Delta + e_l$. Let $e(\Delta)$ be the total execution time of jobs processed on two machines during the period of time between X and T. Since there is only one idle time unit in $\mathcal{B}'$, we have $e(\Delta) = 2\Delta - 1$ and thus

$$E \geq Y + 2\Delta - 1 + e_l \geq \frac{2q+3}{q+3}X - \frac{3}{q+3} + 2\Delta - 1 + e_l.$$

Combining $e_l \leq \Delta$ and $\frac{2q+3}{q+3} \geq \frac{3}{2}$ due to $q \geq 3$, we obtain

$$\frac{C^{LPT}}{C^*} \leq \frac{2X + 2\Delta + 2e_l}{\frac{2q+3}{q+3}X - \frac{3}{q+3} + 2\Delta - 1 + e_l + 2}$$

$$< \frac{2X + 2\Delta + 2e_l}{\frac{2q+3}{q+3}X + 2\Delta + e_l} \leq \frac{2X + 4\Delta}{\frac{3}{2}X + 3\Delta} = \frac{4}{3}.$$

From Lemmas 3 and 5, we obtain the main result in this section.

Theorem 2. *Algorithm LPT has a tight worst case ratio of $\frac{4}{3}$.*

Tight Example. Consider the instance $\mathcal{I}$ with $e_1 = 4$ and $e_2 = e_3 = 3$. It is clear that the first two jobs form a block and $C^{LPT} = 5 + 3 = 8$. In the optimal schedule, the last two jobs are scheduled on one machine at time zero and the first job is scheduled on another machine at time 1. Then we have $C^* = 6$, which implies $\frac{C^{LS}}{C^*} = \frac{4}{3}$.

5 Conclusions

We studied the non-preemptively parallel machine scheduling of minimizing makespan, where each job has to be loaded and unloaded by a single server before and after being processed on one of two machines. We assumed that both loading and unloading take one unit time, and considered the classical algorithms LS and LPT for our problem. We analyzed the structure of list scheduling and showed that the tight worst case ratios of LS and LPT are 12/7 and 4/3, respectively.

References

1. Abdekhodaee, A.H., Wirth, A.: Scheduling parallel machines with a single server: some solvable cases and heuristics. Computers and Operations Research 29, 295–315 (2002)
2. Abdekhodaee, A.H., Wirth, A., Gan, H.S.: Equal processing and equal setup time cases of scheduling parallel machines with a single server. Computers and Operations Research 31, 1867–1889 (2004)
3. Abdekhodaee, A.H., Wirth, A., Gan, H.S.: Scheduling parallel machines with a single server: the general case. Computers and Operations Research 33, 994–1009 (2006)
4. Brucker, P., Dhaenens-Flipo, C., Knust, S., Kravchenko, S.A., Werner, F.: Complexity results for parallel machine problems with a single server. Journal of Scheduling 5, 429–457 (2002)
5. Brucker, P., Knust, S., Wang, G.: Complexity results for flow-shop problems with a single server. European Journal of Operational Research 165, 398–407 (2005)
6. Cheng, T.C.E., Kovalyov, M.: Scheduling a single server in a two-machine flow shop. Computing 70, 167–180 (2003)
7. Graham, R.: Bounds for certain multiprocessing anomalies. Bell System Technical Journal 45, 1563–1581 (1966)
8. Hall, N.G., Potts, C.N., Sriskandarajah, C.: Parallel machine scheduling with a common server. Discrete Applied Mathematics 102, 223–243 (2000)
9. Iravani, S., Teo, C.: Asymptotically optimal schedules for single-server flow shop problems with setup costs and times. Operations Research Letters 33, 421–430 (2005)
10. Jiang, Y., Dong, J., Ji, M.: Preemptive scheduling on two parallel machines with a single server. Computers & Industrial Engineering 66, 514–518 (2013)
11. Kravchenko, S.A., Werner, F.: Parallel machine scheduling problems with a single server. Mathematical and Computer Modelling 26, 1–11 (1997)
12. Ou, J., Qi, X., Lee, C.: Parallel machine scheduling with multiple unloading servers. Journal of Scheduling 13, 213–226 (2010)
13. Su, L., Lee, Y.: The two-machine flowshop no-wait scheduling problem with a single server to minimize the total completion time. Computers & Operations Research 35, 2952–2963 (2008)
14. Su, C.: Online LPT algorithms for parallel machines scheduling with a single server. Journal of Combnatorial Optimization (2012), doi:10.1007/s10878-011-9441-z
15. Wang, G., Cheng, T.C.E.: An approximation algorithm for parallel machine scheduling with acommon server. Journal of the Operational Research Society 52, 234–237 (2001)
16. Werner, F., Kravchenko, S.A.: Scheduling with multiple servers. Automation and Remote Control 71(10), 2109–2121 (2010)
17. Xie, X., Li, Y., Zhou, H., Zheng, Y.: Scheduling parallel machines with a single server. In: Proceeding of International Conference on Measurement, Information and Control, vol. 1, pp. 453–456 (2012)
18. Xie, X., Zheng, Y., Li, Y.: Scheduling parallel machines with a Single Server: A Dedicated Case. In: Proceedings of Fifth International Joint Conference on Computational Sciences and Optimization, pp. 146–149 (2012)
19. Yip, Y., Cheng, C., Low, C.: Sequencing of an M machine flow shop with setup, processing and removal times separated. International Journal of Advanced Manufacturing Technology 30, 286–296 (2006)
20. Zhang, L., Andrew, W.: On-line scheduling of two parallel machines with a single server. Computers and Operations Research 36, 1529–1553 (2009)

Prompt Mechanism for Online Auctions with Multi-unit Demands

Xiangzhong Xiang

Department of Computer Science, The University of Hong Kong, Hong Kong
xzxiang@cs.hku.hk

Abstract. We study the following TV ad placement problem: m identical time-slots are on sell within a period of m days and only one time-slot is available each day. Advertisers arrive online to bid for some time-slots to publish their ads. Typically, advertiser i arrives at the a_i'th day and wishes that her ad would be published for at most s_i days. The ad cannot be published after its expiration time, the d_i'th day. If the ad is published for $x_i \leq s_i$ days, the total value of the ad for advertiser i is $x_i \cdot v_i$; otherwise, the value of the ad to be published for each day diminishes and the total value is always $s_i \cdot v_i$. Our goal is to maximize the social welfare: the sum of values of the published ads. As usual in many online mechanisms, we are aiming to optimize the competitive ratio: the worst ratio between the optimal social welfare and the social welfare achieved by our mechanism.

Our main result is a competitive online mechanism which is truthful and prompt for the TV ad placement problem. In the mechanism, each advertiser is motivated to report her private value v_i truthfully and can learn her payment at the very moment that she wins some time-slots. Before studying the general case where the maximum demands s_i's are non-uniform, we study the special case where all s_i's are uniform and prove that our mechanism achieves a non-trivial competitive ratio of 5. For the general case where the maximum demands s_i's are non-uniform, we prove that our mechanism achieves a competitive ratio of $5 \cdot \lceil s_{max}/s_{min} \rceil$, where s_{max}, s_{min} are the maximum and minimum value of s_i's. Besides, we derive a lower bound of $\min\{\frac{v_{max}+v_{min}}{2v_{min}}, \frac{s_{max}}{s_{min}}\}$ on the competitive ratio for the general case, where v_{max}, v_{min} are the maximum and minimum value of v_i's.

1 Introduction

TV advertising has long been a profitable industry and advertising revenue provides a significant portion of the funding for most privately owned television networks. Advertisers are eager to promote a wide variety of goods, services and ideas by making use of advertisements. From the viewpoint of advertising, one television station, or publisher of advertisements, owns an inventory of time-slots, which are typically between 30 seconds to 120 seconds long and available daily. For an advertiser arriving online, she wishes that her advertisement could be published in a proper time-slot and repeated for a few days before the ad

P. Widmayer, Y. Xu, and B. Zhu (Eds.): COCOA 2013, LNCS 8287, pp. 117–128, 2013.
© Springer International Publishing Switzerland 2013

is expired. As the value of one ad is diminishing when the ad is broadcast for too many times and the advertiser has budget constraint, the repeating times of the ad should have upper limit. In the paper, we design an auction mechanism for publishers to allocate time-slots that maximizes the social welfare while satisfying advertisers' preferences.

1.1   The Problem

We study the following *TV ad placement problem*: m identical time-slots are on sale within a period of m days and only one time-slot is available each day. Advertisers arrive online to bid for some time-slots to publish their ads. Typically, advertiser i arrives at the a_i'th day and wishes her ad could be published for at most s_i consecutive days.[1] The ad cannot be published after its expiration time, the d_i'th day. If the ad is published for $x_i \leq s_i$ days, the total value of the ad for advertiser i is $x_i \cdot v_i$; otherwise, the value of the ad to be published for each day diminishes and the total value is always $s_i \cdot v_i$.[2] The goal is to maximize the social welfare: the sum of values of the published ads.

In this paper, we focus on designing truthful mechanisms. The information of all advertisers arriving in future is unknown in the online auction. However, we assume that when one advertiser i arrives, its arrival time a_i, expiration time d_i and maximum demand s_i are public information. The only private information is the value v_i and selfish advertisers may report false values to the publisher in order to maximize their profits. A truthful mechanism would motivate selfish advertisers to reveal their true values. This goal is usually achieved by means of making the payments collected from advertisers depending on the mechanism's outcome instead of advertisers' reported values.

Besides truthful, we also require our mechanisms to be *prompt*, which means that any advertiser that wins some time-slots could always learn her payment immediately after winning these time-slots. Prompt mechanisms are firstly proposed in [10]. For mechanisms that are not prompt, three main disadvantages are discussed in [10]: (1) A winning advertiser does not know how much money she has spent and thus does not know how much money she has left. She cannot use her remaining money to take part in another auction. (2) A winning advertiser may pay long after she won her time-slots. If she is not honest, she can deny paying the money while her ad has already been published. (3) A winning advertiser essentially provides the publisher with a "blank check" in exchange for time-slots. It is hard for advertisers to verify the exact calculation of their payments. In prompt mechanisms, all these disadvantages are avoided, as any advertiser can learn her payment when her ad begins to be displayed.

---

[1] In the whole paper, we consider the scenario that each advertiser is only interested in publishing her ad on some consecutive days. Note that even if advertisers can publish ads on days which are not consecutive, our prompt mechanism still works and has the same competitive ratio; but our lower bound on the competitive ratio does not hold in such model.

[2] To utilize time-slots efficiently, any rational publisher will not allocate advertiser i more than s_i time-slots.

1.2 Our Results

Our main result is an online mechanism which is truthful and prompt for the TV ad placement problem. In the mechanism, we partition all time-slots into groups evenly and all time-slots in one group can only be allocated to one advertiser. One advertiser can win at most one group even though she may demand more. We prove that each advertiser is motivated to reveal her true value in order to win one group of time-slots. Once an advertiser wins a group of time-slots, the price she pays for each time-slot in the group can be determined to be the least value she can report to win one group.

Before studying the general case where the maximum demands s_i's are non-uniform, we study the special case where all s_i's are uniform and prove that our mechanism achieves a non-trivial competitive ratio of 5. The crucial technique we use is to construct a novel mapping from the optimal solution to the solution produced by our online mechanism. For the general case where the maximum demands s_i's are non-uniform, we prove that our mechanism achieves a competitive ratio of $5 \cdot \lceil s_{max}/s_{min} \rceil$, where s_{max}, s_{min} are the maximum and minimum value of s_i's. If s_{max} is comparable with s_{min}, our mechanism is still very competitive. Besides, we derive a lower bound of $\min\{\frac{v_{max}+v_{min}}{2v_{min}}, \frac{s_{max}}{s_{min}}\}$ on the competitive ratio for the general case, where v_{max}, v_{min} are the maximum and minimum value of v_i's.

Remarks: Besides applied in TV ad placement problem, our online auction mechanism can solve other problems. Instead of time-slots in TV stations, the ad space may be a physical newspaper sheet with ads being published on it daily or a billboard that displays a set of ads on a fixed space with changes every specific time period. Our mechanism can also be used to solve the on-demand data broadcast problem [7, 13], in which clients make requests for data and all requests have deadlines. The server broadcasts the requested data at some time.

1.3 Related Work

The advertisement placement problem has been widely studied in recent years. In [12], the auction system used by Google for allocation and pricing of TV ads is introduced. The system uses a simultaneous ascending auction to generate a schedule of ads for TV companies daily. Online keyword advertising among multiple bidders with limited budgets is studied in [1, 6, 15]. In [1], bidders are offline while the ad places arrive online and an optimal $\frac{e}{e-1}$-competitive randomized algorithm is introduced. Another important branch about advertisement auction is designing truthful mechanisms, which is studied in [2, 3].

One classical technique used in many truthful mechanisms is the VCG payment scheme where each advertiser is charged the harm she causes to other advertisers and bidding the true value is the dominant strategy [14]. However, VCG cannot be applied to online problems because it requires computing the social welfare of the optimal allocation which cannot be computed in an online fashion. Moreover, even when the optimal allocation is known, the payment of a winner cannot be determined by VCG at the time when she wins her time-slots; her payment may depend on future events.

One special case of the TV ad placement problem is studied in [10]. They assume that advertisers arrive and depart over time. In contrast to our problem, each advertiser is interested in winning only one time-slot to publish her ad before she departs. For this special case, they show a 2-competitive prompt and truthful mechanism. The proof of truthfulness and the analysis of the competitive ratio are somewhat straightforward as any advertiser can win only one time-slot in both the optimal solution and the solution produced by their prompt mechanism. When analyzing the competitive ratio, they match at most two advertisers that win one time-slot in the optimal solution to exactly one advertiser that wins one time-slot in their prompt mechanism. In our problem, one advertiser can win more than one group of time-slots in the optimal solution. We need to design a novel matching from the optimal solution to the solution produced by our mechanism without collision. In the analysis, we map at most 5 advertisers in the optimal solution to one advertiser with higher value winning one group in our online mechanism. One difficulty in the analysis is that several advertisers may share one group of time-slots in the optimal solution. We adjust advertisers' arrival time and expiration time slightly and prove a competitive ratio of 5 successfully.

Azar $et\,al.$ [4] studies a problem similar to that in [10]. In their auction problem, each ad can be published for at most once before the ad is expired. But each ad has an arbitrary size no greater than one and several ads can be published in one time-slot on condition that the total size does not exceed one. They design a truthful and prompt mechanism. The mechanism treats ads with size $< \frac{1}{2}$ and size $\geq \frac{1}{2}$ separately. They maintain a tentative schedule of ads for each day, and always prefers ads with higher density (i.e., the ratio of value to size). Their mechanism is proved to be 6-competitive.

In the full information setting, our TV ad placement problem is similar to the online scheduling problem with jobs arriving over time and having deadlines to be finished. Online scheduling with unit-length jobs has been studied in [5,8,9,11]. The best deterministic algorithm achieves a competitive ratio of $2\sqrt{2} - 1$ [11] and no deterministic algorithm can be better than $\frac{\sqrt{5}+1}{2}$-competitive [8]. The on-demand broadcasting problem is studied in [7,13], which can be reduced to online scheduling problem with jobs in different lengths. In [7], Chan $et\,al.$ show an upper bound of $4\Delta + 3$ and a lower bound of $\Delta/\ln\Delta$ on the competitive ratio, where Δ is the ratio between the length of the longest and shortest jobs.

2   Preliminaries

Consider that m identical time-slots are on sale within a period of m days. Only one time-slot is available each day. There are advertisers arriving online and each advertiser has an ad to be published for a period before the ad is expired. One advertiser i can be represented by a tuple (s_i, v_i, a_i, d_i), where $s_i \in \mathbb{N}^+$ is the maximum number of consecutive time-slots the advertiser demands her ad would be published for, $v_i \in \mathbb{R}^+$ refers to the value of the ad if it is published for one day, and $a_i, d_i \in \mathbb{N}^+$ are the arrival and expiration time $(d_i - a_i + 1 \geq s_i)$. The lower bound of all s_i's are s_{min} and known ahead of time. For any advertiser i, we assume that her value v_i is private while the other information is public. Advertiser i wishes that her ad is published in time window $W_i = [a_i, d_i]$. If $x \leq s_i$ consecutive time-slots in W_i are assigned to advertiser i and her payment for these time-slots is p_i, she gains a total value of $x \cdot v_i$ and a net profit of $x \cdot v_i - p_i$.

Our goal is to maximize the social welfare which is the total value of all the published ads. The auction mechanism should be:

(a) **Incentive compatible:** each advertiser i is incentive to reveal her true value v_i in the auction.
(b) **Prompt:** each advertiser can learn her total payment at the very moment that her ad begins to be published.

We say a mechanism is c−competitive if it can always achieve social welfare which is at least $\frac{1}{c}$ times of the optimal social welfare.

The further structure of the paper is as follows: in section 3 we show a lower bound on the competitive ratio of the problem. In section 4, we introduce a mechanism which is truthful and prompt. In section 5, we show the mechanism achieves a competitive ratio of 5 when all advertisers' demands are uniform and a competitive ratio of $5 \cdot \lceil s_{max}/s_{min} \rceil$ when their demands are non-uniform.

3   A Lower Bound on the Competitive Ratio

Before showing our main result, a prompt mechanism for the TV ad placement problem, we derive a lower bound on the competitive ratio for this problem. Assume that for any advertiser i, $s_{min} \leq s_i \leq s_{max}$ and $v_{min} \leq v_i \leq v_{max}$. For the case where all s_i's are one, a lower bound of 2 is shown in [10]. For the general case where s_i's are not uniform, we show that the lower bound of the competitive ratio is $\min\{\frac{v_{max}+v_{min}}{2v_{min}}, \frac{s_{max}}{s_{min}}\}$.

Theorem 1. *The competitive ratio of the TV ad placement problem is at least* $\min\{\frac{v_{max}+v_{min}}{2v_{min}}, \frac{s_{max}}{s_{min}}\}$.

Proof. To prove the lower bound of the competitive ratio, we measure the performance of any prompt mechanism against an adversary that knows all information and adjusts the input sequence according to the decisions made by the prompt mechanism. On the 1'st day, advertiser $u_1 : (s_{max}, v_{min}, 1, s_{max})$ arrives. Wlog, all the x time-slots in time window $[x_0, x_0 + x - 1]$ are allocated to u_1 at the x_0'th day in the prompt mechanism. These x time-slots will not be available to any advertiser arriving after the x_0'th day.

– If $x \leq s_{min}$, the adversary stops the input sequence. The social welfare of the prompt mechanism is: $ALG = x \cdot v_{min} \leq s_{min} \cdot v_{min}$. In the optimal solution, all time-slots in window $[1, s_{max}]$ are allocated to u_1 and the optimal social welfare is: $OPT = s_{max} \cdot v_{min}$. So: $OPT/ALG \geq s_{max}/s_{min}$.

– Else, $x > s_{min}$, the adversary sends another advertiser $u_2 : (x-1, v_{max}, x_0 + 1, x_0 + x - 1)$ and then stops the input sequence. No time-slot in u_2's time window $W_2 = [x_0 + 1, x_0 + x - 1]$ is available for u_2 and $ALG = x \cdot v_{min}$. One feasible solution is to allocate u_2 all the $x - 1$ time-slots in window W_2 and allocate u_1 all the x_0 time-slots in windows $[1, x_0]$. Then we get: $OPT \geq (x - 1) \cdot v_{max} + x_0 \cdot v_{min} \geq (x - 1) \cdot v_{max} + v_{min}$. So

$$\frac{OPT}{ALG} \geq \frac{(x-1) \cdot v_{max} + v_{min}}{x \cdot v_{min}} = \frac{v_{max} - \frac{v_{max} - v_{min}}{x}}{v_{min}} \geq \frac{v_{max} + v_{min}}{2 \cdot v_{min}},$$

the last inequality is true as $x \geq 2$.

No matter what the value of x is, we get that $OPT/ALG \geq \min\{\frac{v_{max} + v_{min}}{2 v_{min}}, \frac{s_{max}}{s_{min}}\}$.

4   A Prompt and Truthful Mechanism

In this section, we introduce a prompt and truthful mechanism for the TV ad placement problem. In section 5, we will continue to analyze the competitive ratio of the mechanism.

In the auction, advertisers arrive online and it is known that s_{min} is the lower bound of all s_i's. When advertiser i arrives, we are not clear about the advertisers arriving later. As shown in the analysis of the section 3, we cannot allocate i either too many or too few time-slots to achieve a low competitive ratio. In our prompt mechanism, no matter what s_i is, we allocate each advertiser 0 or $s = \lceil s_{min}/2 \rceil$ time-slots. We partition all the m time-slots into $M = \lceil m/s \rceil$ groups and call all the s time-slots in time window $[(j - 1) \cdot s + 1, j \cdot s]$ as *group* G_j $(1 \leq j \leq M)$[3]. In our mechanism, each group can be allocated to only one advertiser and each advertiser can win only one group which is totally included in her time window[4].

Our mechanism is implemented by the HALF-algorithm, as shown in Algorithm 1. In the HALF-algorithm, we maintain one candidate advertiser c_j for each group G_j. Whenever a new advertiser i arrives, look at the candidates for groups totally included in W_i and let c_j be the candidate with the lowest value (we say i *competes* on group G_j). If $v_{c_j} < v_i$, i will replace c_j as the candidate of G_j; otherwise, i is rejected irrevocably. On day $(k - 1) \cdot s + 1$, the group G_k is allocated to its current candidate c_k and the payment for the group is calculated. The price that any winner pays for each time-slot in her winning group equals her critical value: the minimum value she can declare and still win one group.

In the HALF-algorithm, any advertiser i can only be allocated s time-slots or 0 slots, although she bids for as many as s_i time-slots. Before proving the truthfulness and promptness of HALF-algorithm, we will prove an important property shown in Lemma 2.

---

[3] When M is not a multiple of s, introduce some dummy slots which will not be used by any advertiser.
[4] Group G_j is totally included in time window $W_i = [a_i, d_i]$ if and only if $a_i \leq (j - 1) \cdot s + 1$ and $d_i \geq j \cdot s$.

Data: Advertisers arriving online
Result: Allocation of time-slots
Set $s := \lceil s_{min}/2 \rceil$ and $t := 1$;   /* t means it is the t'th day now */
Initialize all candidates for groups as dummy advertisers with value of 0;
while $t \leq m$ **do**
   while *there is a new advertiser* $u_i : (s_i, v_i, a_i, d_i)$ *arriving on day* t **do**
      Let S be the set of candidate advertisers for groups which are totally
      included in windows W_i;
      Let c_j be the candidate with the lowest value in S (if there are more
      than one such candidate, choose one arbitrary) ; /* We say i competes
      on group G_j. */
      if $v_{c_j} < v_i$ **then**
       | Make i be the new candidate for group G_j.
      end
   end
   if $t == (k-1) \cdot s + 1$ **then**
      Allocate group G_k to its current candidate advertiser c_k;
      Let p be the minimum value that advertiser c_k can declare and still win
      one group;
      The payment of advertiser c_k is $s \cdot p$;
   end
   $t := t + 1$;
end

Algorithm 1. HALF-algorithm

Lemma 2. *Assume that one advertiser wins a group in the HALF-algorithm. If she has reported a higher bid and others' bids are unchanged, she can still win one group.*

Proof. Suppose that advertiser $i : (s_i, v_i, a_i, d_i)$ wins one group G_j in the HALF-algorithm. We will prove that if she reports $v'_i > v_i$ and others' bids are unchanged, she can still win one group to publish her advertisement. First, note that advertiser i will compete on the same group G_j, no matter what value she reports. Second, compare two runs of the HALF-algorithm in two cases: i reports v_i in case 1 and v'_i in case 2, and we can show that at any time the candidate for any group is the same in these two cases (this implies that i can win the same group in both cases). Before i arrives, these two cases are identical. Look at the next advertiser r arriving after i. For a contradiction, assume that the candidate for some group changes after r arrives in case 2. This can only happen when r competes on group G_j in case 1 and competes on another group G_h in case 2. Assume c_h is the candidate of G_h before r arrives. In case 1, both i and r compete on G_j and i wins. Thus, $v_i \geq v_r$. r competes on G_j instead of G_h so $v_{c_h} \geq v_i$. It follows that $v_{c_h} \geq v_i \geq v_r$. In case 2, r competes on G_h instead of G_i. But as $v_r \leq v_{c_h}$, r cannot become the candidate of G_h. The candidates of G_j and G_h are unchanged and so do the candidates of all the other groups. A contradiction occurs. Thus, the candidates of all groups are unchanged after advertiser r arrives. To finish the proof of monotonicity, we observe all the advertisers arriving after i one by one and use the same analysis.

Theorem 3. *The HALF-algorithm is truthful and prompt.*

Proof. We prove the truthfulness first. The true value of advertiser i is v_i. Let u_i, u_i' be the net profits that advertiser i gains when bidding v_i, v_i' respectively. We argue that $u_i \geq u_i'$ in each of the following cases, as a result bidding truthfully is a dominant strategy.

1. i wins one group when bidding either v_i or v_i'. In these two cases, i competes on one identical group G_j and then wins that group. The price p that i pays for each time-slot in G_j equals her critical value. So p is independent of i's bidding values and her total payment is $p_i = s \cdot p$. Thus, $u_i = s \cdot v_i - s \cdot p = u_i'$.
2. i wins one group when bidding v_i and no group when bidding v_i'. From lemma 2, we know that $v_i \geq v_i'$. When bidding v_i, i wins group G_j and the price paid for each time-slot is p. As p is the minimum value that i can bid to win group G_j, we get $p \leq v_i$. Hence, $u_i = s \cdot v_i - s \cdot p \geq 0 = u_i'$.
3. i wins one group when bidding v_i' and no group when bidding v_i. When bidding v_i', i wins group G_j and the price paid for each time-slot is p. As p is the minimum value that i can bid to win group G_j and i wins no group when bidding v_i, we have $p \geq v_i$. Thus, $u_i' = s \cdot v_i - s \cdot p \leq 0 = u_i$.
4. i wins no group when bidding either v_i or v_i'. Thus, $u_i = u_i' = 0$.

Now we prove the promptness. Recall that regardless what value advertiser i reports, she will compete on one identical group G_j. Moreover, the winner of group G_j cannot be advertisers arriving after time-slot $(j - 1) \cdot s + 1$. As the algorithm is monotone, the payment of i for group G_j is well defined and can be calculated at the very moment when G_j is allocated to i, which is time-slot $(j - 1) \cdot s + 1$. Thus the algorithm is prompt.

5   Competitive Ratios

We have shown a lower bound of $\min\{\frac{v_{max}+v_{min}}{2v_{min}}, \frac{s_{max}}{s_{min}}\}$ on the competitive ratio for the ad placement problem in section 3. We analyze the competitive ratio of the HALF-algorithm in this section. For the case where all demands s_i's are uniform, the HALF-algorithm is proved to be 5-competitive. For the general case where s_i's are non-uniform, the algorithm is proved to be $5 \cdot \lceil s_{max}/s_{min} \rceil$-competitive. Note that when s_{max} is comparable with s_{min}, the algorithm is still very competitive.

5.1   Competitive Ratio When Demands Are Uniform

Assume that all demands s_i's have the same value of s_{min}. Then in HALF-algorithm, $s = \lceil s_{min}/2 \rceil$ and each advertiser i can only win one group (s time-slots) which is totally included in her time window W_i. However, in the optimal solution with the optimal social welfare OPT, i can win any time-slot in W_i. To compare the social welfare of HALF-algorithm, ALG, with OPT, we need to define an intermediate variable:

Definition 4. *OPT' is the optimal social welfare when the maximum number of time-slots that any advertiser i demands is 2s and her time window is* $W_i' = [a_i', d_i'] = [\lfloor \frac{a_i - 1}{s} \rfloor \cdot s + 1, \lceil \frac{d_i}{s} \rceil \cdot s]$.

We call W_i' as i's *extended time window*. As in the optimal solution, any advertiser i can win at most $s_{min} \leq 2s$ time-slots in windows $W_i \subseteq W_i'$, we can get that $OPT \leq OPT'$. We will compare ALG with OPT'. Consider one solution which achieves the social welfare of OPT' now. In the solution, each advertiser bids for at most $2s$ time-slots in time window W'. Note that there are s time-slots in one group and any a_i' is the beginning of one group while any d_i' the end of one group. Without loss of generality, we can find one solution O' in which the social welfare is OPT', each group is allocated to one advertisers and each advertiser i wins 0, 1 or 2 groups in her extended window W_i'. In the following theorem, we will study this solution O' in detail.

Theorem 5. *The HALF-algorithm is 5-competitive when maximum demands are uniform.*

Proof. Let $A = (p_1, \ldots, p_M)$ be the solution of HALF-algorithm where advertiser p_j wins group G_j. Let $O' = (o_1, \ldots, o_M)$ be the solution which achieves the social welfare of OPT'. In O', advertiser o_j wins group G_j and some advertisers may appear twice in O' (e.g. $o_j = o_{j+1}$). We will match each o_j in O' to exactly one advertiser ℓ in A where $v_{o_j} \leq v_\ell$. Each advertiser in A is associated with at most 5 members of O'. In this way, $OPT \leq OPT' \leq 5 \cdot ALG$ and the competitive ratio of 5 is proved.

The matching is constructed as follows. Let $(o_{j_1}, \ldots, o_{j_{k_j}})$ be the members of O' that compete on time-slot j in HALF-algorithm (ordered by their arrival time). Note that o_{j_r} wins group G_{j_r} in O' and G_{j_r} should be in o_{j_r}'s extended time window $W_{o_{j_r}}'$. The number of groups in the extended time window $W_{o_{j_r}}'$ may be one or two more than that in the original time window $W_{o_{j_r}}$. Before showing the matching, we define function P mapping o_{j_r} to one member in A which wins one group in $W_{o_{j_r}}$:

1. $P(o_{j_r}) = p_{j_r}$ if group G_{j_r} is totally included in $W_{o_{j_r}}$;
2. $P(o_{j_r}) = p_{j_r+1}$ if group G_{j_r} is not totally included in $W_{o_{j_r}}$ and is the first group in $W_{o_{j_r}}'$;
3. $P(o_{j_r}) = p_{j_r-1}$ if group G_{j_r} is not totally included in $W_{o_{j_r}}$ and is the last group in $W_{o_{j_r}}'$;

In case (2) group G_{j_r+1} should be totally included in $W_{o_{j_r}}$ and in case (3) group G_{j_r-1} should also be totally included in $W_{o_{j_r}}$ as $s = \lceil s_{min}/2 \rceil$ and $d_{o_{j_r}} - a_{o_{j_r}} + 1 \geq s_{min}$. Now we show the rules of matching:

1. If $o_{j_{k_j}-1} = o_{j_{k_j}}$, match both $o_{j_{k_j}-1}$ and $o_{j_{k_j}}$ to p_j (denoted by $o_{j_{k_j}-1}, o_{j_{k_j}} \to p_j$); otherwise, $o_{j_{k_j}} \to p_j$.
2. If $r < k_j$ and $o_{j_r} \neq o_{j_r+1}$ and $o_{j_r} \neq o_{j_r-1}$, then $o_{j_r} \to P(o_{j_r+1})$.
3. If $r < k_j - 1$ and $o_{j_r} = o_{j_r+1}$, then:
   (a) If $o_{j_r+2} = o_{j_r+3}$, then $o_{j_r} \to P(o_{j_r+2})$ and $o_{j_r+1} \to P(o_{j_r+3})$;

(b) Else if there exists t s.t. $t > r$ and $o_{j_t} = o_{j_{t+1}}$, then choose the minimum t, $o_{j_r} \to P(o_{j_{r+2}})$ and $o_{j_{r+1}} \to P(o_{j_{t+1}})$;

(c) Else, $o_{j_r} \to P(o_{j_{r+2}})$ and $o_{j_{r+1}} \to p_j$;

Rule (1) is used to deal with the last advertiser $o_{j_{k_j}}$ competing on time-slot j. Rule (2) is for advertisers appearing once in O' and rule (3) is for those appearing twice in O'. As one advertiser can appear at most twice in O', our matching covers all cases and each element in O' can be matched to exactly one advertiser in A. Another important fact we will use later is that there are no two elements, $o_{j'_1}$, $o_{j'_2}$ s.t. $o_{j'_1} \to P(o_{j'_3})$ and $o_{j'_2} \to P(o_{j'_3})$ in our matching.

Firstly, we prove that any advertiser p_j is associated with at most 5 elements in O'. There are two possible cases: (a) $o_{j_{k_j}-1} = o_{j_{k_j}}$, which implies that rule (3c) is not applicable. Rule (1) matches two elements, $o_{j_{k_j}-1}$ and $o_{j_{k_j}}$, to p_j. p_j can also appear in matching like $o_{j'_r} \to P(o_{j'_{r'}})$, where $P(o_{j'_{r'}}) = p_j$. Note that function P can map at most three elements in O' to p_j (they are o_{j-1}, o_j, o_{j+1}) , and it does not happen that there are two elements, $o_{j'_1}$, $o_{j'_2}$ s.t. $o_{j'_1} \to P(o_{j'_3})$ and $o_{j'_2} \to P(o_{j'_3})$. So rule (2) and (3) can match at most three elements in O' to p_j. (b) $o_{j_{k_j}-1} \neq o_{j_{k_j}}$. Rule (1) and (3c) matches two elements, $o_{j_{k_j}}$ and $o_{j_{r+1}}$, to p_j ($o_{j_{r+1}}$ may not exist). Similar to the former case, p_j can also appear at most three times in matching like $o_{j'_r} \to P(o_{j'_{r'}})$, where $P(o_{j'_{r'}}) = p_j$.

It remains to be proved that any element in O' is always matched to an advertiser with higher or equal value. In rule (1), since both $o_{j_{k_j}}$ and p_j compete on G_j and p_j wins, $v_{o_{j_{k_j}}} \leq v_{p_j}$. In rule (2), when $o_{j_{r+1}}$ arrives, she competes on G_j rather than the group advertiser $P(o_{j_{r+1}})$ wins. At this moment, o_{j_r} has already arrived, thus the current candidate h for the group advertiser $P(o_{j_{r+1}})$ wins has value at least $v_{o_{j_r}}$, i.e., $v_{o_{j_r}} \leq v_h$. As $v_{P(o_{j_{r+1}})}$ should be no less than v_h, $v_{o_{j_r}} \leq v_h \leq v_{P(o_{j_{r+1}})}$. In rule (3a), when $o_{j_{r+2}}$ arrives, she competes on G_j rather than the group advertiser $P(o_{j_{r+2}})$ or $P(o_{j_{r+3}})$ wins. Similarly, $v_{o_{j_r}} \leq v_{P(o_{j_{r+2}})}$ and $v_{o_{j_{r+1}}} \leq v_{P(o_{j_{r+3}})}$. In rule (3b) and (3c), we can get similar results.

5.2 Competitive Ratio When Demands Are Non-uniform

Now we consider the general case where s_{max} is not necessarily equal to s_{min}. In this case, let $ALG2$ be the social welfare achieved by the HALF-algorithm. Let $O2$ be the optimal solution and $OPT2$ be the optimal social welfare. In $O2$, advertisers may win more than s_{min} time-slots. We will use $O2$ to construct one new solution $O2'$ in which any advertiser wins no more than s_{min} time-slots. The social welfare of $O2'$ is $OPT2'$ and it can be proved that $OPT2 \leq OPT2' \cdot \lceil s_{max}/s_{min} \rceil$. Then we will compare $OPT2'$ with $ALG2$ and a competitive ratio of $5 \cdot \lceil s_{max}/s_{min} \rceil$ is proved.

Theorem 6. *The HALF-algorithm is $5 \cdot \lceil s_{max}/s_{min} \rceil$-competitive when maximum demands are non-uniform.*

Proof. Let $ALG2$ be the social welfare achieved by the HALF-algorithm. Let $O2$ be the optimal solution and $OPT2$ be the optimal social welfare. We use $O2$ to construct a new solution $O2'$. For any advertiser i who wins x_i time-slots in $O2$, we choose the first $x_i' = \lceil x_i / \lceil s_{max}/s_{min} \rceil \rceil$ time-slots from all these x_i time-slots and allocate them to i in $O2'$. The social welfare of $O2'$ is $OPT2'$. As $x_i \leq x_i' \cdot \lceil s_{max}/s_{min} \rceil$ for any i, we get:

$$OPT2 \leq OPT2' \cdot \lceil s_{max}/s_{min} \rceil.$$

In the auction, each advertiser i demands at most s_i time-slots. Now consider another scenario where each advertiser i's maximum demand is s_{min} instead of s_i and her other information remains the same as before. In this scenario, the social welfare achieved by HALF-algorithm is $ALG3$. The optimal solution is $O3$ and the optimal social welfare is $OPT3$. As all maximal demands are uniform, by Theorem 5, we can get:

$$OPT3 \leq 5 \cdot ALG3.$$

Recall that advertiser i wins x_i' slots in $O2'$. As $x_i' \leq \lceil s_{max}/\lceil s_{max}/s_{min} \rceil \rceil \leq s_{min}$, advertiser i wins no more than s_{min} slots in $O2'$. In solution $O3$, any advertiser i can also win no more than s_{min} slots. As $O3$ is the optimal solution, $OPT3$ should be the maximum social welfare and $OPT2' \leq OPT3$. On the other hand, note that the only difference between the two scenarios we have considered is advertisers' maximum demands s_i. However, no matter what the value of s_i is, the HALF-algorithm will only allocate each advertiser 0 or s time-slots. In these two scenarios, the HALF-algorithm has the same output and then $ALG2 = ALG3$. Thus,

$$OPT2 \leq \lceil s_{max}/s_{min} \rceil \cdot OPT2' \leq \lceil s_{max}/s_{min} \rceil \cdot OPT3$$
$$\leq 5 \cdot \lceil s_{max}/s_{min} \rceil \cdot ALG3 = 5 \cdot \lceil s_{max}/s_{min} \rceil \cdot ALG2.$$

References

1. Aggarwal, G., Goel, G., Karande, C., Mehta, A.: Online vertex-weighted bipartite matching and single-bid budgeted allocations. In: Proceedings of the Twenty-Second Annual ACM-SIAM Symposium on Discrete Algorithms, pp. 1253–1264. SIAM (2011)
2. Aggarwal, G., Hartline, J.D.: Knapsack auctions. In: Proceedings of the Seventeenth Annual ACM-SIAM Symposium on Discrete Algorithm, pp. 1083–1092. ACM (2006)
3. Archer, A., Tardos, É.: Truthful mechanisms for one-parameter agents. In: Proceedings of the 42nd IEEE Symposium on Foundations of Computer Science, pp. 482–491. IEEE (2001)
4. Azar, Y., Khaitsin, E.: Prompt mechanism for ad placement over time. In: Persiano, G. (ed.) SAGT 2011. LNCS, vol. 6982, pp. 19–30. Springer, Heidelberg (2011)
5. Bartal, Y., Chin, F.Y.L., Chrobak, M., Fung, S.P.Y., Jawor, W., Lavi, R., Sgall, J., Tichý, T.: Online competitive algorithms for maximizing weighted throughput of unit jobs. In: Diekert, V., Habib, M. (eds.) STACS 2004. LNCS, vol. 2996, pp. 187–198. Springer, Heidelberg (2004)

6. Borgs, C., Chayes, J., Etesami, O., Immorlica, N., Jain, K., Mahdian, M.: Dynamics of bid optimization in online advertisement auctions. In: Proceedings of the 16th International Conference on World Wide Web, pp. 531–540. ACM (2007)
7. Chan, W.-T., Lam, T.-W., Ting, H.-F., Wong, P.W.H.: New results on on-demand broadcasting with deadline via job scheduling with cancellation. In: Chwa, K.-Y., Munro, J.I. (eds.) COCOON 2004. LNCS, vol. 3106, pp. 210–218. Springer, Heidelberg (2004)
8. Chin, F.Y.L., Fung, S.P.Y.: Online scheduling with partial job values: Does timesharing or randomization help? Algorithmica 37(3), 149–164 (2003)
9. Chrobak, M., Jawor, W., Sgall, J., Tichý, T.: Improved online algorithms for buffer management in qoS switches. In: Albers, S., Radzik, T. (eds.) ESA 2004. LNCS, vol. 3221, pp. 204–215. Springer, Heidelberg (2004)
10. Cole, R., Dobzinski, S., Fleischer, L.K.: Prompt mechanisms for online auctions. In: Monien, B., Schroeder, U.-P. (eds.) SAGT 2008. LNCS, vol. 4997, pp. 170–181. Springer, Heidelberg (2008)
11. Englert, M., Westermann, M.: Considering suppressed packets improves buffer management in qos switches. In: Proceedings of the Eighteenth Annual ACM-SIAM Symposium on Discrete Algorithms, pp. 209–218. Society for Industrial and Applied Mathematics (2007)
12. Nisan, N., Bayer, J., Chandra, D., Franji, T., Gardner, R., Matias, Y., Rhodes, N., Seltzer, M., Tom, D., Varian, H., Zigmond, D.: Google's auction for tv ads. In: Albers, S., Marchetti-Spaccamela, A., Matias, Y., Nikoletseas, S., Thomas, W. (eds.) ICALP 2009, Part II. LNCS, vol. 5556, pp. 309–327. Springer, Heidelberg (2009)
13. Ting, H.-F.: A near optimal scheduler for on-demand data broadcasts. In: Calamoneri, T., Finocchi, I., Italiano, G.F. (eds.) CIAC 2006. LNCS, vol. 3998, pp. 163–174. Springer, Heidelberg (2006)
14. Vickrey, W.: Counterspeculation, auctions, and competitive sealed tenders. The Journal of Finance 16(1), 8–37 (1961)
15. Zhou, Y., Chakrabarty, D., Lukose, R.: Budget constrained bidding in keyword auctions and online knapsack problems. In: Papadimitriou, C., Zhang, S. (eds.) WINE 2008. LNCS, vol. 5385, pp. 566–576. Springer, Heidelberg (2008)

Using Basis Dependence Distance Vectors to Calculate the Transitive Closure of Dependence Relations by Means of the Floyd-Warshall Algorithm

Włodzimierz Bielecki, Krzysztof Kraska, and Tomasz Klimek

Faculty of Computer Science and Information Technology
West Pomeranian University of Technology, ul.Żołnierska 49, 71–210 Szczecin, Poland
{wbielecki,kkraska,tklimek}@wi.zut.edu.pl

Abstract. In this paper, we present a modified Floyd-Warshall algorithm, where the most time-consuming part – calculating transitive closure describing self-dependences for each loop statement – is computed by means of basis dependence distance vectors derived from all vectors describing self-dependences. We demonstrate that the presented approach reduces the transitive closure calculation time for parameterized graphs representing all dependences in the loop in comparison with techniques implemented in the Omega and ISL libraries. This increases the applicability scope of techniques based on transitive closure of dependence graphs. Experimental results for NASA Parallel Benchmarks are discussed.

Keywords: basis dependence vectors, transitive closure, Floyd-Warshall algorithm, arbitrarily nested loop, parallelizing compiler.

1 Introduction

Resolving many problems is based on calculating transitive closures of graphs. In this paper, we deal with parameterized graphs whose number of vertices is represented with an expression including parameters. Such graphs can be represented by parameterized relations whose tuples represent vertices while constraints are responsible for defining edges [11]. Transitive closure calculated for such relations can be used in optimizing compilers : to remove redundant synchronization [11], test the legality of iteration reordering transformations [11], apply iteration space slicing [3], form schedules for statement instances of program loops [4]. In general, calculating transitive closure of parameterized graphs is time-consuming [2,11,13]. Sometimes the time of transitive closure calculation prevents applying techniques for extracting coarse- and fine-grained parallelism because this time is not acceptable in practice (several hours and even several days [3,4]. This is why improving transitive closure algorithms aimed at reducing their time complexity is an actual task.

P. Widmayer, Y. Xu, and B. Zhu (Eds.): COCOA 2013, LNCS 8287, pp. 129–140, 2013.
© Springer International Publishing Switzerland 2013

In this paper, we demonstrate how to reduce the time of calculating transitive closure describing self-dependences. For this purpose, we propose to find basis distance dependence vectors from all distance vectors describing self-dependences and then demonstrate how these vectors can be used for calculating transitive closure. For extracting such distance vectors, dependence relations, extracted with a dependence analyzer, are used. Such relations (describing selfdependences) are characterized by the same arity (the number of tuple elements) of input and output tuples. Finaly, we present experimental results showing how the time of transitive closure calculation is reduced for NAS benchmarks [15].

2   Background

In this paper, we deal with the following definitions concerned program loops: iteration vector, loop domain (index set), parameterized loops, perfectly-nested and arbitrarily-nested loops. The explanations of them are given in papers [3,4].

Definition 1 (Dependence). *Two statement instances $S_1(I)$ and $S_2(J)$, where I and J are the iteration vectors, are dependent if both access the same memory location and if at least one access is a write* [4].

Definition 2 (Dependence distance set, dependence distance vector). *We define a dependence distance set $D_{S,T}$ as a set of differences between all such vectors of the same size that stand for a pair of dependent instances of statement T and S. We call each element of set $D_{S,T}$ a (dependence) distance vector and denote it as $d_{S,T}$.*

Definition 3 (Dependence relation). *A dependence relation is a tuple relation of the form $\{[input\_list] \to [output\_list] : constraints\}$, where input_list and output_list are the lists of variables used to describe input and output tuples and constraints is a Presburger formula describing the constraints imposed upon input_list and output_list.*

The general form of a dependence relation is as follows [11]*:*

$$R = \{[s_i, \ldots, s_k] \to [t_i, \ldots, t_k] : \bigvee_{i=1}^{n} \exists \alpha_{i1}, \ldots, \alpha_{im_i} \ s.t. \ \mathcal{F}_i\},$$

where $\mathcal{F}_i, i = 1, 2, \ldots, n$ are represented by Presburger formulas, i.e., they are conjunctions of affine equalities and inequalities on the input variables $s_1, \ldots, s_k$, the output variables $t_1, \ldots, t_k$, the existentially quantified variables $\alpha_{i1}, \ldots, \alpha_{im_i}$, and symbolic constants.

The following concepts of linear algebra are used in the algorithm presented in this paper: vector, unit normal vector, vector space, field, linear combination, linear independence. Details can be found in paper [12].

Definition 4 (Column Space of a Matrix). *Let A be an $m \times n$ matrix. The space spanned by the columns of A is called the column space of A, denoted $C(A)$* [12].

Definition 5 (Basis). *A basis B of a vector space V over a field F (such as $\mathbb{R}$ or $\mathbb{Z}$) is a linearly independent subset of V that spans (or generates) V. Every finite-dimensional vector space V has a basis* [12]*.*

Definition 6 (Positive transitive closure). *Let R be an affine integer tuple relation, then the positive transitive closure R^+ of R is the union of all positive powers of R,*

$$R^+ = \bigcup_{k \geqslant 1} R^k, \quad \text{with} \quad R^k = \begin{cases} R & \text{if } k = 1 \\ R \circ R^{k-1} & \text{if } k \geqslant 2. \end{cases} \tag{1}$$

Definition 7 (Transitive closure). *Transitive closure, R^*, is defined as follows* [11]*:*

$$R^* = R^+ \cup I,$$

where I is the identity relation. R^* describes the same connections in a dependence graph (represented by R) that R^+ does plus connections of each vertex with itself.

To check whether output returned by an algorithm represents exact transitive closure, we can use the well-known fact [11] that for an acyclic relation R (for such a relation $R \cap I = \varnothing$, where I is the identity relation) the following is true:

- if R^+ is exact transitive closure, then:

$$R^+ = R \cup (R \circ R^+),$$

- if R^+ is an over–approximation, then:

$$R^+ \subset R \cup (R \circ R^+).$$

3   Calculating Transitive Closure

To compute the transitive closure of a dependence relation representing all the dependences exposed for an arbitrarily nested loop, we use a modified form of the Floyd-Warshall algorithm (see Algorithm 1). It is not difficult to see that Algorithm 1 has the special key expression $\mathcal{D}_{kj} \circ (\mathcal{D}_{kk}^*) \circ \mathcal{D}_{ik}$, where 'o' denotes the composition operator applied to a pair of relations, $\mathcal{D}_{ik}$ describes all dependences between instances of statement s_i and statement s_k. This means that if there is a dependence from iteration i_1 of statement s_i to iteration i_2 of statement s_k and a chain of self dependences from iteration i_2 to iteration i_3 $\mathcal{D}_{kk}^*$ and finally a dependence from iteration i_3 of statement s_k to iteration i_4 of statement s_j (where $\mathcal{D}_{kj}$ describes all dependences between instances of statement s_k and statement s_j) then there is a transitive dependence from iteration i_1 to iteration i_4. It should be clear that the objective of this technique is to update all the dependences through statements 1,2,...,n in an iteration of each k-loop.

Algorithm 1. The modified Floyd-Warshall algorithm [11]

1:
   Input : $\mathcal{D}^{N \times N}$ array whose element i, j represents a dependence relation describing all direct dependences exposed for instances of statement i and statement j, where N is the total number of statements in the loop, if there exists a dependence from instances of statement $i \leqslant N$ to instances of statement $j \leqslant N$ then $\mathcal{D}_{ij} \neq \emptyset$, otherwise $\mathcal{D}_{ij} = \emptyset$.

2:
   Output : $\mathcal{D}^{N \times N}$ array, where each element $\mathcal{D}_{ij}$ represents a relation describing transitive closure between instances of statement i and instances of statement j.

3:
4: **Method:**
5:
6: **for** each statement k
7:     **for** each statement i
8:         **for** each statement j
9:             $\mathcal{D}_{ij} = \mathcal{D}_{ij} \cup \mathcal{D}_{kj} \circ (\mathcal{D}_{kk}^*) \circ \mathcal{D}_{ik}$
10:

In the presented algorithm, we propose to calculate $\mathcal{D}_{kk}^*$ using a finite linear combination of basis dependence distance vectors [5].

Given a set D of m dependence distance vectors in the n-dimensional integer space derived from a union of dependence relations $\mathcal{D}_{kk}$ (it describes a chain of self dependences of statement s_k in the loop), we first replace all parameterized vectors with constant vectors.

Let v_p be a vector in Z^d and p_i are its parameterized coordinates in the i-positions. We may replace vector v_p with a linear combination of a constant vector v_c, $v_c \in \mathbb{Z}^d$, and unit normal vectors e_i, $e_i \in \mathbb{Z}^d$, where p_i are coefficients, as follows:

$$v_p = v_c + \sum_i p_i \times e_i. \tag{2}$$

If $v_c = 0$ then it can be rejected from (2).

Proof. Without loss of generality, we may assume that the first n positions of v_p have constant coordinates and the last q positions have parameterized ones. Then, we can write:

$$
\begin{pmatrix} c_1 \\ \vdots \\ c_n \\ p_{n+1} \\ \vdots \\ p_d \end{pmatrix}
=
\begin{pmatrix} c_1 \\ \vdots \\ c_n \\ 0 \\ \vdots \\ 0 \end{pmatrix}
+
\begin{pmatrix} 0 \\ \vdots \\ 0 \\ p_{n+1} \\ \vdots \\ p_d \end{pmatrix},
\tag{3}
$$

where here and further $d - n = q$, the second vector can be written as the linear combination of unit normal vectors e_k and parameterized coefficients $p_{n+1}, \ldots, p_d$ in the last d positions:

$$
\begin{pmatrix} 0 \\ \vdots \\ 0 \\ p_{n+1} \\ \vdots \\ p_d \end{pmatrix} = \begin{pmatrix} 0 \\ \vdots \\ 0 \\ p_{n+1} \\ \vdots \\ 0 \end{pmatrix} + \ldots + \begin{pmatrix} 0 \\ \vdots \\ 0 \\ 0 \\ \vdots \\ p_d \end{pmatrix} = p_{n+1} \times \begin{pmatrix} 0 \\ \vdots \\ 0 \\ 1 \\ \vdots \\ 0 \end{pmatrix} + \ldots + p_d \times \begin{pmatrix} 0 \\ \vdots \\ 0 \\ 0 \\ \vdots \\ 1 \end{pmatrix}. \tag{4}
$$

Substituting (4) into (3), we obtain:

$$
\begin{pmatrix} c_1 \\ \vdots \\ c_n \\ p_{n+1} \\ \vdots \\ p_d \end{pmatrix} = \begin{pmatrix} c_1 \\ \vdots \\ c_n \\ 0 \\ \vdots \\ 0 \end{pmatrix} + p_{n+1} \times e_{n+1} + \ldots + p_d \times e_d. \tag{5}
$$

It is obvious that if $v_c = \mathbf{0}$, then v_c can be rejected without affecting the result. □

Property 1. Replacing parameterized vectors with a linear combination of vectors with constant coordinates can be done in a polynomial time.

Proof. To check each position in vector v_p, $v_p \in \mathbb{Z}^d$, the algorithm requires d operations. In the worst case, all d positions can be parameterized coordinates, hence d unit normal vectors e_k, $e_k \in \mathbb{Z}^d$ must be created. This defines $\mathcal{O}\left(d^2\right)$ time complexity of replacing parameterized vectors. □

Let us consider the parameterized dependence distance vector $(N, 2)$. It can be represented as the linear combination of the two linearly independent vectors $(0, 2) + \alpha \times (1, 0)$, where $\alpha \in \mathbb{Z}$.

As a result, we get k, $k \geq m$, dependence distance vectors with constant coordinates. This allows us to get rid of parameterized vectors and to form an integer matrix A, $A \in \mathbb{Z}^{n \times k}$, by inserting dependence distance vectors with constant coordinates into columns of A. The columns of A span vector space V.

To decrease the complexity of further computations, redundant dependence distance vectors are eliminated from matrix A by finding a subset of l, $l \leq k$, linearly independent columns of A. This subset of dependence distance vectors forms the basis B, $B \in \mathbb{Z}^{n \times l}$, of A and generates the same vector space V as A does [12]. Every element of vector space V can be expressed uniquely as a finite linear combination of the basis dependence distance vectors belonging to B. When B is completed, we can calculate the relation $\mathcal{D}_{kk}^*$, as follows:

$$
\mathcal{D}_{kk}^* = \left\{ [x] \to [y] \mid \exists z \; s.t. \; y = x + B^{n \times l} \times z \; \wedge \; y - x \succ 0, \; z \in \mathbb{Z}^l \wedge \atop \wedge \; y \in range\,(\mathcal{D}_{kk}) \; \wedge \; x \in domain\,(\mathcal{D}_{kk}) \right\} \cup I, \tag{6}
$$

where : $\mathcal{D}_{kk}^*$ describes a chain of self dependences of statement s_k in the loop, $B^{n \times l} \times z$ represents a linear combination of the basis dependence distance vectors d_i (the columns of $B^{n \times l}$, $1 \le i \le l$), $y - x \succ 0$ imposes the lexicographically forward constraints on the tuples of $\mathcal{D}_{kk}^*$, I is the identity relation.

For each vertex x in the data dependence graph (where x is the source of a dependence(s), $x \in domain(\mathcal{D}_{kk})$), we can identify all vertices y (the destinations of the dependence(s), $y \in range(\mathcal{D}_{kk})$) that are connected with x by a path of the length equal or more than 1, where y is calculated as x plus a linear combination of the basis dependence distance vectors B, i.e. $y = x + B \times z$, $z \in \mathbb{Z}^l$.

The resulting relation $\mathcal{D}_{kk}^*$ represents the exact transitive closure of relation $\mathcal{D}_{kk}$ or its over-approximation. To prove this, let us note that relation $\mathcal{D}_{kk}^+$ represents all possible paths between vertices x (standing for dependence sources, $x \in domain(\mathcal{D}_{kk})$) and vertices y (standing for dependence destinations, $y \in range(\mathcal{D}_{kk})$) in the dependence graph, represented with relation $\mathcal{D}_{kk}$. Indeed, a linear combination of the base dependence distance vectors $B^{n \times l} \times z$:

o reproduces all dependence distance vectors exposed for the loop,
o describes all existing (true) paths between any pair of x and y as a linear combination of all dependence distance vectors exposed for the loop,
o can describe not existing (false) paths in the dependence graph represented by relation $\mathcal{D}_{kk}^*$.

The last case is possible when on a path between x and y, being described by $\mathcal{D}_{kk}^*$, there exists a vertex w such that $w \in range(\mathcal{D}_{kk}) \wedge w \notin domain(\mathcal{D}_{kk})$. Such a case is presented in Figure 1, where $x_2 \in range(\mathcal{D}_{kk}) \wedge x_2 \notin domain(\mathcal{D}_{kk})$. Relation $\mathcal{D}_{kk}^*$, built according to (6), describes the false path between x_1 and x_4 depicted by the dotted line.

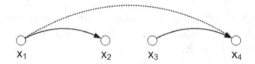

x_1 x_2 x_3 x_4

Fig. 1. False path in the dependence graph

Summing up, we conclude that relation $\mathcal{D}_{kk}^*$ describes all existing paths in the dependence graph represented by relation $\mathcal{D}_{kk}$ and can describe not existing paths, i.e., $(\mathcal{D}_{kk}^*)_{exact} \subseteq \mathcal{D}_{kk}^*$; when relation $\mathcal{D}_{kk}^*$ does not represent not existing paths, $\mathcal{D}_{kk}^* = (\mathcal{D}_{kk}^*)_{exact}$.

4   Related Work

Numerous algorithms for calculating the transitive closure of affine integer tuple relations have been proposed [1,2,6,7,9,11,13]. However, in most of them authors focus on relations whose domain and range are non-parametric polyhedra [1,7,9]. The second limitation of known algorithms is that they require that the arity of input and output tuples (the number of tuple elements) of relations has to be the same [2]. This is why we limit related work only to techniques dealing with parameterized relations whose tuple arities are different in general and relations can describe dependences available in program loops.

On a different line of work, Bozga et al. [7] have studied the computation of transitive closure for the analysis of counter automata (register machines) and they have implemented their method in the tool called FLATA [7]. In this context, relation $R(x, x')$ is a relation that can be written as the finite number of conjunctions of terms of the form $\pm x_i \pm x_j \leqslant a_{i,j}$, $\pm x_i' \pm x_j \leqslant b_{i,j}$, $\pm x_i \pm x_j' \leqslant c_{i,j}$, $\pm x_i' \pm x_j' \leqslant d_{i,j}$, $\pm 2x_i \leqslant e_{i,j}$ or $2x_i' \leqslant f_{i,j}$, where x and y describe counter values, either at the current step, or at the next step, $a_{i,j}, b_{i,j}, c_{i,j}, d_{i,j}, e_{i,j}, f_{i,j} \in \mathbb{Z}$ are integer constants and $1 \leqslant i, j \leqslant n$, $i \neq j$. As we can see, this class of relation does not involve parameters, existentially quantified variables or unions, i.e., it cannot represent dependences in program loops. This is why we do not compare this technique with ours.

To our best knowledge, techniques for computing the transitive closure of parameterized affine integer tuple relations with different input and output arities of tuples were the subject of the investigation of a few papers only [11,13,14]. Kelly et al. [11] proposed a modified Floyd-Warshall algorithm but they have not implemented it in the Omega library [17]. Fourteen years later Verdoolaege has improved and implemented his version of the Floyd-Warshall algorithm in the ISL library [16], but that algorithm and implementation is not the same as ours.

Verdoolaege [13,14] treats each of input relations $R_{i \leqslant m}$ as vertices $V_{i \leqslant m}$ of the directed graph G, where m is the total number of input relations. There exists a directed path E_{ij} from vertex $V_{i \leqslant m}$ to vertex $V_{j \leqslant m}$ (R_j can immediately follow R_i) if the range of R_i intersects the domain of R_j, i.e., if

$$R_j \circ R_i \neq \emptyset. \tag{7}$$

In order to calculate the transitive closure of a dependence relation R, Verdoolaege [13,14] constructs m^2 relations

$$R_{ij} = \bigcup_{i,j \ s.t. \ R_j \circ R_i \neq \emptyset}^{m} R_j \circ R_i. \tag{8}$$

Then he applies Algorithm 1 and returns the union of all output $R_{i,j}$ as transitive closure. In our algorithm, we use information gathered with the Petit dependence analyzer [10] to insert a dependence relation describing dependences between instances of statements i and j as element i, j of array $\mathcal{D}$ (element $\mathcal{D}_{ij}$). Then we call Algorithm 1 to get transitive closure. Information provided with Petit permits us to reduce the time complexity of the Floyd-Warshall algorithm implementation due to skipping a connection check between each pair of input dependence relations (see formula 7).

Because of differences between our implementation of the Floyd-Warshall algorithm and that of Verdoolaege [13,14], in this paper we investigate only how different concepts of calculating the relation $\mathcal{D}_{kk}^*$ impact the time complexity of the Floyd-Warshall algorithm. For this purpose, we have chosen for calculating $\mathcal{D}_{kk}^*$ algorithms implemented in the in ISL [16] and Omega [17] libraries. Those algorithms are based on computing parametric powers R^k and then projecting out the parameter k by making it existentially quantified. As a trivial example, consider the relation $R = \{[x] \to [x+1]\}$. The kth power of R for arbitrary k is $R^k = \{[x] \to [x+k] \mid k \geq 1\}$ and the transitive closure is then $R^+ = \{[x] \to [y] \mid \exists k \in \mathbb{Z}_{\geq 1} : y = x + k\} = \{[x] \to [y] \mid y \geq x+1\}$. Both the algorithms consider the difference set ΔR of the relation, but in the ISL library [16] if all the differences Δ_is are singleton sets, i.e., $\Delta_i = \{\delta_i\}$ with $\delta_i \in Z^d$, then R^+ is calculated as follows :

$$R^+ = \{x \longrightarrow y \mid \exists k_i \in \mathbb{Z}_{\geq 0} : y = x + \sum_i k_i \delta_i \wedge \sum_i k_i = k > 0 \quad (9)$$

which is essentially the same as that of Beletska et al. [2]. If some of the Δ_is are parametric, then each offset Δ_i is extended with an extra coordinate $\Delta_i' = \Delta_i \times \{1\}$, that is a constant and equal to one. The paths constructed by summing such extended offsets have the length k encoded as the difference of their final coordinates, so R^+ can then be decomposed into relations R_i^+, one for each Δ_i,

$$R^+ = ((R_m^+ \cup I) \circ ... \circ (R_2^+ \cup I) \circ (R_1^+ \cup I)) \cap \{x' \to y' \mid \exists k_{>0} : y_{d+1} - x_{d+1} = k\}, \quad (10)$$

with

$$R_i^+ = s \mapsto \{x' \to y' \mid \exists k \in \mathbb{Z}_{\geq 1}, \delta \in k\Delta_i'(s) : y' = x' + \delta\}. \quad (11)$$

Each non-parametric constraint $A_1 x + c_1 \geq 0$ of $\Delta_i'(s)$ from (11) is transformed into the form $A_1 x + k c_1 \geq 0$ and the rest of constraints are rewritten without any changes. For more details see [13,14].

While the algorithms implemented by Verdoolaege [13,14] in the ISL library [16] are designed to compute overapproximations, Kelly et all. [11] in the Omega library [17] propose a heuristic algorithm to compute an underapproximation that does not guarantee calculating exact transitive closure.

5   Experimental Results

The goals of experiments were to evaluate the effectiveness and time complexity of the proposed approach for calculating relation $\mathcal{D}_{kk}^*$ for loops provided by the well-known NAS Parallel Benchmark (NPB) Suite from NASA [15] and compare received results with the effectiveness and time complexity of techniques implemented in the ISL [16] and Omega [17] tools. We have implemented the presented algorithm as an ANSI-C++ software module. The source code of the module was compiled using the gcc compiler v4.3.0 and can be download from: http://www.sfs.zut.edu.pl/files/mfw-omega.tar.gz. Experiments were conducted using an Intel Core2Duo T7300@2.00GHz machine with the Fedora Linux v12 32bit operating system.

The implementation calculates transitive closure according to Algorithm 1 and permits to choose the three options for producing the relation $\mathcal{D}_{kk}^*$ by means of: (i) formula (6), (ii) Omega, and (iii) ISL. Under our experiments, we have examined only such loops provided by NPB that expose dependences. There exist 58 imperfectly-nested loops in NPB that expose dependences. The results of our experiments are collected in Table 1, where time is presented in seconds. The columns "proposed algorithm", "ISL", and "Omega" present the time of calculating transitive closure by means of the Floyd-Warshal algorithm, where relations $\mathcal{D}_{kk}^*$ were calculated by means of applying formula (6), the ISL, and Omega tools, respectively.

Table 1. The results of the experiments on the proposed approach to computing transitive closure (ex: 1 – exact result, 0 – over-approximation; Δt: difference between the transitive closure calculation time of a known correspondent technique and that of the presented approach)

#	Source loop name	Number of relations	Proposed algorithm		ISL[1]		Omega[2]	
			ex	$t\ [s]$	ex	$\Delta t\ [s]$	ex	$\Delta t\ [s]$
	Imperfectly-nested loops							
1)	BT_error.f2p_2	107	1	2.971272	1	8.893896	1	9.152150
2)	BT_exact_rhs.f2p_2	1553	0	32.231114	0	61.389895	0	73.309154
3)	BT_exact_rhs.f2p_3	1553	0	31.938555	0	68.586695	0	74.396025
4)	BT_exact_rhs.f2p_4	1553	0	32.185643	0	61.133555	0	73.996971
5)	BT_initialize.f2p_2	42	1	0.283660	1	0.394805	1	3.007654
6)	BT_initialize.f2p_3	42	1	0.288850	1	0.396405	1	2.998300
7)	BT_initialize.f2p_4	42	1	0.288210	1	0.393468	1	3.052299
8)	BT_initialize.f2p_5	42	1	0.313911	1	0.428361	1	3.003586
9)	BT_initialize.f2p_6	42	1	0.318144	1	0.394464	1	3.008207
10)	BT_initialize.f2p_7	42	1	0.298637	1	0.396687	1	3.018918
11)	BT_rhs.f2p_3	702	0	26.190911	0	268.752587	0	38.544168
12)	BT_rhs.f2p_4	510	0	16.442628	0	236.826475	0	26.149409
13)	LU_blts.f2p_1	4885	1	3632.80771	1	4267.0205	1	5078.63173
14)	LU_buts.f2p_1	5640	1	4010.86541	1	5673.09815	1	5612.88391
15)	LU_erhs.f2p_2	66	1	0.166997	1	0.598715	1	5.963691
16)	LU_erhs.f2p_3	640	0	72.333986	0	164.446467	0	107.584855
17)	LU_erhs.f2p_4	640	0	74.697203	0	192.229223	0	104.277494
18)	LU_erhs.f2p_5	640	0	32.549758	0	237.451972	0	58.911653
19)	LU_HP_blts.f2p_1	3232	0	216.569540	0	216.869540	0	218.789512
20)	LU_HP_buts.f2p_1	3593	0	250.428062	0	447.203115	0	267.193005
21)	LU_HP_erhs.f2p_2	66	1	0.164042	1	0.839860	1	6.439332
22)	LU_HP_erhs.f2p_3	640	0	72.560179	0	263.608092	0	115.785944
23)	LU_HP_erhs.f2p_4	640	0	74.509935	0	262.061739	0	116.060227
24)	LU_HP_erhs.f2p_5	640	0	32.928784	0	236.964911	0	57.891926
25)	LU_HP_rhs.f2p_1	17	1	0.214228	1	1.514948	1	1.245564
26)	LU_HP_rhs.f2p_2	640	0	72.552203	0	387.553991	0	115.388039

- *Continued on next page* -

#	Source loop name	Number of relations	Proposed algorithm		ISL[1]		Omega[2]	
			ex	t [s]	ex	Δt [s]	ex	Δt [s]
27)	LU_HP_rhs.f2p_3	640	0	74.303048	0	262.026583	0	115.403235
28)	LU_HP_rhs.f2p_4	640	0	32.469942	0	237.760219	0	57.595624
29)	LU_rhs.f2p_1	17	1	0.217576	1	1.502999	1	1.217024
30)	LU_rhs.f2p_2	640	0	71.902717	0	279.400439	0	115.549878
31)	LU_rhs.f2p_3	640	0	73.664458	0	277.564800	0	114.485482
32)	LU_rhs.f2p_4	1412	0	199.789371	0	968.474450	0	354.928519
33)	MG_mg.f2p_10	18	1	0.004118	1	0.004387	1	0.004734
34)	MG_mg.f2p_5	24	0	0.628591	0	0.792310	0	2.122442
35)	MG_mg.f2p_6	29	0	0.917312	0	0.973950	0	2.164951
36)	MG_mg.f2p_7	510	1	2.063921	1	17.980857	1	5.568026
37)	MG_mg.f2p_8	55	0	2.299966	0	2.306906	0	7.139500
38)	MG_mg.f2p_9	18	1	0.003687	1	0.004398	1	0.004777
39)	SP_error.f2p_2	107	1	2.496292	1	9.258389	1	9.157796
40)	SP_exact_rhs.f2p_2	1553	0	32.093054	0	97.804819	0	81.841293
41)	SP_exact_rhs.f2p_3	1553	0	32.147199	0	106.642391	0	81.035489
42)	SP_exact_rhs.f2p_4	1553	0	32.297781	0	102.465263	0	81.578539
43)	SP_initialize.f2p_2	24	1	0.224286	1	0.545536	1	3.036898
44)	SP_initialize.f2p_3	24	1	0.223405	1	0.398911	1	3.172281
45)	SP_initialize.f2p_4	24	1	0.221491	1	0.397188	1	3.065710
46)	SP_initialize.f2p_5	24	1	0.223924	1	0.394303	1	3.033620
47)	SP_initialize.f2p_6	24	1	0.221686	1	0.410317	1	3.034202
48)	SP_initialize.f2p_7	24	1	0.222733	1	0.393620	1	3.037680
49)	SP_rhs.f2p_3	699	1	10.780842	1	231.733093	1	20.082132
50)	SP_rhs.f2p_4	507	1	14.171069	1	156.553769	1	23.308064
51)	UA_adapt.f2p_1	10	1	0.046903	1	0.064031	1	0.093033
52)	UA_adapt.f2p_10	14	1	0.013661	1	0.016454	1	0.026414
53)	UA_adapt.f2p_11	11	1	0.013425	1	0.016294	1	0.030629
54)	UA_adapt.f2p_9	14	1	0.005895	1	0.016332	1	0.025601
55)	UA_setup.f2p_14	31	1	0.297392	1	0.947406	1	0.356227
56)	UA_setup.f2p_15	15	1	0.261098	1	0.364980	1	0.283646
57)	UA_transfer.f2p_17	17	1	0.023841	1	0.035448	1	0.086729
58)	UA_utils.f2p_12	20	1	0.181688	1	0.163561	1	0.738736

Analyzing the results presented in Table 1, we can derive the following conclusions. All techniques under experiments are able to calculate transitive closure for all NBP loops exposing dependences. The exactness of the presented approach is the same as that of techniques implemented in Omega and ISL. i.e., all techniques under experiments produce exact transitive closure for the same loops. Calculating relation $\mathcal{D}_{kk}^*$ by means of formula (6) is less time-consuming in comparison with techniques implemented in Omega and ISL that reduces the

---

[1] Integer Set Library – a library for manipulating sets and relations of integer points bounded by affine constraints (available at http://repo.or.cz/w/isl.git)

[2] Omega Project – frameworks and algorithms for the analysis and transformation of scientific programs (available at http://www.cs.umd.edu/projects/omega/)

time of calculating the transitive closure of a relation describing all the dependences in the loop by means of the F-W's algorithm. For all loops, we obtained the shortest time of producing transitive closure.

One possible explanation is that each union that we compose in formula (10) consists of two relations. If there are m disjuncts in the input relation, then the direct application of the composition operation may therefore result in a relation with 2^m disjuncts that is computationally expensive. In general, applying formula (6) results in the number of disjuncts that is much less than 2^m. All these facts permit us to conclude that the presented approach is faster than other well-known approaches.

6 Conclusion

In this paper, we presented a modified Floyd-Warshall algorithm, where the most time consuming part (calculating transitive closure describing self-dependences in the program loop) is calculated by means of basis dependence distance vectors. This results in reducing the time of the transitive closure calculation of parameterized graphs representing dependences in program loops. Reducing this time is explained by using a finite linear combination of basis dependence distance vectors to calculate the $\mathcal{D}_{kk}^*$ term as a part of a modified Floyd-Warshall's algorithm. This conclusion was proved by means of numerous experiments with NPB benchmarks.

The presented approach can be used for resolving many optimizing compilers problems: redundant synchronization removal [11], testing the legality of iteration reordering transformations [11], iteration space slicing [3], forming schedules for statement instances of program loops [4]. In our future work we plan to study the application of the presented approach for extracting both coarse- and fine-grained parallelism for different popular benchmarks.

References

1. Ancourt, C., Coelho, F., Irigoin, F.: A modular static analysis approach to affine loop invariants detection. Electronic Notes in Theoretical Computer Science 267, 3–16 (2010)
2. Beletska, A., Barthou, D., Bielecki, W., Cohen, A.: Computing the transitive closure of a union of affine integer tuple relations. In: Du, D.-Z., Hu, X., Pardalos, P.M. (eds.) COCOA 2009. LNCS, vol. 5573, pp. 98–109. Springer, Heidelberg (2009)
3. Beletska, A., Bielecki, W., Cohen, A., Palkowski, M., Siedlecki, K.: Coarse–grained loop parallelization: Iteration space slicing vs affine transformations. Parallel Computing 37(8), 479–497 (2011)
4. Bielecki, W., Palkowski, M., Klimek, T.: Free scheduling for statement instances of parameterized arbitrarily nested affine loops. Parallel Computing (38), 518–532 (2012), http://dx.doi.org/10.1016/j.parco.2012.06.001
5. Bielecki, W., Kraska, K., Klimek, T.: Transitive closure of a union of dependence relations for parameterized perfectly-nested loops. In: Malyshkin, V. (ed.) PaCT 2013. LNCS, vol. 7979, pp. 37–50. Springer, Heidelberg (2013)

6. Boigelot, B.: Symbolic Methods for Exploring Infinite State Spaces. Ph.D. thesis, Université de Liège (1998)
7. Bozga, M., Gîrlea, C., Iosif, R.: Iterating octagons. In: Kowalewski, S., Philippou, A. (eds.) TACAS 2009. LNCS, vol. 5505, pp. 337–351. Springer, Heidelberg (2009)
8. Diestel, R.: Graph Theory, 4th edn., 451 pages. Springer, Heidelberg (2010)
9. Eve, J., Kurki–Suonio, R.: On computing the transitive closure of a relation. Acta Informatica 25. X 8(4), 303–314 (1977)
10. Kelly, W., Maslov, V., Pugh, W., Rosser, E., Shpeisman, T., Wonnacott, D.: New User Interface for Petit and Other Extensions. User Guide (1996)
11. Kelly, W., Pugh, W., Rosser, E., Shpeisman, T.: Transitive closure of infinite graphs and its applications. In: Huang, C.-H., Sadayappan, P., Banerjee, U., Gelernter, D., Nicolau, A., Padua, D.A. (eds.) LCPC 1995. LNCS, vol. 1033, pp. 126–140. Springer, Heidelberg (1996)
12. Schrijver, A.: Theory of Linear and Integer Programming. Series in Discrete Mathematics (1999)
13. Verdoolaege, S., Cohen, A., Beletska, A.: Transitive closures of affine integer tuple relations and their overapproximations. In: Yahav, E. (ed.) SAS 2011. LNCS, vol. 6887, pp. 216–232. Springer, Heidelberg (2011)
14. Verdoolaege, S.: Integer set library – manual, Tech. rep, Version: isl-0.11 (2012), http://www.kotnet.org/skimo/isl/manual.pdf
15. NASA Advanced Supercomputing Division, http://www.nas.nasa.gov
16. Integer Set Library, http://www.kotnet.org/~skimo/isl/
17. Omega Library, http://www.cs.umd.edu/projects/omega/

A Nash Equilibrium Based Algorithm for Mining Hidden Links in Social Networks

Huan Ma[1], Zaixin Lu[2], Lidan Fan[3], Weili Wu[3], Deying Li[1,*], and Yuqing Zhu[3]

[1] School of Information, Renmin University of China, Beijing 100872, China
[2] NSF Center for Research on Complex Networks, Texas Southern University, Houston, Texas, 77004, USA
[3] Dept. of Computer Science, University of Texas at Dallas, Richardson, TX, 75080, USA
deyingli@ruc.edu.cn

Abstract. With the advance of high techniques, more and more connections between individuals in a social network can be identified, but it is still hard to obtain the complete relation information between individuals for complex structure and individual privacy. However, the social networks have communities. In our work, we aim at mining the invisible or missing relations between individuals within a community in social networks. We propose our algorithm according to the fact that the individuals exist in communities satisfying Nash Equilibrium, which is borrowed from game-theoretic concepts often used in economic researches. Each hidden relation is explored through the individual's loyalty to their community. To the best of our knowledge, this is the first work that studies the problem of mining hidden links from the aspect of Nash Equilibrium. Eventually we confirm the superiority of our approach from extensive experiments over real-world social networks.

1 Introduction

Along with increasing popularity of large, complex networks in computer science and physical domains, the social networks have received a considerable amount of attention from researchers for their theoretical interests and practical importance in real-world applications. In the social networks, the nodes represent individuals, and the links (edges) denote the interactions between the nodes. The most significant feature of social network is "community structure", which means the connections between individuals are dense within the same community while sparse across distinct communities, that is individuals in the same community share more common attributes, and identifying community structure and learning the microscopical relation between individuals in a community can help us to understand individuals' behaviors well, which have wide applications in social marketing [1] [2], urban development [3], criminology [4] and so forth.

Take advantage of online networks, a large volume of data can be obtained by people, however, in the process of collecting, gathering or recording information,

---

* Corresponding author.

P. Widmayer, Y. Xu, and B. Zhu (Eds.): COCOA 2013, LNCS 8287, pp. 141–152, 2013.

some information may be lost for the complex relations between individuals, the data of information from real-world networks is often incomplete and inaccurate, here we call the lost or missing information as hidden links. Particularly, most information is distributed within communities, and a few information is distributed among communities, as a result, the hidden links often reside within communities. To get insight into microscopical individual behaviors, it is of great value for us to mine these lost links more accurately. Since the connections across communities are sparse, in our problem, we assume these links will never be lost. In other words, they are observable to the public. Obviously, our problem is established on the existing community structure.

To efficiently explore the hidden links between individuals whin a community, we start from the aspect of community formation, which was interpreted as a game framework by Chen *et al.* in [6]. They explained this phenomenon from human nature. Each agent is rational and selfish, and their behaviors of joining communities are associated with a gain function and a loss function, therefore, they will choose to join a community that maximize their total interest (the gain of joining the community minus the loss of joining the community). Obeying this rule, all agents finally find their corresponding communities (Each agent can join more than one communities at the same time.) with satisfying Nash Equilibrium. Witnessing this phenomenon, we aim to apply the property into our problem by reasonable and proper modifications, which is different from [7], although they find the missing links in community, they don't use the property of community to find the links, and they use the common neighbor method.

In this paper, we assume that the community structure in a social network has been identified through existing techniques in community detection, and these communities reach Nash Equilibrium status. Here, the concept of Nash Equilibrium is a concrete measurement based on the neighbor distribution of each node, which is different from the one in [6]. Moreover, each individual is required to belong one community, that is, the communities are disjoint. In each community, if no link is lost, then every individual has at least the same number of in-community neighbors as that of out-community neighbors. However, when some links between members within communities are missed, some individuals may become active, meaning that they have tendency to leave their current communities and join other communities. Therefore, we try to search the lost links from these active individuals. As for determining which links are the possible ones adjacent to these active individuals, we modify the techniques used in link prediction problem, which is a long-standing problem in social networks and has received extensive attentions from researchers.

Our paper is the first one to mine the hidden links from the aspect of game-theoretic framework. The main contribution is as follows.

- We prove the $\mathcal{NP}$-hardness of finding a Nash equilibrium when a graph has to be partitioned into two communities and the function used to measure the equilibrium of each node is only based on the degrees in communities.
- We propose a novel measurement modified from Nash equilibrium in [6] to determine the loyalty of each individual to their own communities, and define

all the members who are observed having tendency to leave their current community as *active individuals*.

- By adopting link prediction measurements common neighbors [8] and $Act(\cdot)$ of each node which represents the probability that a node will leave its own community, we search possible hidden links adjacent to these *active individuals* in communities.

2   Game-Theoretic Framework for Mining Hidden Links

2.1   Relative Work in Community Identification

Large networks present attractive common properties, such as power-law degree distributions [9], high network transitivity [10] and the outstanding feature "community structure", meaning that edges between vertices in the same community are dense and sparse between different communities [11]. Identifying community structure in a social network provides people information to understand individual behaviors.

The formation of communities in social networks was put in the game theoretic context by Atheyand *et al.* [12] from economics point of view. In their work, they observed that on deciding which community to participate, individuals will definitely concern about the losses and gains that associate themselves joining a community. Unfortunately, the social network their work considered is not in frame. Later, in [6], Chen *et al.* investigated the dynamic formation of communities from the game-theoretic aspect too. They showed that individuals are intrinsically selfish, and they join or leave communities according to their utilities, where the utility is interpreted as the combination of a gain function and a loss function. The gain function is based on the widely used concept " modularity" that was proposed by Newman in [5]. The loss function reflects the intrinsic cost relating to individuals' choices. They demonstrated that each individual in a community structure finally reaches Nash Equilibrium. Before Chen, existing research addresses detecting social communities mainly based on network topological structures, and many algorithms set a global optimization goal like modularity [5], betweenness [13], or conductance [14].

2.2   Community Formation Game

In our problem, given the community structure, we track back to the community formation process. This is helpful for us to explore the potential existing links. Each individual makes their decisions to join a community according to their k-hop community loyalty to a community. An individual will join a community to which it has the highest k-hop community loyalty. At the very beginning, each individual represents a community. Later, some individuals randomly get together to form larger communities, then, the remaining individuals join these communities according to Nash Equilibrium condition. When the community structure in a social network is established, each individual has more in-community neighbors than the neighbors in any of the other neighbors.

Nash Equilibrium. Since Nash Equilibrium is the key factor that we use to locate the existing but not visible links, here we introduce its formal definition in our problem and several relative terms about it. Finally, we prove the $\mathcal{NP}$-hardness of finding Nash Equilibrium for the community structure.

Definition 1. *Social Network: A social network is denoted as $G = (V, E, C)$, where each node $v_i \in V$ represents an individual in the network, and an edge $(v_i, v_j) \in E$ means that individual v_i has relation with v_j, $C = \{C_1, C_2, ..., C_m\}$ is a set of disjoint communities that forms this network, that is $C_i \cap C_j = \emptyset$, where $i \neq j$.*

Definition 2. *k-Hop Community Loyalty (k-HCL): Given a community $C_j \in C$ and a positive parameter integer k, for each individual $u_i \in V(C_j)$, where $V(C_j)$ is the member of C_j, the k-hop Community Loyalty of u_i to C_j is denoted as $CL_k(i, C_j)$:*

$$CL_k(i, C_j) = \alpha_1 w_1 + \cdots + \alpha_p w_p + \cdots + \alpha_k w_k \qquad (1)$$

where

$$\alpha_p = \frac{|N_{i,j,p}^{in}|}{|N_{i,j,p}^{in}| + |N_{i,j,p}^{out}|} \qquad (2)$$

$N_{i,j,p}^{in}$ and $N_{i,j,p}^{out}$ represent the set of p-hop neighbors of u_i inside community C_j and the set of p-hop neighbors of u_i outside community C_j, respectively, w_p is a coefficient that indicates the importance of p-hop neighbors of u_i, and it decreases exponentially [16] with the increase of p , actually it is the probability of edge, such as in [17], $w_1 = 1\%$, $w_2 = 1\%^2$, $w_p = 1\%^p$. And the parameter k is predetermined.

From equation (2), we conclude that the community loyalty of each individual has close relation with its neighbors, implying that the more closer in-community neighbors one vertex has, the more loyal it is to its own community.

Based on the factor of k-HCL, we give the concept of Nash Equilibrium as follows:

Definition 3. *Nash Equilibrium: Given a set of m disjoint communities $C_1, ..., C_m$ which constitute a social network, the communities are said to be in Nash Equilibrium if $\forall$ community C_t and $\forall i \in C_t$,*

$$\forall 1 \leq j \neq t \leq m, CL_k(i, C_j) \leq CL_k(i, C_t),$$

where $CL_k(i, C_j)$ is the k-HCL value of individual i to community C_j.

Seeing from the above definition, obviously, Nash Equilibrium for the community structure in a social network is such a status: each individual is satisfied with their current community and no one has the tendency (desire) to leave it and join other communities. In other words, the community structure is stable. When the inequality above holds strongly (with "$<$" instead of "$\leq$") for all individuals, then Nash Equilibrium is regarded as strong Nash Equilibrium. Otherwise, if there is an individual i, $CL_k(i, C_t)$ equals to $CL_k(i, C_j)$ for some t and j, then this Nash Equilibrium is classified as weak Nash Equilibrium.

2.3 Hardness of Forming Communities Satisfying Strong Nash Equilibrium

Generally, Nash Equilibrium of each node is considered mainly from its k-hop neighborhood information including neighbors inside and outside its own community. Although achieving Nash Equilibrium only needs simple requirements, it is still hard to partition a graph into two subgraphs satisfying with a strong Nash equilibrium. In the following, we prove that it is $\mathcal{NP}$-hard to partition a general graph into two subgraphs satisfying with strong Nash Equilibrium even if only 1-hop neighbors are considered.

Lemma 1. For a set $S = \{s_1, s_2, \cdots, s_m\}$ which is a vertex subset of $V(G)$, if the induced graph of S forms a cycle and only odd vertices $s_1, s_3, \cdots, s_{2i-1}$ are adjacent to $V(G)\backslash S$, then S cannot be partitioned into two communities S_1 and S_2 satisfying with (∗): for each $u \in V(S_1)$, $|N(u)\backslash V(S_2)| > |N(u) \subseteq V(S_2)|$ and $v \in V(S_2)$, $|N(v)\backslash V(S_1)| > |N(v) \subseteq V(S_1)|$, where $N(u)$ denotes the neighbor set of node u.

Based on the above lemma, we obtain the $\mathcal{NP}$-hardness proof as follows.

Theorem 1. For an arbitrary graph $G = (V, E)$, it is $\mathcal{NP}$-hard to partition G into two subgraphs to reach strong Nash Equilibrium with 1-hop community loyalty requirement.

2.4 Models for Link Prediction

In this part, we formally define our problem and introduce several approaches that have been proposed for link prediction problem. Firstly, we give the concept of Hidden Links.

Definition 4. *Hidden Links: In the real-world networks, some relation links between individuals may be lost when gathering or recording information, we call the lost or missing relation links as hidden links.*

Definition 5. *Equilibrium based Hidden Links Mining (EHLM) problem: Given a set of m disjoint communities $C_1, ..., C_m$ which constitute a social network, our aim is to find the active individuals that destroy Nash Equilibrium of the community structure, and mine possible (hidden) links that connect them with other individuals in the same community.*

As for the topic of predicting link problem in social networks, extensive researches have been conducted to infer links from observable data sets like [18], and [19]. All approaches in these works took $score(x, y)$ as a key factor, which is a connection weight assigned to each pair of vertices (x, y) based on input graph G. The higher the score is, the higher probability that x and y are connected. In the following, we introduce several methods that define $score(x, y)$. For simplicity, we denote $N(x)$ as the neighbor set of node x in graph G. A number of measurements proposed are based on the idea that two vertices x and y are more likely to have a hidden link between them if $N(x)$ and $N(y)$ have a large scale overlap.

The most basic approach is to rank node pairs $< u, v >$ according to the length of their shortest path in a network, which is based on the observation obtained by Newman in G_{collab} [8], in which Newman proposed common neighbor as a measurement, only considering the number of neighbors that node x and y have in common, that is, $score(x, y) := |N(x) \cap N(y)|$.Salton $et\ al.$ in [21] introduced the Jaccards coefficient which is a similarity metric often used for information retrieval. It is used to estimate the probability that both of x and y have a randomly selected property p. Actually, when p is viewed as the common neighbors of x and y in G, the measurement becomes

$$score(x, y) := |N(x) \cap N(y)|/|N(x) \cup N(y)|.$$

In [22], Adamic $et\ al.$ considered the measurement

$$score(x, y) := \sum_{p \in N(x) \cap N(y)} 1/\log(N(p))$$

is derived. In [8] and [20], a measure $score(x, y) := |N(x)| \cdot |N(y)|$ is proposed according to the empirical observation, which indicates that the probability of co-authorship between x and y is related to the product of the numbers of authors who work with both x and y. Through refining the definition of shortest-path, Katz in [23] proposed a measure

$$score(x, y) := \sum_{\ell=1}^{\infty} \beta^{\ell} \cdot |paths_{x,y}^{\langle \ell \rangle}|$$

where $paths_{x,y}^{\langle \ell \rangle}$ is the set of all paths with length ℓ from x to y, β is a positive parameter.

3   Methodology

In this section, we propose an algorithm called Two-Stage Link Mining (T-SLM) for the Equilibrium based Hidden Links Mining (EHLM) problem. This algorithm contains two phases. The first phase is to identify *active individuals*, in which we calculate the k-HCL values of each vertex to different communities. Through comparing these k-HCL values that associate with each vertex, we can obtain the community each node belongs to. Then we find the vertices who are not in the communities that they are supposed to be in, and name them as active vertices. The second phase is to search possible links adjacent with those active individuals. For every active individual, we continue to find its adjacent links until it satisfies Nash Equilibrium. Finally, all active individuals are satisfied with their own communities, indicating that all communities are observably stable and the entire community structure accords with Nash equilibrium.

To facilitate the description of our algorithm, in the first place, we define *active individual* based on the definition of k-HCL given in section 2.2.

Definition 6. *Active Individual (AI): Given a set of m disjoint communities $C_1, ..., C_m$ which constitute a social network, an individual u_i in community C_j is said to be active if $\exists\, t$, such that $CL_k(i, C_t) > CL_k(i, C_j)$, where $CL_k(i, C_j)$ is the k-HCL value of individual u_i to community C_j.*

For an individual u_i belonging to community C_j, if $\exists\, p$ such that $CL_k(i, C_j) < CL_k(i, C_p)$, then u_i becomes an active individual.

3.1   Select Active Individual Set

In this part, we determine active individuals appeared in each community based on k-HCL value and Nash Equilibrium of community structure defined in section 2.2. All the obtained active individuals from a set, which we call Active Individual Set (AIS). To simplify the expression of our algorithm, Some notions are introduced first. Assume that a social network is partitioned into m communities, namely $C_1, C_2, \cdots, C_m$. The diameter $D_{i,j}$ of node i in community C_j is the maximum number of hops between i and any vertex belongs to C_j. Therefore, we have the diameter d_j of community C_j as $d_j = \max\{D_{i,j}, i \in C_j\}$. k is computed by $k = \max\{d_i, i = 1, 2, \cdots, m\}$. The concrete description of this location procedure is as follows.

Algorithm 1. EHLM Algorithm-Locate Active Individual Set (SAIS)

INPUT: A social network $G = (V, E, C)$, m communities $C_1, \cdots, C_m, w_1, \cdots, w_k$, where k is predetermined and $S = \phi$.
OUTPUT: Active Individual Set $S \subseteq V(G)$.
For each node $i \in C_j$, construct the $D_{i,j}$-*level Breadth First Search Tree* (DBFST) rooted at i, include all the nodes within $D_{i,j}$-hop of i to the DBFST. Put the corresponding vertex set to $N^{in}_{i,j,r}$ and $N^{out}_{i,j,r}$, respectively, where $r = 1, 2, \cdots, D_{i,j}$. (Notice that $k \geq D_{i,j}$, we set $N^{in}_{i,j,m'} = \phi$ and $N^{out}_{i,j,r'} = \phi$ for $r' = D_{i,j} + 1, \cdots, k$.)
For $1 \leq j \leq m$
  For $i \in C_j$
    According to Formulation (1) and (2), compute $CL_k(i, C_j)$ and $CL_k(i, C_t)$ for $\forall i \in V(G)$ and $\forall t, 1 \leq t \neq j \leq m$, compute $Act(i) = \max_{t \neq j}\{CL_k(i, C_t) - CL_k(i, C_j)\}$ and record it;
  If $Act(i) \neq 0$
    $S = i \cup S$;
  End
End
Return S.

3.2   Search Hidden Links

From Algorithm 3.1, we obtain all the active individuals in communities in G. In this phase, we explore possible but invisible links that are adjacent to these active individuals. Both of the two ends of these links lie in the same community. Under the condition that we have known the active vertices, to find their matching pairs corresponding those links, we compute the number of common neighbors and the total k-HCL values of a pair of nodes. Then we match the active nodes in the same community, if there is not active nodes, we will use the common neighbors method to find the hidden links. And we will update the loyalty after adding a link. In the following, we provide the concrete description of the second phase implementation.

Algorithm 2. EHLM Algorithm-Search Hidden Links

INPUT: The active individual set S, the total number of added links AL, $HL = \phi$, $L = 0$.

OUTPUT: The set of hidden links HL.

if $L \leq AL$

    if $s = \phi$

        Calculate $argmax_{\{u_i, u_k\}}(Act(u_i) + Act(u_k))$, where u_i and u_k belong to the same community, denote the corresponding nodes pair as (u_m, u_n);

    else

        Calculate $argmax_{\{u_i, u_k\}}|N(u_i) \cap N(u_k)|$, where u_i and u_k belong to the the whole graph, denote the corresponding nodes pair as (u_m, u_n);

$W = W \cup (u_m, u_n)$;

Execute step the inner loop in Algorithm 1 to update S;

Return W.

3.3   Algorithm Analysis

Our EHLM algorithm contains two stages. The first step is to determine the active individuals set: for each node, find all of its k-hop neighbors through using the Breadth First Search (BFS) method, then compute the Community Loyalty (CL) of each node to all the different communities through applying eqnarray (1) and (2). Next, we find the active nodes through measure the value of $Act(\cdot)$ with regard to each node, when the value is larger than zero, we place the corresponding node into Active Individuals Set. After finding these active individuals, we find proper pairs among those active individuals to add links between them when Active Individual Set is not null, otherwise, we will use the common neighbor algorithm to find the hidden links in the whole graph. Every time we determine a link, we will update the active individuals set, since the newly added link improves the community loyalty of some individuals.

4   Simulation

Before we show our results we introduce the data sets we use and our comparison algorithms. Four realistic data sets are used: **NetHEPT**, **Enron Email**,

Arenas and **American College Football**. Before we carry out the experiments, to get communities, we adopt the approaches proposed in [24] and [25]. The communities of NetHEPT and Enron are partitioned by [24], and the communities of Arenas and ACF are partitioned by [25]. Next, we delete the links in Table IV. Here, we set the parameter $k = 1$, suppose the maximum number of common neighbors is N. We denote $N(x)$ as the set of neighbors of node x in social networks. The three comparison algorithms we use to find the hidden links between nodes in communities are: *Random method*, *Common neighbors*, and *Jaccard's coefficient*, all of which is introduced in [15].

Table 1. Results of correct probability on Arenas/Email

Predictor	Jaccard's coefficient(%)	common neighbors(%)	our algorithm(%)
Rank=1	5.1	7.7	8.6
Rank=4	5.7	8.2	9.175
Rank=16	3.3	8.88	10.04
Rank=64	4.29	8.225	9.31

Table 2. Results of correct probability on ACF/Team

Predictor	Jaccard's coefficient(%)	common neighbors(%)	our algorithm(%)
Rank=1	78.2	54.2	80.1
Rank=4	77.12	55.275	80.375
Rank=8	75.1	55.8	77.825
Rank=16	73.36	55.83	75.62

Table 3. Results of correct probability on NetHEPT/Author

Predictor	Jaccard's coefficient(%)	common neighbors(%)	our algorithm(%)
Rank=1	24.01	14.23	24.8
Rank=4	22.72	14.75	29.86
Rank=16	21.125	14.81	24.78
Rank=64	21.59	15.39	20.68

4.1 Experimental Results

To compare with other algorithms, we concerned about the probabilities of correct predictions (i.e precisions). We randomly delete a number of links in each graph, and conduct our algorithm and the above three methods on the incomplete graphs. For each comparison, we run each algorithm 1000 times to get the average. We fix the $Rank = 1, 4, 16, 64$.

Table 4. Results of correct probability on Enron/Email

Predictor	Jaccard's coefficient(%)	common neighbors(%)	our algorithm(%)
Rank=1	7.15	21.463	31.002
Rank=4	20.468	49.351	58.125
Rank=16	18.732	32.485	42.163
Rank=64	13.375	13.281	19.952

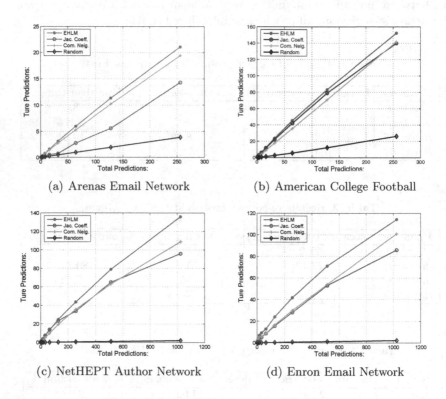

(a) Arenas Email Network　　　　(b) American College Football

(c) NetHEPT Author Network　　　　(d) Enron Email Network

Fig. 1. True Predictions vs. Total Predictions

From Table 1, we can find that the probabilities of correct predictions for EHLM and Common Neighbors (*Com. Neig.*) decrease as the number of total predictions is increasing from 1 to 64. This agrees with our intuition in that top ranked pairs have higher probabilities to be hidden links. In addition, our algorithm always outperforms *Com. Neig.* and Jaccard's Coefficient (*Jac. Coeff.*). When comparing the first 16 predictions, our algorithm is about 15 percent better than *Com. Neig.* and it outperforms *Jac. Coeff.* even much better. When comparing 64 predictions, the gap between our algorithm and *Jac. Coeff.* is smaller. From Table 2, 3 and 4, we can see the results on other three data sets. Note that the precisions not only depend on the total number of predictions but also depend on the number of hidden links. The more the number of hidden links, the higher the probabilities of correct predictions. In Table 2 (i.e. the results on ACF/Author network), about 40 percent of links are hidden, therefore, the precisions of all the algorithms are relatively high; while in Table 1 (i.e. the results on Arenas/Email netowrk), only 10 percent of links are hidden, and the precisions of all three algorithms are relatively low. Fig.1 plots the experimental results of True Predictions vs. Total Predictions. It can be seen that with the increase of total predictions, the number of true predictions increases. In all the test cases, our algorithm works the best.

5   Conclusion

In this paper, we address the problem of mining hidden links between individuals within communities. Instead of directly predicting these links, we consider this problem from viewpoint of community formation process, and model it as a game-theoretic framework. Each individual is assumed to be selfish, and it chooses to join a community with the maximal loyalty value. Generally, the community structure in a social networks satisfies Nash Equilibrium, and we search for the active individuals and mining the existing links hidden. Our work is the first attempt to predict links from the aspect of the community formation. To test the efficiency of our algorithm, we conducted extensive experiments over large-scale real world social networks and compare our algorithm with state-of-art algorithms and our algorithm outperforms them.

Acknowledgment. This work was supported in part by National Natural Science Foundation of China under grants 61070191 and 91124001, and Research Fund for the Doctoral Program of Higher Education of China under grant 20100004110001. It was also supported by the by National Science Foundation of USA under grants CNS-1016320, CCF-082999.

References

1. Kotler, P., Zaltman, G.: Social Marketing: An Approach to Planned Social Change. The Journal of Marketing 35(3), 3–12 (1971)
2. McKenzie Mohr, D., Smith, W.: Fostering Sustainable Behavior: An Introduction to Community Based Social Marketing. New Society Publishers (1999)
3. Kasarda, J.D., Janowitz, M.: Community Attachment in Mass Society. American Sociological Review 39(3), 328–339 (1974)
4. Sampson, R.J., Groves, W.B.: Community Structure and Crime: Testing Social Disorganization Theory. American Journal of Sociology 94(4), 774 (1989)
5. Newman, M.E.J.: Modularity and community structure in networks. Proceedings of the National Academy of Sciences 103(23), 8577–8582 (2006)
6. Chen, W., Liu, Z., Sun, X., Wang, Y.: A game-theoretic framework to identify overlapping communities in social networks. Data Min. Knowl. Discov. 21(2), 224–240 (2010)
7. Yan, B., Gregory, S.: Finding missing edges in networks based on their community structure. Physical Review E 85, 056112 (2012)

8. Newman, M.E.J.: Clustering and preferential attachment in growing networks. Physical Review Letters E 64(025102) (2001)
9. Barabási, A.L., Albert, R.: Emergence of scaling in random networks. Science 286, 509–512 (1999)
10. Watts, D.J., Strogatz, S.H.: Collective dynamics of 'small-world' networks. Nature 393, 440–442 (1998)
11. Girvan, M., Newman, M.E.J.: Community structure in social and biological networks. Proc. Natl. Acad. Sci. USA 99, 7821–7826 (2002)
12. Atheyand, S., Jha, S.: A theory of community formation and social hierarchy, working paper (2006)
13. Gregory, S.: A fast algorithm to find overlapping communities in networks. In: Daelemans, W., Goethals, B., Morik, K. (eds.) ECML PKDD 2008, Part I. LNCS (LNAI), vol. 5211, pp. 408–423. Springer, Heidelberg (2008)
14. Brandes, U., Erlebach, T.: Network Analysis: methodological foundations. Springer (2005)
15. Liben-Nowell, D., Kleinberg, J.: The link-prediction problem for social networks. J. American Society for Information Science and Technology 58(7), 1019–1031 (2007)
16. Lancichinetti, A., Kivel, M., Saramki, J., Fortunato, S.: Characterizing the Community Structure of Complex Networks. PLoS One 5(8), e11976 (2010)
17. Kempe, D., Kleinberg, J.M., Tardos, E.: Maximizing the spread of influence through a social network. In: The 9th ACM SIGKDD Conference on Knowledge Discovery and Data Mining, pp. 137–146 (2003)
18. Goldberg, D.S., Roth, F.P.: Assessing experimentally derived interactions in a small world. Proceedings of the National Academy of Sciences 100(8), 4372–4376 (2003)
19. Popescul, A., Ungar, L.: Statistical relational learning for link prediction. In: Workshop on Learning Statistical Models From Relational Data at the International Joint Conference on Artificial Intelligence, pp. 81–90. ACM Press, New York (2003)
20. Barabási, A.L., Jeong, H., Néda, Z., Ravasz, E., Schubert, A., Vicsek, T.: Evolution of the social network of scientific collaboration. Physica A 311(34), 590–614 (2002)
21. Salton, G., McGill, M.J.: Introduction to modern information retrieval. McGraw-Hill, New York (1983)
22. Adamic, L.A., Adar, E.: Friends and neighbors on the Web. Social Networks 25(3), 211–230 (2003)
23. Katz, L.: A new status index derived from sociometric analysis. Psychometrika 18(1), 39–43 (1953)
24. Blondel, V.D., Guillaume, J., Lambiotte, R., Lefebvre, E.: Fast unfolding of communities in large networks. J. Stat. Mech.: Theory and Experiment (2008)
25. Hu, Y., Chen, H., Zhang, P., Li, M., Di, Z., Fan, Y.: Comparative definition of community and corresponding identifying algorithm. Phys. Rev. E. 78(2), 1–7 (2008)

An Improved Exact Algorithm for Undirected Feedback Vertex Set[*]

Mingyu Xiao[1] and Hiroshi Nagamochi[2]

[1] School of Computer Science and Engineering,
University of Electronic Science and Technology of China, China
myxiao@gmail.com
[2] Department of Applied Mathematics and Physics,
Graduate School of Informatics, Kyoto University, Japan
nag@amp.i.kyoto-u.ac.jp

Abstract. A feedback vertex set in an undirected graph is a subset of vertices removal of which leaves a graph with no cycles. Razgon (SWAT 2006) gave a $1.8899^n n^{O(1)}$-time algorithm for finding a minimum feedback vertex set in an n-vertex undirected graph, which is the first exact algorithm for the problem that breaks the trivial barrier of 2^n. Later, Fomin *et al.* (Algorithmica 2008) improved the result to $1.7548^n n^{O(1)}$. In this paper, we further improve the result to $1.7356^n n^{O(1)}$. Our algorithm is analyzed by using the measure-and-conquer method. After showing some properties of the problem, we get improvements by introducing a new measure scheme on the structure of reduced graphs.

1 Introduction

There are many practical problems that require us to find a minimum number of vertices or edges in a graph that intersect all cycles in the graph. An example is the *deadlock recovery* problem in concurrent programs [15], which has applications in operating system, computer architecture communities, database system and so on. In this problem, the system is represented by a resource allocation graph (RAG) with vertices being processed and a deadlock in the system is represented by a cycle in RAG. We need to abort a set of processes in the system to recover it from deadlocks. The set of processes is corresponding to a minimum set of vertices in RAG intersecting all cycles. Another well-known application is the *rank aggregation* problem [14,7], in which we are asked to find out a ranking minimizing the number of pairs that occur in a different order in the outcome ranking, i.e., to delete a minimum number of arcs in a directed graph so that all cycles in the graph are broken. Some other applications are also mentioned in the literature [9,5,10].

The above kinds of problems of destroying cycles in graphs are known as *feedback set* problems. They have been systematically studied in both practice and theory due to their importance. According to the graph being directed or

[*] Supported by NFSC of China under the Grant 61370071 and Fundamental Research Funds for the Central Universities under the Grant ZYGX2012J069.

P. Widmayer, Y. Xu, and B. Zhu (Eds.): COCOA 2013, LNCS 8287, pp. 153–164, 2013.
© Springer International Publishing Switzerland 2013

undirected, and the elements to be deleted being vertices or edges, we can define four different versions of the problem: Directed Feedback Vertex Set (DFVS), Undirected Feedback Vertex Set (UFVS), Directed Feedback Arc Set (DFAS) and Undirected Feedback Edge Set (UFES). UFES is polynomial-time solvable since it is equal to the problem of finding a maximum spanning tree in a graph, while the other three problems become NP-hard, where UFVS and DFAS are also included in the list of Karp's 21 NP-complete problems [13].

UFVS admits a polynomial-time approximation algorithm with ratio 2 [1], whereas no constant-ratio approximation algorithm has been found for the problems in directed graphs [8]. These problems have also been extensively studied in parameterized and exact algorithms. For parameterized problems with parameter k being the size of the solution, there is a long list of contributions to k-UFVS [12,6,4,3], and it can be solved in $O(3.83^k k n^2)$ time now [3]. Whether k-DFVS admits fixed-parameterized tractable algorithms or not had been an open problem for a long time and finally it was solved affirmatively by Chen $et\ al.$ [5]. Exact algorithms faster than the trivial $2^n n^{O(1)}$ time for both UFVS and DFVS were developed recently. Based on a good choice of the measure to problem instances, a $1.8899^n n^{O(1)}$-time algorithm for UFVS was designed by Razgon [16], and later the running time was improved to $1.7548^n n^{O(1)}$ by Fomin $et\ al.$ [10]. Razgon [17] also broke the barrier 2^n for DFVS by giving a $1.9977^n n^{O(1)}$-time algorithm.

In this paper, we will design a $1.7356^n n^{O(1)}$-time exact algorithm for UFVS, which improves the previous results. Our algorithm also uses some good branching rules introduced in [16] and [10]. But our algorithm adopts a different measure scheme from the previous algorithms. Under this measure scheme, we can get the improvements by showing the worst case of the algorithm in [10] will not always happen. To effectively use the measure scheme, we also analyze some new structural properties of the problem.

2   Preliminaries

In this paper, a graph stands for an undirected graph with possible multiple edges and self-loops. A *simple graph* is a graph without any multiple edges and self-loops. Let $G = (V, E)$ be a graph and $X \subseteq V$ be a subset of vertices. The subgraph induced by X is denoted by $G[X]$, and $G[V \setminus X]$ is also written as $G \setminus X$. Contracting X into a single vertex v^* means to remove X from G after changing the end-vertex $u \in X$ of each edge $uv \in E$ to the new vertex v^*, where we keep any parallel edges between v^* and $v \in V \setminus X$ while we remove all resulting loops incident to v^*. A vertex v in a connected graph is called a *cut-vertex*, if removing it leaves more than one connected component. For a vertex $v \in V$, the set of the neighbors of v is denoted by $N(v)$, and the degree $d(v)$ of v is defined to be the number of edges incident to v, where $d(v) \geq |N(v)|$. In a simple graph, $d(v) = |N(v)|$. After some preprocessing, our algorithm will keep the graph as a simple graph.

A subset $X \subseteq V$ of vertices is called a *feedback vertex set* of a graph G if $G \setminus X$ is a forest. Note that X is a feedback vertex set of G if and only if $V \setminus X$

induces a forest. The problem of finding a minimum feedback vertex set (FVS) is equivalent to the problem of finding a maximum induced forest (MIF). For the purpose of description, we will design an algorithm in terms of MIF. Given a subset $F \subseteq V$ of vertices which induces a forest $G[F]$, a maximum induced forest containing F is called an F-MIF. The problem of finding an F-MIF is called the *forced MIF*. We will use (G, F) to denote an instance of the forced MIF.

3   Reducing the Instance

We review some cases where a given instance (G, F) can be simplified. First, if there is a vertex with a self-loop, then this vertex should be deleted from the graph. Second, if there is a vertex not contained in any cycle in G (including degree-1 vertices), we can add it to F directly without changing the optimality of the instance. Third, if there are multiple edges between a vertex $u \in V \setminus F$ and a vertex $v \in F$, then u needs to be excluded from any F-MIF. Fourth, if the maximum degree of G is 2, then we can solve the instance easily by removing one vertex in $V \setminus F$ from each cycle in G and returning the vertex set of the resulting graph as a solution. The next lemma provides the fifth reduction.

Proposition 1. [17] *Given an instance (G, F), let G' be the graph obtained from G by contracting each connected component T of $G[F]$ into a single vertex v_T, and F' be the corresponding set obtained from F by replacing $V[T]$ of each connected component T of $G[F]$ with the new vertex v_T. Then for a subset $X \subseteq V \setminus F$, a set $F \cup X$ is an F-MIF in (G, F) if and only if $F' \cup X$ is an F'-MIF in (G', F').*

When the set of vertices in F in the current instance (G, F) is not an independent set in the graph G, we apply the procedure in the proposition. We denote by Rd the algorithm that consists of the above five reductions. The algorithm $\mathrm{Rd}(G, F)$ will return an instance (G', F') to which none of the above five reductions is applicable anymore. When there are parallel edges between two vertices not in F, we will show that simply branching on one of the two vertices by including it to F or not is good enough for our analysis. Then we can deal with all parallel edges. We will call an instance a *reduced instance*, if the graph has no parallel edge and none of the above five reductions can be applied.

In this paper, we use the following new lemma to deal with degree-2 vertices in a reduced instance.

Lemma 1. *Let (G, F) be a reduced instance and $v \in V \setminus F$ be a vertex of degree 2. Then there is an F-MIF S containing v.*

Proof. Since a reduced instance has no parallel edges, the two neighbors of v are different. Let $N(v) = \{u_1, u_2\}$. To derive a contradiction, we assume that no F-MIF of (G, F) contains v. Let S be an F-MIF of (G, F). Since S is maximal subject to acyclicity of induced graph $G[S]$, $S \cup \{v\}$ induces a graph $G[S \cup \{v\}]$ with a cycle C passing through $\{u_1, v, u_2\}$. First consider the case where $G[S \cup \{v\}]$ contains exactly one such cycle C. Since F is an independent set, C contains at least one vertex $v' \in V(C) \setminus (F \cup \{v\})$, for which $S' = (S \setminus \{v'\}) \cup \{v\}$ still induces an acyclic graph, and hence S' is another F-MIF of (G, F), contradicting the assumption. Next consider the case where $G[S \cup \{v\}]$ contains at least two distinct cycles C_1 and C_2. In this case, $C_1 - \{v\}$ and $C_2 - \{v\}$ are paths between u_1 and u_2. Since $C_1 - \{v\} \neq C_2 - \{v\}$, there is a cycle in the union of $C_1 - \{v\}$ and $C_2 - \{v\}$, contradicting that $G[S]$ is acyclic. This proves the lemma. ∎

4   A Divide-and-Conquer Algorithm Based on Cut-Vertices

We also use the following new lemma, which leads to a divide-and-conquer method to eliminate cut-vertices in any instances (a proof of this lemma can be found in the full version of this paper).

Lemma 2. *Let (G, F) be a reduced instance of forced MIF with a cut-vertex v, and let H be a connected component in $G \setminus \{v\}$. Let $F_1 = F \cap (V(H) \cup \{v\})$, $G_1 = G[V(H) \cup \{v\}]$, $F_2 = F \setminus V(H)$, $G_2 = G \setminus V(H)$, and $F_i^* = F_i \cup \{v\}$, $i = 1, 2$.*

(i) *Assume that $v \in F$. Then an F-MIF of (G, F) is obtained by the union of any F_1-MIF of $I_1 = (G_1, F_1)$ and F_2-MIF of $I_2 = (G_2, F_2)$.*

(ii) *Assume that $v \notin F$ and $|S_1^*| > |S_1|$ for an F_1^*-MIF S_1^* of $I_1^* = (G_1, F_1^*)$ and an F_1'-MIF S_1 of $I_H = (H, F_1' = F \cap V(H))$. Then an F-MIF of (G, F) is obtained by the union of $S_1^* \setminus \{v\}$ and S_2 for any F_1^*-MIF S_1^* of $I_1^* = (G_1, F_1^*)$ and F_2-MIF S_2 of $I_2 = (G_2, F_2)$.*

(iii) *Assume that $v \notin F$ and $|S_1^*| \leq |S_1|$ for an F_1^*-MIF S_1^* of $I_1^* = (G_1, F_1^*)$ and an F_1'-MIF S_1 of $I_H = (H, F_1' = F \cap V(H))$. Then an F-MIF of (G, F) is obtained by the union any F_1-MIF of $I_H = (H, F_1)$ and F_2'-MIF of $I_H' = (G \setminus (V(H) \cup \{v\}), F_2' = F \setminus (V(H) \cup \{v\}))$.*

If a cut-vertex v is in F, we need to compute I_1 and I_2 according to Lemma 2. On the other hand, if v is not in F, we first compute I_1^* and I_H and then compute either I_2 or I_H' according to the solutions to I_1^* and I_H. After applying this step, we can assume that the graph G is biconnected.

5   The Idea of the Branching Operations

A simple idea to design algorithms for forced MIF is that: we pick up a vertex $t \in F$ and select a neighbor v ($\in V \setminus F$) of it, and then branch on v by either including it to F or deleting it from the graph. Deleting v in the second branch decreases the number of vertices only by one. We look at the first branch. When v is included to F, we delete all vertices in $N(v) \cap N(t)$ from the graph and contract v and t into a new vertex, say t. When $N(v) \cap N(t) = \emptyset$, the number of vertices decreases again only by one in the first branch. It leads to only running time of $2^n n^{O(1)}$. However, in the first branch we observe more information: the vertices in $N(v) \setminus N(t)$ will become a neighbor of t in the new graph. If all the vertices in $V \setminus T$ become neighbors of t, then the problem may become an easier problem. This implies that the vertices in $N(t)$ may be easier to be handled than other vertices in $V \setminus F$. We set a weight to each vertex in our graph to distinguish their contribution to the computational complexity of the problem. Define the weight of each vertex in F to be 0. We select a vertex $t \in F$ as a designated vertex in our algorithm. Then define the weight of each vertex in $V \setminus (F \cup N(t))$ to be 1, and set the weight of each vertex in $N(t)$ to be a value $\alpha \in [0.5, 1.0]$, where the best value of α will be determined according to the analysis of our algorithm described later. We use the sum w of all the vertex weight as the measure of the instance, where $w \leq n$ holds. Suppose that we choose a neighbor v of t for branching, and let d be the number of neighbors of v except t (i.e., $d = |N(v)| - 1$). Here we assume the worst case that $N(v) \cap N(t) = \emptyset$. By using $C(w) = \tau^w$ ($\tau > 1$) to denote the worst size of the search tree generated from an instance with measure w by our algorithm, we get the following recurrence for the above branching operation

$$C(w) \leq C(w - (\alpha + d(1 - \alpha))) + C(w - \alpha), \tag{1}$$

which will be the bottleneck of the algorithm. To get a smaller upper bound on the size of the search tree of the algorithm, we hope that $d(1 - \alpha)$ is as large as possible.

When $d = 1$, we can simply include v to F without branching, because if any F-MIF S not containing v always includes the other neighbor v' ($\neq t$) of v by the maximality and can be modified into another F-MIF $S' = (S \setminus \{v'\}) \cup \{v\}$.

Razgon [16] gave a way of dealing with the case of $d = 2$ effectively so that the worst recurrence becomes

$$C(w) \leq C(w - (\alpha + 3(1 - \alpha))) + C(w - \alpha).$$

We can verify that $C(w) = O(1.8899^w)$ by choosing $\alpha = 0.6370$. That is how Razgon [16] obtained the first exact algorithm that breaks the barrier of 2^n for MIF.

Fomin et al. [10] further analyzed the problem and found a way to deal with the case of $d = 3$, and then improved the worst case recurrence to

$$C(w) \leq C(w - (\alpha + 4(1 - \alpha))) + C(w - \alpha), \tag{2}$$

which solves to $C(w) = O(1.7548^w)$ with $\alpha = 0.5116$. This leads to the current best time bound for exact algorithms for FVS (MIF).

Previous techniques to deal with the cases of $d = 2$ and $d = 3$ come from the following lemma.

Lemma 3. [10] *Let (G, F) be a reduced instance. For a vertex $t \in F$, let $v \in V \setminus F$ be a neighbor of t such that $|N(v)| \geq 3$. If $N(v) \cap F = \{t\}$, then there is an F-MIF S satisfying one of the following properties: For $D(v) = N(v) \setminus \{t\} = \{v_1, v_2, \ldots, v_d\}$*

1. $v \in S$;
2. $(D(v) \cup \{v\}) \cap S = \{v_1, v_2\}$; and
3. $\{v, v_i, v_{i+1}, \ldots, v_d\} \cap S = \{v_i\}$ for some $i \in \{3, \ldots, d\}$, where $|D(v)| \geq 3$.

Based on the lemma, we obtain the following branching rule. When $d = 2$ ($|N(v)| = 3$), we can branch into two instances at v by either including v to F or by including v_1 and v_2 to F and deleting v. When $d = 3$ ($|N(v)| = 4$), we can branch into three instances at v by including v to F, by including v_1 and v_2 to F and deleting v, or by including v_3 to F and deleting v. These two branching rules are more effective than simply branching on v by including it to F or not. Therefore, we get effective ways to deal with the cases of $d = 2$ and $d = 3$. However, Lemma 3 only handles the case where $N(v) \cap F = \{t\}$. When v is also adjacent to other vertices in F, the the analysis will become complicated and some bad cases arise. To extend Lemma 3, Fomin et al. [10] introduced two concepts "active vertex" and "generalized neighbor."

In their algorithm, at most a vertex $t \in F$ is designated as an *active vertex*. When we contract a set of vertices into a single vertex in $\mathrm{Rd}(G, F)$, the new vertex is called an active vertex and denoted by t if the set includes the active vertex t. The algorithm always choose a neighbor of the active vertex t to branch on based on the "generalized degree." For a vertex $v \in V \setminus F$, a vertex $u \in V \setminus F$ is called a *generalized neighbor* of v if u is a neighbor of v or u and v share a common neighbor s in $F \setminus \{t\}$. Denote by $GD(v)$ the set of generalized neighbors of v and $gd(v) = |GD(v)|$ the *generalized degree* of v. Hence $GD(v) = (N(v) \setminus F) \cup \{u \in N(s) \setminus F \mid s \in N(v) \cap (F \setminus \{t\})\}$. For generalized neighbors, the next property holds, which nearly corresponds to Lemma 3.

Lemma 4. *Let (G, F) be a reduced instance with a designated vertex $t \in F$. For a vertex $v \in V \setminus F$, denote $GD(v) = \{v_1, v_2, \ldots, v_d\}$ ($d = gd(v)$). If $gd(v) \leq 1$, then there is an F-MIF containing v. Otherwise ($gd(v) \geq 2$) there is an F-MIF S satisfying one of the following properties:*

1. $v \in S$;
2. $(GD(v) \cup \{v\}) \cap S = \{v_1, v_2\}$; and
3. $\{v, v_i, v_{i+1}, \ldots, v_d\} \cap S = \{v_i\}$ for some $i \in \{3, \ldots, d\}$, where $gd(v) \geq 3$.

Proof. Let S be an F-MIF of (G, F), where we assume that $v \notin S$ (otherwise we are done). By the maximality of S, we see that $G[S' \cup \{v\}]$ contains a cycle C passing through v, and $V(C) \setminus \{t\}$ contains at least one vertex u in $GD(v) = (N(v) \setminus F) \cup \{u \in N(s) \setminus F \mid s \in N(v) \cap (F \setminus \{t\})\}$. If S' contains no other vertex from $GD(v) \setminus \{u\}$, then $S' = (S \setminus \{u\}) \cup \{v\}$ is an F-MIF containing v. Assume that S contains at least two vertices from $GD(v)$. Hence $gd(v) \geq 2$ (which means that there is an F-MIF S' containing v when $gd(v) \leq 1$). Let i (≥ 2) be the largest index such that $v_i \in S$, where $i \geq 2$. If $i = 2$, then we have the second condition. Otherwise for $i \geq 3$ we have the third condition. ∎

Note that in this lemma, we do not require the vertex v to be a neighbor of the active vertex t. But our algorithm only branches on a vertex adjacent to the active vertex based on the lemma.

6   The Algorithm

Our algorithm for forced MIF is described in Fig. 1. In Step 1, the algorithm deals with some parallel edges between two vertices $u, v \in V \setminus F$. We can branch by either including u to F or deleting it from the graph. In the first branch we can also remove v directly since at least one of u and v should be deleted to destroy all cycles containing them. Note that when u is of degree-2, we can simply remove v without branching, because now u is adjacent to only v and v intersects all cycles containing u. After this step, the algorithm will not create parallel edges between two vertices in $v \setminus F$ any more. Step 2 calls $\mathrm{Rd}(G, F)$ to reduce the instance. Lemma 1 can guarantee the correctness of Step 3. Steps 3 and 4 are used to deal with the graph that is not biconnected. After Step 4, the instance satisfies the following properties: the instance is a reduced instance; all vertices in $V \setminus F$ are of degree ≥ 3; and the graph G is biconnected. Based on these properties, we will design the major reduction operations in the algorithm.

A vertex $v \in N(t)$ with $gd(v) = 3$ is called a *good vertex*, if (i) $GD(v) \setminus N(t)$ is not empty; or (ii) when $N(t)$ contains no good vertex of Case (i), any vertex $v \in N(t)$ with $gd(v) = 3$ will do. For any vertex $v \in V \setminus F$, let $F_v = N(v) \cap (F \setminus \{t\})$. A vertex $v \in N(t)$ with $gd(v) \geq 4$ is called an *effective vertex* if (i) it is in a path vav' where $a \in F \setminus \{t\}$ and $v' \in V \setminus (F \cup N(t))$; or (ii) when $N(t)$ contains no effective vertex of Case (i), any vertex $v \in N(t)$ that maximizes $dg(v)$ and then maximizes $|F_v|$ will do. By using Lemma 4, we get Steps 8, 9 and 10 in Fig. 1.

7   The Analysis

We will use the measure-and-conquer method [11] to analyze the algorithm. In this method, we set a weight to each vertex in the graph and use the sum w of the total vertex weight as the measure to scale the size of the instance. In fact, we will set the vertex weight of each vertex at most 1. Then w is at most the number n of vertices and a running time bound with respect to w will imply a running time bound with respect to n.

Fomin *et al.* used the measure-and-conquer method in the analysis of their algorithm [10]. In their analysis, they only considered three kinds of different vertex weight: setting the weight of each vertex in F as 0, the weight of each vertex in $N(t) \cap (V \setminus F)$ as $\alpha = 0.5116$, and the weight of each remaining vertex in $V \setminus (F \cup N(t))$ as 1, where t is the active vertex in the graph. By listing out all recurrences as (1) generated by the algorithm, Fomin *et al.* proved that the algorithm runs in $1.7548^n n^{O(1)}$ time and the worst case is (2).

One way to further decrease the time bound of Fomin *et al.*'s algorithm is to improve the worst case of (2) by using the branching rule from Lemma 4 for $d = 4$. However, this new branching rule will create four instances and it will generate a recurrence even worse than (2). In this paper, we try to improve the result of MIF in a different way. We will use a different measure scheme to improve the worst case analysis.

In most measure-and-conquer algorithms for graph problems, such as the algorithms in [2,11,18], we set different vertex weights for vertices of different

Input: A graph $G = (V, E)$ and a subset of vertices $F \subseteq V$. Initially $F = \emptyset$.
Output: An F-MIF S of G.

1. **If** {There are parallel edges between two vertices $u, v \in V \setminus F$}, if one of u and v, say u is of degree 2, then return $S := \text{mif}(G \setminus \{u, v\}, F) \cup \{u\}$, else return

$$S := \max\{ \text{mif}(G \setminus \{u\}, F \cup \{v\}), \text{mif}(G \setminus \{v\}, F) \}.$$

2. **Elseif** {(G, F) is not a reduced instance}, let $(G', F') := \text{Rd}(G, F)$ and return $S := \text{mif}(G', F')$.

3. **Elseif** {There is a degree-2 vertex v in $V \setminus F$}, select such a vertex v, and return $S := \text{mif}(G, F \cup \{v\})$.

4. **Elseif** {G consists of $k \geq 2$ connected components $G_i = (V_i, E_i)$ ($i = 1, 2, \ldots, k$)}, return

$$S := \bigcup_{i=1}^{k} \text{mif}(G_i, F_i = V_i \cap F).$$

5. **Elseif** {G is not biconnected}, use Lemma 2 to decompose (G, F) into subinstances, where we select H as the component of minimum number of vertices.

6. **Elseif** {$F = \emptyset$}, select a vertex v of maximum degree and return

$$S := \max\{ \text{mif}(G, \{v\}), \text{mif}(G \setminus \{v\}, \emptyset) \},$$

where v is designated as the active vertex t in the instance $(G, \{v\})$.

7. **Elseif** {F contains no active vertex}, choose an arbitrary vertex $t \in F$ as an active vertex. Denote the active vertex by t from now on.

8. **Elseif** {There is a vertex $v \in N(t)$ with $gd(v) \leq 1$}, select such a vertex v, and return $S := \text{mif}(G, F \cup \{v\})$.

9. **Elseif** {There is a vertex $v \in N(t)$ with $gd(v) = 2$}, select such a vertex v, where $GD(v) = \{v_1, v_2\}$, and return

$$S := \max\{ \text{mif}(G, F \cup \{v\}), \text{mif}(G \setminus \{v\}, F \cup \{v_1, v_2\}) \}.$$

10. **Elseif** {There is a vertex $v \in N(t)$ with $gd(v) = 3$} select such a good vertex v, where $GD(v) = \{v_1, v_2, v_3\}$ (v_3 is chosen so that $v_1 \in GD(v_2)$ if possible and $v_3 \notin N(t)$ if v is of Case (i)), and return

$$S := \max \{ \text{mif}(G, F \cup \{v\}), \text{mif}(G \setminus \{v, v_3\}, F \cup \{v_1, v_2\}),$$
$$\text{mif}(G \setminus \{v\}, F \cup \{v_3\}) \}.$$

11. **Else** /* $gd(v) \geq 4$ for each $v \in N(t)$ */ select an effective vertex $v \in N(t)$, and return
$$S := \max\{ \text{mif}(G, F \cup \{v\}), \text{mif}(G \setminus \{v\}, F) \}.$$

Note. In Steps 9 and 10, if adding v_1 and v_2 to F induces a cycle in the second branch, we just ignore this branch.

Fig. 1. Algorithm $\text{mif}(G, F)$

degree. One reason for this would be that vertices of higher degree may have larger contribution to the structural complexity of the graph. However, MIF seems different from the previous measure-and-conquer algorithms in the sense that the current fastest algorithm is obtained by using weights with no difference between vertices of different degree.

We wonder whether the algorithm for MIF can be improved if we distinguish vertices of different degree (without significantly modifying the algorithm). Note that in some branching operations in the algorithm, we will delete some vertices from the graph and the degree of some vertices will decrease. When the maximum degree of the graph is at most 2, the forced MIF can be solved in polynomial time. These properties imply that vertices of different degree have different contribution to the hardness of the problem. After carefully analyzing the algorithm based on the above observation, we improve the running time bound for this problem from $1.7548^n n^{O(1)}$ to $1.7356^n n^{O(1)}$ by using a sophisticated measure scheme.

Our improvement is obtained from the following intuitive observation. The worst case of Fomin et $al.$'s algorithm is to branch on a vertex v of $gd(v) = 4$ (as Step 10 in $mif(G, F)$) with recurrence (2). However, we observe that this case may not always happen. If there is a vertex $u \in GD(v)$ of high degree (such as degree ≥ 6), in the branch of including v to F (where u becomes a vertex adjacent to the new active vertex), we can branch on a vertex u of degree at least 6 in the next step, which may lead to a good performance. If there is a vertex $u \in GD(v)$ of low degree (such as degree ≤ 5), in the other branch of deleting v from the graph, the degree of u will decrease by one and u will become an 'easy' vertex in the future. Then it is possible to decrease some measure from u by setting different vertex weight to vertices of different degree. Both cases imply that we can get a recurrence better then (2). This is the main idea on how our algorithm gets the claimed improvement. Next, we show the details about the weight setting and the recurrences generated by the algorithm.

7.1   Weight Setting and Basic Analysis

Let t be the active vertex in the graph and $U = V \setminus (F \cup N(t))$. For each vertex $v \in V$, we set its vertex-weight $w(v)$ to be

$$
w(v) = \begin{cases} 0 & \text{if } v = t \text{ (v is the active vertex)} \\ \beta & \text{if } v \in F \setminus \{t\} \\ \alpha & \text{if } v \in N(t) \cap (V \setminus F) \\ 0 & \text{if } v \in U \text{ and } d(v) \in \{0, 1\} \\ w_i & \text{if } v \in U \text{ and } d(v) = i \text{ for } i \in \{2, 3, 4\} \\ 1 & \text{if } v \in U \text{ and } d(v) \geq 5, \end{cases}
$$

where w_i is the weight of a vertex $v \in U$ of degree $i \geq 0$.

To simplify some arguments, we require vertex weights satisfy the following constraints

$$0.01 < \beta = w_2 \leq w_3 \leq w_4 \leq 1 \quad \text{and} \quad \max\{\beta, 0.5\} \leq \alpha \leq 1. \tag{3}$$

We also assume that

$$w_3 + \beta \leq 2\alpha. \tag{4}$$

We will determine the best value of β, α, w_3 and w_4 so that the worst recurrence in the algorithm is as good as possible.

Let Δ_i $(i \geq 1)$ denote $w_i - w_{i-1}$, where $w_0 = w_1 = 0$ and $w_j = 1$ for all $j \geq 5$. We see that $\Delta_i \geq 0$ for each $i \geq 1$. Recall that we use $C(w)$ to denote the worst size of the search tree with respect to the measure w generated by the algorithm. We will list out all possible recurrences created by the algorithm. To do so, we analyze the decrease of the measure in each of the generated instances in each recurrence.

In the algorithm, we have two basic operations in branching on a vertex v in a reduced instance: one is to include vertex v to F and the other is to delete vertex v from the graph. We consider the following three cases (1), (2) and (3):

(1) The vertex v is in U, and it is included to F and also selected as the active vertex t. In this case, the weight of v decreases from $w(v)$ to β, the weight of v further decreases from β to 0, and the weight of each neighbor $u \notin F$ of v decreases from $w(u)$ to α. In total, the measure w decreases by at least

$$\Delta w \geq w(v) + \sum_{u \in N(v) \setminus F} (w(u) - \alpha). \tag{5}$$

(2) The vertex $v \in V \setminus F$ is a neighbor of the active vertex t, and it is included to v by the first branch, and $\mathrm{Rd}(G, F)$ is executed in the next step, where t and v (possibly with some other vertices in F adjacent to v) will be contracted into a new active vertex: We analyze the decrease of the measure w in these two operations together. In fact, the decrease of the measure come from three kinds of vertices, which are v itself, vertices in $GD(v)$, and vertices in $F_v = N(v) \cap (F \setminus \{t\})$. The weight of v decreases from α to 0. For vertices $u \in GD(v)$, we distinguish two cases. If u is also in $N(t)$, then after including v to F, vertex u will be removed from the graph by executing $\mathrm{Rd}(G, F)$. Then the weight of each vertex $u \in GD(v) \cap N(t)$ decreases from α to 0, whereas the weight of each vertex $u \in GD(v) \setminus N(t)$ decreases from $w(u) = w_{d(u)}$ to α. We see that the weight of some vertices in F also decreases. After including v to F, all vertices in F_v together with v and t will be contracted into a new active vertex. This decreases the weight of each vertex in F_v by β. In total, the decrease of the measure w is at least

$$\Delta w = \alpha + |GD(v) \cap N(t)|\alpha + \sum_{u \in GD(v) \setminus N(t)} (w_{d(u)} - \alpha) + |F_v|\beta. \tag{6}$$

(3) The vertex v is deleted from the graph: In this case the measure decreases by the deletion of v and the weight decrease of some of its neighbors (since their degrees decrease by 1). Then the weight of each vertex $u \in N(v) \setminus (F \cup N(t))$ decreases by $\Delta_{d(u)}$. Then deleting v decreases the measure w by at least

$$\Delta w = w(v) + \sum_{u \in N(v) \setminus (F \cup N(t))} \Delta_{d(u)}. \tag{7}$$

where $w(v) = 1$ when v is in U and $w(v) = \alpha$ when v is in $N(t)$.

We will frequently use (5), (6) and (7) to derive our recurrences generated by the algorithm.

7.2   Some Techniques

To ease amortization on our analysis, we introduce "*shift*" σ for some recurrences which are not bottlenecks in our algorithm. Suppose that there are two branching operations A and B with recurrences $C(w) \leq C(w - t_{(A1)}) + C(w - t_{(A2)})$ and $C(w) \leq C(w - t_{(B1)}) + C(w - t_{(B2)})$, and that branching operation B is always applied to the subinstance G_1 generated by the first branch of A in the algorithm. The branching operation B may lead to a better recurrence (with a smaller branching factor) than A does. To improve the recurrence for operation A, we will save some decreasing of the measure in operation B to A. Instead, we save $\sigma_B \geq 0$ from B by evaluating the branch rule B with a worse recurrence

$$C(w) \leq C(w - (t_{(B1)} - \sigma_B)) + C(w - (t_{(B2)} - \sigma_B)).$$

The saved weight σ_B will be included to the recurrence for operation A to obtain

$$C(w) \leq C(w - (t_{(A1)} + \sigma_B)) + C(w - t_{(A2)}).$$

The amount of save is also called *shift*. In our algorithm, we introduce only one shift σ when branching on a vertex v with $gd(v) = 4$ in Step 11.

7.3   The Final Result

By using the above techniques, we can list out all the recurrences generated in the algorithm. The detailed analysis and the recurrence list can be found in the full version of this paper. We build up a quasiconvex program according to the list of recurrences. After solve it we get the running time bound of $1.7356^n n^{O(1)}$ by letting $\alpha = 0.5547$, $\beta = 0.0851$, $w_3 = 0.9793$, $w_4 = 0.9919$ and $\sigma = 0.0876$.

Theorem 1. *The minimum feedback vertex set in an n-vertex undirected graph can be solved in $1.7356^n n^{O(1)}$ time.*

8   Concluding Remarks

With new reductions based on biconnectivity and a deeper analysis, we improved the running time bound for the maximum induced forest problem (the feedback vertex set problem) from $1.7548^n n^{O(1)}$ to $1.7356^n n^{O(1)}$. The improvement is analyzed by using a measure scheme where we distinguish vertices of different degree. Previous measure-and-conquer algorithms for MIF did not consider this before. It should be interesting to know that this measure scheme is still helpful for MIF. Furthermore, we introduced many structural properties for the problem, and our algorithm does not need to solve the maximum independent set problem while previous best algorithm needs this as a subroutine.

References

1. Bafna, V., Berman, P., Fujito, T.: A 2-approximation algorithm for the undirected feedback vertex set problem. SIAM J. on Disc. Math. 12(3), 289–297 (1999)
2. Bourgeois, N., Escoffier, B., Paschos, V.T., van Rooij, J.M.M.: Fast algorithms for max independent set. Algorithmica 62(1-2), 382–415 (2012)
3. Cao, Y., Chen, J., Liu, Y.: On feedback vertex set new measure and new structures. In: Kaplan, H. (ed.) SWAT 2010. LNCS, vol. 6139, pp. 93–104. Springer, Heidelberg (2010)
4. Chen, J., Fomin, F., Liu, Y., Lu, S., Villanger, Y.: Improved algorithms for feedback vertex set problems. J. Comput. Syst. Sci. 74, 1188–1198 (2008)
5. Chen, J., Liu, Y., Lu, S., O'Sullivan, B., Razgon, I.: A fixed-parameter algorithm for the directed feedback vertex set problem. J. ACM 55, 1–19 (2008)
6. Dehne, F., Fellows, M.R., Langston, M.A., Rosamond, F.A., Stevens, K.: An $O(2^{O(k)}n^3)$ FPT algorithm for the undirected feedback vertex set problem. In: Wang, L. (ed.) COCOON 2005. LNCS, vol. 3595, pp. 859–869. Springer, Heidelberg (2005)
7. Dwork, C., Kumar, R., Naor, M., Sivakumar, D.: Rank aggregation revisited (2001) (manuscript)
8. Even, G., Naor, J., Schieber, B., Sudan, M.: Approximating minimum feedback sets and multicuts in directed graphs. Algorithmica 20, 151–174 (1998)
9. Festa, P., Pardalos, P.M., Resende, M.G.C.: Feedback set problems. In: Handbook of Combinatorial Optimization, vol. A, pp. 209–258. Kluwer Acad. Publ., Dordrecht (1999)
10. Fomin, F.V., Gaspers, S., Pyatkin, A.V., Razgon, I.: On the minimum feedback vertex set problem: exact and enumeration algorithms. Algorithmica 52(2), 293–307 (2008)
11. Fomin, F.V., Grandoni, F., Kratsch, D.: A measure & conquer approach for the analysis of exact algorithms. J. ACM 56(5), 1–32 (2009)
12. Guo, J., Gramm, J., Huffner, F., Niedermeier, R., Wernicke, S.: Compression-based fixed-parameter algorithms for feedback vertex set and edge bipartization. J. Comput. Syst. Sci. 72, 1386–1396 (2006)
13. Karp, R.M.: Reducibility Among Combinatorial Problems. In: Miller, R.M., Thatcher, J.W. (eds.) Complexity of Computer Computations, pp. 85–103. Plenum Press, Nwe York (1972)
14. Kemeny, J., Snell, J.: Mathematical models in the social sciences. Blaisdell (1962)
15. Silberschatz, A., Galvin, P.: Operating System Concepts, 4th edn. Addison-Wesley (1994)
16. Razgon, I.: Exact computation of maximum induced forest. In: Arge, L., Freivalds, R. (eds.) SWAT 2006. LNCS, vol. 4059, pp. 160–171. Springer, Heidelberg (2006)
17. Razgon, I.: Computing minimum directed feedback vertex set in $O(1.9977^n)$. In: ICTCS 2007, Rome, Italy, pp. 70–81 (2007)
18. Xiao, M., Nagamochi, H.: A refined exact algorithm for edge dominating set. In: Agrawal, M., Cooper, S.B., Li, A. (eds.) TAMC 2012. LNCS, vol. 7287, pp. 360–372. Springer, Heidelberg (2012)

An Inductive Construction of Minimally Rigid Body-Hinge Simple Graphs

Yuya Higashikawa[1], Naoyuki Kamiyama[2,*], Naoki Katoh[1,*], and Yuki Kobayashi[1]

[1] Graduate School of Engineering, Kyoto University,
Kyoto Daigaku Katsura, Nishikyo-ku, Kyoto 615-8540, Japan
{as.higashikawa,naoki,as-kobayashi}@archi.kyoto-u.ac.jp
[2] Institute of Mathematics for Industry, Kyushu University,
744 Motooka, Nishi-ku, Fukuoka 819-0395, Japan
kamiyama@imi.kyushu-u.ac.jp

Abstract. In this paper, we propose an inductive construction of minimally rigid body-hinge simple graphs. Inductive construction is one of well-studied topics in Combinatorics and Combinatorial Optimization. We develop an inductive construction for minimally rigid body-hinge simple graphs in d-dimension with $d \geq 3$ by which we can develop a polynomial-time algorithm for enumerating all minimally rigid body-hinge simple graphs.

Keywords: Body-hinge framework, Panel-hinge framework, Body-hinge graph, Combinatorial rigidity, Rigid realization.

1 Introduction

A d-dimensional body-hinge framework (Fig. 1(a)) is a collection of d-dimensional rigid bodies connected by *hinges*, where a hinge is a $(d-2)$-dimensional affine subspace, i.e., pin-joints in 2-space, line-hinges in 3-space, plane-hinges in 4-space etc. Bodies are allowed to move continuously in $\mathbb{R}^d$ so that the relative motion of any two bodies connected by a hinge is a rotation around it and the framework is called rigid if every motion provides a framework isometric to the original one. We consider a body-hinge framework as a pair $(G, \mathbf{p})$ of a multigraph $G = (V, E)$ and a mapping $\mathbf{p}$ from $e \in E$ to a $(d-2)$-dimensional affine subspace $\mathbf{p}(e)$ in $\mathbb{R}^d$. Namely, $v \in V$ corresponds to a body and $uv \in E$ corresponds to a hinge $\mathbf{p}(uv)$ which joints the two bodies corresponding to u and v. Then, G is said to be *realized* as a generic body-hinge framework $(G, \mathbf{p})$ in $\mathbb{R}^d$, and is called a *body-hinge graph*.

Tay [10] and Whiteley [13] independently proved that the infinitesimal rigidity of a *generic* body-hinge framework $(G, \mathbf{p})$ is determined only by its underlying graph G. Body-hinge framework is generic if its rigidity matrix has a maximum rank on all subgraphs [15]. From basic algebraic geometry results this means

---

* Supported by JSPS Grant-in-Aid for Scientific Research(A)(25240004).

P. Widmayer, Y. Xu, and B. Zhu (Eds.): COCOA 2013, LNCS 8287, pp. 165–177, 2013.

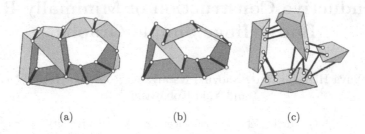

Fig. 1. (a) Body-hinge, (b) panel-hinge and (c) body-bar frameworks

that 'almost all' body-hinge realizations of G are generic in $\mathbb{R}^d$ (see [13,14,16] for details). For generic frameworks, rigidity is equivalent to infinitesimal rigidity [8,14].

Let $D = \binom{d+1}{2}$ and $\widetilde{G}$ denotes the graph obtained from G by replacing each edge by $(D-1)$ parallel edges.

Proposition 1. [10,13] *A graph G can be realized as an infinitesimally rigid body-hinge framework in $\mathbb{R}^d$ if and only if $\widetilde{G}$ contains D edge-disjoint spanning trees.*

Katoh and Tanigawa [4] proved that the generic rigidity of minimally rigid *panel-hinge framework* (Fig. 1(b)) is equivalent to that of body-hinge frameworks.

A necessary and sufficient condition that a generic 2-dimensional framework is rigid is given by Laman [5]. The corresponding graph is called a Laman graph. It is known that Laman graphs are generated via Henneberg operations, and all Laman graphs can be constructed by these operations [3,14]. Furthermore, by using the matroid property, we can efficiently enumerate all Laman graphs by the algorithm of [12]. On the other hand, it remains open to derive combinatorial characterizations for 3-dimensional bar-joint frameworks [14].

However, for a special class of generic 3-dimensional bar-joint frameworks such as generic body-bar and body-hinge frameworks, a combinatorial charac-terization was developed by Tay and Whiteley [10,13]. A body-bar framework (Fig. 1(c)) is a collection of rigid bodies connected by rigid bars with joints which is represented by a body-bar graph where a rigid body corresponds to a vertex and a bar corresponds to an edge. A graph (a minimally rigid body-bar graph) corresponding to a minimally generically rigid body-bar framework in d-dimension can be characterized as one that contains D edge-disjoint spanning trees. D edge-disjoint spanning trees can be viewed as a union of six graphic matroid, and thus we can easily develop a polynomial-time algorithm for enu-merating all minimally rigid body-bar graphs by using the algorithm of [12]. Moreover, an inductive construction for all minimally rigid body-bar graphs in d-space is known [2]. We investigate the combinatorial property of graphs G such $\widetilde{G}$ contains D edge-disjoint spanning trees. Such graphs are called *body-hinge rigid graphs*. Similarly to a minimally body-bar graph, we can define a minimally rigid body-hinge graph which corresponds to a minimally generically rigid body-hinge framework. However, for minimally rigid body-hinge graphs, inductive operations that create one with larger size from a smaller one are not known. Notice that Katoh and Tanigawa [4] showed two operations (i.e., contraction of minimally rigid proper subgraph, and splitting-off operation at a vertex of degree two) that create a minimally rigid body-hinge graph with a smaller size when a minimally rigid body-hinge graph is given. However, this does not directly imply that we can develop an algorithm for enumerating all minimally rigid body-hinge graphs that runs in polynomial-time per output.

Our Results: We develop five operations that inductively construct minimally rigid body-hinge simple graphs for d-dimension with $d \geq 3$. More precisely, we prove the following theorems.

Theorem 1. *For any given minimally rigid body-hinge simple graph G, we apply at least one of five operations to G so that the resulting graph is also a minimally rigid body-hinge simple graph. Also, every operation can be executed in polynomial-time to apply operations.*

Theorem 2. *Any minimally rigid body-hinge simple graph can be constructed by a sequence of operations of five types starting from the triangle graph.*

In this paper, we focus on body-hinge simple graphs, because it is not interesting to consider body-hinge multigraphs from the viewpoint of engineering applications.

The rest of this paper is organized as follows. In Section 2, we introduce necessary notations and facts that are needed to prove Theorems 1 and 2. In Section 3, we introduce five operations, and give a proof of Theorem 1. In Section 4, we give a proof of Theorem 2.

2   Preliminaries

Let $G = (V, E)$ be a multigraph which may contain parallel edges but no self-loops. For $X \subseteq V$, let $G[X]$ be the graph induced by X. For $X \subseteq V$, let $\delta_G(X) = \{uv \in E \mid u \in X, v \notin X\}$. For $X = \{v\}$, we shall omit set brackets when describing singleton sets, e.g., $\delta_G(v)$ implies $\delta_G(\{v\})$. Throughout the paper, a *partition* $\mathcal{P}$ of V is a collection $\{V_1, V_2, \ldots, V_m\}$ of vertex subsets for some positive integer m such that $V_i \neq \emptyset$ for $1 \leq i \leq m, V_i \cap V_j = \emptyset$ for any $1 \leq i, j \leq m, i \neq j$, and $\cup_{i=1}^m V_i = V$. Let $\delta_G(\mathcal{P})$ denote the set of edges of G connecting distinct subsets of $\mathcal{P}$. Let $\widetilde{E}$ denote the edge set of $\widetilde{G}$. Also, with $\tilde{e}$ denote the *set* of corresponding $D - 1$ parallel copies of e in $\widetilde{E}$. We index the edges of $\tilde{e}$ by $1 \leq i \leq D - 1$, and e_i, or $(e)_i$, denotes th ith element in $\tilde{e}$. The following Tutte-Nash-Williams disjoint tree theorem is well known [7,11].

Proposition 2 (Tutte, Nash-Williams). *A multigraph $G = (V, E)$ contains k edge-disjoint spanning trees if and only if $|\delta_G(\mathcal{P})| \geq k(|\mathcal{P}| - 1)$ holds for every partition $\mathcal{P}$ of V.*

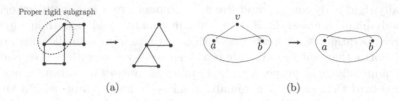

Fig. 2. (a) Contraction and (b) splitting off

Proposition 2 implies that a body-hinge graph can be realized as a rigid body-hinge framework if and only if $\widetilde{G}$ has D edge-disjoint spanning trees. Katoh and Tanigawa [4] showed two operations which produce a minimally rigid body-hinge graph of a smaller size from a minimally rigid body-hinge graph $G = (V, E)$.

The first operation is the contraction of a proper rigid subgraph; $G' = (V', E')$ is called a proper rigid subgraph if it is a rigid subgraph of G satisfying $1 < |V'| < |V|$ (Fig. 2(a)). The second operation is a so-called *splitting off* operation (Fig. 2(b)). For a vertex v of a graph G, we denote by $N_G(v)$ the set of vertices adjacent to v in G. A splitting off is defined only at a vertex v of degree two. Let $N_G(v) = \{a, b\}$. We denote by G_v^{ab} the graph obtained from G by removing v (and the edges incident to v) and then inserting a new edge ab. So, the resulting graph G_v^{ab} is isomorphic to that obtained by contracting either va or vb. Note that it is not straightforward to construct a polynomial-time algorithm for enumerating all minimally rigid body-hinge graphs based on these two operations. The following six lemmas are known [4].

Lemma 1. [4] *Let G be a rigid body-hinge graph. Then, G is 2-edge-connected.*

Lemma 2. [4] *Let $G = (V, E)$ be a minimally rigid body-hinge multigraph, and let $G' = (V', E')$ be a rigid subgraph of G. Then, the graph obtained from G by contracting E' is a minimally rigid body-hinge multigraph.*

Lemma 3. [4] *Let $G = (V, E)$ be a multigraph which contains no proper rigid subgraph. Then, the following holds.*

$$(D - 1)|E| < D(|V| - 1) + D - 1. \tag{1}$$

Lemma 4. [4] *Let $G = (V, E)$ be a minimally rigid body-hinge graph which contains no proper rigid subgraph. Then, either G is a cycle graph of at most D vertices or it contains a chain $v_0, v_1, \ldots, v_d$ of length d such that $v_i v_{i+1} \in E$ for $0 \leq i \leq d - 1$ and $|\delta_G(v_i)| = 2$ for $1 \leq i \leq d - 1$.*

Lemma 5. [4] *Let $G = (V, E)$ be a minimally rigid body-hinge graph. Then, for any vertex v of degree two with $N_G(v) = \{a, b\}$, G_v^{ab} is a rigid body-hinge graph.*

Lemma 6. [4] *Let $G = (V, E)$ be a minimally rigid body-hinge graph that contains no proper rigid subgraph. Then, for any vertex v of degree two with $N_G(v) = \{a, b\}$, G_v^{ab} is a minimally rigid body-hinge graph.*

3   Five Operations Which Inductively Construct Minimally Rigid Body-Hinge Simple Graphs

Consider a minimally rigid body-hinge simple graph $G = (V, E)$, where the number of vertices is n with $n \geq 3$. Notice that for $n = 3$, G is a triangle (a simple cycle of length 3). We define five operations which construct a minimally rigid body-hinge simple graph $G' = (V', E')$ of a larger size.

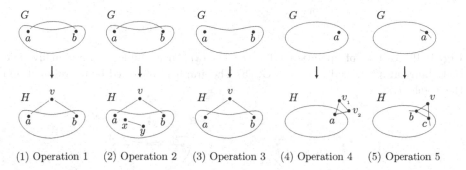

(1) Operation 1   (2) Operation 2   (3) Operation 3   (4) Operation 4   (5) Operation 5

Fig. 3. Operations which inductively construct a minimally rigid body-hinge simple graph

Operation 1 (edge-split): Choose an edge ab, insert a new vertex v on ab if the resulting graph is rigid (Fig. 3(1)).

Operation 2 (edge-split plus 1-addition): There exists a partition of V denoted by $\mathcal{P}$ such that $(D-1)|\delta_G(\mathcal{P})| = D(|\mathcal{P}| - 1)$. Let $\mathcal{P} = \{V_1, V_2, \ldots, V_m\}$. For an edge $ab \in \delta_G(\mathcal{P})$, split ab to add a new vertex v. Find vertices x, $y \in V$ such that $x \in V_i$ and $y \in V_j$ with $1 \leq i, j \leq m$, $i \neq j$, and the graph H obtained from G by adding an edge xy to $E \cup \{va, vb\}\backslash\{ab\}$ is minimally rigid. Then add an edge xy (Fig. 3(2)).

Operation 3 (vertex 2-addition): Add a new vertex v, choose two existing vertices a and b, and add edges va, vb if the resulting graph is minimally rigid (Fig. 3(3)).

Operation 4 (triangle-addition): Choose an arbitrary vertex a, add two new vertices v_1 and v_2 as well as edges v_1a, v_1v_2, v_2a (Fig. 3(4)).

Operation 5 (triangle-expansion): Choose an arbitrary vertex a. Suppose that a has degree d. Let G' be created from G by replacing vertex a with a triangle v, b, c where b and c are connected to at least one neighbor of a, and v is connected to b and c. Formally, let $2 \leq d' \leq d$, $v_1, \ldots, v_d$ be the neighbors of a and $G' = (V', E')$ with $V' = V \cup \{b, c\}\backslash a$, $E' = E \cup \{v_ib : 1 \leq i \leq d'-1\} \cup \{v_ic : d' \leq i \leq d\}\backslash\{v_ia : 1 \leq i \leq d\}$. We apply this operation only if the resulting graph is minimally rigid (Fig. 3(5)).

3.1   Proof of Theorem 1

Let $H = (V', E')$ be the graph obtained by applying one of the five operations.

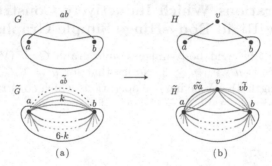

Fig. 4. Illustration of operation 1 for $D = 6$. (a) An example of a minimally rigid body-hinge graph G and the graph $\widetilde{G}$. (b) The graph H obtained by operation 1 and the graph $\widetilde{H}$.

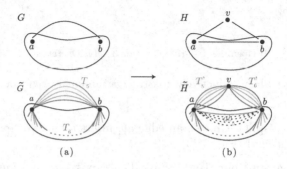

Fig. 5. Illustration of operation 2 for $D = 6$. (a) An example of a minimally rigid body-hinge graph G and the graph $\widetilde{G}$. (b) The graph H obtained by operation 2 and the graph $\widetilde{H}$.

(1) edge-split (Fig. 3(1))

First, We shall show that H is rigid. We denote by T_j $(1 \le j \le D)$ D edge-disjoint spanning trees of $\widetilde{G}$. Let us show that we can construct edge-disjoint spanning trees $T'_1, \ldots, T'_D$ of $\widetilde{H}$ from each of $T_1, \ldots, T_D$. From the assumption of operation 1, let us assume that each T_j $(1 \le j \le k)$ with $k \le D - 2$ uses an edge ab_j while T_j $(k + 1 \le j \le D)$ does not use any of $\widetilde{ab}$ (See Fig. 4(a)). For a spanning tree T_j $(1 \le j \le k)$, let $T'_j = T_j \setminus \{ab_j\} \cup \{va_j, vb_j\}$ (Fig. 4(b)). For each T_j $(k + 1 \le j \le D)$, we create a spanning tree T'_j by adding one of $\widetilde{va}, \widetilde{vb}$, so that T'_j is a spanning tree. This is possible because $2(D - 1 - k) \ge D - k$. Then it is clear that T'_j $(1 \le j \le D)$ are edge-disjoint spanning trees.

Next, suppose, for a contradiction, that H is not minimal, i.e., there exists a redundant edge f in H. Therefore, H is rigid even if we remove f. $f \ne va$ or vb since v is of degree two. Thus, $f \in E$ holds. Since $H \setminus \{f\}$ is rigid, $H^{ab}_v \setminus \{f\}$ is also rigid by Lemma 5. Since $H^{ab}_v = G$, this implies that G is redundantly rigid, contradicting the minimal rigidity of G. So, H is a minimally rigid body-hinge graph.

(2) edge-split plus 1-addition (Fig. 3(2))
First, we shall show that H is rigid. Notice that the graph obtained by splitting the edge ab is no longer rigid since by $(D-1)|\delta_G(\mathcal{P})| = D(|\mathcal{P}| - 1)$, $\mathcal{P}' = (V_1, V_2, \ldots, V_m, V_{m+1} (= \{v\}))$ does not satisfy the condition of Proposition 2. We shall first show that there always exists an edge xy with $x \in V_i$ and $y \in V_j$ with $1 \leq i, j \leq m$, $i \neq j$ such that the graph H obtained from G by adding an edge xy to $E \cup \{va, vb\}\backslash ab$ is rigid. Such edge xy, if it exists, is called *eligible*. To show this, it suffices to show that ab is eligible. For this, let T_j $(1 \leq j \leq D)$ be D edge-disjoint spanning trees in $\widetilde{G}$ such that T_j $(1 \leq j \leq D-1)$ is a spanning tree which uses edge ab_j, and T_D is the one which does not use $\widetilde{ab}$ (Fig. 5(a)). For $T_1, \ldots, T_D$, $T_j' = T_j \cup \{va_j, vb_j\}\backslash\{ab_j\}$ for $j = 1, 2, \ldots, D-2$, $T_{D-1}' = T_{D-1} \cup \{va_{D-1}\}$, and $T_D' = T_D \cup \{vb_{D-1}\}$ (Fig. 5(b)). Then $T_1', \ldots, T_D'$ are clearly edge-disjoint spanning trees in $\widetilde{H}$. Thus H is rigid. However, it may not be minimal. Therefore, we apply this operation only if the resulting graph is minimally rigid, which can be checked by applying the pebble game algorithm [6], for instance.

(3) vertex 2-addition (Fig. 3(3))
We shall show that H is rigid. We shall show that we can construct D edge-disjoint spanning trees T_j' $(1 \leq j \leq D)$ in $\widetilde{H}$ from D edge-disjoint spanning trees T_j $(1 \leq j \leq D)$ of $\widetilde{G}$ by adding a single edge va_j or $vb_{j'}$ (Fig. 6(b)). Since the number of edges $\widetilde{va} \cup \widetilde{vb}$ is $2(D-1)$, this is always possible. Then, the graph H obtained by vertex 2-addition is also rigid. However, it may not be minimal. Therefore, we apply this operation only if the resulting graph is minimally rigid.

(4) triangle-addition (Fig. 3(4))
Let us prove that H is a minimally rigid body-hinge graph. Since the graph G is a minimally rigid body-hinge graph, there exist D edge-disjoint spanning trees in $\widetilde{G}$. Let F be a triangle graph where F is a minimally rigid body-hinge graph. Then there exist D edge-disjoint spanning trees in $\widetilde{F}$. Since G and F share a single vertex v, the union of a spanning tree of $\widetilde{G}$ and that of $\widetilde{F}$ is clearly a

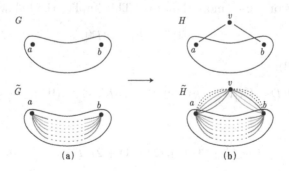

Fig. 6. Illustration of operation 3 for $D = 6$. (a) An example of a minimally rigid body-hinge graph G and the graph $\widetilde{G}$ obtained from G. (b) The graph H obtained by operation 3 and the graph $\widetilde{H}$ obtained from H.

spanning tree of $\widetilde{H}$. By the minimal rigidity of G and F, the minimality of H is obvious. Therefore, H is a minimally rigid body-hinge graph.

(5) triangle-expansion (Fig. 3(5))

We shall show that H is rigid. Let F be a triangle graph. Notice that F is a minimally rigid body-hinge graph. Then there exist D edge-disjoint spanning trees in $\widetilde{F}$. The union of a spanning tree of $\widetilde{G}$ and that of $\widetilde{F}$ is clearly a spanning tree of $\widetilde{H}$. Then, the graph H obtained by triangle-expansion is also rigid. However, it may not be minimal. Therefore, we apply this operation only if the resulting graph is minimally rigid.

Finally we will show that all five operations can be executed in polynomial-time. Notice that we can find $\mathcal{P}$ satisfying the condition of operation 2 in polynomial-time (more precisely, $O(n)$ applications of max-flow computation [9]). Checking whether the graph obtained by adding an edge xy in applying operation 2 is rigid or not can be done in $O(n^2)$ time by applying the pebble game algorithm [6]. It is not difficult to see that the other four operations can be done in polynomial-time. □

4   Proof of Theorem 2

In this section, we prove Theorem 2. For this purpose, we first prove the following three lemmas.

Lemma 7. *Let $G' = (V', E')$ be a minimally rigid body-hinge simple graph which contains no proper rigid subgraph such that $|V'| \geq 3$. Then, the number of vertices of degree two is at least $(D-3)|V'|/(D-1) + 2/(D-1)$.*

Proof. By Lemma 3, we have the following:

$$(D-1)|E'| < D(|V'| - 1) + (D-1). \tag{2}$$

Let k denote the number of the vertices of degree two. Then, there exist $|V'| - k$ vertices of degree more than two. This implies the following:

$$2k + 3(|V'| - k) \leq 2|E'|. \tag{3}$$

By (2) and (3),

$$2(D-1)k + 3(D-1)(|V'| - k) \leq 2(D-1)|E'| \leq 2D(|V'| - 1) + 2(D-1).$$

As a result,

$$k \geq (D-3)|V'|/(D-1) + 2/(D-1). \qquad \square$$

Lemma 8. *Let $G = (V, E)$ be a minimally rigid body-hinge multigraph. (a) G does not contain more than two parallel edges. (b) Contraction of two parallel edges into a single vertex never produces new parallel edges.*

Fig. 7. The case where two parallel edges are produced after the contraction of two parallel edges

Proof. (a) Omitted. (b) After the contraction of parallel edges, the minimal rigidity is preserved by Lemma 2. Suppose for a contradiction that new parallel edges are produced in the graph obtained by the contraction of parallel edges. This implies that there exist parallel edges e and e' between vertices a and $b \in V$, and moreover vertices a and b have a common adjacent vertex v (Fig. 7). Then the edge e is redundant, contradicting the minimality of G. Therefore, new parallel edges are not produced due to the contraction of parallel edges.
□

The following lemma plays a key role in proving Theorem 2.

Lemma 9. *There exists at least one vertex of degree two in a minimally rigid body-hinge simple graph $G = (V, E)$ with $|V| \geq 3$.*

Suppose, for a contradiction, that there exists a minimally rigid body-hinge simple graph G such that the degree of each vertex is at least three. Let us consider the following two cases.

Case 1. G does not have a proper rigid subgraph.

Since the degree of each vertex is at least three, we have

$$3|V| \leq 2|E|. \tag{4}$$

By Lemma 3 and (4), we have

$$3(D-1)|V| \leq 2(D-1)|E| \leq 2D(|V|-1)+2(D-1). \tag{5}$$

As a result, we have $(D-3)|V| \leq -2$, contradicting $|V| \geq 3$.
The rest of the proof is omitted.

4.1   Proof of Theorem 2

The proof is by induction on the number of vertices in a minimally rigid body-hinge simple graph. For the base case, a triangle graph is obviously a minimally rigid body-hinge simple graph and thus satisfies the theorem. Suppose that any minimally rigid body-hinge graph of $n - 1$ or less vertices for $n \geq 4$ can be constructed by a sequence of the five operations starting from the triangle graph. Suppose we have a minimally rigid body-hinge graph $G = (V, E)$ of n vertices.

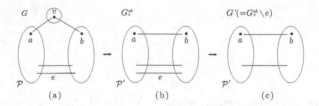

Fig. 8. The reverse direction of operation 2

Then, we prove that we can construct a minimally rigid body-hinge graph of $n-1$ or $n-2$ vertices by applying one of the five operations on G in the reverse direction. Let E_1 and E_2 be a partition of E such that E_1 and E_2 are both nonempty. Let $G[E_1]$ and $G[E_2]$ be the subgraphs edge-induced by E_1 and E_2, respectively. Suppose that $G[E_1]$ and $G[E_2]$ share a single vertex v, i.e., v is a cut point, and $G[E_2]$ is a triangle. Then, G is obtained by applying operation 4 to $G[E_1]$. Therefore, we assume that G does not have a partition into E_1 and E_2 such that $G[E_2]$ is a triangle. Furthermore, if G has a proper rigid subgraph of the triangle graph and the degree of one of vertices of the triangle graph is two, we apply operation 5 in the reverse direction. In this case, the resulting graph is minimally rigid by Lemma 2. Therefore, in the subsequent discussion, we assume that G does not have a triangle graph as its subgraph. In the following, we prove that it is possible to apply at least one of operations 1, 2 and 3 in the reverse direction. We divide the proof into two cases depending on whether G has a proper rigid subgraph or not.

Case 1. G does not have a proper rigid subgraph.
By Lemma 9, there exists at least one vertex of degree two. In this case, by Lemma 5, a minimally rigid body-hinge graph of $n-1$ vertices can be constructed by the reverse direction of operation 1 (i.e., splitting off operation).

Case 2. G has a proper rigid subgraph.
There exists a vertex v of degree two by Lemma 9. Let us consider applying operation 1 in the reverse direction at a vertex v. Notice that this operation is in fact equivalent to a splitting off operation, In this case, the resulting graph is rigid by Lemma 5. Thus, the resulting graph is denoted by G_v^{ab} where a and b are adjacent vertices of v. Notice that there does not exist an edge ab because we assume that G does not have a triangle graph as its subgraph. Thus, G_v^{ab} is a simple graph. However, the minimality is not guaranteed. If the resulting graph is minimally rigid, we can obtain a minimally rigid body-hinge graph of $n-1$ vertices. So, suppose otherwise. If the edge ab is redundant, the graph obtained by removing ab from G_v^{ab} is rigid. The resulting graph is isomorphic to that obtained by applying operation 3 in the reverse direction (vertex 2-addition). Then, by the minimal rigidity of G, we remark that the graph obtained by applying operation 3 in the reverse direction is always minimally rigid. Therefore, if there exists a vertex v of degree two and the edge ab is redundant for G_v^{ab}, the reverse direction of operation 3 for G produces a minimal rigid body-hinge graph of $n-1$ vertices.

Thus, consider the case where for G_v^{ab}, ab is not redundant but there exists a redundant edge $e \in E \setminus \{av, bv\}$ (see Fig. 8). Let $G' = G_v^{ab} \setminus \{e\}$. Since G is a minimally rigid body-hinge graph, and by Proposition 2,

$$(D - 1)|\delta_G(\mathcal{P})| \geq D(|\mathcal{P}| - 1) \tag{6}$$

holds for any partition $\mathcal{P}$. By the minimal rigidity of G, if we remove edge e, $G \setminus \{e\}$ is not rigid. Then, by Proposition 2, there exists a partition $\mathcal{P}$ satisfying $e \in \delta_G(\mathcal{P})$ such that

$$(D - 1)|\delta_{G \setminus \{e\}}(\mathcal{P})| < D(|\mathcal{P}| - 1) \tag{7}$$

holds. Thus,

$$(D - 1)|\delta_G(\mathcal{P})| < D|\mathcal{P}| - 1 \tag{8}$$

follows. Let $\mathcal{P} = \{V_1, V_2, \ldots, V_m\}$ be such that $v \in V_1$. Then, let us consider the following cases depending on whether $|V_1| \geq 2$ or $|V_1| = 1$.

Subcase 2A. $|V_1| \geq 2$. Let $\mathcal{P}'$ be the partition obtained by removing v from V_1 in $\mathcal{P}$. Then, we have

$$|\mathcal{P}'| = |\mathcal{P}|, \ |\delta_{G'}(\mathcal{P}')| \leq |\delta_G(\mathcal{P})| - 1. \tag{9}$$

By (9), we have

$$(D - 1)|\delta_{G'}(\mathcal{P}')| - D(|\mathcal{P}'| - 1) \leq (D - 1)(|\delta_G(\mathcal{P})| - 1) - D(|\mathcal{P}| - 1)$$
$$= (D - 1)|\delta_G(\mathcal{P})| - D|\mathcal{P}| + 1 < 0. \tag{10}$$

Because this fact contradicts the rigidity of G', there does not exist such $\mathcal{P}$.

Subcase 2B. $|V_1| = 1$. Let $\mathcal{P}' = \mathcal{P} \setminus \{\{v\}\}$. (i) a and b belong to V_i for some $i \neq 1$. Then, we have

$$|\mathcal{P}'| = |\mathcal{P}| - 1, \ |\delta_{G'}(\mathcal{P}')| = |\delta_G(\mathcal{P})| - 3. \tag{11}$$

Then,

$$(D - 1)|\delta_{G'}(\mathcal{P}')| - D(|\mathcal{P}'| - 1) = (D - 1)(|\delta_G(\mathcal{P})| - 3) - D(|\mathcal{P}| - 2)$$
$$= (D - 1)|\delta_G(\mathcal{P})| - D|\mathcal{P}| + 1 - D + 2 < 0. \tag{12}$$

Since this fact contradicts the rigidity of G', there does not exist such $\mathcal{P}$.

(ii) a and b belong to V_i and V_j for some $i \neq j$ with $i > 1, j > 1$. Let $\mathcal{P}' = \mathcal{P} \setminus V_1$. Then, we have

$$|\mathcal{P}'| = |\mathcal{P}| - 1, \ |\delta_{G'}(\mathcal{P}')| = |\delta_G(\mathcal{P})| - 2. \tag{13}$$

Because G_v^{ab} is rigid and edge e is redundant, G' is also a rigid body-hinge graph. Then, by Proposition 2, we have

$$(D - 1)|\delta_{G'}(\mathcal{P}')| \geq D(|\mathcal{P}'| - 1). \tag{14}$$

By (13) and (14), we have

$$(D-1)|\delta_G(\mathcal{P})| \geq D|\mathcal{P}| - 2. \tag{15}$$

By (8) and (15),

$$(D-1)|\delta_G(\mathcal{P})| = D|\mathcal{P}| - 2 \tag{16}$$

follows.

By (13) and (16), we have

$$(D-1)(|\delta_{G'}(\mathcal{P}')| + 2) = D(|\mathcal{P}'| + 1) - 2$$
$$(D-1)|\delta_{G'}(\mathcal{P}')| = D(|\mathcal{P}'| - 1). \tag{17}$$

By (17), G' is minimally rigid. Furthermore, notice that the equation (17) is equivalent to the condition of operation 2. Hence, a minimally rigid body-hinge graph of $n-1$ vertices can be constructed by applying operation 2 in the reverse direction for a minimally rigid body-hinge graph G. Therefore, for any given minimally rigid body-hinge simple graph $G = (V, E)$ of n vertices, a minimally rigid body-hinge graph G' of $n-1$ or $n-2$ vertices can be constructed by applying one of the five operations in the reverse direction.   □

As a result, it is possible to construct an algorithm for enumerating minimally rigid body-hinge simple graphs by applying the reverse search [1]. The running time is polynomial per output ($O(n^6)$ time). The details are omitted. Moreover it is possible to define operations to generate all minimally rigid body-hinge multigraphs. In this case, we need three operations instead of operations 4 and 5. The details are also omitted.

References

1. Avis, D., Fukuda, K.: Reverse search for enumeration. Discrete Applied Mathematics 65(1), 21–46 (1996)
2. Frank, A., Szegǫ, L.: Constructive characterizations for packing and covering with trees. Discrete Applied Mathematics 131(2), 347–371 (2003)
3. Henneberg, L.: Die graphische statik der starren system. Leipzig (1911)
4. Katoh, N., Tanigawa, S.: A proof of the molecular conjecture. Discrete and Computational Geometry 45, 647–700 (2011)
5. Laman, G.: On graphs and rigidity of plane skeletal structures. Journal of Engineering Mathematics 4(4), 331–340 (1970)
6. Lee, A., Streinu, I.: Pebble game algorithms and sparse graphs. Discrete Mathematics 308(8), 1425–1437 (2008)
7. Nash-Williams, C.: Edge-disjoint spanning trees of finite graphs. Journal of the London Mathematical Society 36, 445–450 (1961)
8. Roth, A.: The rigidity of graphs. AMS 245, 279–289 (1979)
9. Schrijver, A.: Combinatorial Optimization, vol. B, p. 881, Corollary 51.3b. Springer (2003)
10. Tay, T.: Linking $(n-2)$-dimensional panels in n-space ii:$(n-2, 2)$-frameworks and body and hinge structures. Graphs and Combinatorics 5(1), 245–273 (1989)

11. Tutte, W.T.: On the problem of decomposing a graph into n connected factors. Journal of the London Mathematical Society 36, 221–230 (1961)
12. Uno, T.: A new approach for speeding up enumeration algorithms and its application for matroid bases. In: Asano, T., Imai, H., Lee, D.T., Nakano, S.-I., Tokuyama, T. (eds.) COCOON 1999. LNCS, vol. 1627, pp. 349–359. Springer, Heidelberg (1999)
13. Whiteley, W.: The union of matroids and the rigidity of frameworks. SIAM Journal on Discrete Mathematics 1(2), 237–255 (1988)
14. Whiteley, W.: Some matroids from discrete applied geometry. Contemporary Mathematics 197, 171–311 (1996)
15. Whiteley, W.: Rigidity of molecular structures: generic and geometric analysis. In: Thorpe, M.F., Duxbury, P.M. (eds.) Rigidity Theory and Applications, pp. 21–46 (1999)
16. Whiteley, W.: Rigidity and scene analysis. In: Goodman, J., ORourke, J. (eds.) Handbook of Discrete and Computational Geometry, 2nd edn. ch. 60, pp. 1327–1354. Chapman Hall/CRC Press, Boca Raton, FL (2004)

On Complexities of Minus Domination

Luérbio Faria[1], Wing-Kai Hon[2], Ton Kloks, Hsiang-Hsuan Liu[2],
Tao-Ming Wang[3], and Yue-Li Wang[4]

[1] Instituto de Matemática e Estatística
Universidade do Estado do Rio de Janeiro, Brazil
`luerbio@cos.ufrj.br`
[2] Department of Computer Science
National Tsing Hua University, Taiwan
`{wkhon,hhliu}@cs.nthu.edu.tw`
[3] Department of Applied Mathematics
Tunghai University, Taichung, Taiwan
`wang@go.thu.edu.tw`
[4] Department of Information Management
National Taiwan University of Science and Technology
`ylwang@cs.ntust.edu.tw`

Abstract. A function $f : V \to \{-1, 0, 1\}$ is a minus-domination function of
a graph $G = (V, E)$ if the values over the vertices in each closed neighborhood sum to a positive number. The weight of f is the sum of $f(x)$ over all
vertices $x \in V$. In the minus-domination problem, one tries to minimize the
weight of a minus-domination function. In this paper, we show that (1) the
minus-domination problem is fixed-parameter tractable for d-degenerate
graphs when parameterized by the size of the minus-dominating set and by
d, where the size of a minus domination is the number of vertices that are
assigned 1, (2) the minus-domination problem is polynomial for graphs of
bounded rankwidth and for strongly chordal graphs, (3) it is NP-complete
for splitgraphs, and (4) unless P = NP there is no fixed-parameter algorithm for minus-domination.

1 Introduction

A fresh breeze seems to be blowing through the area of domination problems.
This research area is aroused anew by the recent fixed-parameter investigations
(see, e.g., [2,7,26,27]).

Let $G = (V, E)$ be a graph and let $f : V \to S$ be a function that assigns some
integer from $S \subseteq \mathbb{Z}$ to every vertex of G. For a subset $W \subseteq V$ we write

$$f(W) = \sum_{x \in W} f(x).$$

The function f is a domination function if for every vertex x, $f(N[x]) > 0$, where
$N[x] = \{x\} \cup N(x)$ is the closed neighborhood of x. The *weight* of f is defined as
the value $f(V)$.

P. Widmayer, Y. Xu, and B. Zhu (Eds.): COCOA 2013, LNCS 8287, pp. 178–189, 2013.

In this manner, the ordinary domination problem is described by a domination function that assigns a value of $\{0, 1\}$ to each element of V. A signed domination function assigns a value of $\{-1, 1\}$ to each vertex x. The minimal weight over all dominating and signed dominating functions are denoted by $\gamma(G)$ and $\gamma_s(G)$, respectively. In this paper we look at the minus-domination problem.

Definition 1. *Let* $G = (V, E)$ *be a graph. A function* $f : V \to \{-1, 0, 1\}$ *is a minus-domination function if* $f(N[x]) > 0$ *for every vertex* x.

In the minus-domination problem one tries to minimize the weight of a minus-domination function. The minimal weight over all minus-domination functions is denoted as $\gamma^-(G)$. Notice that the weight may be negative. For example, consider a K_4 and add one new vertex for every edge, adjacent to the endpoints of that edge. Assign a value 1 to every vertex of the K_4 and assign a value -1 to each of the six other vertices. This is a valid minus-domination function and its weight is -2.

The problem to determine the value of $\gamma^-(G)$ is NP-complete, even when restricted to bipartite graphs, chordal graphs and planar graphs with maximal degree four [3,5]. Sharp bounds for the minimum weight are obtained in, e.g., [21].

Damaschke shows that, unless $P = NP$, the value of γ^- cannot be approximated in polynomial time within a factor $1 + \epsilon$, for some $\epsilon > 0$, not even for graphs with maximum degree at most four [3, Theorem 7].

In this paper we settle the complexity of the minus-domination problem for planar graphs, d-degenerate graphs, cographs, strongly chordal graphs and splitgraphs.

2   Planar Graphs and d-Degenerate Graphs

In this section, we show that the minus-domination problem is fixed-parameter tractable for planar graphs and d-degenerate graphs.

2.1   Planar Graphs

Determining the smallest weight over all minus-dominating functions is NP-complete, even when restricted to planar graphs [5].

Let $G = (V, E)$ be a graph and let $f : V \to S$ be a domination function. Following Zheng et al. [26], we define the *size* of f as the number of vertices $x \in V$ with $f(x) > 0$. We denote the size of a minus-dominating function f as size(f).

Consider signed-domination functions of size at most k. It is easy to see that $|V(G)| = O(k^2)$ (see [26]). It follows that the signed domination problem parameterized by the size is fixed-parameter tractable. This is not so clear for the minus domination problem. For example, consider a star and assign to the center a value of 1 and to every leaf a value of zero. This is a valid minus-domination function with size 1 but the number of vertices is unbounded.

Theorem 1. *For planar graphs the minus-domination problem, parameterized by the size, is fixed-parameter tractable.*

Proof. Let $f : V \to \{-1, 0, 1\}$ be a minus-domination function. Let

$$D = \{ x \mid x \in V \quad \text{and} \quad f(x) = 1 \}.$$

Then D is a dominating set in G. It follows that, for all graphs G,

$$\gamma^-(G) \leqslant \gamma(G) \leqslant \min \{ \text{size}(f) \mid f \text{ is a minus-dominating function} \}.$$

The first subexponential fixed-parameter algorithm for domination in planar graphs appeared in [1]. In [1], the authors prove that, if G is a planar graph with $\gamma(G) \leqslant k$, then the treewidth of G is $O(\sqrt{k})$. Using a tree decomposition of bounded treewidth one can solve the domination problem in $O(2^{15.13\sqrt{k}} \cdot k + n^3 + k^4)$ time (or conclude that $\gamma(G) > k$). The results were generalized to some nonplanar classes of graphs by Demaine et al. [4].

The minus-domination problem with size bounded by k can be formulated in monadic second-order logic. By Courcelle's theorem, any such problem can be solved in linear time on graphs of bounded treewidth (see, e.g., [13,15]). This proves the theorem.   □

2.2   d-Degenerate Graphs

Definition 2. *A graph is* d-*degenerate if each of its induced subgraphs has a vertex of degree at most* d.

Graphs with bounded degeneracy contain, e.g., graphs that are embeddable on some fixed surface, families of graphs that exclude some minor, graphs of bounded treewidth, etc. [24].

In this subsection we show that, for each fixed d, the minus domination problem, parameterized by the size, is fixed-parameter tractable for d-degenerate graphs.

In the minus domination problem one searches for a partition of the vertices into three parts, say red, white and blue. The red vertices are assigned -1, white are 1 and blue are 0. Notice that, when considering a partition of the vertices, we allow that some parts of the partition are empty. Zheng et al. proved a lemma similar to the one below for the signed domination problem in [27, Theorem 2] and [26, Lemma 6].

Lemma 1. *Assume that* $G = (V, E)$ *has a minus-dominating function with size at most* k. *Let* R, W *and* B *be the coloring of the vertices into red, white and blue, defined by this minus-domination function. Then*

$$|W \cup R| = O(k^2).$$

Proof. By assumption, the minus-domination function colors at most k vertices white. Consider the subgraph G' induced by the red and white vertices. Consider a vertex x of G'. Then at least half of its neighbors are colored white, otherwise its closed neighborhood has weight at most zero. Since there are at most k white vertices, each vertex of G' has degree less than 2k.

Notice also that each red vertex has at least two white neighbors. Since there are only k white vertices, and each white vertex has degree less than 2k, the number of red vertices is less than $2k^2$. This proves the lemma.   □

For algorithmic purposes one usually considers the following generalization of the domination problem. Consider graphs in which each vertex is either colored black or white. In the parameterized black-and-white domination problem the objective is to find a set D of at most k vertices such that

> for each black vertex x, $N[x] \cap D \neq \emptyset$.

Obviously, the domination problem is a special case, in which each vertex is black.

For the minus-domination problem we describe an algorithm for a black-and-white version, where the vertices with a 0 or -1 are black and such that each closed neighborhood of a black vertex has a positive weight. To see that this solves the minus-domination problem, just consider the case where all vertices are black.

Alon and Gutner prove, in their seminal paper, that the domination problem is fixed-parameter tractable for d-degenerate graphs [2]. The main ingredient of their paper is the following lemma.

Lemma 2. *Let* $G = (V, E)$ *be a d-degenerate black-and-white colored graph. Let B and W be the set of black and white vertices. If* $|B| > (4d + 2)k$ *then the set*

$$\Omega = \left\{ x \mid x \in V \quad and \quad |N[x] \cap B| \geqslant \frac{|B|}{k} \right\} \quad satisfies \quad |\Omega| \leqslant (4d + 2)k.$$

To prove that the minus-domination problem, parameterized by the size, is fixed-parameter tractable for d-degenerate graphs, we adapt the proof of [2, Theorem 1].

Theorem 2. *For each d and k, there exists a linear algorithm for finding a minus-domination of size at most k in a d-degenerate black-and-white graph, if such a set exists.*

Proof. Let B and W be the set of black and white vertices. First assume that $|B| \leqslant (4d + 2)k$. If there is a minus-domination function of size at most k then there are k vertices (assigned 1) that dominate all vertices in B.

The algorithm tries all possible subsets $R \subseteq B$ for the set of red vertices (those are assigned -1). Number the closed neighborhoods of the vertices in $R \cup B$, say

$$N_1, \ldots, N_t,$$

where $t = |B \cup R| \leqslant (4d + 2)k$. Define an equivalence relation on the vertices of $V \setminus R$ by making two vertices equivalent if they are contained in exactly the same subsets N_i. For each equivalence class that contains more than k vertices which are not red, remove all of them except at most k vertices. This kernelization reduces the graph to an instance H with at most $g(k, d)$ vertices, for some function g.

Consider all subsets of $V(H)$ with at most k vertices of which none is red. Give these vertices the value 1 and the remaining vertices that are not red the value 0. Check if this is a valid minus-domination.

Now assume that $|B| > (4d + 2)k$. Then, by Lemma 2, $|\Omega| \leqslant (4d + 2)k$. Notice that at least one vertex of Ω is assigned 1 in any minus-domination function of size k. In that case the algorithm grows a search tree of size at most $(4d + 2)^k \cdot k!$ before it arrives at the previous case (see [2]).   □

3   Cographs

A minus domination with bounded size can be formulated in monadic second-order logic without quantification over subsets of edges. It follows that there is a linear-time algorithm to solve the problem for graphs of bounded treewidth or rankwidth (or cliquewidth) [17]. It is less obvious that γ^- is computable for bounded rankwidth when there is no restriction on the size. In this section we adapt a method of Yeh and Chang [25] to show this.

It is well-known that the graphs of rankwidth one are the distance-hereditary graphs. We first analyze the complexity of the minus-domination problem for the class of cographs. Cographs form a proper subclass of the class of distance-hereditary graphs.

We denote a path with four vertices by P_4.

Definition 3. *A cograph is a graph without induced* P_4.

Cographs are characterized by the property that each induced subgraph with at least two vertices is either a join or a union of two smaller cographs. It follows that cographs admit a decomposition tree (T, f) where T is a rooted binary tree and where f is a bijection from the vertices of G to the leaves of T. Each internal node is labeled as $\otimes$ or $\oplus$. When the label is $\otimes$ then all vertices of the left subtree are adjacent to all vertices of the right subtree. A node that is labeled as $\otimes$ is called a join-node. When the label is $\oplus$ there is no edge between vertices of the right and left subtree. A node that is labeled as $\oplus$ is called a union-node. One refers to a decomposition tree of this type as a cotree. A cotree for a cograph can be obtained in linear time.

Theorem 3. *There exists a polynomial algorithm that computes* γ^- *for cographs.*

Proof. Let $G = (V, E)$ be a cograph. We assume that a cotree for G is part of the input. Consider a subtree T' and let $W \subseteq V$ be the set of vertices that are mapped to the leaves in T'.

For three numbers (a, b, c), an (a, b, c)-function is a function $f : W \to \{-1, 0, 1\}$ such that f assigns a vertices the value -1, b vertices the value 0 and c vertices the value 1. Obviously, we have that $a + b + c = |W|$.

For an integer t, let

$$\zeta(t, a, b, c) = \max |\{ x \mid x \in W \quad \text{and} \quad f(N[x] \cap W) + t > 0 \quad \text{and}$$
$$\text{where } f \text{ is an } (a, b, c)\text{-function} \}|.$$

When the set is empty we let $\zeta(t, a, b, c) = -\infty$.

Notice that a minus-domination function with minimum weight can be computed when ζ is known for the root node, that is, when $W = V$. Namely,

$$\gamma^-(G) = \min \{ -a + c \mid a + b + c = n \quad \text{and} \quad \zeta(0, a, b, c) = n \}.$$

We show how the values $\zeta(t, a, b, c)$ can be computed. Assume that G is the union of two cographs $G_1 = (V_1, E_1)$ and $G_2 = (V_2, E_2)$. We denote the ζ-values for G_1 and G_2 by ζ_1 and ζ_2. Then

$$\zeta(t, a, b, c) = \max \{ \zeta_1(t, a_1, b_1, c_1) + \zeta_2(t, a_2, b_2, c_2)$$
$$\text{where} \quad a_1 + a_2 = a \quad b_1 + b_2 = b \quad c_1 + c_2 = c \}.$$

Now assume that G is the join of G_1 and G_2. Then

$$\zeta(t, a, b, c) = \max \{ \zeta_1(t - c_2 + a_2, a_1, b_1, c_1) + \zeta_2(t - c_1 + a_1, a_2, b_2, c_2)$$
$$\text{where} \quad a_1 + a_2 = a \quad b_1 + b_2 = b \quad c_1 + c_2 = c \}.$$

This proves the theorem.   □

Remark 1. Notice that complete multipartite graphs are cographs. Formulas for the signed and minus domination number of complete multipartite graphs appear in a recent paper by Liang [19].

3.1 Distance-Hereditary Graphs

Definition 4. *A graph G is distance hereditary if for every pair of nonadjacent vertices x and y and for every connected induced subgraph H of G which contains x and y, the distance between x and y in H is the same as the distance between x and y in G.*

Distance-hereditary graphs are the graphs of rankwidth one (see, e.g., [15]). They were introduced in 1977 by Howorka [12] as those graphs in which, for every pair of nonadjacent vertices, all the chordless paths between them have the same length. They have a decomposition tree (T, f) where T is a rooted binary tree and f is a bijection from the vertices to the leaves of T. For each branch, the 'twinset' of that branch is defined as those vertices in the leaves that have neighbors in leaves outside that branch. Each twinset induces a cograph. Each internal node of T is labeled as $\oplus$ or $\otimes$. When the label is $\otimes$ then all the vertices in the twinset of the left branch are adjacent to all the vertices in the twinset of the right branch. When the label is $\oplus$ there are no edges between vertices mapped to different branches. The twinset of a parent is either empty, or the twinset of one of the two children or the union of the twinsets at the two children.

Theorem 4. *There exists a polynomial algorithm that computes γ^- for distance-hereditary graphs.*

Proof. See ArXiv1307.6663.

□

Remark 2. It is not hard to see that similar results can be derived for graphs of bounded rankwidth, that is, γ^- is computable in polynomial time for graphs of bounded rankwidth (see, e.g., [15]). The rankwidth appears as a function in the exponent of n. Graphs of bounded treewidth are contained in the class of bounded rankwidth and so a similar statement holds for graphs of bounded treewidth. At the moment we do not believe that there is a fixed-parameter algorithm, parameterized by treewidth or rankwidth, to compute γ^-. The results of [27, Section 4.2] seem wrong.[1]

4   Strongly Chordal Graphs

The minus domination problem is NP-complete for chordal graphs. In this section we show that the problem can be solved in polynomial time for strongly chordal graphs.

A graph is chordal if it has no induced cycle of length more than three. A chord in a cycle is an edge that runs between two vertices that are not consecutive in the cycle. Let $C = [x_1, \ldots, x_{2k}]$ be an even cycle of length $2k$. A chord $\{x_i, x_j\}$ in C is odd if the distance in C between x_i and x_j is odd.

Definition 5. *A chordal graph G is strongly chordal if each cycle in G of even length at least 6 has an odd chord.*

There are many characterizations of strongly chordal graphs [6,14]. Perhaps the best known examples of strongly chordal graphs are the interval graphs.

---

[1] We communicated with the authors of [27] and our ideas about it are now in agreement.

In strongly chordal graphs the domination number is equal to the 2-packing number (see, e.g., [23, Theorem 7.4.4]). It follows that the domination number for strongly chordal graphs is polynomial [6].

Theorem 5. *The minus domination problem for strongly chordal graphs can be solved in* $O(\min\{n^2, m \log n\})$ *time. Here* n *is the number of vertices and* m *is the number of edges of the graph.*

Proof. Farber [6] describes a linear programming formulation for the domination problem. In this linear programming formulation we can change the variables from x_i to $z_i = x_i + 1$. This changes the constraints $-1 \leqslant x_i \leqslant 1$ into $0 \leqslant z_i \leqslant 2$. The linear program becomes

$$\text{Minimize} \quad \sum_{i=1}^{n} z_i$$

$$\text{subject to} \quad \sum_{i \in N[k]} z_i \geqslant b_k \qquad \text{for each } k$$

$$\text{and} \quad 0 \leqslant z_i \leqslant 2 \qquad \text{for each } i.$$

In our case, the variable b_k is equal to $|N[k]| + 1$.

The closed neighborhood matrix of a strongly chordal graph is totally balanced. By [9,11,16] (see also, e.g., [23, Theorem A.3.4]) the integer program and its linear relaxation have the same value.

To deal with the constraint $z_i \leqslant 2$ we write the LP as

$$\text{Minimize} \quad j^T \cdot z$$

$$\text{subject to} \quad \begin{pmatrix} A \\ -I \end{pmatrix} z \geqslant \begin{pmatrix} b \\ -2 \cdot j \end{pmatrix} \qquad \text{and} \quad z \geqslant 0.$$

Here, the matrix A is the closed neigborhood matrix, and the vector b is equal to

$$b = j + Aj.$$

The dual of this LP is

$$\text{Maximize} \quad b^T \cdot y_1 - 2j^T \cdot y_2$$

$$\text{subject to} \quad Ay_1 \leqslant j + y_2 \quad \text{and} \qquad y_1 \geqslant 0 \quad \text{and} \quad y_2 \geqslant 0.$$

Notice that

$$y_{2,k} = \max\left\{ 0, -1 + \sum_{i \in N[k]} y_i \right\} \quad \text{for all } k.$$

The linear program and its dual can be solved via Farber's method in linear time, when a strong elimination ordering of the graph is part of the input. We omit the details; see also remark 4 below. □

Remark 3. When G is strongly chordal then so is G^2 [20]. A simple vertex of G is simplicial in G^2. The weighted 2-packing problem in G asks for the maximal weight independent set in G^2. This can be solved in linear time [8]. It uses the fact that in any chordal graph, with integer weights on the vertices, the maximal weight over all independent sets equals the minimal number of cliques that have the property that every vertex is covered at least as many times by cliques as its weight.

Corollary 1. *There exists a linear-time algorithm that solves minus domination on interval graphs.*

Remark 4. After the publication of our draft on arXiv, one of the authors of their paper, quoted in the footnote, drew our attention to their result. The authors claim a linear algorithm for minus domination on strongly chordal graphs. (Here, they assume that a strong elimination ordering is a part of the input). [2]

5 Splitgraphs

In this section we show that the minus-domination problem is NP-complete for splitgraphs. We reduce the $(3, 2)$-hitting set problem to the minus-domination problem. The $(3, 2)$-hitting set problem is defined as follows (see, e.g., [22]).

Instance: Let $\mathcal{C}$ be a collection of sets, each containing exactly three elements from a universe $\mathcal{U}$.
Question: Find a smallest set $\mathcal{U}' \subseteq \mathcal{U}$ such that for each $C \in \mathcal{C}$,

$$|C \cap \mathcal{U}'| \geqslant 2.$$

Lemma 3. *The $(3, 2)$-hitting set is NP-complete.*

Proof. The reduction is from vertex cover, i.e., $(2, 1)$-hitting set. The $(2, 1)$-hitting set is defined similar as above, except that in this case every subset has two elements and the problem is to find a subset $\mathcal{U}'$ which hits every subset at least once.

Consider an instance of $(2, 1)$-hitting set. Let $\mathcal{C}$ be the collection of 2-element subsets of a universe $\mathcal{U}$. Add four vertices to the universe, say α, β, γ and δ. Add α to every subset of $\mathcal{C}$ and add two subsets, $\{\alpha, \beta, \gamma\}$ and $\{\alpha, \beta, \delta\}$. We claim that any solution of this $(3, 2)$-hitting set problem has α in the hitting set. If not, then $\{\beta, \gamma, \delta\}$ is a subset of the $(3, 2)$-hitting set. In that case we may replace the elements β, γ and δ with α and β. Then we obtain a $(3, 2)$-hitting set with fewer elements.

Thus, we may assume that α is in the $(3, 2)$-hitting set. But now the problem is equivalent to the $(2, 1)$-hitting set, since every adapted subset contains α. □

Theorem 6. *The minus-domination problem is NP-complete for splitgraphs.*

Proof. See ArXiv1307.6663.

□

---

[2] C. Lee and M. Chang, Variations of Y-dominating functions on graphs, *Discrete Mathematics* **308** (2008), pp. 4185–4204.

5.1   Minus Domination Is Not FPT

Consider the following problem.

Instance: A graph G.
Question: Does G have a minus domination of weight at most 0?

Following Hattingh et al. [10], we call this 'the zero minus-domination problem.'
Consider the graph L in Figure 1.

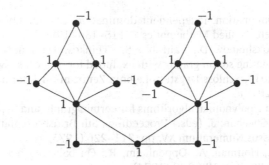

Fig. 1. The graph L. It has $\gamma^-(L) = -1$

Lemma 4. *The graph* L *has minus-domination weight* $\gamma^-(L) = -1$. *The minus-domination function that achieves this weight is unique; it is the one depicted in Figure 1.*

Theorem 7. *The zero minus-domination problem is NP-complete.*

Proof. Let H be a graph and let G be the union of H and k disjoint copies of L. Obviously

$$\gamma^-(G) = \gamma^-(H) + k \cdot \gamma^-(L) = \gamma^-(H) - k.$$

It follows that $\gamma^-(G) \leqslant 0$ if and only if $\gamma^-(H) \leqslant k$. By Theorem 6, given a graph H and a positive k it is NP-complete to decide whether $\gamma^-(H) \leqslant k$.   □

Theorem 8. *The minus-domination problem is para-NP-complete unless* $P = NP$.

Proof. Assume there exists an algorithm which runs in time $O(f(k) \cdot n^c)$ and that determines whether a graph G has a minus domination of weight at most k. Then the zero minus-domination problem would be solvable in polynomial time.   □

References

1. Alber, J., Bodlaender, H., Fernau, H., Kloks, T., Niedermeier, R.: Fixed-parameter algorithms for dominating set and related problems on planar graphs. Algorithmica 33, 461–493 (2002)

2. Alon, N., Gutner, S.: Linear time algorithms for finding a dominating set of fixed size in degenerated graphs. Algorithmica 54, 544–556 (2009)
3. Damaschke, P.: Minus domination in small-degree graphs. Discrete Applied Mathematics 108, 53–64 (2001)
4. Demaine, E., Formin, F., Hajiaghayi, M., Thilikos, D.: Fixed-parameter algorithms for (k, r)-center in planar graphs and map graphs. ACM Transactions on Algorithms 1, 33–47 (2005)
5. Dunbar, J., Goddard, W., Hedetniemi, S., Henning, M., McRae, A.: The algorithmic complexity of minus domination in graphs. Discrete Applied Mathematics 68, 73–84 (1996)
6. Farber, M.: Domination, independent domination, and duality in strongly chordal graphs. Discrete Applied Mathematics 7, 115–130 (1984)
7. Fomin, F., Lokshtanov, D., Saurabh, S., Thilikos, D.: Linear kernels for (connected) dominating set on graphs with excluded topological subgraphs. In: Proceedings STACS 2013, Schloss Dagstuhl-Leibniz-Zentrum für Informatik. LPIcs, vol. 20, pp. 92–103 (2013)
8. Frank, A.: Some polynomial algorithms for certain graphs and hypergraphs. In: Nash-Williams, C., Sheehan, J. (eds.) Proceedings 5th British Combinatorial Conference 1975. Congressus Numeratium XV, pp. 211–226 (1975)
9. Fulkerson, D., Hoffman, A., Oppenheim, R.: On balanced matrices. Mathematical Programming Study 1, 120–132 (1974)
10. Hattingh, J., Henning, M., Slater, P.: The algorithmic complexity of signed domination in graphs. Australasian Journal of Combinatorics 12, 101–112 (1995)
11. Hoffman, A., Kolen, A., Sakarovitch, M.: Totally-balanced and greedy matrices. SIAM Journal on Algebraic and Discrete Methods 6, 721–730 (1985)
12. Howorka, E.: A characterization of distance-hereditary graphs. The Quarterly Journal of Mathematics 28, 417–420 (1977)
13. Kloks, T.: Treewidth. LNCS, vol. 842. Springer, Heidelberg (1994)
14. Kloks, T., Liu, C., Poon, S.: Feedback vertex set on chordal bipartite graphs. Manuscript on arXiv: 1104.3915 (2012)
15. Kloks, T., Wang, Y.: Advances in graph algorithms (2013) (manuscript)
16. Kolen, A.: Location problems on trees and in the rectilinear plane, PhD Thesis, Mathematisch centrum, Amsterdam (1982)
17. Langer, A., Rossmanith, P., Sikdar, S.: Linear-time algorithms for graphs of bounded rankwidth: A fresh look using game theory. In: Ogihara, M., Tarui, J. (eds.) TAMC 2011. LNCS, vol. 6648, pp. 505–516. Springer, Heidelberg (2011)
18. Lee, C., Chang, M.: Variations of Y-dominating functions on graphs. Discrete Mathematics 308, 4185–4204 (2008)
19. Liang, H.: Signed and minus domination in complete multipartite graphs. Manuscript on arXiv: 1205.0343 (2012)
20. Lubiw, A.: Γ-Free matrices, Master's Thesis, University of Waterloo, Canada (1982)
21. Matoušek, J.: Lower bound on the minus-domination number. Discrete Mathematics 233, 361–370 (2001)
22. Mellor, A., Prieto, E., Mathieson, L., Moscato, P.: A kernelisation approach for multiple d-hitting set and its application in optimal multi-drug therapeutic combinations. PLoS One 5, e13055 (2010)
23. Scheinerman, E., Ullman, D.: Fractional graph theory. Wiley (1997)
24. Thomason, A.: The extremal function for complete minors. Journal of Combinatorial Theory, Series B 81, 318–338 (2001)

25. Yeh, H., Chang, G.: Algorithmic aspects of majority domination. Taiwanese Journal of Mathematics 1, 343–350 (1997)
26. Zheng, Y., Wang, J., Feng, Q.: Kernelization and lowerbounds of the signed domination problem. In: Fellows, M., Tan, X., Zhu, B. (eds.) FAW-AAIM 2013. LNCS, vol. 7924, pp. 261–271. Springer, Heidelberg (2013)
27. Zheng, Y., Wang, J., Feng, Q., Chen, J.: FPT results for signed domination. In: Agrawal, M., Cooper, S.B., Li, A. (eds.) TAMC 2012. LNCS, vol. 7287, pp. 572–583. Springer, Heidelberg (2012)

A Linear-Time Algorithm for Reconciliation
of Non-binary Gene Tree and Binary Species Tree

Yu Zheng[1], Taoyang Wu[2], and Louxin Zhang[1]

[1] Department of Mathematics, National University of Singapore, Singapore 119076
{matzhyu,matzlx}@nus.edu.sg
[2] School of Computing Sciences, University of East Anglia, Norwich NR4 7TJ U.K.
taoyang.wu@uea.ac.uk

Abstract. Tree reconciliation approach to inferring the duplication history for a gene family poses challenging problems when input gene and species trees are non-binary. We present the first linear-time algorithm that outputs a reconciliation of a non-binary gene tree and a binary species tree that minimizes the gene loss cost under the constraint of having the smallest gene duplication cost. As a part of a method for reconciling two non-binary trees, this algorithm has been implemented in a software package (http://phylotoo.appspot.com).

1 Introduction

Millions of genes in current species belong to only thousands of gene families. A gene family includes the instances of the same gene in different species and duplicate genes in the same species. Two genes found from different species are orthologous if they arose by speciation in the most recent common ancestor of the species [9]. Since orthologous genes tend to retain similar biological functions, they are often used to infer the pattern of gene gain and loss, the mode of signaling pathway evolution, and the relationship between genotype and phenotype. Therefore, ortholog identification is often the first task of a genome study.

Genes are gained through duplication and horizontal gene transfer, and get lost via deletion and pseudogenization in evolution. Identifying orthologs is essentially to find out how a gene family has evolved. A key method for this problem is to use an explicit model of the evolutionary history of a gene family, in the form of gene tree. It compares the gene tree and the evolutionary history of the species where the genes are found – the species tree – using a procedure known as tree reconciliation [14]. Species tree reconciliation formalizes the following intuition: If the offspring of a binary gene tree node is distributed in the same set of species as that of one of its child nodes, then the node corresponds to a duplication event [10,19]. The tree reconciliation approach is less prone to error than a sequence-match-based method, particularly in the situation when gene loss events are not rare [14].

Standard reconciliation between a binary gene tree and a binary species tree is linear-time computable [6,22]. However, tree reconciliation presents challenging

P. Widmayer, Y. Xu, and B. Zhu (Eds.): COCOA 2013, LNCS 8287, pp. 190–201, 2013.

problems when either the input gene tree or the input species tree is not binary [7,21]. Non-binary gene tree nodes are called soft polytomies because the true pattern of gene divergence is binary [12], but there is not enough signal in the data to time the true diverging events. On top of ambiguity in gene tree, there are also uncertainties in a species tree. The NCBI taxonomy database and other reference species trees are often non-binary due to unsolved species diverging order [13]. Notung, the best packages for tree reconciliation, requests that one of the two reconciled trees has to be binary [7,21].

Here, we focus on the reconciliation of a non-binary gene tree and a binary species tree. Quadratic algorithms are known for this problem [4,7]. It was also independently studied for arbitrary species trees in [3], where the optimality criterion used is inferring the smallest set of gene duplication events that poses the least gene loss events. Such an optimality criterion is used since a gene lose event is often attributed to an absence in the species when many species have sparse sampling of genes [3]. Unfortunately, the heuristic method proposed in the work might stop before an optimal solution is found. Hence, in general, it overestimates the number of loss events.

In the present paper, we present the first linear-time algorithm that outputs a reconciliation that minimizes the gene loss cost under the constraint of having the smallest gene duplication cost. Most importantly, it is a natural extension of the standard reconciliation procedure from binary to non-binary gene trees.

2   Tree Reconciliation Problem

2.1   Gene Trees and Species Trees

Both gene and species trees are rooted trees in which only the "root" node is of degree two and all the other nodes have degree 1 or at least 3. Both gene and species trees are also leaf–labeled. A species tree represents the evolutionary history of a set of modern species. In a species tree, a leaf is labeled by a unique modern species, whereas each internal node naturally corresponds to a unique subset of modern species.

A gene tree represents the evolutionary relationship of the members of a gene family, in which a leaf represents a unique gene. A gene family having multiple members found in a species is often a product of a series of gene duplication events. Since the gene tree of a gene family does not represent its explicit duplication history, one has to reconcile it and the corresponding species tree to infer the duplication history of the gene family.

In the study of tree reconciliation, a gene tree leaf is labeled with the species in which the corresponding gene is found. Clearly, the gene tree of a gene family is not uniquely leaf labeled if multiple gene members are found in the same modern species. Since a gene duplication event might be inferred above the root of a species tree, we draw a branch entering its root (see Fig. 1 for example). For a gene or species tree T, we use $r(T)$ to denote its root; we use $V(T)$, $Lf(T)$,

$\overset{\circ}{V}(T)$, and $E(T)$ to denote the sets of nodes, leaves, internal (or non-leaf) nodes and all the branches of T, respectively. Clearly, $V(T) = \mathrm{Lf}(T) \cup \overset{\circ}{V}(T)$.

For any $u \in V(T)$, all the nodes in the path from $r(T)$ to u are the *ancestors* of u. If w is an ancestor of u, we also call u a *descendant* of w. The induced subtree of T on all the descendants of u is written $T(u)$. We use $L(u)$ to denote the set of the labels of the leaves of $T(u)$.

We say w is the parent of v or v is a child of w if w is an ancestor of v and a branch connects w and v. The parent of v is denoted by $p(v)$. A tree node is *binary* if it has exactly two children; it is *non-binary* otherwise. A tree is *binary* if all its internal nodes are binary (Fig. 1A) and *non-binary* otherwise.

Finally, for each $u \in \mathrm{Lf}(T)$, we use $l_T(u)$ to denote the label of u. In addition, $l_T^{-1}(c)$ denotes the unique leaf having label c if T is a species tree.

2.2   Tree Reconciliation

Consider a *binary* species tree S and a *binary* gene tree G such that $L(r_G) = L(r_S)$. A *reconciliation* f between G and S is a map from $V(G)$ to $V(S)$ having the following properties:

(i) (Leaf-preservation) For each $x \in \mathrm{Lf}(G)$, $f(x) \in \mathrm{Lf}(S)$, having the same label as x.

(ii) (Order-preservation) For $g, g' \in V(G)$ such that g' is a descendant of g, then, either $f(g') = f(g)$ or $f(g')$ is a descendant of $f(g)$.

For a subset $X \subseteq V(S)$, we use $\mathrm{lca}(X)$ to denote the unique node y that satisfies the following two properties: (i) y is a common ancestor of nodes of X, that is, an ancestor of every node of X; (ii) no child of y is a common ancestor of the nodes of X. This $\mathrm{lca}(X)$ is called the most recent common ancestor of the nodes in X in S. Clearly, the map λ defined by:

$$\lambda(u) = \begin{cases} l_S^{-1}(l_G(u)), & u \in \mathrm{Lf}(G), \\ \mathrm{lca}\left(\{\lambda(v)|p(v) = u\}\right), & u \in \overset{\circ}{V}(G) \end{cases} \tag{1}$$

is a reconciliation between G and S. Note that λ is the minimum one in the sense that, for any reconciliation f between G and S, $\lambda(u) = f(u)$ or $\lambda(u)$ is a descendant of $f(u)$ for every $u \in V(G)$. Hence, λ is called the *lca reconciliation* between G and S [10].

A $u \in \overset{\circ}{V}(G)$ is a duplication node if $\lambda(u) = \lambda(v)$ for some child v of u. The number of inferred duplications is defined as the *duplication cost* of the reconciliation of G and S, denoted by $D(G, S)$ [19]. A hypothetical evolutionary history of the gene family will be obtained if, for each duplication node in G, a duplication is assumed to occur in the branch entering to $\lambda(u)$ in S.

The resulting evolutionary history of the gene family may contain gene loss events in general. For example, for a non-root $v \in V(G)$, if $\lambda(p(v)) \neq \lambda(v)$, we have to assume the corresponding ancestral gene copy had lost in each species represented by a branch off the lineage path from $\lambda(p(v))$ to $\lambda(v)$. Overall, we have to assume

$$l(G, S) = \sum_{u \in \overset{\circ}{V}(G)} \sum_{v:p(v)=u} [n(u, v) + d(u) - 2]$$

gene losses, where $d(u) = 1$ if u is a duplication node under λ and 0 otherwise, and $n(u, v)$ denotes the number of nodes in the path between u and v inclusively. The $l(G, S)$ is called the *gene loss cost* of the λ between G and S. For example, $D(G', S) = 4$, $D(G'', S) = 3$, and $l(G', S) = l(G'', S) = 7$ for G', G'' and S given in Fig. 1.

The weighted linear combination of the gene duplication and loss costs (called the affine reconciliation cost here) is also used to study gene and species tree reconciliation.

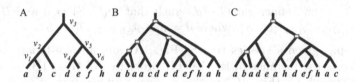

Fig. 1. An illustration of binary tree reconciliation. **A.** A binary species tree S of 7 species a, b, c, d, e, f, h. **B.** A binary gene tree G' of a gene family over the species. The reconciliation between G' and S produces an evolutionary history of 4 duplication (indicated by square) and 7 gene loss events (not shown). **C.** Another gene tree G'' over the same gene family. The reconciliation between G'' and S produces an evolutionary history of 3 duplication and 7 gene loss events, shown in Fig. 3.

2.3   Reconciliation of Two Non-binary Trees

In graph theory, an edge contraction is an operation that removes an edge from a graph while simultaneously merging the two nodes previously connected by the edge. Given two rooted trees T and T', T' *refines* T if T can be obtained from T' via a series of edge contractions. Clearly, we can map each $v \in V(T)$ to a unique node $v' \in V(T')$ such that $L_T(v) = L_{T'}(v')$. Moreover, if T' is a binary tree, T' is called a *binary refinement* of T. We use $\mathcal{BR}(T)$ to denote the set of all binary refinements of T.

We shall study non-binary tree reconciliation in the binary refinement model [8]. The reconciliation problem is formally defined as a discrete optimization problem:

General Reconciliation (GR)
INPUT: A species tree S, a set of gene trees G_i $(1 \le i \le k)$ and a reconciliation cost $c(\,,)$.
OUTPUT: A $\hat{S} \in \mathcal{BR}(S)$ and $\hat{G}_i \in \mathcal{BR}(G_i)$ for every i such that $\sum_{1 \le i \le k} c(\hat{G}_i, \hat{S})$ is minimized.

When the input species and gene trees are binary, the solution is the lca reconciliation defined earlier for the duplication, loss, or affine reconciliation cost [5,11] and hence linear-time computable [6,22].

A star tree is a rooted tree in which the all the leaves are the children of the root. Clearly, every binary tree is a binary refinement of the star tree over the same set of species. When the input species tree S is a star tree and the input gene trees G_i are binary, a solution to the GR problem is just a binary species tree $\hat{S}$ that minimizes $\sum_i c(G_i, \hat{S})$ over all the binary species trees. Hence, the species tree problem becomes a special case of the GR problem. Recall that the species tree problem is NP-hard [17] and even is hard to be approximated within a logarithmic factor [2]. Therefore, the GR problem is an important, but difficult problem in computational biology. Here, we shall work on the following special case of the GR problem:

Non-binary Gene Tree Refinement (NGTR)
INPUT: A binary species tree S and a non-binary gene tree G.
OUTPUT: A binary refinement $\hat{G}$ of G such that $D(\hat{G}, S) = d$ and $l(\hat{G}, S) = \min_{G' \in BR(G):D(G',S)=d} l(G', S)$, where $d = \min_{G' \in BR(G)} D(G', S)$.

For example, given a species tree S (Fig. 1A) and an non-binary gene tree G (Fig. 2A) as the instance of the NGTR problem, G'' (Fig. 1B) is a desired solution. Although G' (Fig. 1C) is also a binary refinements of G, it is not a solution, as it does not give the smallest duplication cost.

3   Algorithm

Consider a species tree S and a gene tree G such that $L(G) = L(S)$. For each $G' \in BR(G)$, every $g \in V(G)$ and the corresponding node in G' are mapped to the same node in S under the lca reconciliations λ_G and $\lambda_{G'}$. Hence, the refinement of G can be achieved by refining each star subtree consisting of a non-binary node and its children separately.

Our goal is to find $\hat{G} \in BR(G)$ by resolving every non-binary node of G using S such that $\hat{G}$ has the smallest duplication cost d when $\hat{G}$ and S are reconciled. Moreover, the reconciliation of $\hat{G}$ and S also has the minimum loss cost over $G' \in BR(G)$ such that $D(G', S) = d$.

Consider a non-binary internal node $g \in \overset{\circ}{V}(G)$ with k children g_i, $k \geq 3$. We set $I(g) = \{s \in V(S) : s \in P(\lambda(g), \lambda(g_i))$ for some $i\}$, where $P(\lambda(g), \lambda(g_i))$ denotes the set of nodes in the path from $I(g)$ forms a subtree rooted at $\lambda(g)$ (see Fig. 2B). For simplicity, we also use $I(g)$ to represent the resulting subtree. Clearly, in $I(g)$, every leaf is $\lambda(g_i)$ for some i. However, $I(g)$ may not be a binary subtree because some internal nodes may have a child not belonging to $I(g)$ (Fig. 2B). We use $I^+(g)$ to denote the binary tree obtained by including all the children of the non-leaf nodes of $I(g)$. For each $x \in V(I^+(g))$, we define $\omega(x)$ to be the number of the children (of g) that are mapped to x under λ. We further define:

$$m(x) = \begin{cases} \omega(x), \text{if } x \in \text{Lf}(I^+(g)), \\ \omega(x) + \max\{m(x_1), m(x_2)\}, \text{otherwise,} \end{cases}$$

where x_1 and x_2 are the children of x in $I^+(g)$.

For the non-binary tree g in G in Fig. 2A and the species tree S in Fig. 1A, we have:

$$\omega(a) = \omega(v_i) = 1, \ 1 \le i \le 5, \ m(b) = m(c) = m(v_6) = 0,$$
$$m(a) = m(v_4) = 1, \ m(v_1) = m(v_5) = 2, \ m(v_2) = 3, \ m(v_3) = 4.$$

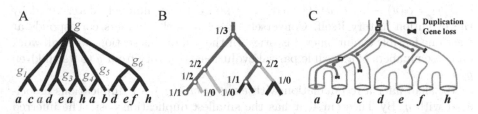

Fig. 2. An illustration of computing $m(\ ), \alpha(\ ), \beta(\ )$ for refining a non-binary gene tree node. **A.** A gene tree G with non-binary root g. **B.** $I(g)$ and $I^+(g)$ in the reconciliation of G (panel A) and S (Figure 1.A). $I(g)$ consists of v_1–v_5 and the leftmost leaf labeled with species a (red circles), on which $\alpha(x)$ and $\beta(x)$ is written beside each node x in the format of $\alpha(x)/\beta(x)$. $I^+(g)$ is obtained from $I(g)$ by adding v_6 and the leaves labeled with species b and c. The edges in $I^+(g)$ but not in $I(g)$ are drawn in gray. **C.** The duplication history of the children of g induced by refining g. After g is refined, the resulting binary refinement of G is G'' in Fig. 1.C.

Theorem 1. *At least $m(\lambda(g)) - 1$ duplication events are required to produce the ancestral genes represented by $g_1, g_2, \cdots, g_k$ in the reconciliation of S and any $G' \in \mathcal{BR}(G)$.*

Proof. Consider the following partial order set:

$$\mathcal{O} = (\{L(\lambda(g_1)), L(\lambda(g_2)), \cdots, L(\lambda(g_k))\}, \subseteq),$$

where $L(\lambda(g_i))$ is the set of labels of the leaves below in $\lambda(g_i)$ in S and $\subseteq$ is the set inclusion. For the example presented in Fig 2, $\mathcal{O}$ contains the following six subsets:

$$\{a, b, c\}, \ \{a\}, \ \{a, b\}, \ \{d, e\}, \ \{a, b, c, d, e, f, h\}, \ \{d, e, f, h\}.$$

Note that all the image nodes appearing in a path from the root to a leaf in $I^+(g)$ form a chain in $\mathcal{O}$. We have that $m(\lambda(g))$ is the largest size of such a chain. An $\mathcal{A} \subset \mathcal{O}$ is an antichain if for any $x, y \in \mathcal{A}$, they are incomparable, i.e.,

$x \not\subseteq y$ and $y \not\subseteq x$. For any $i \neq j$, if $L(\lambda(g_i))$ and $L(\lambda(g_j))$ are incomparable, they are disjoint since they correspond to two tree nodes in S. Hence, an antichain consists of disjoint elements in $\mathcal{O}$. Let M be the smallest number of antichains into which $\mathcal{O}$ may be partitioned. In [3], it is proved that $M-1$ is a lower bound on the number of duplications needed to produce $g_1, g_2, \cdots, g_k$. By a dual of Dilworth's theorem [18], M is equal to $m(\lambda(g))$, the size of the longest chain. $\square$

Consider a hypothetical evolution of a gene family in the corresponding species tree S as shown in Fig. 3. Thick branches represent ancestral species in S. Two numbers are naturally associated with each branch $e = (p(u), u)$: the number $\alpha(u)$ of ancestral gene copies flowing into e, and the number $\beta(u)$ of ancestral gene copies flowing out e. Clearly, if t duplication events occurred inside e, $\beta(u) = \alpha(u) + t > \alpha(u)$, where we assume each duplication event produces exactly one copy of only one gene. Similarly, if l gene loss events occurred inside e, $\beta(u) = \alpha(u) - l < \alpha(u)$. Clearly, $\alpha(u)$ and $\beta(u)$ are uniquely determined by the evolution history itself. Conversely, a set of such numbers corresponds at least one feasible evolutionary histories. In the rest of this section, we shall work on these numbers of a feasible partial evolutionary history from g to its children g_i's.

We shall infer a reconciliation with exactly $m(\lambda(g)) - 1$ duplications associated with g. By Theorem 1, it has the smallest duplication cost. The inferred duplication events have to be postulated on proper branches of $I^+(g)$ to minimize gene losses simultaneously. To this end, we define $\alpha(u)$ and $\beta(u)$ for each node u of $I^+(g)$ as follows. In the rest of this section, u_1 and u_2 denote the two children of u if it is an internal node of S.

Let $r = \lambda(g)$, the root of the subtree $I^+(g)$. We set

$$\alpha(r) = 1, \tag{2}$$

$$\beta(r) = \max\{\min(m(r_1), m(r_2)), 1\} + \omega(r). \tag{3}$$

For any $u \neq r$ with parent $p(u)$ and sibling u' in $I^+(g)$,

$$\alpha(u) = \alpha(u') = \beta(p(u)) - \omega(p(u)), \tag{4}$$

$$\beta(u) = \begin{cases} m(u), & \text{if } \alpha(u) \geq m(u), \\ m(u), & \text{if } u \text{ is a leaf of } I^+(g), \\ \gamma(u), & \text{otherwise}, \end{cases} \tag{5}$$

where $\gamma(u)$ is:

$$\max\{\alpha(u), \min(m(u_1), m(u_2)) + \omega(u), 1 + \omega(u)\}. \tag{6}$$

Continue the example given in Fig. 1–2, the computation of $\alpha()$ and $\beta()$ is shown in Fig. 2B.

If $\alpha(u) < \beta(u)$, we postulate $\beta(u) - \alpha(u)$ duplication events on the branch $(p(u), u)$; if $\alpha(u) > \beta(u)$, we postulate $\alpha(u) - \beta(u)$ gene loss events on the branch instead. In total, we postulate $\sum_{u \in I^+(g)} \max(\beta(u) - \alpha(u), 0)$ duplication events and $\sum_{u \in I^+(g)} \max(\alpha(u) - \beta(u), 0)$ gene loss events.

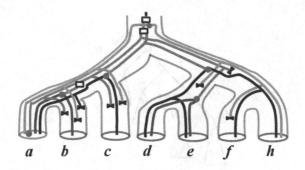

Fig. 3. A schematic view of the duplication history of the gene family inferred from the lca reconciliation of S (Fig. 1A) and G (Fig. 2A). It is extended from the partial history in Fig. 2.C by reconciling the children of g, having 3 duplication and 7 loss events.

For the example given in given in Fig. 1–2, we infer two duplication events above the root r and one inside the branch from v_2 to v_1 in S (Fig. 1A), resulting in the binary refinement G'' (Fig. 1C) and the duplication history of the gene family shown in Fig. 3.

Theorem 2. *(1) The reconciliation described above has the smallest duplication cost (that is, $m(\lambda(g)) - 1$) for resolving g.*
(2) It also has the minimum gene loss cost over all the reconciliations that induce a binary refinement of g and have duplication cost $m(\lambda(g)) - 1$.

The idea of its proof is clear although the proof is long. Recall that $\lambda(g)$ is the root of $I^+(g)$. In the subtree $I^+(g)$, by the definition of $m(\ )$, at most $m(\lambda(g))$ children of g are mapped onto a root-to-leaf path from $\lambda(g)$; furthermore, there is a path P containing exactly $m(\lambda(g))$ children images. By setting $\alpha(u)$ and $\beta(u)$ with formulas (2)-(5), we only postulate duplication events in P and push-down duplication events away from the root as much as possible by postulating a duplication event whenever it is necessary. By doing so, we guarantee that the resulting reconciliation has the minimum gene loss cost while keeping the duplication cost unchanged. For the example considered above, P is the leftmost path from the root to the leaf labeled with a (Fig. 2B). We postulate all three duplication events in P and three loss events on the branches off P (Fig. 2C). To prove Theorem 2, we first establish three facts about $\alpha()$ and $\beta()$. Due to limit space, the proof of the following lemma is omitted.

Lemma 1. *Let u be an internal node of $I^+(g)$.*
(1) If $\alpha(u) < m(u)$, then $\alpha(u) \le \beta(u)$.
(2) If $\alpha(u) \ge m(u)$, then, $\alpha(w) \ge m(w) = \beta(w)$ for any descendant w of u.
(3) $\omega(u) + \min\{m(u')|p(u') = u\} \le \beta(u) \le m(u)$, and $1 \le \alpha(u)$.

Proof of Theorem 2. (1). Assume that $m(r)$ is computed through the nodes on the following path P in $I^+(v)$ (Fig 4):

$$P : u_0 = r, u_1, u_2, \cdots, u_t, \tag{7}$$

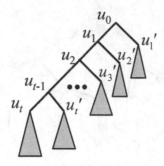

Fig. 4. The path P defined in the proof of Theorem 2.3, along which $m(r)$ is computed and $m(r)$ duplication events have to be postulated, where $r = \lambda(g)$, the root of $I^+(g)$

i.e., u_t is a leaf of $I^+(g)$ and $m(u_{i-1}) = \omega(u_{i-1}) + m(u_i)$ for $i = t, t-1, \cdots, 1$. Let u_i' be the sibling of u_i in $I^+(g)$. Then, $m(u_i') \leq m(u_i)$ for $i \geq 1$. By the fact (3) in Lemma 1, $\beta(u_{i-1}) \geq \omega(u_{i-1}) + m(u_i')$, $i \geq 1$. Hence, $\alpha(u_i') = \beta(u_{i-1}) - \omega(u_{i-1}) \geq m(u_i')$, $i \geq 1$. By the fact (2) in Lemma 1, $\alpha(w) \geq \beta(w)$ for any w in the subtree rooted at u_i' ($i \geq 1$). This implies that there are no duplication events postulated in the subtree below u_i', $i \geq 1$. Therefore, all the duplication events are postulated on P. Since u_t is a leaf, by definition, $\beta(u_t) = m(u_t)$. Let j be the smallest index satisfying $\beta(u_j) = m(u_j)$. Then, for any $i < j$, $a(u_i) < m(u_i)$ and hence $\alpha(u_i) \leq \beta(u_i)$ by the fact (1) in Lemma 1. Thus, the number of the inferred duplication events is equal to

$$\sum_{i=0}^{j-1}(\beta(u_i) - \alpha(u_i))$$

$$= (\beta(u_0) - 1) + \sum_{i=1}^{j-1}(\beta(u_i) - \beta(u_{i-1}) + \omega(u_{i-1}))$$

$$= \sum_{i=0}^{j-2}\omega(u_i) - 1 + \beta(u_{j-1}) \leq \sum_{i=0}^{j-2}\omega(u_i) - 1 + m(u_{j-1}) = m(r) - 1.$$

where the last inequality follows from the fact (3) in Lemma 1.

(2). Assume R is a partial reconciliation that uses $m(r) - 1$ duplication events for resolving g. We shall prove that R cannot have less gene loss cost than the reconciliation defined above.

For each $u \in I^+(g)$, we use $\hat{\alpha}(u)$ and $\hat{\beta}(u)$ to denote the number of genes entering and leaving the branch from $p(u)$ to u for R. Consider P given in Eqn. (7). Since $m(r)$ children of g are mapped to the nodes on P and a duplication event can only produce one more gene, all the duplication events must occur on P and, additionally, no gene loss occurs on P. This is because the duplication history induced by R has exactly $m(r) - 1$ duplication events responsible for the children of g. Thus, $\hat{\beta}(u_i) \geq \hat{\alpha}(u_i)$ for any u_i on P. Moreover, we have the following results. Their proofs can be found in the final version of this work.

Fact 1. For any u_i on P, $\hat{\beta}(u_i) \geq \beta(u_i)$.

Fact 2. For any descendant w of u'_i in $I^+(g)$, $\hat{\alpha}(w) \geq \alpha(w)$.

Using the two fact given above, we continue to prove the part 2 of Theorem 2. For each u'_i, the subtree $S(u'_i)$ (rooted at u'_i) can be decomposed into the union of disjoint paths

$$\bar{P} : w_1, w_2, \cdots, w_{\bar{t}}$$

such that $m(w_{i-1}) = m(w_i) + \omega(w_{i-1})$ and $w_{\bar{t}}$ is a leaf. In each of these disjoint paths $\bar{P}$, R has $l_R = \sum_{j=1}^{\bar{t}}(\hat{\alpha}(w_j) - \hat{\beta}(w_j)) = \hat{\alpha}(w_1) - m(w_1)$ gene loss events, whereas our reconciliation has $l_O = \alpha(w_1) - m(w_1)$ gene loss events. By Fact 2, $l_R \geq l_O$. □

4   Efficient Implementation of the Algorithm

Now, we briefly discuss how to implement the algorithm in linear time. Consider an internal node u in $I^+(g)$ with children u_1 and u_2. Assume $m(u_1) > m(u_2) = 0$ and $\omega(u) = 0$, that is, no child of g is mapped to u and u_2. By definition, u_2 has to be a leaf of $I^+(g)$. By Eqn. (6), $\gamma(u) = \alpha(u)$ since $\alpha(u) \geq 1$ (Lemma 1). This implies that $\alpha(u) = \beta(u) = \alpha(u_1) = \alpha(u_2)$. In other words, we only need to update $\alpha()$ and $\beta()$ for all the nodes that are the child images or have degree-3 in $I^+(g)$. From the proof of the fact (1) of Theorem 2, we only need to compute the values of $\alpha()$ and $\beta()$ on the path P defined in Eqn. (7) for the purpose of inferring duplication events. We call such a path is the 'dominant' path. We need to build the condensed version $\hat{P}$ from P by removing the degree-2 nodes u such that $\omega(u) = 0$ to have linear-time implementation. This is because $\hat{P}$ has at most $|C(g)|$ nodes, but P may not, where $C(g)$ is the set of the children of g in the gene tree.

Based on the above discussion, we implement the algorithm as follows:

1. Compute $\omega()$ and $m()$ in the species tree and build the condensed dominant paths $\hat{P}_g$ using doubly linked list for all non-binary gene tree nodes g simultaneously.
2. Resolve each non-binary gene tree node g by inferring duplication events on the corresponding path $\hat{P}_g$.
3. Obtain the binary refinement of G by assembling all the inferred duplication events together.

The details of Step 1 is omitted due to space limit. It can be done in time $O(n + m)$ for input gene and species trees of n and m nodes, respectively, using the techniques appearing in [22]. More specifically, we need to pre-process the species tree S so that we have the postorder traversal of S and can find $\text{lca}(\{u, v\})$ for any $u, v \in V(S)$ using constant operations [20].

Step 2 can be done in time $O(n)$ since the length of each condensed dominant path $\hat{P}_g$ is less than or equal to the number of the children of g.

Step 3 can clearly be done in $O(n + m)$. If one would like to output the gene loss cost, we could simply find it by reconciling the resulting binary refinement and the species tree in $O(n + m)$ time [22].

As a part of a program called TxT for reconciling two non-binary trees, the algorithm has been implemented. TxT is available on http://phylotoo.appspot.com.

5   Conclusion

We have presented the first linear-time algorithm for the NGTR problem. All the existing algorithms for this problem have quadratic-time complexity [4,7]. Our algorithm benefits from an elegant theorem in partial order theory [18]. Unlike [3], to resolve non-binary gene tree nodes, we focus on the longest chain instead of the disjoint partitions of the partial order set defined in the proof of Theorem 1.

Our study takes the same approach as in [3]. We compute in linear time a reconciliation with the optimal duplication cost. Moreover, it has the smallest gene loss cost over all reconciliations with the optimal duplication cost. When two binary trees are reconciled, the lca reconciliation has not only the best duplication cost [11], but also the optimal gene loss cost [5]. However, such a reconciliation minimizing the both costs may not exist for some non-binary gene and species trees. Our proposed algorithm is identical to the standard duplication inference procedure when applied to binary gene tree nodes. Thus, our algorithm can be considered as a natural generalization of the standard reconciliation to non-binary gene trees. Hence, our study advances tree reconciliation approach.

Finally, techniques developed in this work are powerful for studying the NGTR problem. After our work [23], Lafond et al. presented an $O(|G||S|)$-time algorithm for computing a reconciliation that has the smallest sum of the gene duplication and loss costs for an non-binary gene tree G and a binary gene tree S [15]. By refining our techniques here, we have obtained an $O(|G| + |S|)$-time algorithm for for the same problem recently.

Acknowledgments. LX Zhang would like to thank Daniel Huson for suggestion of working on non-binary tree reconciliation. He would also like to thank David A. Liberles for comments on the preliminary version of this paper. This work was financially supported by Singapore MOE tier-2 grant R-146-000-134-112 and tier-1 grant R-146-000-177-112.

References

1. Arvestad, L., Lagergren, J., Sennblad, B.: The gene evolution model and computing its associated probabilities. J. ACM 56, 1–44 (2009)
2. Bansal, M.S., Shamir, S.: A note on the fixed parameter tractability of the gene-duplication problem. IEEE-ACM Trans. Comput. Biol. Bioinform. 8, 848–850 (2010)

3. Berglund-Sonnhammer, A., et al.: Optimal gene trees from sequences and species trees using a soft interpretation of parsimony. J. Mol. Evol. 63, 240–250 (2006)

4. Chang, W.-C., Eulenstein, O.: Reconciling gene trees with apparent polynomies. In: Chen, D.Z., Lee, D.T. (eds.) COCOON 2006. LNCS, vol. 4112, pp. 235–244. Springer, Heidelberg (2006)

5. Chauve, C., El-Mabrouk, N.: New perspectives on gene family evolution: losses in reconciliation and a link with supertrees. In: Batzoglou, S. (ed.) RECOMB 2009. LNCS, vol. 5541, pp. 46–58. Springer, Heidelberg (2009)

6. Chen, K., Durand, D., Farach-Colton, M.: NOTUNG: a program for dating gene duplications and optimizing gene family trees. J. Comput. Biol. 7, 429–447 (2000)

7. Durand, D., Halldorsson, B., Vernot, B.: A hybrid micro-macroevolutionary approach to gene tree reconstruction. J. Comput. Biol. 13(2), 320–335 (2005)

8. Eulenstein, O., Huzurbazar, S., Liberles, D.: Reconciling Phylogenetic Trees. In: Dittmar, K., Liberles, D. (eds.) Evolution After Duplication, pp. 185–206. Wiley-Blackwell, New Jersey (2010)

9. Fitch, W.M.: Distinguishing homologous from analogous proteins. Syst. Zool. 19, 99–113 (1970)

10. Goodman, M., et al.: Fitting the gene lineage into its species lineage, a parsimony strategy illustrated by cladograms constructed from globin sequences. Syst. Zool. 28, 132–163 (1979)

11. Górecki, P., Tiuryn, J.: DLS-trees: a model of evolutionary scenarios. Theoret. Comput. Sci. 359, 378–399 (2006)

12. Hudson, R.: Gene genealogies and the coalescent process. In: Oxford Surveys in Evolutionary Biology, vol. 7, pp. 1–44. Oxford University Press (1990)

13. Koonin, E.V.: The origin and early evolution of eukaryotes in the light of phylogenomics. Genome Biol. 11, 209 (2010)

14. Kristensen, D.M., Wolf, Y.I., Mushegian, A.R., Koonin, E.V.: Computational methods for gene orthology inference. Briefings Bioinform. 12, 379–391 (2011)

15. Lafond, M., Swenson, K.M., El-Mabrouk, N.: An optimal reconciliation algorithm for gene trees with polytomies. In: Raphael, B., Tang, J. (eds.) WABI 2012. LNCS, vol. 7534, pp. 106–122. Springer, Heidelberg (2012)

16. Maddison, W.: Reconstructing character evolution on polytomous cladograms. Cladistics 5, 365–377 (1989)

17. Ma, B., Li, M., Zhang, L.X.: From gene trees to species trees. SIAM J. Comput. 30, 729–752 (2000); Also in Proc. RECOMB 1998, pp. 182–191 (2000)

18. Mirsky, L.: A dual of Dilworth's decomposition theorem. Amer. Math. Monthly 78, 876–877 (1971)

19. Page, R.: Maps between trees and cladistic analysis of historical associations among genes, organisms, and areas. Syst. Biol. 43, 58–77 (1994)

20. Schieber, B., Vishkin, U.: On finding lowest common ancestors: simplification and parallelization. SIAM J. Comput. 17, 1253–1262 (1988)

21. Vernot, B., Stolzer, M., Goldman, A., Durand, D.: Reconciliation with non-binary species trees. J. Comput. Biol. 15(8), 981–1006 (2008)

22. Zhang, L.X.: On a Mirkin-Muchnik-Smith conjecture for comparing molecular phylogenies. J. Comput. Biol. 4, 177–187 (1997)

23. Zheng, Y., Wu, T., Zhang, L.X.: Reconciliation of Gene and Species Trees With Polytomies, arXiv:1201.3995, arxiv.org (2012)

On Some Proximity Problems of Colored Sets*

Chenglin Fan[1], Jun Luo[1,2], and Farong Zhong[3]

[1] Shenzhen Institutes of Advanced Technology
Chinese Academy of Sciences, Shenzhen, China
[2] Huawei Noah's Ark Laboratory, Hong Kong
[3] College of Math, Physics and Information Sciences
Zhejiang Normal University, Jinhua, China
{cl.fan,jun.luo}@siat.ac.cn,zrf@zjnu.cn

Abstract. The maximum diameter color-spanning set problem (MaxDCS) is defined as follows: given n points with m colors, select m points with m distinct colors such that the diameter of the set of chosen points is maximized. In this paper, we design an optimal $O(n \log n)$ time algorithm using rotating calipers for MaxDCS problem in the plane. Our algorithm can also be used to solve the maximum diameter problem of imprecise points modeled as polygons. We also give an optimal algorithm for the all farthest foreign neighbor problem (AFFN) in the plane, and propose algorithms to answer the farthest foreign neighbor query (FFNQ) of colored sets in two and three dimensional space. Furthermore, we study the problem of computing the closest pair of color-spanning set (CPCS) in d dimensional space, and remove the factor $\log m$ of the best known time bound if we treat d as a constant.

1 Introduction

Computing the diameter of a set of n points in a d-dimensional space ($d = 1, 2, 3....$) has a long history of research [2,4]. The diameter of a point set is the maximum Euclidean distance between any two points of the set. By a reduction to set disjointness, the computation requires $\Omega(n \log n)$ operations in the algebraic computation tree model [1]. The paper [14] is completely devoted to this problem and several efficient algorithms are proposed.

However, these algorithms are based on the assumption that the positions of input points are precise. If a point may randomly appear at one of the many candidate positions, which are painted with the same color, how to compute the maximum possible diameter of the point set with different colors? The problem is called the Maximum Diameter Color-spanning Set (MaxDCS) problem. The solution to the problem is useful in large computer networks. For example, a large company tries to pool resources to solve a certain computational task. But some uncertain factors interfere with the accuracy of the data and the company want to know the worst (or the best) cost based on those imprecise data.

---

* This research is partially funded by the International Science & Technology Cooperation Program of China (2010DFA92720-24) and NSF of China under project 11271351.

P. Widmayer, Y. Xu, and B. Zhu (Eds.): COCOA 2013, LNCS 8287, pp. 202–213, 2013.
© Springer International Publishing Switzerland 2013

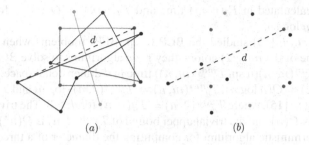

Fig. 1. Example of maximum diameter of imprecise points modeled as (a) polygons and (b) point set

In 2007, Löffler *et al.* [19] studied the largest diameter problem and the smallest diameter problem based on imprecise data. But they used a continuous region model, such as disc and square. They showed that the largest diameter problem can be solved in $O(n \log n)$ time based on the square and the disc models. But their algorithm [19] cannot be adapted to the case when imprecise data are modeled as general polygons. The algorithm in our paper can be used to compute the maximum diameter of imprecise points in $O(n \log n)$ time, which can be modeled as polygons (see Figure 1), as the two points forming the largest diameter must be among the vertices of two polygons. On the other hand, the smallest diameter problem under the disc model is more complex and they proposed an $(1 + \epsilon)$-approximation and $O(n^{c\epsilon^{-\frac{1}{2}}})$ time algorithm for the problem, where $c \approx 6.66$ [19].

The minimum diameter color-spanning set (MDCS) problem is firstly studied by Zhang *et al.* in spatial databases [21]. But Zhang *et al.* only proposed an $O(n^m)$ time algorithm for the problem. It is unfortunately a brute force algorithm. Then Chen *et al.* implemented the algorithm in a geographical tagging system [22]. Finally, Fleischer *et al.* [23] proved that the problem is NP-Complete in L_p $(p \geq 2)$ metric and proposed an efficient constant factor approximation algorithm for the MDCS problem. Fan *et al.* [24] recently studied several other color-spanning problems; they gave an efficient randomized algorithm to compute a maximum diameter color-spanning set, and they showed that it is NP-hard to compute a largest closest pair color-spanning set and a planar minimum color-spanning tree.

While all-pairs nearest neighbors in any fixed dimension d can be computed in optimal $O(n \log n)$ time [5], no algorithm with similar efficiency is known for the all-pairs farthest neighbors. Agarwal *et al.* [9] showed that the three-dimensional all-pairs farthest neighbors can be computed in $O(n^{4/3} \log^{4/3} n)$ expected time, and posed closing the gap between this and the only lower bound of $\Omega(n \log n)$ as a challenging open problem. Cheong [17] *et al.* studied the all farthest pair problem when all the points are at the convex positions in R^3, and gave an expected $O(n \log^2 n)$ time algorithm to compute it.

The bichromatic closest (resp. farthest) pairs (BCP, resp. BFP) [6] is formulated as follows: Given a set n red and m blue points in R^d, find a red point p and a blue point q such that the distance between p and q is minimum among all red-blue pairs, which can be calculated in $\Gamma_d(n, m)$ time and $\Gamma_d(n, n) = O(n^{1+\varepsilon})$ for $d \geq 3$ and $\Gamma_2(n, n) = O(n \log n)$.

Dumitrescu et $al.$ [15] studied the BCP (resp. BFP) problems when each point is colored with one of the $m (\geq 2)$ colors, they gave algorithms to solve BCP (resp. BFP) problems in $T_d^{min}(m, n)$(resp. $T_d^{max}(m, n)$) time in d dimensional space. They showed that $T_2^{min}(m, n) = O(n \log n)$, $T_d^{min}(m, n) = T_d^{min}(2, n) \cdot \log m$ and $T_d^{max}(m, n) = T_d^{max}(2, n) \cdot \log m$ [15], where $T_d^{min}(2, n) = \Gamma_d(n, n)$ for $d \geq 3$. The trivial low bound of $T_d^{min}(2, n)$ is $\Omega(n)$ and the trivial upper bound of $T_d^{min}(2, n)$ is $O(n^2)$. Ramos gave an optimal deterministic algorithm for computing the diameter of a three-dimensional point set, which can also be used to solve the bichromatic diameter in three-dimensional space in $O(n \log n)$ time [13]. Combining the above two results, we can solve BFP of m colors in $O(n \log^2 n)$ time.

In paper [10], Agarwal et $al.$ gave an $O(n \log n)$ time algorithm for the following problem: Given a collection of sets with total of n points in the plane, find for each point a closest neighbor that does not belong to the same set, which is the earliest version to compute all foreign nearest pairs of points in the plane.

In this paper, all the problems we study are based on Euclidean distance. We propose an optimal time algorithm for MaxDCS. To the best of our knowledge, this is the first $O(n \log n)$ time algorithm for MaxDCS. We also give an optimal algorithm for all farthest foreign neighbors problem (AFFN) in the plane, and propose algorithms to answer the the farthest foreign neighbor query (FFNQ) of colored sets in two and three dimensional space. At last we study the problem of computing the closest pair of color-spanning set (CPCS) in d dimensional space, and remove the factor $\log m$ off the best known time bound if we treat d as a constant.

Table 1 lists the results of our algorithms and previous algorithms.

Table 1. An overview of the time complexity of various problems of colored sets. The number in parenthesis is the number of dimension, and k denotes the number of colors of point set.

Problem	Previous results	Our results
MaxDCS(2)	$O(n \log^2 n)$ [13, 15]	$O(n \log n)$
AFFN(2)	none	$O(n \log n)$
FFNQ(2)	none	$O(\log n)$
FFNQ(3)	none	$O(\log^2 n)$
CPCS(d)	$T_d^{min}(2, n) \log m$ [15]	$T_d^{min}(2, n)$

2   The Algorithm for MaxDCS

Problem 1. Suppose that we are given n points with m colors. How to select m points with m different colors such that the diameter of the m selected points is maximized?

Let CH^* be the convex hull of all n input points $\{p_1, p_2, ..., p_n\}$ and let $\langle v_1, v_2, ..., v_t \rangle$ be the vertices of CH^* in clockwise order. We can divide $\langle v_1, v_2, ..., v_t \rangle$ into r sequences $G_1, G_2, ..., G_r$ such that all vertices in one sequence have the same color and appear consecutively on CH^*. Denote the first vertex in G_i as v_i^s and the last vertex

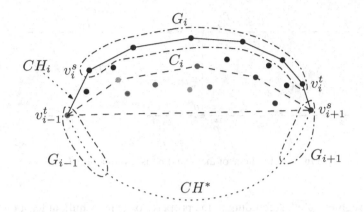

Fig. 2. Illustration of *Associate Convex Hull* CH_i and *Associate Chain* C_i

in G_i as v_i^t (in clockwise order). Let the convex hull constructed from v_{i-1}^t, v_{i+1}^s and all the vertices in G_i be CH_i, which is called the *Associate Convex Hull* of G_i. Let D_i be the set of points inside CH_i with colors different from the color of the vertices in G_i. We can construct a convex chain C_i such that it starts from v_{i-1}^t, ends at v_{i+1}^s and encloses all the points of D_i. We call C_i the *Associate Chain* of G_i (see Figure 2).

Lemma 1. *All Associate Chains C_i ($i = 1, 2, ..., r$) can be computed in $O(n \log n)$ time.*

Proof. First of all, we can construct CH^* in $O(n \log n)$ time. Then the vertices of CH^* are separated in r groups $G_1, G_2, ..., G_r$ by traversing the vertices of CH^* in $O(n)$ time. At the same time all CH_i's can be constructed during the traversal. For each point p_k ($k = 1, 2, ..., n$), we can decide which CH_i it is located in $O(\log n)$ time (note that each point p_k could belong to at most two neighboring *Associate Convex Hulls*). Thus this step in total also takes $O(n \log n)$ time. For all points in CH_i, we can get rid of those points with the same color as those in G_i to obtain D_i which can be done in linear time. Since one point can only belong to at most two neighboring *Associate Convex Hulls*, we have $\sum_{i=1}^{r} |D_i| = O(n)$. Therefore, we can construct all *Associate Chains* in $O(n \log n)$ time.   □

Let the two points realizing MaxDCS be p' and p''. For two points p_i and p_j, the distance between them is denoted as $d(p_i, p_j)$ and the color of p_i and p_j is denoted as $Col(p_i)$ and $Col(p_j)$ respectively. We have the following lemmas:

Lemma 2. *At least one of p' and p'' is a vertex of CH^*.*

Proof. Assume that neither of p' and p'' is a vertex of CH^* as is shown in Figure 3. We draw two parallel lines l_1 and l_2 through p' and p'' respectively which are perpendicular

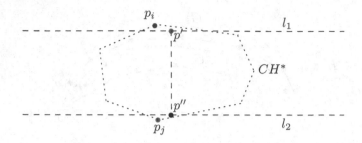

Fig. 3. At least one of the p' and p'' is a vertex of CH^*

to the line segment $\overline{p'p''}$. According to the property of convex hull, at least one vertex, say p_i, must lie above or on l_1 and another vertex, say p_j, must lie below or on l_2. Therefore, $d(p', p_j) \geq d(p', p'')$, $d(p'', p_i) \geq d(p', p'')$ and $d(p_i, p_j) \geq d(p', p'')$. Since $Col(p') \neq Col(p'')$ according to our assumption, $Col(p') \neq Col(p_j)$ or $Col(p'') \neq Col(p_i)$ or $Col(p_i) \neq Col(p_j)$. Then at least one pair of (p', p_j), (p'', p_i) and (p_i, p_j) is a better candidate for realizing MaxDCS than the pair of (p', p''). This is a contradiction and the lemma is proved. □

Now, without loss of generality, let p' be a vertex of CH^*.

Lemma 3. *If p'' is not a vertex of CH^*, then it must be the vertex of some* Associate Chain.

Proof. We can draw two parallel lines l_1 and l_2 through p' and p'' respectively which are perpendicular to $\overline{p'p''}$, similar to the proof of lemma 2 (see Figure 4). Since p' is a vertex of CH^*, there is no point above l_1. For the points below l_2, the color of these points must be the same as the color of p'. (Otherwise, any one of these points together with p' is the better candidate for realizing MaxDCS than the pair (p', p''). Assume that the vertices of CH^* below l_2 are in G_i, then v_{i-1}^t and v_{i+1}^s are between l_1 and l_2. Therefore, p'' must be a vertex of C_i since all points below l_2 have the same color as p' and they do not belong to C_i. The lemma is proved. □

According to Lemma 2 and Lemma 3, there are two cases for p' and p'':

1. p' and p'' are the vertices of CH^*. Actually, we need to find l_1 and l_2 with the maximum distance. We can use the rotating calipers method [4] to compute the diameter of a point set ((p', p'') is called an antipodal pair in the rotating calipers method). The only difference is that in our algorithm, we do not need to record the distance between an antipodal pair with the same color.

2. p' is a vertex of CH^* and p'' is a vertex of some C_i. In this case, we can rotate l_1 along CH^* and at the same time rotate l_2 along the associate chain. However, an associate chain may intersect with its neighboring associate chain and it may cause trouble when rotating l_2. Fortunately, associate chains that are not neighbors do not intersect. We can construct convex hull CH_{even} which is the convex hull of all C_i's where i is an even number (see Figure 5). Observe that all vertices of C_i (i is even)

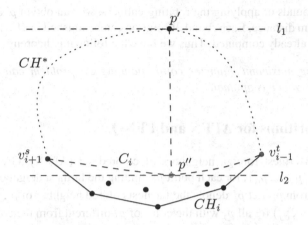

Fig. 4. p'' is the vertex of an *Associate Chain*

are the vertices of CH_{even}. Similarly, we can construct convex hull CH_{odd} which includes all the vertices of C_i (i is odd). Then l_2 needs to be rotated twice, once along CH_{even} and the other along CH_{odd}. For l_1, it just rotates along CH^* twice accordingly. Note that there is one minor case: when the number of the associate chains is odd, the first and the last chain of CH_{odd} intersect. We can deal with this case easily by just rotating l_2 along the odd associate chains from the first one to the last one and l_1 along CH^* accordingly as before. The only difference is that l_1 does not rotate exactly one round. It rotates a little bit more than one round but definitely no more than two rounds.

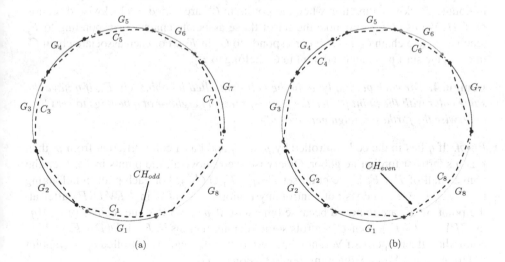

Fig. 5. Illustration of (a) CH_{odd} and (b) CH_{even}

After three rounds of applying the rotating calipers, we can obtain p' and p'' which has the maximum distance. Each round of rotation can be finished in $O(n)$ time if CH^* and all C_i's are already computed. Thus we have the following theorem:

Theorem 1. *The maximum diameter color-spanning set problem can be solved in $O(n \log n)$ time, which is optimal.*

3   The Algorithms for AFFN and FFNQ

Problem 2. All farthest foreign neighbors of colored set (AFFN): Given n colored points $P = \{p_1, p_2, ..., p_n\}$, for each point p_i, finding the farthest neighbor which has a color different from p_i. Let p_i^f denote the farthest colored neighbor of p_i, which satisfy $\{d(p_i, p_i^f) \geq d(p_i, p_j)\}$ for all p_j with the color of p_j different from the color of p_i.

We construct the farthest-point Delaunay triangulation (the dual graph of farthest-point Voronoi diagram) of all the points P in the plane, $FPDT(P)$. The farthest-point Voronoi diagram $FPVD(P)$ can be computed in $O(n \log n)$ time. A point of P has a cell in the farthest-point Voronoi diagram if and only if it is a vertex of the convex hull of P. What is more, let $cw(p)$ and $ccw(p)$ denote the adjacent point of p in clockwise and counterclockwise direction of convex hull respectively, then the cell will come in between the cells of $cw(p)$ and $ccw(p)$ [20].

We construct the convex hull CH for the point set P, and let the point set on the convex hull be P_{CH}. We then construct the associate chains C_i and groups G_i as in Lemma 1 in $O(n \log n)$ time.

Then we divide $FPDT(P)$ into the same colored sets $F_j (1 \leq j \leq m)$. Let F_j' denote the set of points adjacent to points in F_j. The point set F_j is composed of several groups of G_i. All the points in each group G_i have the same color. If one point in G_i belong to F_j, then all the points in G_i belong to F_j. The cells controlled by G_i are sorted in counterclockwise direction when the points in G_i are sorted in clockwise direction on CH. We use CC_j to denote the set of those associate chains corresponding to F_j (each associate chain C_i in CC_j corresponds to G_i in F_j). For each associate chain C_i in CC_j, the start point and endpoint in C_i belong to F_j'.

Lemma 4. *For each point q lying in the cell controlled by point p in F_j, if q have the same color with the point p, then the farthest foreign neighbor of q belongs to set CC_j, otherwise the farthest foreign neighbor of q is p.*

Proof. If q lies in the cell controlled by p in F_j and has a color different from p, then p is q's farthest foreign neighbor. Otherwise we remove all the points in F_j, then the convex hull of $P \setminus F_j$ is the point set $P_{CH} \setminus F_j \cup CC_j$. For each point p belonging to $P_{CH} \setminus F_j \setminus F_j'$, p does not control any region of $cell(F_j)$ in $FPVD(P)$ after all the points in F_j are removed because otherwise if p controls some regions of $cell(q)$ in $FPVD(P \setminus F_j)$, then P controls some disjoint regions in $FPVD(P \setminus F_j)$ which contradicts the property of Voronoi diagram, that is the region controlled by each point in farthest point Voronoi diagram should be connected.

Therefore q belongs to the cell controlled by point p in CC_j in $FPVD(P \setminus F_j)$, and all the points in CC_j have some different colors from q. Then the farthest foreign neighbor of q belongs to set CC_j.   □

Theorem 2. *The all farthest foreign neighbors of colored set problem in the plane can be solved in $O(n \log n)$ time and $O(n)$ space, which is optimal.*

Proof. The steps to compute AFFN in the plane are as follows:

1. $FPVD(P)$ and $FPDT(P)$ can be constructed in $O(n \log n)$ time [20]. $FPDT(P)$ has $O(n)$ edges. All the groups of G_i and C_i can be computed in $O(n \log n)$ time, and $\sum (G_i + C_i) = O(n)$ according to Lemma 1.
2. Find the same colored set F_j and F'_j of $FPDT(P)$, which takes $O(n)$ time as $\sum (|F_j| + |F'_j|) = O(n)$.
3. For each set CC_j, construct the farthest point Voronoi diagram of CC_j. The total time cost is $\sum |CC_j| \log(|CC_j|) = O(n \log n)$.
4. For the point p located in $cell(p')$ in $FPVD(P)$, if p' has a color different from p, then p' is the farthest foreign neighbor of p. Otherwise we locate p in $FPVD(CC_j)$ to find its farthest foreign neighbor according to Lemma 4, where the color of p is the same as the color of F_j. Hence the location for each point takes $O(\log n)$ time, and total time is $O(n \log n)$.

□

The above theorem can be easily extended to the farthest foreign neighbor query (FFNQ) of a point:

Theorem 3. *Given a colored sets P of n points in the plane and a point q (q might not be in P) with some color, the farthest foreign neighbor query (FFNQ) of q can be answered in $O(\log n)$ time in the plane with $O(n \log n)$ preprocessing time and $O(n)$ space.*

Now we consider the FFNQ problem in three dimensions. We first compute the three dimensional convex hull $CH(3)$ in $O(n \log n)$, and let O be a vertex of $CH(3)$. Let $P_{CH(3)}$ denote the point set of $CH(3)$.

The $CH(3)$ of n points in space consists of $n_f \leq 2n - 4$ faces and $n_e \leq 3n - 6$ edges [20]. We assume that all the faces of $CH(3)$ are triangles (namely simplicial polytope), otherwise we just add some edges to triangulate it.

For each face f_i of $CH(3)$, we use f_i and point the O to construct a tetrahedron t_i. Then the space surround by $CH(3)$ consist of n_f tetrahedra.

For each point p inside $CH(3)$, we located it to find which tetrahedron t_i it belongs to in $O(\log^2 n)$ using $O(n \log n)$ space [16]. Let S_i denote the point set of P located in t_i.

If we remove a vertex p_j ($p_j \neq O$) from $CH(3)$, those faces with vertex p_j will disappear. Let $face_j$ denote the set of those disappeared faces, T_j denote the set of tetrahedra with one face in $face_j$, and T'_j denote those points of P inside T_j but without those points of the same color as p_j. For each face f_i, since f_i has only three vertices, f_i appears on at most three different tetrahedra sets like T_j. Let T''_j denote those point set of $\{T'_i | p_i \in F_j\}$, hence $\sum |T''_j| = O(n)$. Let FT_j denote those point set $F'_j \cup T''_j$.

Lemma 4 still holds in three dimensional space when FT_j replaces CC_j. The remaining part of the algorithm is similar to the two dimensional case. The time complexity analysis is as follows:

The complexity of $FPVD(P)$ and $FPDT(P)$ in d dimension is $O(n^{\lceil d/2 \rceil})$ [3], which can be constructed in $O(n^{\lceil d/2 \rceil} + n \log n)$ time [8]. Hence the $FPDT(P)$ in three dimensional space can be constructed in $O(n^2)$ time and the complexity of $FPDT(P)$ is $O(n^2)$. However, the size of $FPDT(P)$ is smaller than the bound $O(n^{\lceil d/2 \rceil})$ in general case [7]. Chan *et al* [12] give an output sensitive algorithm to compute $FPDT(P)$ in three dimensional space in $O((n+f)\log^2 f)$ time, where f $(f \in [n, n^2])$ is the size of $FPDT(P)$.

We compute the Voronoi diagram for each FT_j. Then the total time is

$$\sum_{i=1}^{M} |FT_j|^2 \leq \sum_{k=1}^{f/n} |U_k'|^2 = O(f/n \times n^2) = O(fn)$$

where U_k' is the union of several FT_j such that $n \leq |U_k'| \leq 2n$. Since $|FT_j| \leq n$ and $\sum |FT_j| = O(f)$, then $k = O(f/n)$).

The total size of all Voronoi diagrams is also $O(fn)$ using above analysis. In three-dimensional space, it is possible to answer point location queries in $O(\log^2 N)$ using $O(N \log N)$ space and $O(N \log N)$ preprocessing time of size N [16]. Therefore we can answer the nearest foreign neighbor query in $O(\log^2 n)$ using $O(fn \log n)$ space and $O(fn \log n)$ preprocessing time .   □

Therefore we have the following theorems:

Theorem 4. *Given a colored set P of n points in three dimensions and a point q (q might not be in P) with color, the farthest foreign neighbor query of q can be answered in $O(\log^2 n)$ time with $O(fn \log n)$ preprocessing time and $O(fn \log n)$ preprocessing space, where f is the size of $DT(P)$.*

4   The Algorithm for $CPCS(d)$

Problem 3. The closest pair of color-spanning set in d dimensional space (CPCS(d)): Given n input points P in d dimensions, find a pair (p, q), satisfying $\{d(p,q) \leq d(p', q'), Col(p) \neq Col(q), Col(p') \neq Col(q')\}$ for any p', q' in the space.

We use the well separated pairs decomposition (WSPD) method together with a compressed quadtree to deal with this problem. Well separated pairs decomposition was defined by Callahan and Kosaraju [11]. We use the version of WSPD (very roughly) from [18].

The steps of our algorithm are as follows:

1. Construct the smallest enclosing box of P in d dimensional space. Using quadtree subdivision to divide the box into smaller boxes (child boxes) until the points in each disjoint box have the same color.

2. Find the smallest box B of the subdivision in which there exist at least two points in B with different colors. Let d_0 denote the diameter of B and b_0 denote the edge length of the side of B. Then the distance of $CPCS(d)$ is less than or equal to d_0.

Obviously, the box B is divided into child boxes following step 1 and the points in each child box have the same color which has a side length $b_0/2$.

3. For the box obtained thus far whose diameter is larger than $d_0/2$, we divide it into smaller boxes until its side length is $b_0/2$. Now the side length of any such base box is $b_0/2$.

4. For two disjoint base boxes u, v, let $dis(u, v) = min||p-q||$, where $p \in u, q \in v$. If two boxes contain points of the same color, then we can ignore them. Otherwise we compute the $dis(u, v)$ and the distance of $CPCS(d)$ is the minimum of all those $dis(u, v)$.

Lemma 5. *For each base box u, there are at most $O(2^d d^{d/2})$ disjoint base boxes v satisfying $dis(u, v) \leq d_0$.*

Proof. For any base box b, the length of the side of b is $b_0/2$ and $d_0 = b_0 * d^{1/2}$. Hence there are at most $O(d_0/(b_0/2))^d = O(2^d d^{d/2})$ disjoint base boxes whose distance from b is less than or equal to d_0. $\qquad\square$

Theorem 5. *The time complexity for computing the distance of $CPCS(d)$ is $T_d^{min}(m, n) = O(2^{2d} d^d) * T_d^{min}(2, n) + O(n \log n + 2^d n) = T_d^{min}(2, n)$, if d is a constant.*

Proof. Let the quadtree after step 3 be T, and the compressed quadtree corresponding to T be $\mathbb{T}$. Construct a ϵ^{-1} WSPD W for $\mathbb{T}$. Let (u, v) be a pair of boxes in W and $\epsilon = 1/2$, then there are only two possible cases:

1. $max\{diam(u), diam(v)\} > d_0/2$. Since we know u, v are ϵ^{-1} well separated, then $dis(u, v) \geq \epsilon^{-1} * max\{diam(u), diam(v)\} > \epsilon^{-1} d_0/2 = d_0$, that means we do not need to compute the distance of those pairs.
2. $diam(u) = diam(v) = d_0/2$. Those are the pairs we need to compute at step 4.

According to the Lemma 5.1 in [18], one can construct a 2-WSPD of size $O(2^d n)$ with the construction time being $O(2^d n \log n + 2^d n)$. Of course, there is a little difference in our algorithm, as the box with diameter $d_0/2$ does not need to be divided further, but that does not affect the time complexity of our algorithm.

At step 4, let the time to compute the distance between the box pair (u, v) be $CP(|u|, |v|)$, where $|u|$ and $|v|$ denote the number of points in u and v respectively. Then $CP(|u|, |v|) = T_d^{min}(2, |u| + |v|)$ since u contains the points of one color and v contains the points of the other color. Then this problem is exactly the BCP problem. Because $T_d^{min}(2, n) = \Omega(n)$, we have $T_d^{min}(2, |x|) + T_d^{min}(2, |y|) \leq T_d^{min}(2, |x| + |y|)$. Because we only need to compute the box pair whose distance is less than or equal to d_0 and according to Lemma 5, each box appears at most $O(2^d d^{d/2})$ times in those pairs, then $\sum_{dis(u,v) \leq d_0}(|u| + |v|) = O(2^d d^{d/2} n)$. So we have $\sum_{dis(u,v) \leq d_0} CP(|u|, |v|) \leq T_d^{min}(2, \sum_{dis(u,v) \leq d_0}(|u| + |v|)) \leq T_d^{min}(2, O(2^d d^{d/2} n)) = O((2^d d^{d/2})^2) T_d^{min}(2, n) = O(2^{2d} d^d) T_d^{min}(2, n)$ as $T_d^{min}(2, n) = O(n^2)$. If we treat d as a constant, then the time to compute the distance of $CPCS(d)$ is $T_d^{min}(m, n) = O(2^{2d} d^d) T_d^{min}(2, n) + O(n \log n + 2^d n) = T_d^{min}(2, n)$. $\qquad\square$

5   Conclusions

In this paper, we propose an optimal $O(n \log n)$ time algorithm for the maximum diameter color-spanning set problem. Our algorithm can also be used to solve the maximum diameter problem of imprecise points modeled as polygons since the candidate pair of points must be vertices of two polygons, and the vertices of each polygons are painted in the same color.

We also give $O(n \log n)$ time and $O(n)$ space algorithms for AFFN problems in the plane . For the query of the farthest foreign neighbor in two dimension, we propose $O(\log n)$ query time algorithms with $O(n \log n)$ preprocessing time and $O(n)$ preprocessing space. For the three dimension query problems, we give $O(\log^2 n)$ query time algorithms with $O(fn \log n)$ preprocessing time and $O(fn \log n)$ preprocessing space, where f is the size of Farthest point Delaunay triangulation of P. We also give an algorithm to improve the best known bound of the CPCS(d) problem, and conclude that the $CPCS(d)$ of m colors can be computed in the same time with $CPCS(d)$ of two colors when d is a constant. In the future, we will focus on the problems of computing the farthest foreign pair in higher dimensional space, and approximate nearest neighbor query of color point set.

References

1. Preparata, F.P., Shamos, M.I.: Computational geometry: an introduction. Springer, New York (1985)
2. Shamos, M.I.: Computational geometry, Ph.D. thesis, Yale University (1978)
3. Klee, V.: On the complexity of d-dimensional Voronoi diagrams. Archiv der Mathematik 34, 75–80 (1980)
4. Toussaint, G.: Solving geometric problems with the rotating calipers. In: Proc. MELECON 1983 (1983)
5. Vaidya, P.M.: An $O(n \log n)$ algorithm for the all-nearest-neighbors problem. Discrete Comput. Geom., 101–115 (1989)
6. Agarwal, P.K., Edelsbrunner, H., Schwarzkopf, O., Welzl, E.: Euclidean minimum spanning trees and bichromatic closest pairs. In: Proceedings of the Sixth Annual Symposium on Computational Geometry, pp. 203–210 (1990)
7. Rex, A.: Dwyer, Higher-dimensional voronoi diagrams in linear expected time. Discrete Comput. Geom. 6(1), 343–367 (1991)
8. Chazelle, B.: An optimal convex hull algorithm and new results on cuttings. In: Proc. 32nd Annu. IEEE Sympos. Found. Comput. Sci., pp. 29–38 (1991)
9. Agarwal, P.K., Matousek, J., Suri, S.: Farthest Neighbors, Maximum Spanning Trees and Related Problems in Higher Dimensions. Comput. Geom. Theory Appl. 1(4), 189–201 (1992)
10. Aggarwal, A., Edelsbrunner, H., Raghavan, P., Tiwari, P.: Optimal Time Bounds for Some Proximity Problems in the Plane. Information Processing Letters 42, 55–60 (1992)
11. Callahan, P.B., Kosaraju, S.R.: A decomposition of multidimensional point sets with applications to k-nearest-neighbors and n-body potential fields. J. Assoc. Comput. Mach. 42, 67–90 (1995)
12. Chan, T.M., Snoeyink, J., Yap, C.-K., Dividing, P., Pruning, D.: Output-Sensitive Construction of Four-Dimensional Polytopes and Three-Dimensional Voronoi Diagrams. Discrete Comput. Geom. 18(4), 433–454 (1997)

13. Ramos, E.A.: An Optimal Deterministic Algorithm for Computing the Diameter of a Three-Dimensional Point Set. Discrete Comput. Geom. 26, 233–244 (2001)
14. Malandain, G., Boissonnat, J.: Computing The Diameter of a Point set. International Journal of Computational Geometry and Applications 12(6), 489–509 (2002)
15. Dumitrescu, A., Guha, S.: Extreme Distances in Multicolored Point Sets. In: Sloot, P.M.A., Tan, C.J.K., Dongarra, J., Hoekstra, A.G. (eds.) ICCS 2002, Part III. LNCS, vol. 2331, pp. 14–25. Springer, Heidelberg (2002)
16. Snoeyink, J.: Point location. In: Goodman, J.E., O'Rourke, J. (eds.) Handbook of Discrete and Computational Geometry, 2nd edn. ch. 34 (2004)
17. Cheong, O., Shin, C.S., Vigneron, A.: Computing farthest neighbors on a convex polytope. Theor. Comput. Sci. 296(1), 47–58 (2003)
18. Har-Peled, S., Mendel, M.: Fast construction of nets in low dimensional metrics and their applications. SIAM J. Comput. 35(5), 1148–1184 (2006)
19. Löffler, M., van Kreveld, M.: Largest bounding box, smallest diameter, and related problems on imprecise points. In: Dehne, F., Sack, J.-R., Zeh, N. (eds.) WADS 2007. LNCS, vol. 4619, pp. 446–457. Springer, Heidelberg (2007)
20. Berg, M., Kreveld, M., Overmars, M., Schwarzkopf, O.: Computational Geometry, 3rd edn. Springer (2008)
21. Zhang, D., Chee, Y.M., Mondal, A., Tung, A.K.H., Kitsuregawa, M.: Keyword search in spatial databases: Towards searching by document. In: Proceedings of the 25th IEEE International Conference on Data Engineering (ICDE 2009), pp. 688–699 (2009)
22. Chen, Y., Shen, S., Gu, Y., Hui, M., Li, F., Liu, C., Liu, L., Ooi, B.C., Yang, X., Zhang, D., Zhou, Y.: MarcoPolo: A community system for sharing and integrating travel information on maps. In: Proceedings of the 12th International Conference on Extending Database Technology (EDBT 2009), pp. 1148–1151 (2009)
23. Fleischer, R., Xu, X.: Computing Minimum Diameter Color-Spanning Sets. In: Lee, D.-T., Chen, D.Z., Ying, S. (eds.) FAW 2010. LNCS, vol. 6213, pp. 285–292. Springer, Heidelberg (2010)
24. Fan, C., Ju, W., Luo, J., Zhu, B.: On Some Geometric Problems of Color-Spanning Sets. Journal of Combinatorial Optimization 26(2), 266–283 (2013)

An Extended Strange Planet Protocol*

Jin Liu, Zhenhua Duan**, and Cong Tian

ICTT and ISN Laboratory, Xidian University, Xi'an, 710071, P.R. China

Abstract. Strange planet protocol is about a special way that three species mate. This paper extends the original strange planet protocol to m species and analyzes conditions for secure communications with it. To do so, conditions of must-fail, might-fail and cannot-fail states are formalized and the number of cannot-fail states is discussed. Finally, an example is given to show how the extended strange planet protocol works.

Keywords: Automata, protocols, secure communication.

1 Introduction

The original strange planet problem is described as follows. On a distant strange planet, there are three species a, b, and c. Any two different species can mate to produce two children of the third species but the participants die. As you can see, the number of individuals never changes. The planet fails if at the some point all individuals are of the same species. That is, no more breeding can take place. Now a question arises: "Is it possible that no matter what breeding choices are made, the planet will not fail? ". It mirrors real one about protocols [2–4]. "The planet fails" means that a protocol enters an error state.

To describe the problem, we need some notations below. A state is defined as a tuple (a,b,c) in italic type, denoting the numbers of individuals of species a, b, and c, respectively. We write abc instead of (a,b,c) as a state for brevity. After a breeding, a new successor state is produced. A state abc can reach a state $a'b'c'$ if $a'b'c'$ is a descendant of abc. We say a state is a final state if there exist two species with 0 numbers at the state, and a might-fail state if the state can reach a final state, and a cannot-fail state if the state can never reach a final state. For two numbers x and y, we denote $(x \mod 3) = (y \mod 3)$ by $x \overset{3}{=} y$.

For the original strange planet problem, some conclusions are achieved in [1]:

1. State $a'b'c'$ is reachable from a non-final state abc iff genetic factor (gene) condition **G1** holds.

$$a' + b' + c' = a + b + c$$
$$(a' - b') \overset{3}{=} (a - b)$$
$$(a' - c') \overset{3}{=} (a - c) \qquad \textbf{(G1)}$$

---

* This research is supported by the NSFC Grant Nos. 61133001, 61272118, 61272117, 61202038, 61322202, 91218301 and National Program on Key Basic Research Project (973 Program) Grant No. 2010CB328102.

** Corresponding author.

P. Widmayer, Y. Xu, and B. Zhu (Eds.): COCOA 2013, LNCS 8287, pp. 214–225, 2013.

$$(b' - c') \overset{3}{=} (b - c)$$

2. Two non-final states $a'b'c'$ and abc are reachable from each other iff condition **G1** holds.
3. A state abc $(a, b, c \geq 0)$ is a might-fail state iff it satisfies condition **M1**, **M2**, or **M3** given below.

$$\begin{cases} (b - c) \overset{3}{=} 0, and \\ (a + b + c) \overset{3}{=} (a - b) \overset{3}{=} (a - c) \end{cases} \tag{M1}$$

$$\begin{cases} (a - c) \overset{3}{=} 0, and \\ (a + b + c) \overset{3}{=} (a - b) \overset{3}{=} (b - c) \end{cases} \tag{M2}$$

$$\begin{cases} (a - b) \overset{3}{=} 0, and \\ (a + b + c) \overset{3}{=} (a - c) \overset{3}{=} (b - c) \end{cases} \tag{M3}$$

4. For n individuals, there are totally $2n + \frac{n(n-3)}{3}$ cannot-fail states.

In this paper, we extend the problem with 3 different species to m different species. Accordingly, the rule for mating is extended such that any $m - 1$ different species can mate. If they do, the participants die and $m - 1$ children of the mth species are born. Similarly, the number of individuals never changes just as the case for three species. Whenever the numbers of any two species decrease to 0, the planet will fail, i.e. no more breeding can take place. We investigate the conditions of cannot-fail, might-fail and must-fail states, as well as the number of cannot-fail states.

The paper is organized as follows. In the next section, the extended strange planet problem with m species is introduced. The conditions of cannot-fail, might-fail and must-fail states are presented in Section 3, and the number of cannot-fail states is calculated in Section 4. Section 5 gives an example to show how the m species strange planet protocol can be used in secure communications [5]. Finally, in Section 6, conclusion and future research are presented.

2 m Species Strange Planet Protocol

Let N_0 denote the set of all non-negative integers. Given m species $a_0, a_1, \cdots,$ and a_{m-1}, and total n of numbers of all m species. We model the extended strange planet problem with automata [6] and use the notations introduced in [1]. Further, let a_i denote the number of individual specie a_i $(0 \leq i \leq m - 1)$, thus, $a_0 + \cdots + a_{m-1} = n$. We use $(a_0, a_1, a_2, \cdots, a_{m-1})$ $(a_i \in N_0)$ to represent a state, denoting the numbers of individuals of species $a_0, a_1, \cdots,$ and a_{m-1}, respectively. For the sake of brevity, we use $a_0 a_1 a_2 \cdots a_{m-1}$ to replace $(a_0, a_1, a_2, \cdots, a_{m-1})$. An a_i-event occurs if individuals of species $a_0, a_1, \cdots, a_{i-1}, a_{i+1}, \cdots,$ and a_{m-1} breed and produce $m - 1$ a_i's $(0 \leq i \leq m - 1)$. The definitions of the final state and cannot-fail state are the same as the case of three species. A might-fail state is a state which can reach not only a final state, but also itself (i.e, in a circle). Final states and the states that can only reach final states are called must-fail states.

As an example, we assume that there are five species a, b, c, d, and e. Any four can mate. For example, a, b, c, and d mate to produce children e. Each kind of the participants decreases by one while the number of species e increases by 4.

Fig. 1 shows the model of the strange planet with four species and six individuals where $1230, 0123, 3012$ and 2301 are cannot-fail states, and all others must-fail states. Considering the symmetry, we just take the combination of the numbers of the four species as a sample.

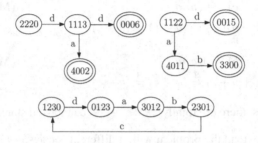

Fig. 1. Strange planet with four species and six individuals

3 Principles in Transitions

Fact 1. *A state $a_0 a_1 a_2 \cdots a_{m-1}$ is a non-final state iff **D2** holds.*

$$\text{For any } i, j \in N_0, 0 \leq i, j \leq m - 1, \text{ if } i \neq j, a_i + a_j \neq 0. \quad (\textbf{D2})$$

That is, there exists at most one 0 in $a_0, a_1, a_2, \cdots,$ and a_{m-1}.   □

Fact 2. *A state $a_0 a_1 a_2 \cdots a_{m-1}$ has a precursor state if there is at least one number in $a_0, a_1, a_2, \cdots,$ and a_{m-1} which is no less than $m - 1$.*   □

The proofs of the above Facts are straightforward.

Lemma 1. *If $a'_0 a'_1 a'_2 \cdots a'_{m-1}$ is a direct successor of a non-final state $a_0 a_1 a_2 \cdots a_{m-1}$, condition **G2**, called non-final genetic factor, holds.*

$$a'_0 + a'_1 + \cdots + a'_{m-1} = a_0 + a_1 + \cdots + a_{m-1}$$
$$(a'_0 - a'_1) \overset{m}{=} (a_0 - a_1)$$
$$(a'_0 - a'_2) \overset{m}{=} (a_0 - a_2)$$
$$\cdots \qquad\qquad (\textbf{G2})$$
$$(a'_0 - a'_{m-1}) \overset{m}{=} (a_0 - a_{m-1})$$

$$(a_1' - a_2') \overset{m}{=} (a_1 - a_2)$$
$$(a_1' - a_3') \overset{m}{=} (a_1 - a_3)$$
$$\cdots$$
$$(a_1' - a_{m-1}') \overset{m}{=} (a_1 - a_{m-1})$$
$$\cdots$$
$$(a_{m-2}' - a_{m-1}') \overset{m}{=} (a_{m-2} - a_{m-1})$$

Proof. For a non-final state $a_0 a_1 a_2 \cdots a_{m-1}$, there are at most m possible transitions at state $a_0 a_1 a_2 \cdots a_{m-1}$ as depicted in Fig. 2.

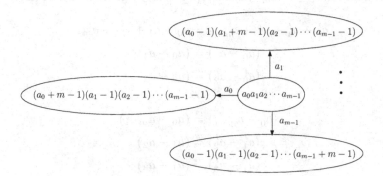

Fig. 2. Possible transitions

When a_0-event occurs: we have $a_0' = a_0 - 1 + m$, $a_1' = a_1 - 1$, $\cdots$, and $a_{m-1}' = a_{m-1} - 1$. Thus,

$$a_0' + a_1' + \cdots + a_{m-1}' = a_0 + a_1 + \cdots + a_{m-1}$$
$$(a_0' - a_1') \overset{m}{=} (a_0 - a_1 + m) \overset{m}{=} (a_0 - a_1)$$
$$(a_0' - a_2') \overset{m}{=} (a_0 - a_2 + m) \overset{m}{=} (a_0 - a_2)$$
$$\cdots$$
$$(a_0' - a_{m-1}') \overset{m}{=} (a_0 - a_{m-1} + m) \overset{m}{=} (a_0 - a_{m-1})$$
$$(a_1' - a_2') \overset{m}{=} (a_1 - a_2)$$
$$(a_1' - a_3') \overset{m}{=} (a_1 - a_3)$$
$$\cdots$$
$$(a_1' - a_{m-1}') \overset{m}{=} (a_1 - a_{m-1})$$
$$\cdots$$

$$(a'_{m-2} - a'_{m-1}) \stackrel{m}{=} (a_{m-2} - a_{m-1})$$

It can be proved similarly in the case in which event a_1, a_2, $\cdots$, or a_{m-1} occurs. □

Theorem 1. If a state $a'_0 a'_1 a'_2 \cdots a'_{m-1}$ is reachable from a non-final state $a_0 a_1 a_2 \cdots$ a_{m-1}, **G2** holds. On the contrary, if **G2** holds, it can be at least obtained that $a_0 a_1 a_2 \cdots$ a_{m-1} can reach $a'_0 a'_1 a'_2 \cdots a'_{m-1}$ or $a'_0 a'_1 a'_2 \cdots a'_{m-1}$ can reach $a_0 a_1 a_2 \cdots a_{m-1}$, under the condition that the precursor state is a non-final one.

Proof. Suppose state $a'_0 a'_1 a'_2 \cdots a'_{m-1}$ is reachable from the non-final state $a_0 a_1 a_2 \cdots$ a_{m-1}. By Lemma 1, **G2** holds.

For two states $a_0 a_1 a_2 \cdots a_{m-1}$ and $a'_0 a'_1 a'_2 \cdots a'_{m-1}$, if **G2** holds, we have:

$$a'_0 + a'_1 + \cdots + a'_{m-1} = a_0 + a_1 + \cdots + a_{m-1}$$
$$(a'_0 - a'_1) \stackrel{m}{=} (a_0 - a_1)$$
$$(a'_0 - a'_2) \stackrel{m}{=} (a_0 - a_2)$$
$$\cdots$$
$$(a'_0 - a'_{m-1}) \stackrel{m}{=} (a_0 - a_{m-1})$$
$$(a'_1 - a'_2) \stackrel{m}{=} (a_1 - a_2)$$
$$(a'_1 - a'_3) \stackrel{m}{=} (a_1 - a_3)$$
$$\cdots$$
$$(a'_1 - a'_{m-1}) \stackrel{m}{=} (a_1 - a_{m-1})$$
$$\cdots$$
$$(a'_{m-2} - a'_{m-1}) \stackrel{m}{=} (a_{m-2} - a_{m-1})$$

Without loss of generality, we assume $a'_0 - a_0 \geq a'_1 - a_1 \geq \cdots \geq a'_{m-1} - a_{m-1}$. Thus,

$$a'_0 - a'_1 = a_0 - a_1 - mk_1$$
$$a'_0 - a'_2 = a_0 - a_2 - mk_2$$
$$\cdots$$
$$a'_0 - a'_{m-1} = a_0 - a_{m-1} - mk_{m-1}$$

where k_1, k_2, $\cdots$, and k_{m-1} are integers. Equivalently, we have

$$a_1 - a'_1 + mk_1 = a_0 - a'_0$$
$$a_2 - a'_2 + mk_2 = a_0 - a'_0$$
$$\cdots$$
$$a_{m-1} - a'_{m-1} + mk_{m-1} = a_0 - a'_0$$

Then,

$$a_0 - a_0' = a_0 - a_0'$$
$$a_0 - a_0' = a_1 - a_1' + mk_1$$
$$a_0 - a_0' = a_2 - a_2' + mk_2$$
$$\cdots$$
$$a_0 - a_0' = a_{m-1} - a_{m-1}' + mk_{m-1}$$

Equivalently,

$$a_0' = a_0 - (a_0 - a_0')$$
$$a_1' = a_1 - (a_0 - a_0') + mk_1$$
$$a_2' = a_2 - (a_0 - a_0') + mk_2$$
$$\cdots$$
$$a_{m-1}' = a_{m-1} - (a_0 - a_0') + mk_{m-1}$$

Owing to $a_0' - a_0 \geq a_1' - a_1 \geq \cdots \geq a_{m-1}' - a_{m-1}$, we have $k_{m-1} \geq \cdots \geq k2 \geq k1 \geq 0$. Consequently, we can find some reasonable breeding choices such that $a_0 a_1 a_2 \cdots a_{m-1}$ can reach $a_0' a_1' a_2' \cdots a_{m-1}'$. During the process, all the species numbers are non-negative.

a_0		a_1		a_2	$\cdots$	a_{m-1}	
$-$	-1	$-$	-1	-1		$-1+m$	$-$
					$\cdots$		
$\vert$	-1	$\vert$	-1	$-1+m$	$\vert$	-1	$\vert$
$\vert$	-1	$\vert$	$-1+m$	-1	$\vert$	-1	$\vert$
$\vert$		plus m			$\cdots$		$\vert$
$\vert$	-1	for	-1	-1	$\vert$	$-1+m$	$\vert$
$\vert$		k_1 times			$\cdots$		$\vert$
$\vert$	-1	$\vert$	-1	$-1+m$ plus m		-1	$\vert$
$\vert$	-1	$-$	$-1+m$	-1 for		-1	$\vert$
$\vert$	-1		-1	-1 k_2 times		$-1+m$	$\vert$
$(a_0 - a_0')$				$\cdots$			plus m
times	-1		-1	$-1+m$	$\vert$	-1	for
$\vert$				$\cdots$			k_{m-1} times
$\vert$	-1		-1	-1	$\vert$	$-1+m$	$\vert$
$\vert$				$\cdots$			$\vert$
$\vert$	-1		-1	$-1+m$	$-$	-1	$\vert$
$\vert$				$\cdots$			$\vert$
$\vert$	-1		-1	-1		$-1+m$	$\vert$
				$\cdots$			
$-$	-1		-1	-1		$-1+m$	$-$

Accordingly, state $a_0 a_1 a_2 \cdots a_{m-1}$ can reach state $a_0' a_1' a_2' \cdots a_{m-1}'$.   □

We point out that if **G2** holds, state $a_0' a_1' a_2' \cdots a_{m-1}'$ is reachable from non-final state $a_0 a_1 a_2 \cdots a_{m-1}$, but not vice versa. For instance, states 2221 and 1150 satisfy condition **G2**, however, 2221 can reach 1150 but 1150 cannot reach 2221.

Corollary 1. Two cannot-fail states or might-fail states $a'_0 a'_1 a'_2 \cdots a'_{m-1}$ and $a_0 a_1 a_2 \cdots a_{m-1}$ are reachable from each other iff condition **G2** holds. □

In the following, some conditions are given to show how to decide whether a state is a must-fail, might-fail or cannot-fail state.

1. must-fail states: Must-fail states consist of two kinds of states, final states and some specific non-final states which cannot reach themselves (named tailed states for convenience). Thus, to decide whether a given state is a must-fail state, we need to check whether it is a final state or a tailed state. By the definition, a final state $a_0 a_1 a_2 \cdots a_{m-1}$ can be distinguished by checking whether there exist at least two 0 in $a_0, a_1, a_2, \cdots$, and a_{m-1}. Further, to decide whether a state $a_0 a_1 a_2 \cdots a_{m-1}$ is a tailed state, we first sort $a_0 a_1 a_2 \cdots a_{m-1}$ in ascending order and obtain $a'_0 a'_1 a'_2 \cdots a'_{m-1}$. Then we compare each a'_i, $0 \le i \le m-1$, with the ith number in $0, 1, 2, \cdots, m-1$. Consequently, state $a_0 a_1 a_2 \cdots a_{m-1}$ is a tailed state iff there exists at least one a'_i such that $a'_i < i$. As an example, for a state $(2, 1, 2, 1, 6, 4)$ in 6 species strange planet protocol, we sort it in ascending order first and obtain a sequence $(a'_0 = 1, a'_1 = 1, a'_2 = 2, a'_3 = 2, a'_4 = 4, a'_5 = 6)$. Thus, $(2, 1, 2, 1, 6, 4)$ is a tailed state since the a'_3, i.e. 2, is smaller than 3.

The following Lemma shows the correctness of the above approach for deciding whether a state is a tailed state.

Lemma 2. Let $a'_0 a'_1 a'_2 \cdots a'_{m-1}$ be the sequence obtained by sorting the numbers of state $a_0 a_1 a_2 \cdots a_{m-1}$ in ascending order. $a_0 a_1 a_2 \cdots a_{m-1}$ is a tailed state iff there exists at least one a'_i such that $a'_i < i$.

Proof. ($\Rightarrow$): Since sequence $a'_0 a'_1 a'_2 \cdots a'_{m-1}$ is in ascending order, it can be obtained that $a_k \le a_{k+1}$, $0 \le k \le m-2$. Suppose i is the smallest integer in $0, 1, \cdots$, and $m-1$ such that $a_i < i$. Then we have $i + 1 - a_i \ge 2$. This means that after at most a_i^{th} mating, there are at least two species a_l and a_k, $0 \le l, k \le i$, which have never been added by m. Some reasonable breeding choices are shown below:

a_0	a_1	a_2	$\cdots$	a_l	$\cdots$	a_k	$\cdots$	a_i	a_{i+1}	$\cdots$	a_{m-1}	—
-1	$-1+m$	-1		-1		-1		-1	-1		-1	|
$-1+m$	-1	-1		-1		-1		-1	-1		-1	|
-1	-1	-1		-1		-1		-1	$-1+m$		-1	*at most*
			$\cdots$									a_i *times*
-1	-1	-1		-1		-1		-1	-1		$-1+m$	|
-1	-1	$-1+m$		-1		-1		-1	-1		-1	|
-1	-1	-1		-1		-1	$-1+m$		-1	-1		-1	|
-1	-1	$-1+m$		-1		-1		-1	-1		-1	—

Since $a_l, a_k \le a_i$, according to the mating rules, a_l and a_k must be 0. This leads to $a_0 a_1 \cdots a_i \cdots a_{m-1}$ being a tailed state.

($\Leftarrow$): We prove the sufficiency by contradiction. Without loss of generality, we assume each $a'_i \ge i$, $0 \le i \le m-1$. Then the state can be equally denoted as

$(0+b_0, 1+b_1, \cdots, m-1+b_{m-1})$, $b'_i \geq 0, 0 \leq i \leq m-1$. Thus we have the following breeding sequences:

$$0 + b_0, \ 1 + b_1, \ 2 + b_2, \ \cdots, \ m-1+b_{m-1}$$
$$\downarrow$$
$$m - 1 + b_0, \ 0 + b_1, \ 1 + b_2, \ \cdots, \ m - 2 + b_{m-1}$$
$$\downarrow$$
$$m - 2 + b_0, \ m - 1 + b_1, \ 0 + b_2, \ \cdots, \ m - 3 + b_{m-1}$$
$$\downarrow$$
$$\cdots$$
$$\downarrow$$
$$1 + b_0, \ 2 + b_1, \ 3 + b_2, \ \cdots, \ 0 + b_{m-1}$$
$$\downarrow$$
$$0 + b_0, \ 1 + b_1, \ 2 + b_2, \ \cdots, \ m - 1 + b_{m-1}$$

It is evident that the state is in a loop, which is contradictory with the definition of tailed states. So, the hypothesis is false. Therefore, the Lemma holds. □

2. might-fail states: A might-fail state is one that can reach not only final states, but also itself. This means that it is in a circle and can reach a final state. Thus, for a given state $a_0 a_1 a_2 \cdots a_{m-1}$, we first check whether it can reach a final state and then decide whether it is in a loop.

If there are at least two numbers in $\{a_0 \bmod m, a_1 \bmod m, \cdots, a_{m-1} \bmod m\}$ which are equal, state $a_0 a_1 a_2 \cdots a_{m-1}$ can reach a final state. Further, state $a_0 a_1 a_2 \cdots a_{m-1}$ is in a loop if it is not a tailed state or a final state.

Lemma 3. Let $a'_0 a'_1 a'_2 \cdots a'_{m-1}$ be the sequence obtained by sorting the numbers in state $a_0 a_1 a_2 \cdots a_{m-1}$ in ascending order. $a_0 a_1 a_2 \cdots a_{m-1}$ is a might-fail state iff the following two conditions hold together:

(1) $\exists i, j, 0 \leq i, j \leq m - 1, i \neq j$, such that $a_i \overset{m}{=} a_j$,
(2) $a'_i \geq i$, for $\forall i \in \{0, 1, 2, \cdots, m - 1\}$.

Proof. In fact, condition (1) ensures that a state can reach a final state and condition (2) guarantees that a state is in a loop.
(1) ($\Rightarrow$): For m strange planet protocol with n individuals, i.e. $a_0 + a_1 + a_2 + \cdots + a_{m-1} = n$, we have:

$$a_0 = mk_0 + q_0$$
$$a_1 = mk_1 + q_1$$
$$a_2 = mk_2 + q_2$$
$$\cdots$$
$$a_{m-1} = mk_{m-1} + q_{m-1}$$

where $k_0, k_1, k_2, \cdots$, and $k_{m-1} \geq 0$, $q_0, q_1, q_2, \cdots$, and $q_{m-1} = 0, 1, 2, \cdots$, or $m-1$. Without loss of generality, we assume $q_0 = q_1$, $k_0 \geq k_1$. Thus, $a_0 - a_1 = m(k_0 - k_1)$. We can find the following breeding choices:

	a_0		a_1	a_2	$\cdots$	a_{m-1}
$-$	-1	$-$	$-1+m$	-1		-1
$\mid$	-1	$\mid$	-1	$-1+m$		-1
$t\ times$		$\mid$		$\cdots$		
$\mid$	-1	$\mid$	-1	-1		$-1+m$
$\mid$	-1	$\mid$	$-1+m$	-1		-1
$\mid$	-1	plus m	-1	$-1+m$		-1
$\mid$		for (k_0-k_1)		$\cdots$		
$\mid$	-1	$times$	-1	-1		$-1+m$
$\mid$	-1	$\mid$	$-1+m$	-1		-1
$-$		$-$		$\cdots$		

Now we have $a_0' = a_1 + m(k_0 - k_1) - t$ and $a_1' = a_1 + m(k_0 - k_1) - t$. That is $a_0' = a_1'$. Then after a_0' (or a_1') times non-a_0-event or non-a_1-event occur, a_0 and a_1 will reduce to 0 simultaneously. Therefore, state $a_0a_1\cdots a_{m-1}$ can reach a final state.

($\Leftarrow$): Now we prove that if a state can reach a final state, condition (1) holds. Suppose state $a_0a_1a_2\cdots a_{m-1}$ can reach the final state $00b_2\cdots b_{m-1}$. **G2** holds between them. Then we have $(a_0 - a_1) \stackrel{m}{=} (b_0 - b_1) \stackrel{m}{=} (0 - 0) \stackrel{m}{=} 0$ and $a_0 \stackrel{m}{=} a_1$. Thus, condition (1) holds.

(2) A state can reach itself iff it is not a tailed state or a must-fail state. By Lemma 2, state $a_0a_1\cdots a_{m-1}$ is in a loop iff for each a_i', $0 \le i \le m-1$, it has $a_i' \ge i$.   □

3. cannot-fail states: Cannot-fail states are the ones that cannot reach final states. This means that it is a state in a loop but cannot reach a final state. Accordingly, we have the following corollary.

Corollary 2. Let $a_0'a_1'a_2'\cdots a_{m-1}'$ be the sequence obtained by sorting the numbers in state $a_0a_1a_2\cdots a_{m-1}$ in ascending order. $a_0a_1a_2\cdots a_{m-1}$ is a cannot-fail state if the following two conditions hold:

(1) $\not\exists\, i$ and j, $0 \le i, j \le m-1$, $i \ne j$, such that $a_i \stackrel{m}{=} a_j$,
(2) $a_i' \ge i$, for $\forall\, i \in \{0, 1, 2, \cdots, m-1\}$.   □

In Corollary 2, condition (1) confirms that state $a_0a_1a_2\cdots a_{m-1}$ cannot reach a final state, and (2) indicates that state $a_0a_1a_2\cdots a_{m-1}$ occurs in a loop.

Equivalently, a state $a_0a_1\cdots a_{m-1}$, is a cannot-fail state if $(a_0\ mod\ m, a_1\ mod\ m, \cdots, a_{m-1}\ mod\ m)$ is an arbitrary permutation of 0, 1, 2, $\cdots$, and $m-1$. So we also have the corollary below.

Corollary 3. State $a_0a_1a_2\cdots a_{m-1}$ is a cannot-fail state iff there do not exist two numbers, i and j, $0 \le i, j \le m-1$, $i \ne j$, such that $a_i \stackrel{m}{=} a_j$.   □

4   Properties with m Strange Planet Protocol

Let S be the set of all states. For any $x, y \in S$, xRy, if x can reach y. With respect to relation R, S is partitioned into three sets, cannot-fail, must-fail, and might-fail states as shown in Fig. 3.

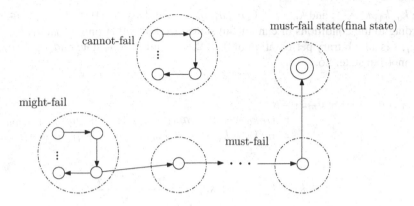

Fig. 3. Possible Subsets

Theorem 2. The binary reachable relation R over the set of cannot-fail states (or the set of might-fail states) is an equivalence relation.

Proof. Let $x(x_0x_1 \cdots x_{m-1})$ and $y(y_0y_1 \cdots y_{m-1})$ be two different states in the set of cannot-fail states (or the set of might-fail states) and xRy. It is evident that condition **G2** holds. Namely, $(x_i - x_j) \overset{m}{=} (y_i - y_j), 0 \leq i < j \leq m - 1$.

(1) Suppose that there exists a state $z(z_0z_1 \cdots z_{m-1})$ such that yRz. Then, $(y_i - y_j) \overset{m}{=} (z_i - z_j), 0 \leq i < j \leq m - 1$, it follows that $(x_i - x_j) \overset{m}{=} (z_i - z_j), 0 \leq i < j \leq m - 1$. By non-final genetic factor and Theorem 1, **G2** also holds between x and z, and xRz. So relation R is transitive.
(2) Since **G2** holds between x and y, by Corollary 1 we have yRx. Thus, relation R is symmetric.
(3) Let $x'(x'_0x'_1 \cdots x'_{m-1})$ be the sorted state of x. Then according to Lemma 3 and Corollary 2, $x'_i \geq i, 0 \leq i \leq m - 1$, which guarantees that x' is in a loop. That is to say, x can reach itself, namely xRx. Hence, relation R is reflexive.

Therefore, the theorem holds.   □

Theorem 3. For m species strange planet protocol with n individuals, there are totally

$$\frac{(m + \frac{n - \sum_{i=0}^{m-1} i}{m} - 1)!}{(\frac{n - \sum_{i=0}^{m-1} i}{m})! \times (m - 1)!} \times m!$$

cannot-fail states.

Proof. For any $a_0 + a_1 + a_2 + \cdots + a_{m-1} = n$, we have

$$a_0 = mk_0 + q_0$$
$$a_1 = mk_1 + q_1$$
$$\cdots$$
$$a_{m-1} = mk_{m-1} + q_{m-1}$$

where $k_0, k_1, k_2, \cdots$, and $k_{m-1} \geq 0$, $q_0, q_1, q_2, \cdots$, and $q_{m-1} = 0, 1, 2, \cdots$, or $m - 1$. According to the conditions of cannot-fail states, we know that only if $q_0, q_1, q_2, \cdots$, and q_{m-1} is an arbitrary permutation of $0, 1, 2, \cdots$, and $m - 1$, state $a_0 a_1 a_2 \cdots a_{m-1}$ is a cannot-fail state. So,

$$a_0 + a_1 + a_2 + \cdots + a_{m-1} = n$$
$$= (mk_0 + q_0) + (mk_1 + q_1) + \cdots + (mk_{m-1} + q_{m-1})$$
$$= m(k_0 + k_1 + \cdots + k_{m-1}) + (q_0 + q_1 + \cdots + q_{m-1})$$

Thus,

$$k_0 + k_1 + k_2 + \cdots + k_{m-1} = \frac{n - \sum_{i=0}^{m-1} i}{m}$$

The number of the combinations of $k_0, k_1, \cdots$, and k_{m-1} is equal to the problem of putting $r = \frac{n - \sum_{i=0}^{m-1} i}{m}$ indistinguishable objects into m distinguishable buckets, namely C_{m+r-1}^r [7]. Consequently, the number of the cannot-fail states will be:

$$C_{m+r-1}^r \times m \times (m - 1)! = \frac{(m + \frac{n - \sum_{i=0}^{m-1} i}{m} - 1)!}{(\frac{n - \sum_{i=0}^{m-1} i}{m})! \times (m - 1)!} \times m!$$

Therefore, the theorem holds.   □

Note that by the proof of Theorem 3, it can be inferred that only if $n \geq \sum_{i=0}^{m-1} i$ and $n \bmod m = 0$, there exist cannot-fail states in the extended strange planet problem. Therefore, for m species and n individuals, the number of cannot-fail states is:

$$\begin{cases} \frac{(m + \frac{n - \sum_{i=0}^{m-1} i}{m} - 1)!}{(\frac{n - \sum_{i=0}^{m-1} i}{m})! \times (m-1)!} \times m!, & n \geq \sum_{i=0}^{m-1} i \text{ and } n \bmod m = 0, \\ 0, & otherwise. \end{cases}$$

5 Applications

We use a small password generator example to show how the m strange planet protocol can be used in secure communications. A password of an important file constitutes five parts, which is held by five different persons, respectively. Let $Pass_1, Pass_2, Pass_3, Pass_4$ and $Pass_5$ be the data held by $worker_1, worker_2, worker_3, worker_4$, and $worker_5$, respectively. In order to open the file, the five parties are required to partici-pant in simultaneously.

To keep the password safe, whenever the password is used successfully, $Pass_1, Pass_2, Pass_3, Pass_4$ and $Pass_5$ will be changed according to rules in 5 strange planet protocol automatically. Note that the sum of the 5 data never changes. A new generated password is any successor of the former one. In that way, the password generator will fail if the current password is a might-fail state or a must-fail state, because both can lead to a final state which has no successors. That is to say, only cannot-fail states can make the password generator work permanently. We point out that here $worker_1$,

$worker_2$, $worker_3$, $worker_4$, and $worker_5$ do not know the rule. Each of them only holds their data assigned by the generator. To ensure that new data will always be generated, the initially assigned data should be a cannot-fail state in the extended strange planet protocol.

For instance, let $Pass_1 = 66$, $Pass_2 = 58$, $Pass_3 = 14$, $Pass_4 = 20$, $Pass_5 = 32$ be the initial data held by the 5 workers, respectively. The numbers modulo m are 1, 3, 4, 0, 2, which exactly satisfies the condition of cannot-fail states. The password generator will work forever and a possible infinite sequence will be produced as shown below:

$Pass_1$	$Pass_2$	$Pass_3$	$Pass_4$	$Pass_5$
66	58	14	20	32
65	62	13	19	31
64	61	12	18	35
68	60	11	17	34
72	59	10	16	33
71	58	14	15	32

$\cdots$

6 Conclusion

This paper presents a general extension of the original strange planet problem and reveals the essential principles in it. An intelligent password generator is designed to show how the protocol can be used in practice. In the near future, we will further study more general mating rules among species of the extended protocol as well as the essence under it. How the discovered properties can be used in secure communications of real world will also be investigated.

References

1. Tian, C., Duan, Z., Liu, J.: Secure Communication with Strange Planet Protocol. Optimization Letters, 1–9 (2012)
2. Holzmann, G.J.: Design and Validation of Computer Protocols. Pretentice Hall (1991) ISBN 0-13-539925-4
3. Perlman, R.: Interconnections: Bridges, Routers, Switches, and Internetworking Protocols, 2nd edn. Addison-Wesley (1999) ISBN 0-201-63448-1
4. Comer, D.E.: Internetworking with TCP/IP - Principles, 4th edn. Protocols and Architecture. Prentice Hall (2000) ISBN 0-13-018380-6
5. Agrawal, D.P., Zeng, Q.-A.: Introduction to Wireless and Mobile Systems, 2nd edn. Thomson (April 2005) ISBN. 978-0534493035
6. Hopcroft, J.E., Motwani, R., Ullman J.D.: Introduction to Automata Theory, Languages, and Computation, 2nd edn. Pearson Education (2000) ISBN 0-201-44124-1
7. Rosen, K.H., Krithivasan, K.: Discrete mathematics and its applications. McGraw-Hill, New York (1999)

Online Bin Covering: Expectations vs. Guarantees*

Marie G. Christ, Lene M. Favrholdt, and Kim S. Larsen

University of Southern Denmark, Odense, Denmark
{christm,lenem,kslarsen}@imada.sdu.dk

Abstract. Bin covering is a dual version of classic bin packing. As usual, bins have size one and items with sizes between zero and one must be packed. However, in bin covering, the objective is to cover as many bins as possible, where a bin is covered if the sizes of items placed in the bin sum up to at least one. We are considering the online version of bin covering. Two classic algorithms for online bin packing that have natural dual versions are HARMONIC$_k$ and NEXT-FIT. Though these two algorithms are quite different in nature, competitive analysis does not distinguish these bin covering algorithms.

In order to understand the combinatorial structure of the algorithms better, we turn to other performance measures, namely relative worst order, random order, and max/max analysis, as well as analyses under restricted input assumptions or uniformly distributed input. In this way, our study also supplements the ongoing systematic studies of the relative strengths of various performance measures.

We make the case that when guarantees are needed, even under restricted input sequences, the dual HARMONIC$_k$ algorithm is preferable. In addition, we establish quite robust theoretical results showing that if items come from a uniform distribution or even if just the ordering of items is uniformly random, then dual NEXT-FIT is the right choice.

1 Introduction

Bin covering [1] is a dual version of classic bin packing. As usual, bins have size one and items with sizes between zero and one must be packed. However, in bin covering, the objective is to cover as many bins as possible, where a bin is covered if the sizes of items placed in the bin sum up to at least one. We are considering the online version of bin covering. A problem is online if the input sequence is presented to the algorithm one item at a time, and the algorithm must make an irrevocable decision regarding the current item without knowledge of future items.

Bin covering algorithms have numerous important applications. For instance when packing or canning food items guaranteeing a minimum weight or volume, reductions in the overpacking of even a few percent may have a large economic

---

* Supported in part by the Danish Council for Independent Research.

P. Widmayer, Y. Xu, and B. Zhu (Eds.): COCOA 2013, LNCS 8287, pp. 226–237, 2013.
© Springer International Publishing Switzerland 2013

impact. If items arrive on a conveyor belt, for instance, the problem becomes online.

Classic algorithms for online bin packing are NEXT-FIT and HARMONIC$_k$ [21]. NEXT-FIT is a very simple and natural algorithm, and HARMONIC$_k$ was designed to obtain a competitive ratio [24,19] better than any Any-Fit algorithm (First-Fit and Best-Fit are examples of Any-Fit algorithms for bin packing, and the competitive ratio of Next-Fit is worse than both these algorithms). HARMONIC$_k$ and variations of it have been analyzed extensively [22,25,23]. We consider the obvious dual version of these, DNF [1] and DH$_k$ [12]. These algorithms are quite different in nature and the bin packing versions are clearly separated, having competitive ratios of 2 and approximately 1.691, respectively. However, for bin covering, competitive analysis does not separate them! In fact, for bin covering, competitive analysis categorizes both algorithms as being worst possible (among reasonable algorithms). This is unlike the situation in bin packing, and in general, results from bin packing do not transfer directly to bin covering.

To understand the algorithmic differences better, it is therefore necessary to employ different techniques, and we turn to other generally applicable performance measures, namely relative worst order analysis, random order analysis, and max/max analysis. As for almost all performance measures, the idea is to abstract away some details of the problem to enable comparisons. Without some abstraction, it is hard to ever, analytically, claim that one algorithm is better than another, since almost any algorithm performs better than any other algorithm on at least one input sequence. For all the measures considered here, the abstraction can be viewed as being defined via first a partitioning of the set of input sequences of a given length and then an aggregation of the results from each partition. For each sequence length, competitive analysis, for instance, considers all the ratios of the online performance to the optimal offline performance obtained for each sequence of that length, and then takes the worst ratio of all of these. The measures above employ a less fine-grained partition of the input space. Worst order and random order analysis group permutations of the same sequence together instead of considering each sequence separately, considering worst-case or average-case performance, respectively, within each partition. With max/max analysis the partitioning of the input space is even coarser: for each sequence length n, the online worst-case behavior over all sequences of length n is compared to the worst-case optimal offline behavior over all sequences of length n. There is no one correct way to compare algorithms, but since these measures focus on different aspects of algorithmic behavior, considering all of the ones above lead to a very broad analysis of the problem. Extensive motivational sections can be found in the papers introducing these measures and in the survey [13]. As a further supplement, we analyze restricted input sequences, where items have similar size, which is likely to happen in practice if one is packing products with an origin in nature, for instance. Finally, we consider input sequences containing items having uniformly distributed sizes.

Relative worst order analysis [3,4] has been applied to many problems; a recent list can be found in [15]. In [16], bin covering was analyzed, but using a version of the problem allowing items of size 1. We analyze the more commonly studied version for bin covering, where all items are strictly smaller than 1. Since worst-case sequences from [16] contain items of size 1, this leads to slightly different results. For completeness, we include these results. Random order analysis [20] was introduced for classic bin packing, but has also been used for other problems; a server problem, for instance [7]. Max/max analysis [2] was introduced as an early step towards refining the results from competitive analysis for paging and a server problem.

Relative worst order analysis emphasizes the fact that there exist multisets of input items where DNF can perform $\frac{3}{2}$ times as poorly as DH_k. On the other hand, DH_k's method of limiting the worst case also means that it has less of an opportunity to reach the best case, as opposed to DNF. This is reflected in the random order analysis, where DNF comes out at least as well as DH_k. Another way of approaching randomness is to analyze a uniform distribution. We establish new results on DH_k showing that its performance here is slightly worse than that of DNF, in line with the random order results. With the max/max analysis, a distinction between the two algorithms can only be achieved, when the item sizes are limited, and DH_k is the algorithm selected as best by this measure. With respect to competitive analysis, we also consider restricted input in the sense that item sizes may only vary across one or two consecutive DH_k partitioning points. This is a formal way of treating the case where items are of similar size, while allowing greater variation when this size is large. We show that with this restricted form of input, considering the worst case measures of competitive analysis, DH_k is deemed better than DNF, as DNF is vulnerable to worst-case sequences where DH_k can organize the packing differently.

This study also contributes to the ongoing systematic studies of the relative strengths of various performance measures, initiated in [7]. Up until that paper, most performance measures were introduced for a specific problem to overcome the limitations of competitive analysis. In [7], comparisons of performance measures different from competitive analysis were initiated, and this line of work has been continued in [5,8,6], among others. Our results supplement results in [11], showing that no deterministic algorithm for the bin covering problem can be better than $\frac{1}{2}$-competitive and giving an asymptotically optimal algorithm for the case of items being uniformly distributed on $(0, 1)$. For DNF, [10] established an expected competitive ratio of $\frac{2}{e}$ under the same conditions.

Due to space restrictions, several proofs have been omitted or shortened. Refer to [9] for all the details.

Bin Covering

In the one dimensional bin covering problem, the algorithm gets an input sequence $I = \langle i_1, i_2, \ldots \rangle$ of item sizes, where for all j, $0 < i_j < 1$. The goal is to pack the items in a maximum number of bins, each having size 1, such that the sum of the sizes of the items within each bin is at least one, i.e., the bin is *covered*. Requiring items to be strictly smaller than 1 corresponds to assuming that items of size 1 are treated separately. This makes sense, since there is no advantage in combining an item of size 1 with any other items in a bin. In other words, any algorithm not giving special treatment to items of size 1 could trivially be improved by doing so.

In algorithms for bin packing and covering, it is standard to use the terminology that a bin is *open* if it is one of the bins that an algorithm is currently considering for the next item, and *closed* if the bin has received items, but the algorithm will not consider that bin again for future items.

Thus, the objective for a bin covering algorithm A is to maximize the number of bins covered as a result of processing an input sequence I. We let $A(I)$ denote this number of covered bins. We let OPT denote an optimal offline algorithm. Thus, $\text{OPT}(I)$ is the largest number of bins that can be covered by any algorithm processing I.

Assmann, Johnson, Kleitman, and Leung [1] introduced the Dual NEXT-FIT algorithm (DNF), an adaption of the NEXT-FIT algorithm for bin packing. DNF always keeps a single bin open. The arriving items are packed into the open bin until the open bin has a content of at least one. Then the open bin is closed and a new empty bin becomes the open bin.

HARMONIC$_k$ was introduced for bin packing by Lee and Lee [21]. This algorithm partitions the interval $(0, 1]$ into k subintervals, with the partitioning points at $\frac{1}{2}, \frac{1}{3}, \ldots, \frac{1}{k}$, resulting in the intervals $(0, \frac{1}{k}], (\frac{1}{k}, \frac{1}{k-1}], \ldots, (\frac{1}{2}, 1)$. For each of these k subintervals, HARMONIC$_k$ keeps one open bin into which the items belonging to this subinterval are packed at their arrival. This means that each closed bin for the interval $(\frac{1}{j}, \frac{1}{j-1}]$ contains exactly j items. The natural adaptation to the bin covering problem is to use $(0, \frac{1}{k}), [\frac{1}{k}, \frac{1}{k-1}), \ldots, [\frac{1}{2}, 1)$. The resulting algorithm, DHARMONIC$_k$ (DH$_k$), uses exactly j items from the interval $[\frac{1}{j}, \frac{1}{j-1})$ to cover a bin. All through the paper we assume that $k \geq 2$, since for $k = 1$, DH$_k$ becomes DNF.

2   Competitive Analysis

In competitive analysis [24,19], the performance of an online algorithm is compared to that of an optimal offline algorithm OPT. An algorithm A for a maximization problem is called *c-competitive* if there exists a fixed constant b such that for any input sequence I, it holds that $A(I) \geq c\,\text{OPT}(I) + b$. The supremum over all such c is the *competitive ratio* CR(A) of A. Note that some authors reverse the order of the algorithm and OPT to get ratios larger than one.

For bin covering, Csirik and Totik [11] showed that no deterministic online algorithm can be better than $\frac{1}{2}$-competitive. DNF was shown to be $\frac{1}{2}$-competitive in [1], and the same result for DH$_k$ was noted in [16]. For completeness, to show that this result is tight for a large class of algorithms, we define a *reasonable* algorithm to be one that closes bins as soon as they are covered, does not close bins before they are covered, and does not have more than a constant number of open bins at any point.

Theorem 1. *Any deterministic reasonable algorithm has competitive ratio $\frac{1}{2}$.*

2.1 Limiting the Item Sizes

In some applications of the bin covering problem it is likely that the sizes of the items contained in an input sequence differ only slightly, e.g., packing similar food items into a container, guaranteeing the consumer a minimum weight. In the following, we investigate the performance of DNF and DH_k on sequences with similar-sized items. Since it seems reasonable to allow larger variance in size when the considered sizes are large, we consider sequences containing item sizes from consecutive DH_k intervals.

We first consider intervals $(a, b) \subseteq (0, 1)$ that contain exactly one DH_k partitioning point. Afterwards, we consider sequences with exactly two DH_k partitioning points. We emphasize that there are no restrictions on the endpoints a and b, which can be any real numbers, as long as the interval between them contains exactly one or two DH_k partitioning points. In both cases, DH_k turns out to have the better ratio.

For any $(a, b) \subseteq (0, 1)$, we let $\mathrm{CR}_{a,b}$ denote the competitive ratio on sequences where all item sizes are in (a, b).

If (a, b) does not contain at least one of the interval borders used by DH_k, then DH_k behaves exactly like DNF. If (a, b) contains a DH_k border, then we define $\frac{1}{p} = \max\left\{\frac{1}{l} \,\middle|\, l \in \mathbb{N}, \frac{1}{l} < b\right\}$, and refer to $\frac{1}{p}$ as the *maximal border in* (a, b).

Theorem 2. *If* $\frac{1}{p+1} \leq a < \frac{1}{p}$, *then* $\mathrm{CR}_{a,b}(\mathrm{DNF}) = \frac{p}{p+1}$.

Theorem 3. *If* $\frac{1}{p+1} \leq a < \frac{1}{p}$ *and* $k \geq p$, *then* $\mathrm{CR}_{a,b}(\mathrm{DH}_k) = \frac{p^2+1}{p(p+1)}$.

It follows that if (a, b) contains exactly one DH_k partitioning point, $\frac{1}{p}$, and $k \geq p$, then DH_k has a better competitive ratio than DNF:

Corollary 1. *If* $\frac{1}{p+1} \leq a < \frac{1}{p}$ *and* $k \geq p$, *then* $\mathrm{CR}_{a,b}(\mathrm{DH}_k) > \mathrm{CR}_{a,b}(\mathrm{DNF})$.

We now consider intervals $(a, b) \subseteq (0, 1)$ that contain exactly two DH_k partitioning points. For the following theorem, note that $\frac{1}{p} < \frac{p+2}{p(p+1)} < \frac{1}{p-1}$.

Theorem 4. *If* $a < \frac{1}{p+1}$, *then*

$$\mathrm{CR}_{a,b}(\mathrm{DNF}) \leq \begin{cases} \dfrac{p+1}{p+2}, & \text{if } b \leq \frac{p+2}{p(p+1)} \\[2mm] \dfrac{p(p+1)}{p^2+2p+2}, & \text{otherwise} \end{cases}$$

Proof. Replacing p by $p+1$ in Theorem 2, we get an upper bound of $\frac{p+1}{p+2}$, since the upper bound of Theorem 2 only assumes $a < \frac{1}{p} < b$. This proves the upper bound for $b \leq \frac{p+2}{p(p+1)}$.

If $b > \frac{p+2}{p(p+1)}$, we choose ε, $0 < \varepsilon < \min\left\{\frac{1}{2(p-1)(p+1)n}\left(\frac{1}{p+1} - a\right), b - \frac{p+2}{p(p+1)}\right\}$, the only purpose of this complicated expression being that we should ensure that all items below belong to (a, b). Now, we consider a sequence consisting of the following subsequences:

$$- \langle (\frac{1}{p})^{p-1}, \frac{1}{p} - 2\varepsilon, \frac{p+2}{p(p+1)} + \varepsilon \rangle^{(p+1)(p-2)n}$$

$$- \langle \frac{1}{p+1} + i(p-1)\varepsilon, \frac{1}{p+1} - (i+1)(p-1)\varepsilon, \langle \frac{1}{p+1} + \varepsilon \rangle^{p-2}, \frac{1}{p+1} - \varepsilon, \frac{p+2}{p(p+1)} + \varepsilon \rangle$$
for $i = 1, 2, \ldots, (p+1)n$

$$- \langle \frac{1}{p+1} + i(p-1)\varepsilon, \frac{1}{p+1} - (i+1)(p-1)\varepsilon, \langle \frac{1}{p+1} \rangle^{p-2}, \frac{1}{p+1} - \varepsilon, \frac{p+2}{p(p+1)} + \varepsilon \rangle$$
for $i = (p+1)n + 1, (p+1)n + 2, \ldots, 2(p+1)n - 1$

$$- \langle \frac{1}{p+1} + 2(p+1)n(p-1)\varepsilon, \langle \frac{1}{p+1} \rangle^{p-2}, \frac{1}{p+1} - \varepsilon, \frac{p+2}{p(p+1)} + \varepsilon \rangle$$

$$- \langle \frac{1}{p+1} - (p-1)\varepsilon \rangle$$

Giving the items in this order, DNF covers $(p+1)(p-2)n + (p+1)n + (2(p+1)n - 1 - (p+1)n) + 1 = p(p+1)n$ bins. In the full version it is shown that OPT covers $(p^2 + 2p + 2)n$ bins. □

Theorem 5. *If $\frac{1}{p+2} \le a < \frac{1}{p+1}$ and $k \ge p+1$, then*

$$CR_{a,b}(DH_k) = \begin{cases} \dfrac{p^3 + 2p^2 + p + 2}{p(p+1)(p+2)}, & \text{if } b \le \frac{p+2}{p(p+1)} \\[2mm] \dfrac{p^3 + 2p^2 + 2}{p(p+1)(p+2)}, & \text{otherwise} \end{cases}$$

Proof. We only sketch the proof of the lower bound here.

Items of size less than $\frac{1}{p+1}$ are called *small*, items of size at least $\frac{1}{p}$ are called *large*, and the remaining items are called *medium*. Let s, m, and ℓ denote the number of small, medium, and large items, respectively.

Consider an optimal packing. For $i = 1, 2, 3$, let n_i denote the number of bins with exactly $p+i-1$ items. Then, $n = n_1 + n_2 + n_3$ is the number of bins covered by OPT. Since DH_k covers exactly $\lfloor \frac{s}{p+2} \rfloor + \lfloor \frac{m}{p+1} \rfloor + \lfloor \frac{\ell}{p} \rfloor$ bins, we can consider items from the three types of bins separately. The contribution to the number of bins covered by DH_k from the n_i items is at least $d_i - 3$, where

$$d_i \ge \begin{cases} \dfrac{p^3 + 2p^2 + p + 2}{p(p+1)(p+2)} n_i, & \text{if } b \le \frac{p+2}{p(p+1)} \\[2mm] \dfrac{p^3 + 2p^2 + 2}{p(p+1)(p+2)} n_i, & \text{otherwise} \end{cases}$$

□

It follows that if (a, b) contains exactly two DH_k partitioning points, then DH_k has a better competitive ratio than DNF:

Corollary 2. *If $\frac{1}{p+2} \le a < \frac{1}{p+1}$, then $CR_{a,b}(DH_k) > CR_{a,b}(DNF)$.*

3 Relative Worst Order Analysis

Relative worst order analysis was introduced by Boyar and Favrholdt [3] and compares the performance of two algorithms A and B directly instead of via the comparison to OPT. Algorithms are compared on the same input sequence I, but on the worst possible permutation of I for each algorithm.

Formally, if n is the length of I, and σ is a permutation on n elements, then $\sigma(I)$ denotes I permuted by σ, and we define $A_W(I) = \min_\sigma A(\sigma(I))$. If there exists a fixed constant b such that, for any input sequence I, $A_W(I) \geq B_W(I) - b$, then A and B are *comparable* and the *relative worst order ratio* of A to B is defined as follows: $WR(A, B) = \sup\{c \mid \exists b \, \forall I : A_W(I) \geq c \, B_W(I) - b\}$.

Note that since the performance of DH_k does not depend on the order in which the items are given, relative worst order analysis of DNF versus DH_k gives the same result as simply comparing the two algorithms on each sequence separately, just as competitive analysis with OPT replaced by DH_k.

In [16], a relative worst order analysis of DH_k and DNF is given for the model that allows items of size 1. It is shown that, for $i < j$, $WR(H_j, H_i) = \frac{i+1}{i}$. Hence, in this model, $WR(DH_k, DNF) = 2$, for $k \geq 2$, since DNF and DH_1 are equivalent. Note that, for $i \geq 2$, the result from [16] holds for our model too, since the lower bound sequences for these cases do not contain items of size 1.

We first show that DH_k and DNF are comparable. This is a special case of the corresponding result in [16].

Lemma 1. *For any $k \geq 1$ and any input sequence I, $DH_{kW}(I) \geq DNF_W(I) - (k - 1)$.*

Thus, according to relative worst order analysis, DH_k is at least as good as DNF. The next lemma establishes a separation between the two algorithms.

Lemma 2. *For any $k \geq 2$, $WR(DH_k, DNF) \geq \frac{3}{2}$.*

By providing a matching upper bound, we determine the exact relative worst order ratio of the two algorithms.

Theorem 6. $WR(DH_k, DNF) = \frac{3}{2}$.

Thus, we conclude that according to relative worst order analysis, DH_k is a better algorithm than DNF.

4   The Random Order Ratio

The random order ratio was introduced by Kenyon [20] as the worst ratio obtained over all sequences I, comparing the expected value of an algorithm A, with respect to a uniform distribution of all permutations, σ, of I, to the value of OPT on I:

$$RR(A) = \liminf_{OPT(I) \to \infty} \frac{E_\sigma[A(\sigma(I))]}{OPT(I)}$$

Note that OPT is still assumed to know the entire sequence in advance, so there is no expectation involved in computing $OPT(I)$.

The following theorem gives a bound on how well DNF can perform with respect to the random order ratio.

Theorem 7. *The random order ratio of* DNF *is at most* $\frac{4}{5}$.

Proof. Let S^n denote all sequences of length n with item sizes from $\mathcal{I}$, where $\mathcal{I} = \{\varepsilon, 1 - \varepsilon\}$ for an $0 < \varepsilon < \frac{1}{n}$. Define

$$S_i^n = \{I \in S^n \mid I \text{ contains } i \text{ items of size } \varepsilon \text{ and } n - i \text{ items of size } 1 - \varepsilon\}$$

Then we can consider the following disjoint partitioning $S^n = \bigcup_{0 \leq i \leq n} S_i^n$. We let R^n denote the set of all sequences of length n.

The first inequality below follows from two facts:

- For any pair of sequences, $I, I' \in S_i^n$, $\mathrm{OPT}(I) = \mathrm{OPT}(I')$.
- For two sums $A = \sum_{i=1}^n a_i$ and $B = \sum_{i=1}^n b_i$, $\frac{A}{B} \geq \min_{1 \leq i \leq n} \frac{a_i}{b_i}$.

$$\frac{\mathrm{E}_{I \in S^n}[\mathrm{DNF}(I)]}{\mathrm{E}_{I \in S^n}[\mathrm{OPT}(I)]} \geq \min_{0 \leq i \leq n} \frac{\mathrm{E}_{I \in S_i^n}[\mathrm{DNF}(I)]}{\mathrm{OPT}(I_i^n)}, \text{ where } I_i^n \in S_i^n$$

$$= \min_{I \in S^n} \frac{\mathrm{E}_\sigma[\mathrm{DNF}(\sigma(I))]}{\mathrm{OPT}(I)} \geq \min_{I \in R^n} \frac{\mathrm{E}_\sigma[\mathrm{DNF}(\sigma(I))]}{\mathrm{OPT}(I)}$$

Hence,

$$\lim_{n \to \infty} \frac{\mathrm{E}_{I \in S^n}[\mathrm{DNF}(I)]}{\mathrm{E}_{I \in S^n}[\mathrm{OPT}(I)]} \geq \liminf_{\mathrm{OPT}(I) \to \infty} \frac{\mathrm{E}_\sigma[\mathrm{DNF}(\sigma(I))]}{\mathrm{OPT}(I)} = \mathrm{RR}(\mathrm{DNF}).$$

In the rest of the proof, we compute the leftmost expression from the above, which then gives us an upper bound on the random order ratio of DNF.

There is no difference between choosing some element from S^n uniformly at random and generating a length n sequence iteratively by choosing the next item from $\mathcal{I}$ with equal probability. Thus, we can analyze the behavior of DNF by considering a Markov chain, where the state of the system after i items have been processed is determined by the state of the open bin. The Markov chain is finite and has just three states: either there is no open bin (N – for "No"), one open bin containing one large item of size $1 - \varepsilon$ (L – for "Large"), or one bin with a number of small items, each of size ε (S – for "Small"). Note that since $\varepsilon < \frac{1}{n}$, there is room for all the small items in one bin, if necessary.

This is an irreducible chain, where all states are positive recurrent, which implies that it has a stationary (equilibrium) distribution, and the probability of ending up in each of the states converges independently of the starting state [14]. The probability of being in one of the states N, L, or S can be calculated from the following equations:

$$1 = \mathrm{Prob}[N] + \mathrm{Prob}[L] + \mathrm{Prob}[S]$$
$$\mathrm{Prob}[N] = \mathrm{Prob}[L] + \mathrm{Prob}[S]/2$$
$$\mathrm{Prob}[L] = \mathrm{Prob}[N]/2$$
$$\mathrm{Prob}[S] = \mathrm{Prob}[N]/2 + \mathrm{Prob}[S]/2$$

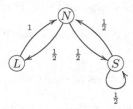

Fig. 1. A Markov chain describing DNF's behavior on the considered sequences

This system has the solution $\mathrm{Prob}[N] = \mathrm{Prob}[S] = \frac{2}{5}$ and $\mathrm{Prob}[L] = \frac{1}{5}$. From this it follows that $\mathrm{E}_{I \in S^n}[\mathrm{DNF}(I)]$ tends to $\mathrm{Prob}[N]n = \frac{2}{5}n$.

For the optimal algorithm, note that its result only depends on the number of items of each size. In particular, after n items, it can cover $\lfloor \frac{n}{2} \rfloor$ bins, unless there are more small than large items. All the small items would be wasted.

Using random walks, it is easy to see that the expected difference between the number of large and small items is a low order term compared with n, and therefore does not affect the limit.

A sequence of independent stochastic variables $\{X_i\}_{i \geq 1}$, where $\mathrm{Prob}[X_i = 1] = \mathrm{Prob}[X_i = -1] = \frac{1}{2}$, is called a *simple random walk* [14]. It is well known that if we define $T_n = \sum_{i=1}^{n} X_i$, then $\lim_{n \to \infty} \frac{\mathrm{E}[|T_n|]}{\sqrt{n}} = \sqrt{\frac{2}{\pi}}$ [17]. Hence, $\mathrm{E}[|T_n|] \in O(\sqrt{n})$, and then $\mathrm{E}_{I \in S^n}[\mathrm{OPT}(I)] = \frac{n}{2} - O(\sqrt{n})$.

In conclusion, we get $\lim_{n \to \infty} \frac{\mathrm{E}_{I \in S^n}[\mathrm{DNF}(I)]}{\mathrm{E}_{I \in S^n}[\mathrm{OPT}(I)]} = \lim_{n \to \infty} \frac{\frac{2}{5}n}{\frac{n}{2} - O(\sqrt{n})} = \frac{4}{5}$. □

Theorem 8. *The random order ratio of* DH_k *is* $\frac{1}{2}$.

Proof. The performance of DH_k does not depend on the order of the items in the sequence. Given a sequence containing n items of size $1 - \varepsilon$ and n items of size ε, where $\varepsilon < \frac{1}{n}$, DH_k will always cover $\frac{n}{2}$ bins, while OPT will cover n bins. The lower bound is given by Theorem 1, since the random order ratio of a bin covering algorithm is never worse than its competitive ratio. □

Thus, according to random order analysis, DNF is at least as good as DH_k. Though it seems hard to raise the lower bound on the random order ratio for DNF above $\frac{1}{2}$, and thereby separate the two algorithms, we conjecture that DNF is in fact strictly better than DH_k with respect to this measure. We discuss this further in the conclusion.

5   The Max/Max Ratio

The max/max ratio was introduced by Ben-David and Borodin [2] and compares an algorithm's worst-case behavior on any sequence of length n with OPT's worst-case behavior on any sequence of length n. The max/max ratio was introduced for the minimization problems paging and K-server. Since bin covering is a maximization problem, we actually need a min/min ratio. Additionally, since the input items can be arbitrarily small, letting the sequence length approach infinity does not give interesting results. Thus, we modify the measure to consider the volume, $vol(I)$, of a sequence I, where $vol(I)$ is the sum of the sizes of all the items in I:

$$MR_{vol}(A) = \frac{\liminf_{v \to \infty} \min_{vol(I)=v} A(I)/v}{\liminf_{v \to \infty} \min_{vol(I)=v} OPT(I)/v}$$

This measure cannot distinguish between DNF and DH_k in the general case:

Theorem 9. *Both DNF and DH_k have a min/min ratio of 1.*

If the item sizes are restricted to be from an interval $(a, b) \subseteq (0, 1)$, the min/min ratio can distinguish between DNF and DH_k. If (a, b) does not contain at least one of the interval borders used by DH_k, then DH_k behaves exactly like DNF. If (a, b) contains a DH_k border, then we define, as in Section 2, $\frac{1}{p}$ as the *maximal border in (a, b).*

Theorem 10. *With item sizes in $(a, b) \subseteq (0, 1)$, where $\frac{1}{p} \in (a, b)$, DH_k has a min/min ratio of 1 and DNF has a min/min ratio of $\max\left\{\frac{1+\frac{1}{p}}{1+b}, \frac{pb}{1+b}\right\}$.*

Note that $\frac{1+\frac{1}{p}}{1+b} < 1$ is equivalent to $\frac{1}{p} < b$, which follows from the definition and maximality of $\frac{1}{p}$. Furthermore, $\frac{pb}{1+b} < 1$ is equivalent to $b < \frac{1}{p-1}$, which is satisfied as long as b is not equal to $\frac{1}{p-1}$. Thus, according to min/min analysis, DH_k is better than DNF when item sizes are restricted to an interval $(a, b) \in (0, 1)$ containing a DH_k border, and $b \neq \frac{1}{p-1}$ where $\frac{1}{p}$ is the maximal border.

6 Uniform Distribution

In this section, we study the expected performance ratio of DNF and DH_k on sequences containing items drawn uniformly at random from the interval $(0, 1)$.

The expected performance ratio $ER_U(A)$ is the ratio between the expected performance of the algorithms A and OPT on sequences of length n, containing items drawn uniformly at random from the interval $(0, 1)$:

$$ER_U(A) = \lim_{n \to \infty} \frac{E_{I \in U_n(0,1)}[A(I)]}{E_{I \in U_n(0,1)}[OPT(I)]}.$$

Theorem 11. *On a sequence containing items drawn uniformly at random from the interval $(0, 1)$,*

$$ER_U(DH_2) = \frac{1}{2} + \frac{1}{e^2 - e} \approx 0.7141 \text{ and}$$

$$\lim_{k \to \infty} ER_U(DH_k) = \frac{12 - \pi^2}{3} \approx 0.7101.$$

This should be compared with a result from [10], showing that on a uniform distribution, DNF has an expected performance ratio of $\frac{2}{e} \approx 0.7358$. Thus, under this assumption, DNF is a little better than DH_k.

7   Concluding Remarks

Our starting point was the fact that the very different bin covering algorithms, DNF and DH_k, are not separated by competitive analysis. Thus, the question is which algorithm to use in different scenarios. DH_k was designed to guard against worst-case sequences, and since these are often made up using pathological input, mixing very large and very small items, we have carried out analyses using the worst-case performance, but on restricted input of items of similar size. The comparison is still in DH_k's favor, though less so. Under similar conditions, Max/max analysis and relative worst order analysis also point to DH_k.

In contrast, DNF is a little better than DH_k when considering expected performance under a uniform distribution. This seems fairly robust; even if we add an element of worst-case requirements in the form of random order analysis, DNF does not appear worse than DH_k. Thus, even if an adversary gets to choose the worst sequence for the algorithm, just the fact that the items are received in the order of a random permutation removes DH_k's advantage over DNF.

Thus, unless guarantees are desired or it is known that items do not arrive in a random order, it is worth considering DNF as the algorithm of choice.

DH_k has a random order ratio of $\frac{1}{2}$, which is worst possible, whereas the upper bound we have on DNF is $\frac{4}{5}$. We conjecture that these two algorithms can be separated, and discuss this issue in rest of the section. It seems intuitively almost obvious that DNF would always get a ratio larger than $\frac{1}{2}$. The difficulty in establishing this formally stems from problems handling the size aspects using probability theory. In the hardest case, there are a linear number of very large items such that if they end up on top of each other pairwise, we get the ratio of $\frac{1}{2}$. Thus, we need to prove that some fraction of these large items do not end up pairwise on top of each other. The small items that would be packed with the large items in an optimal packing can be cut into very small pieces so there are orders of magnitude more small items than large items—but still of possibly dramatically varying size, relatively. Whereas we have strong theoretical tools for bounding the deviation from the expected number of items in certain locations in the form of Chebyshev's inequality, for instance, it is much harder to reason regarding deviations from the expected size, and it is exactly the sum of sizes of small items surrounding a large item that decides whether or not two large items end up on top of each other.

Results on the random order ratio are often difficult to establish. An exceptionally tight result appears in [18], where it is shown that the random order ratio of Next-Fit for bin packing is exactly 2. Note, however, that this result does not give indication that the random order ratio of DNF for bin covering should be $\frac{1}{2}$. The sequence establishing the lower bound of 2 consists of n items of size $\frac{1}{2}$ and kn items of size $\epsilon < \frac{1}{kn}$, for some large k. For a random ordering of these items, each item of size $\frac{1}{2}$ has a high probability of being combined with at least one of the small items. For bin covering, the problem is reversed; we must prove that each large item has a significant probability of being surrounded by a sufficient volume of small items so that it will not go into the same bin as a neighboring large item.

References

1. Assmann, S.F., Johnson, D.S., Kleitman, D.J., Leung, J.Y.-T.: On a dual version of the one-dimensional bin packing problem. J. Algorithms 5(4), 502–525 (1984)
2. Ben-David, S., Borodin, A.: A new measure for the study of on-line algorithms. Algorithmica 11(1), 73–91 (1994)
3. Boyar, J., Favrholdt, L.M.: The relative worst order ratio for on-line algorithms. ACM Trans. Algorithms 3(2) (2007)
4. Boyar, J., Favrholdt, L.M., Larsen, K.S.: The relative worst order ratio applied to paging. J. Comput. Sys. Sci. 73(5), 818–843 (2007)
5. Boyar, J., Gupta, S., Larsen, K.S.: Access graphs results for LRU versus FIFO under relative worst order analysis. In: Fomin, F.V., Kaski, P. (eds.) SWAT 2012. LNCS, vol. 7357, pp. 328–339. Springer, Heidelberg (2012)
6. Boyar, J., Gupta, S., Larsen, K.S.: Relative interval analysis of paging algorithms on access graphs. In: Dehne, F., Solis-Oba, R., Sack, J.-R. (eds.) WADS 2013. LNCS, vol. 8037, pp. 195–206. Springer, Heidelberg (2013)
7. Boyar, J., Irani, S., Larsen, K.S.: A comparison of performance measures for online algorithms. In: Dehne, F., Gavrilova, M., Sack, J.-R., Tóth, C.D. (eds.) WADS 2009. LNCS, vol. 5664, pp. 119–130. Springer, Heidelberg (2009)
8. Boyar, J., Larsen, K.S., Maiti, A.: A comparison of performance measures via online search. In: Snoeyink, J., Lu, P., Su, K., Wang, L. (eds.) AAIM 2012 and FAW 2012. LNCS, vol. 7285, pp. 303–314. Springer, Heidelberg (2012)
9. Christ, M., Favrholdt, L.M., Larsen, K.S.: Online bin covering: Expectations vs. guarantees. arXiv:1309.6477(cs.DS) (2013)
10. Csirik, J., Frenk, J.B.G., Galambos, G., Kan, A.H.G.R.: Probabilistic analysis of algorithms for dual bin packing problems. J. Algorithms 12(2), 189–203 (1991)
11. Csirik, J., Totik, V.: Online algorithms for a dual version of bin packing. Discrete Appl. Math. 21(2), 163–167 (1988)
12. Csirik, J., Woeginger, G.: On-line packing and covering problems. In: Fiat, A., Woeginger, G.J. (eds.) Online Algorithms 1996. LNCS, vol. 1442, pp. 147–177. Springer, Heidelberg (1998)
13. Dorrigiv, R., López-Ortiz, A.: A survey of performance measures for on-line algorithms. SIGACT News 36(3), 67–81 (2005)
14. Durrett, R.: Probability: Theory and Examples. Dixbury Press (1991)
15. Ehmsen, M.R., Kohrt, J.S., Larsen, K.S.: List factoring and relative worst order analysis. Algorithmica 66(2), 287–309 (2013)
16. Epstein, L., Favrholdt, L.M., Kohrt, J.S.: Comparing online algorithms for bin packing problems. J. Scheduling 15(1), 13–21 (2012)
17. Hoffmann-Jørgensen, J.: Probability with a View towards Statistics, vol. I. Chapman & Hall (1994)
18. Coffman Jr., E.G., Csirik, J., Rónyai, L., Zsbán, A.: Random-order bin packing. Discrete Appl. Math. 156, 2810–2816 (2008)
19. Karlin, A.R., Manasse, M.S., Rudolph, L., Sleator, D.D.: Competitive snoopy caching. Algorithmica 3, 79–119 (1988)
20. Kenyon, C.: Best-fit bin-packing with random order. In: SODA, pp. 359–364 (1996)
21. Lee, C.C., Lee, D.T.: A simple on-line bin-packing algorithm. J. ACM 32(3), 562–572 (1985)
22. Ramanan, P.V., Brown, D.J., Lee, C.C., Lee, D.T.: On-line bin packing in linear time. J. Algorithms 10(3), 305–326 (1989)
23. Seiden, S.S.: On the online bin packing problem. J. ACM 49(5), 640–671 (2002)
24. Sleator, D.D., Tarjan, R.E.: Amortized efficiency of list update and paging rules. Comm. ACM 28(2), 202–208 (1985)
25. Woeginger, G.: Improved space for bounded space, on-line bin-packing. SIAM J. Disc. Math. 6(4), 575–581 (1993)

Map of Geometric Minimal Cuts for General Planar Embedding[*]

Lei Xu[1], Evanthia Papadopoulou[2], and Jinhui Xu[1]

[1] Department of Computer Science and Engineering
State University of New York at Buffalo
Buffalo, NY 14260, USA
{lxu,jinhui}@buffalo.edu
[2] Faculty of Informatics
Università della Svizzera italiana
Via Giuseppe Buffi 13
CH 6904 Lugano, Switzerland
evanthia.papadopoulou@unisi.ch

Abstract. In this paper, we consider the problem of computing the map of geometric minimal cuts (MGMC) induced by a general planar embedding (i.e., the edge orientation is either rectilinear or diagonal) of a subgraph $H = (V_H, E_H)$ of an input graph $G = (V, E)$. The MGMC problem is motivated by the critical area extraction problem in VLSI layout and finds applications in several other areas. In this paper, we extend an earlier result for planar rectilinear embedding to its more general case. The increased freedom on edge orientation in the embedding imposes new challenges, mainly due to the fact that the inducing region of a geometric minimal cut is no longer unique. We show that the MGMC problem can be solved by computing the L_∞ Hausdorff Voronoi diagram of a set of rectangle families, each containing an infinite number of axis-aligned rectangles. By exploiting the geometric properties of these rectangle families, we present an output-sensitive algorithm for computing the Hausdorff Voronoi diagram in this general case which runs in $O((N + K) \log^2 N \log \log N)$ time, where K is the complexity of the Hausdorff Voronoi diagram and N is the number of geometric minimal cuts.

1 Introduction

In this paper, we consider the following problem, called *Map of Geometric Minimal Cuts or MGMC* problem: Given a graph $G = (V, E)$ and an planar embedding of a subgraph $H = (V_H, E_H)$ of G with rectilinear or diagonal edges, compute a map $\mathcal{M}$ of the embedding plane P of H so that for every point $p \in P$, the cell in $\mathcal{M}$ containing p is associated with the "closest" *geometric cut* (in G) to p, where the distance between a point p and a cut C is defined as the maximum distance between p and any individual element of C. A geometric cut C

---

[*] The research of the third author was supported in part by NSF under grant IIS-1115220.

P. Widmayer, Y. Xu, and B. Zhu (Eds.): COCOA 2013, LNCS 8287, pp. 238–249, 2013.
© Springer International Publishing Switzerland 2013

of G is a set of edges and vertices in H that overlap a given geometric shape S in P and whose removal from G disconnects G. In this paper we consider the case where geometric cuts are induced by axis-aligned rectangles and distances are measured by the L_∞ metric. The main objective of the MGMC problem is to compute the map $\mathcal{M}$ of all geometric minimal (or canonical) cuts (the exact definition of geometric minimal cuts will be given in next section) of the planar embedding of H.

The MGMC problem was introduced in [5] motivated by the VLSI critical area computation problem. The critical area problem for various types of faults can be reduced to different variants of Voronoi diagrams that lead to accurate critical area extraction (see e.g., [4,6] and references therein). A VLSI net can be modeled as a graph $G = (V, E)$ with a subgraph embedded on every conducting layer. A subgraph $H = (V_H, E_H)$ on a layer X is vulnerable to random defects associated with layer X. Defects on layer X may create *cuts* on graph G that result in disconnecting the net N. The Voronoi framework for critical area extraction asks for a subdivision of layer X into regions that reveal for every point p the radius of the smallest disk centered at p inducing a cut of G.

The MGMC problem was first addressed in [5,6], based on higher order Voronoi diagrams and an iterative process to determine min-cuts that resulted in the L_∞ Hausdorff Voronoi diagram of all geometric min-cuts. In [7] the rectilinear version of the problem was considered and an output sensitive approach was proposed that first computed all possible geometric min-cuts and then directly computed the L_∞ Hausdorff Voronoi diagram, where each geometric min-cut induced an axis-aligned rectangle representing its minimum inducing region. The MGMC problem is also closely related to the farthest line-segment Voronoi diagram which has constant complexity for non-crossing line segments [2].

2   Geometric Cuts

Let $G = (V, E)$ be the undirected graph in an MGMC problem and $H = (V_H, E_H)$ be its planar subgraph embedded in the plane P with $|V| = N_G$, $|E| = M_G$, $|V_H| = n$, and $|E_H| = m$. Due to the planarity of H, $m = O(n)$. Edges in H are straight line segments with rectilinear or diagonal orientation. A pair of vertices u and v in a graph is connected if there is a path in this graph from u to v, and disconnected otherwise. A graph is connected if every pair of its distinct vertices is connected. Without loss of generality (WLOG), we assume that G is connected. A cut C of G is a subset of edges in G whose removal disconnects G. A cut C is minimal if removing any edge from C no longer forms a cut.

Definition 1 (see [7]). *Let R be a connected region in P, and $C = R \cap H$ be the set of edges in H intersected by R. C is called a geometric cut induced by R if the removal of C from G disconnects G. A geometric cut C is called a 1-D geometric cut (or a 1-D cut) if $R(C)$ is a segment. If $R(C)$ is an axis-aligned rectangle, then C is called a 2-D geometric cut (or a 2-D cut). A geometric cut*

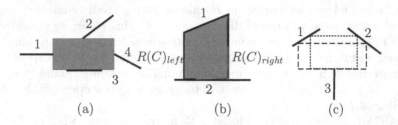

Fig. 1. (a) R(C) of a 2-D cut C is bounded by 4 edges. (b) Each vertical line segment between $R(C)_{left}$ and $R(C)_{right}$ (gray region) forms a 1-D cut $\{1, 2\}$. (c) Two minimum reducing regions (dashed and dotted rectangles) of geometric cut $\{1,2,3\}$.

C is a geometric minimal cut if the set of edges intersected by any rectangle shrinking from R(C) is no longer a cut.

When there is no ambiguity of the region R, we often call the cut induced by R as a geometric cut for simplicity. For a given cut C, its *minimum inducing region* $R(C)$ is the minimum axis-aligned rectangle which intersects every edge of C. For some geometric cut C, its $R(C)$ could be degenerated into a horizontal or vertical line segment, or even a single point. If $R(C)$ is not a point, it may not be fixed for a given geometric minimal cut C (see Figure 1).

Let $B(C)$ denote the set of edges bounding a geometric minimal cut C (i.e., the set of edges in C intersecting $R(C)$). Due to the minimality nature of C, removing any edge in $B(C)$ will lead to a non-cut. This means that any edge in $B(C)$ is necessary for forming the cut. However, this is not necessarily true for edges in $C \setminus B(C)$. Thus, a geometric minimal cut may not be a minimal cut. This also explains why the number of geometric minimal cuts is polynomial and the number of minimal cuts is exponential [7].

Clearly edges in $B(C)$ define the boundary position of $R(C)$. However, it is not true that all edges in $B(C)$ are needed to define $R(C)$.

Lemma 1. *For any 1-D (or 2-D) geometric minimal cut, the number of edges in $B(C)$ needed to define $R(C)$ is at most two (or four) (see Figure 1(a)).*

For simplicity, we assume thereafter that $B(C)$ contains only those edges which are barely sufficient to define $R(C)$.

For a 1-D cut C, the location of $R(C)$ may not be fixed, since there may be an infinite number of 1-D cuts cutting the same set of edges (see Figure 1(b)). For a 2-D cut C, it is also possible that $R(C)$ is not fixed due to the appearance of diagonal oriented edge(s) in $B(C)$. For example, if the vertex of $R(C)$ incident to a diagonal oriented edge $e \in B(C)$, moving the vertex along e continuously could generate an infinite number of different $R(C)$. In Figure 1(c), two minimum inducing regions represented by dotted and dashed rectangles are induced by the geometric cut $\{1, 2, 3\}$.

Lemma 2. *For a given geometric minima cut C, if $R(C)$ is not fixed, the number of $R(C)$ is infinite.*

Note that in the presence of non-fixed inducing region, the computation of map $\mathcal{M}$ is quite different. In this case, if point $p \in P$ falls in the cell of a geometric minimal cut C, p is closer to one of its $R(C)$ than to all $R(C')$ of any other cut C'. Thus the MGMC problem is to construct a Hausdorff Voronoi diagram of which each cell corresponding to a geometric minimal cut C is a union of its $R(C)$. The main challenge is to efficiently deal with those Voronoi cell owned by an infinite number of rectangles corresponding to the same geometric minimal cut.

3 Identifying Geometric Minimal Cuts and Minimum Inducing Regions

To compute the map $\mathcal{M}$ of geometric minimal cuts, we first identify all possible geometric minimal cuts and then construct the Hausdorff Voronoi diagram of their infinite number of minimum inducing regions.

3.1 Computing Geometric Minimal Cuts

To identify all 1-D and 2-D geometric minimal cuts, we adopt the algorithm proposed in [7]. In [7], it has shown that all geometric minimal cuts induced by a planar rectilinear embedding of H can be identified in a worst case $O(n^3 \log n (\log \log n)^3)$ time and in $O(n \log n (\log \log n)^3)$ time if the maximum size of the cut is bounded by a constant.

3.2 Computing Minimum Inducing Regions

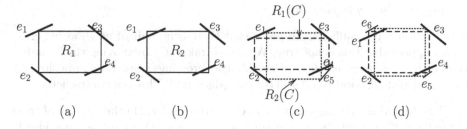

Fig. 2. (a) R_1 with 2 neighboring crossings $\{e_3, e_4\}$ is not an R(C). (b) R_2 with 2 opposite crossings $\{e_1, e_4\}$ is an R(C). (c) Compute e_5 by $\{e_1, e_2, e_3\}$. (d) Compute $\{e_5, e_6\}$ by $\{e_1, e_3\}$.

First, we emphasize that the methods of computing minimum inducing regions described in this section actually can be applied to arbitrary planar embedding of H. Given a set $\mathcal{C}$ of geometric minimal cuts of H, we need to first identify their minimum inducing regions before computing the Hausdorff Voronoi diagram. For a given cut $C \in \mathcal{C}$, from our previous discussion we know that $R(C)$ may not be unique. If $R(C)$ is fixed, $R(C)$ is bounded by the edges in $B(C)$ and can be computed in $O(1)$ time. If $R(C)$ is not fixed, we know (from Lemma 2) that there are infinite number of $R(C)$s. Thus it is impossible to compute the Hausdorff Voronoi diagram for all such $R(C)$s. To overcome this difficulty, our main idea is to find a discrete representation to capture the behaviors of all possible $R(C)$s. In other words, we need to find a small set of extreme $R(C)$s to represent the infinite number of $R(C)$s. To achieve this goal, our idea is to analyze the geometric properties of all $R(C)$s. For instance, if $R(C)$ is not fixed for a given 1-D cut C, it is easy to see that each $R(C)$ is bounded by the two extreme 1-D cuts $R(C)_{left}$ and $R(C)_{right}$ (or $R(C)_{top}$ and $R(C)_{bottom}$), and the two bounding edges in $B(C)$ (see Figure 1(b)). For a 2-D cut C, it is more complicated since (1) $B(C)$ contains up to 4 edges and (2) one or more edges could be arbitrarily orientated. From now on, we assume that $B(C)$ consists of 4 non-rectilinear edges. We focus on this case since all the other cases are simpler and can be handled similarly. Thus we omit the details for other cases in this extended abstract.

Definition 2. $B(C)$ *is general if it contains 4 non-rectilinear edges.*

Definition 3. *Given an edge e of a general $B(C)$, e is crossing (or tangent to) a rectangle R if e intersects R twice (or once). Each intersection is called a crossing (or tangency) between $B(C)$ and R. (see Figure 2(b))*

To better understand these concepts, consider the geometric minimal cut $C = \{e_1, e_2, e_3, e_4\}$ shown in Figure 2. R_1 is not a minimum inducing region of C since shrinking it a little bit still cuts C. e_2 (e_3) is tangent to R_2. R_2 is an $R(C)$ with 2 crossings $\{e_1, e_4\}$.

Lemma 3. *Given a geometric minimal cut C, if $R(C)$ is not unique, the number of crossing between $B(C)$ and any $R(C)$ is at most 2. If it is 2, the two crossings are not neighboring to each other.*

Proof. Clearly it is sufficient to prove that there is no pair of neighboring crossings. Suppose that this is not true. We can shrink $R(C)$ by moving the boundary edge of $R(C)$ connecting the two neighboring crossings toward its opposite edge by a small distance and $R(C)$ still cuts all edges in C. This is a contradiction.   □

Thus, to find all $R(C)$s, we have two cases to consider, (1) the number of crossing is 1 and (2) the number of crossings is 2. For case (1), we explain our idea by an example. In Figure 2(c), a general $B(C)$ contains 4 edges $\{e_1, e_2, e_3, e_4\}$ with slopes $\{\kappa_1, \kappa_2, \kappa_3, \kappa_4\}$ respectively. A rectangle (shown by dashed line) tangent to $\{e_1, e_2, e_3\}$ respectively and crossed by e_4 is a minimum inducing region $R_1(C)$. Another rectangle (shown by dotted line) tangent to the same set of edges as $R_1(C)$ forms another minimum inducing region $R_2(C)$. $R_2(C)$ is also crossed by e_4. For all $R(C)$s tangent to e_1, e_2, e_3, their corner points which are not on any edge of $B(C)$ induce a new edge e_5 (i.e., union of all such corner points forms an edge), called *skating edge*. The skating edge e_5 can be easily computed from $\{e_1, e_2, e_3\}$. Given a set of 4 edges $\{e_1, e_2, e_3, e_5\}$, if we move a point p along the valid interval of e_1, each p corresponds to exactly one minimum inducing region $R(C)$. Note that the valid interval in which $R(C)$ exists can be easily computed. Thus we call $\{e'_1, e'_2, e'_3, e_5\}$ a configuration of $R(C)$, where e'_i is the valid interval of e_i for $i \in \{1, 2, 3\}$. For case (2), as shown in Figure 2(d), the dotted and dashed rectangles are tangent to e_2 and e_3 respectively. Similar to case (1), we can also compute two skating edges e_5 and e_6 for all $R(C)$s tangent to e_2 and e_3. For any three edges of $B(C)$, the corresponding configuration can be computed in either case (1) or (2). Thus we have the following lemma.

Lemma 4. *Given a geometric minimal cut C, $R(C)$ has at most 4 configurations and each configuration is a set of 4 edges.*

For a general $B(C)$, we only need to find a set of 4 configurations to represent all its $R(C)$s.

Lemma 5. *Given a geometric minimal cut C, the representation of $R(C)$ (i.e., all configurations if $R(C)$ is not unique) can be computed in $O(|B(C)|)$ time.*

4   Generating Map of Geometric Minimal Cuts

Given a set $\mathcal{C}$ of geometric minimal cuts of H, the Hausdorff Voronoi diagram of $\mathcal{C}$ is a partition of the embedding plane P of H into regions (or cells) so that the Hausdorff Voronoi cell of a cut $C \in \mathcal{C}$ is the union of all points whose Hausdorff distance to some $R(C)$ is closer than to any minimum inducing region of other cuts in $\mathcal{C}$.

In our MGMC problem, we have four types of objects, the fixed and non-fixed minimum inducing regions of 1-D geometric minimal cuts and the fixed and non-fixed minimum inducing regions of 2-D geometric minimal cuts. As it is well known, the Hausdorff Voronoi diagram can be viewed as the intersections of wavefronts propagating from each object with unit speed. Thus our construction of the Hausdorff Voronoi diagram uses the wave propagation concept. We focus on the discussion of 2-D $R(C)$s since the same idea can be applied to the 1-D case. More specifically, for non-fixed $R(C)$, we assume that $B(C)$ is general. We have two types of objects to consider, the fixed rectangle $R(C)$ and the union of non-fixed rectangle $UR(C)$. To visualize the whole growing process, we can lift the waves to 3D with time being the third dimension and thus each object corresponds to a 3D cone. We will discuss the properties of 3D cones of $R(C)$ and $UR(C)$ in the next section.

Lemma 6 (see [7]). *The Hausdorff Voronoi diagram can be obtained by projecting the lower envelope of the 3D facet cones to the xy plane.*

4.1   Properties and Plane Sweep Approach

Lemma 7. *Let C be a 2-D geometric minimal cut with a fixed $R(C)$. At any moment, the wavefront of $R(C)$ is either empty or an axis-aligned rectangle. Furthermore, the wavefront in 3D is a facet cone apexed at a segment and with each facet forming a 45 degree angle with the xy plane. (see Figure 3)*

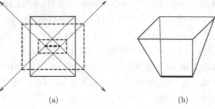

Fig. 3. (a) The wavefront (dashed line) of a fixed $R(C)$ (solid line). (b) Its corresponding 3D V-cone. Note that the bold dashed line in (a) is corresponding to the bold solid line in (b).

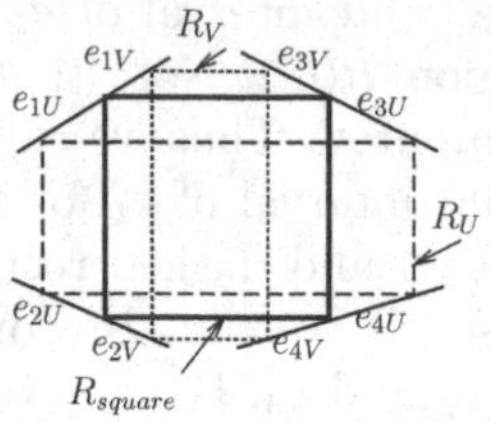

Fig. 4. R_{square} (bold line), R_U (dashed line) and R_V (dotted line)

Definition 4. *Given a fixed $R(C)$, a 3D facet cone $\partial W(C)$ is a U-cone (or V-cone) if its apex segment s_C is parallel to the y (or x) axis. (see Figure 3(b))*

Next we discuss the properties of the wavefront of $UR(C)$ of a general $B(C)$. By Lemma 4, we know that $UR(C)$ is represented by at most 4 configurations $UR_1(C)$, $UR_2(C)$, $UR_3(C)$ and $UR_4(C)$, with each corresponding to a 3D wavefront. The 3D wavefront of $UR(C)$ is simply the lower envelope of the wavefronts of the four configurations (looking from $-\infty$ of the z axis). Since the property of each wavefront is the same, we only need to focus on one configuration $UR_1(C) = \{e_1, e_2, e_3, e_4\}$.

$UR_1(C)$ is the union of an infinite number of $R(C)$s. It is possible that some of them have U-cones as their 3D wavefronts and the others have V-cones as their 3D wavefronts. To distinguish these $R(C)$s, we further classify $UR_1(C)$ into two sub-configurations $UR_{1U}(C) = \{e_{1U}, e_{2U}, e_{3U}, e_{4U}\}$ and $UR_{1V}(C) = \{e_{1V}, e_{2V}, e_{3V}, e_{4V}\}$ such that any rectangle from $UR_{1U}(C)$ (or $UR_{1V}(C)$) generates only U-cone (or V-cone). The computation of $UR_{1U}(C)$ and $UR_{1V}(C)$ can be done in $O(1)$ time since we only need to check the position of the square (denoted by R_{square} in Figure 4) of $UR_1(C)$ if it exists. All the rectangles of $UR_1(C)$ with length bigger (or smaller) than the width form U-cones (or V-cones). In Figure 4, R_U (or R_V) is a minimum inducing region corresponding to an U- (or V-) cone in 3D. Now we analyze the property of the wavefront of $UR_{1U}(C)$ and $UR_{1V}(C)$ respectively.

To better illustrate the whole growing process of the wavefront of $UR_{1U}(C) = \{e_{1U}, e_{2U}, e_{3U}, e_{4U}\}$, we first choose 3 rectangles $\{R_{min}, R_{mid}, R_{max}\}$ such that R_{min} (or R_{max}) is an extreme rectangle of $UR_{1U}(C)$ in which the difference between the length and width is minimum (or maximum). R_{mid} is any rectangle in between. We first analyze the wavefront of these three rectangles and then generalize the idea to all rectangles in $UR_{1U}(C)$.

Since the Hausdorff distance to a rectangle R_{min} is determined by the four corner points, an equivalent view is to propagate 4 separated waves from the 4

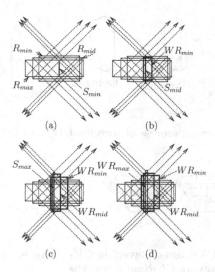

Fig. 5. The growing process of the wavefront for 3 rectangles

corner points of R_{min} with each being an L_∞ ball. Let $B1_{min}$, $B2_{min}$, $B3_{min}$ and $B4_{min}$ be the 4 L_∞ balls of R_{min}. The common intersection of the 4 balls are the wave WR_{min} of R_{min}. We grow the 4 corner balls of R_{mid} and R_{max} in the same way and denote their waves by WR_{mid} and WR_{max} respectively. Initially, both of them are empty. We call it stage 1. Once the size of the 4 balls of R_{min} reaches the minimum Hausdorff distance to R_{min}, their common intersection forms a segment s_{min} located at the center of R_{min} and parallel to the shorter side of R_{min} (see Figure 5(a)). As Bi_{min} grows, WR_{min} becomes a rectangle. Later, s_{mid} appears when the size of the 4 balls of R_{mid} reaches the minimum Hausdorff distance to R_{mid} and intersects WR_{min} (see Figure 5(b)). After that, WR_{mid} grows in the same way as WR_{min} does. Finally, s_{max} appears and intersects the above two rectangles (see Figure 5(c)). We call the above procedure stage 2. In stage 3, all 3 rectangles grows simultaneously (see Figure 5(d)).

To analyze the wavefront $WUR_{1U}(C)$ of $UR_{1U}(C)$, we generalize the above discrete process by replacing R_{mid} with the union of all rectangles between R_{mid} and R_{max}. It is easy to see the shape of $WUR_{1U}(C)$ at each stage. In stage 1, $WUR_{1U}(C)$ is empty. In stage 2, $WUR_{1U}(C)$ is a segment s_{min}. $WUR_{1U}(C)$ keeps the shape as a hexagon (see Figure 6(a)) until s_{max} appears. Each non-rectilinear edge of the hexagon is parallel to $\{e_{1U}, e_{2U}, e_{3U}, e_{4U}\}$ respectively. In stage 3, $WUR_{1U}(C)$ is an octagon, since each endpoint of s_{max} grows to an segment paralleled to the x-axis. We denote the two endpoints of s_{min} (or s_{max}) as pa_{min} and pb_{min} (or pa_{max} and pb_{max}). e_a (or e_b) is the straight line segment between pa_{min} and pa_{max} (or pb_{min} and pb_{max}) (see Figure 6(b)). The property of wavefront of $UR_{1V}(C)$ is the same except that everything in each stage is rotated counterclockwise by 90-degree.

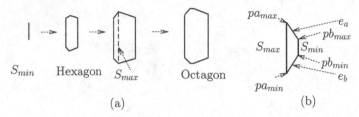

S_{min} 　Hexagon　S_{max} 　　Octagon

(a)

(b)

Fig. 6. (a) The growing process of wavefront $WUR_{1U}(C)$. (b)e_a and e_b.

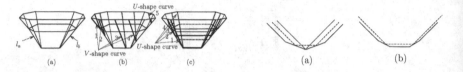

(a)　　　(b)　　　(c)　　　　　　　(a)　　　　　(b)

Fig. 7. (a) CV-cone. (b) 5 stages of sweeping CV- cone. (c) 4 stages of sweeping CU-cone (view the process by rotated 90-degree).

Fig. 8. (a) stage 4 (b) stage 5

To better understand the whole growing process, we lift the wavefront $WUR_{1U}(C)$ to 3D, with time being the third dimension (see Figure 7(a)). The following lemma summarizes the main properties of the growing process.

Lemma 8. *Let C be a 2-D geometric minimal cut with non-fixed $R(C)$. At stage 1, the wavefront $WUR_{1U}(C)$ (or $WUR_{1V}(C)$) is empty. At stage 2 (or 3), $WUR_{1U}(C)$ (or $WUR_{1V}(C)$) is a hexagon (or an octagon) with the property discussed above. Furthermore, the wavefront in 3D is a facet cone apexed at a segment and with each facet forming a 45 degree angle with the xy plane. It is a 6-sided (or 8-sided) facet cone in stage 2 (or 3). Let l_a and l_b be the top and bottom (or left and right) edge of the 6-sided facet cone of $WUR_{1U}(C)$ ($WUR_{1V}(C)$) at stage 2. Then e_a (or e_b) is the projection of l_a (or l_b) on the xy plane.*

Definition 5. *Given a non-fixed $R(C)$ and one of its sub-configuration I, a 3D facet cone $\partial WI(C)$ is a CU-cone (or CV-cone) if its apex segment s_{min} is parallel to the y (or x) axis.*

By the above lemma, CU-cone (or CV-cone) has 6 facets at stage 2 and 8 facets at stage 3. Thus we totally have 4 types of objects in 3D, U-cone, V-cone, CU-cone and CV-cone.

Lemma 9. *The wavefront of $UR(C)$ can be represented by at most 4 CU-cones and 4 CV-cones.*

To efficiently construct the Hausdorff Voronoi diagram $HVD(\mathcal{C})$, we follow the spirit of Fortune's plane sweep algorithm for points [3], and sweep along the

x axis direction a tilted plane Q in 3D which is parallel to the y axis and forms a 45 degree with the xy plane. Q intersects the xy plane at a sweep line L parallel to the y axis.

Since every facet of a 3D facet cone forms a 45 or 135-degree angle with the xy plane and apexed at either a horizontal or vertical segment, at each moment, the intersection of Q and a cone $\partial WI(C)$ is either a V-shape curve (i.e., consisting of a 45-degree ray and a 135-degree ray on Q) or a U-shape curve (i.e., consisting of a 45-degree ray, a segment parallel to L, and a 135-degree ray). When the cone is first encountered, it introduces either a V-shape curve or a U-shape curve to Q. When L (or Q) moves, the curve grows and its shape may change from a V-shape to a U-shape. In addition, the height of the apex of a V-shape curve could change due to the existence of CV-cones. For both CU-cone and CV-cone, the 45-degree and 135-degree ray of a U-shape curve could move along the y direction on Q. Next, we discuss the intersection between each type of cones and Q in details.

First we consider U cones. Let $\partial W(C)$ be any U cone with apex segment s_C, and v_1 and v_2 be the two endpoints of s_C. When the sweep plane Q first encounters $\partial W(C)$, it introduces a U-shape curve C_u to Q. Let r_l, r_r, and s_m be the left and right rays and the middle segment of C_u respectively. Initially s_m is the apex segment s_C, and r_l and r_r are the two edges of facet cone. When Q (or L) moves, C_u grows and always maintains its U-shape.

Lemma 10 (see [7]). *Let $\partial W(C)$, C_u, r_l, r_r and s_m be defined as above. When Q moves in the direction of the x axis, C_u is always a U-shape curve. The supporting lines of r_l and r_r remain the same on Q, and the two endpoints of s_m (the fixed points of r_l and r_r) moves upwards in unit speed along the two supporting lines.*

Let $\partial WI(C)$ be any CU-cone with apex segment s_{min}, and pa_{min} and pb_{min} be the two endpoints of s_{min}. When the sweep plane Q first encounters $\partial WI(C)$, it introduces a U-shape curve C_{cu} to Q. Let r_l, r_r, and s_m be the left and right rays and the middle segment of C_{cu} respectively.

Lemma 11. *Let $\partial WI(C)$, C_{cu}, r_l, r_r and s_m be defined as above. When Q moves in the direction of the x axis, C_{cu} is always a U-shape curve (i.e., at each moment, r_l and r_r remain 45-degree and 135-degree respectively.). The whole process can be divided into 4 stages (see Figure 7(c)). At stage 1, s_m is the apex segment s_{MM} which is the edge of rectangle RI parallel to y-axis with smaller x coordinate. RI is the rectangle growing from s_{min} at the moment when s_{max} appears. At stage 2, s_m moves in unit speed downwards from s_{MM} to s_{min} with its two endpoints staying on l_a and l_b respectively. At stage 3, s_m moves in unit speed upwards from s_{min} to S_{NN} with its two endpoints staying on l_a and l_b respectively. S_{NN} is the edge of rectangle RI parallel to s_{MM}. At stage 4, two endpoints of s_m moves upwards in unit speed along r_l and r_r.*

For an arbitrary V cone $\partial W(C')$, let $s_{C'}$ be its apex segment, and v_1' and v_2' be its two endpoints (or left and right endpoints). When Q first touches $\partial W(C')$

at v_1', it generates a V-shape curve C_v'. C_v' remains a V-shape curve before encountering v_2'. After that, C_v' becomes a U-shape curve.

Lemma 12 (see [7]). *Let r_l and r_r be the two rays of C_v', and s_m be the middle segment of the U-shape curve C_v' after Q visiting v_2'. During the whole sweeping process, the supporting lines of r_l and r_r are fixed lines on Q. C_v' remains the same V-shape curve on Q before encountering v_2'. s_m moves upwards in unit speed along the supporting lines of r_l and r_r after Q encounters v_2'.*

For an arbitrary CV-cone $\partial WI(C')$, let s_{min}' be its apex segment, and pa_{min}' and pb_{min}' be its two endpoints (or left and right endpoints).

Lemma 13. *The whole sweeping process of a CV-cone can be divided into 5 stages (see Figure 7(b)). At stage 1, when Q first touches $\partial WI(C')$ at pa_{max}', it generates a V-shape curve C_{cv}. At stage 2, C_{cv} is still a V-shape curve moving in unit speed downwards from pa_{max}' to pa_{min}' with its apex staying on l_a'. At stage 3, C_{cv} remains the same V-shape curve on Q before encountering pb_{min}'. At stage 4, C_{cv} is a 4-edge V-shape curve (see solid lines in Figure 8 (a)) moving in unit speed upwards from pb_{min}' to pb_{max}' with its apex staying on l_b'. At stage 5, after Q encounters pb_{max}', C_{cv} becomes an 5-edge U-shape curve (see solid lines in Figure 8 (b)) with the middle segment s_m' moves upwards in unit speed along the fixed supporting lines of r_l' and r_r' on Q.*

The 4-edge V-shape curve (5-edge U-shape curve) is essentially a union of U-shape curves if a V-shape is considered as a degenerate case of U-shape curve (See dotted lines in Figure 8 (a)). Two of those U-shape curves are shown at dotted lines in Figure 8. For each 4-edge V-shape curve (5-edge U-shape curve), we divide it to two curves, one curve with 45 and 135 degrees lines and another curve with the rest. Instead of working on the 4-edge V-shape curve and 5-edge U-shape curve directly, we convert each of them to the normal U or V-shape curve by computing the intersections with other curves on the beach line. Since any two curves may intersect at most twice, the same complexity will be kept of the beach line as [7]. Thus, even though there are 4 types of cones, at each moment, we only have U and V-shape curves on Q. We have two cases in which a hidden U or V-shape curve could appear in the beach line instead of only one case as in [7].

Lemma 14. *Let $\partial W(C_1)$ be either a U or V cone and $\partial W(C_2)$ be a V cone with its left endpoint v_1 of s_{C_2} being inside of $\partial W(C_1)$ and its right endpoint v_2 being outside of $\partial W(C_1)$. If $\partial W(C_2)$ is not entirely contained by the union $\cup_{C_i \in C; C_i \neq C_2} \partial W(C_i)$, the V-shape curve C_2 introduced by $\partial W(C_2)$ will be hidden by the beach line at the beginning and then becomes part of the beach line later. This is the first case in which a hidden U or V-shape curve could appear in the beach line.*

Lemma 15. *Let $\partial W(C_1')$ be any type of cone and $\partial W(C_2')$ be a CV-cone with pa_{max}' or pa_{min}' being inside of $\partial W(C_1')$ and its right endpoint pb_{min}' being outside of $\partial W(C_1')$. If $\partial W(C_2')$ is not entirely contained by the union*

$\cup_{C_i \in \mathcal{C}; C_i \neq C_2'} \partial W(C_i)$, the V-shape curve C_2' introduced by $\partial W(C_2')$ will be hidden by the beach line at the beginning and then becomes part of the beach line later. This is the second case in which a hidden U or V-shape curve could appear in the beach line.

Lemma 16. *Let $\mathcal{C}$ be a set of N minimal geometrical cuts. The edges of $HVD(\mathcal{C})$ are either segments or rays, and the vertices of the $HVD(\mathcal{C})$ are either the vertices of bisectors or the intersections of bisectors.*

Lemma 17. *The size K of the L_∞ Hausdorff Voronoi diagram of N minimum geometrical cuts is $O(N + M')$, where M' is the number of intersecting minimum inducing region pairs. The bound is tight in the worst case.*

4.2   Events, Data Structures and Algorithm

To implement the plane sweep algorithm, we use similar data structures as in [7] with one modification for handling V events. To efficiently detect all possible V events, our idea is to process the apex points of all V-shape curves, including (1) the left endpoint of a V-cone's apex segment and (2) pa_{max} and pa_{min} of a CV-cone into the 3D dynamic range search tree data structure MD. Thus we are able to handle all events efficiently in a similar way as [7].

Theorem 1. *The L_∞ Hausdorff Voronoi diagram $HVD(\mathcal{C})$ of a set $\mathcal{C}$ of geometric minimal cuts can be constructed by a plane sweep algorithm in $O((N + K)\log^2 N \log\log N)$ time, where $N = |\mathcal{C}|$ and K is the complexity of the Hausdorff Voronoi diagram.*

References

1. Abellanas, M., Hernandez, G., Klein, R., Neumann-Lara, V., Urrutia, J.: A Combinatorial Property of Convex Sets. Discrete & Computational Geometry 17, 307–318 (1997)
2. Dey, S.K., Papadopoulou, E.: The $L_\infty(L_1)$ Farthest Line-Segment Voronoi diagram. In: The Ninth International Symposium on Voronoi Diagrams in Science and Engineering, pp. 49–55 (2012)
3. Fortune, S.: A sweepline algorithm for Voronoi diagrams. Algorithmica 2, 153–174 (1987)
4. Papadopoulou, E.: Critical area computation for missing material defects in VLSI circuits. IEEE Transactions on Computer-Aided Design 20(5), 583–597 (2001)
5. Papadopoulou, E.: Higher order Voronoi diagrams of segments for VLSI critical area extraction. In: The Eighteenth International Symposium on Algorithms and Computation, pp. 716–727 (2007)
6. Papadopoulou, E.: Net-aware critical area extraction for opens in VLSI circuits via high-order Voronoi diagram. IEEE Transactions on Computer-Aided Design 20(5), 583–597 (2011)
7. Xu, J., Xu, L., Papadopoulou, E.: Computing the Map of Geometric Minimal Cuts. In: The Twentieth International Symposium on Algorithms and Computation, pp. 244–254 (2009)
8. Xu, J., Xu, L., Papadopoulou, E.: Map of Geometric Minimal Cuts with Applications. In: Handbook of Combinatorial Optimization, 2nd edn. Springer (2013)

A New Approach to the Upper Bound on the Average Distance from the Fermat-Weber Center of a Convex Body[*]

Xuehou Tan[1,2] and Bo Jiang[1]

[1] Dalian Maritime University, Linghai Road 1, Dalian, China
[2] Tokai University, 4-1-1 Kitakaname, Hiratsuka 259-1292, Japan
tan@wing.ncc.u-tokai.ac.jp

Abstract. We show that for any convex body Q in the plane, the average distance from the Fermat-Weber center of Q to the points in Q is at most $\frac{99-50\sqrt{3}}{36} \cdot \Delta(Q) < 0.3444 \cdot \Delta(Q)$, where $\Delta(Q)$ denotes the diameter of Q. This improves upon the previous bound of $\frac{2(4-\sqrt{3})}{13} \cdot \Delta(Q) \approx 0.3490 \cdot \Delta(Q)$, due to Dumitrescu, Jiang and Tòth. Our new method to evaluate the average distance from the Fermat-Weber center of Q is to transform Q into a circular sector of radius $\Delta(Q)/2$. Some points of Q may decrease their distances to the Fermat-Weber center in Q after the transformation, but the total amount of varied distances can be well controlled. Our work sheds more light on the conjectured upper bound $\Delta(Q)/3$.

1 Introduction

The Fermat-Weber center of a measurable planar set Q with positive area is a point in the plane, such that the average distance from it to the points in Q is minimal. Clearly, the Fermat-Weber center gives the ideal location, say, for a fire station that serves the region Q. The classical Fermat-Weber problem is to find a point in a set F of feasible facility locations, which minimizes the average distance to the points in a set D of (possibly weighted) demand locations. For a survey of the Fermat-Weber problem, see [10]. Related work on the Weber problem can also be found in [7].

Let $\|pq\|$ denote the Euclidean distance between two points p and q in the plane, and pq the line segment with two endpoints p and q. For a measurable set Q with positive area and a point y in the plane, we denote by $\mu_Q(y)$ the average distance between y and the points x in Q, that is, $\mu_Q(y) = \int_{x \in Q} \|xy\| dx / area(Q)$, where $area(Q)$ denotes the area of the body Q. Let $\mathcal{FW}_Q$ be a point for which this average distance is minimal, namely, $\mu_Q(\mathcal{FW}_Q) = \min_y \mu_Q(y)$. We simply write $\mu_Q^* = \mu_Q(\mathcal{FW}_Q)$. The point $\mathcal{FW}_Q$ is a Fermat-Weber center of Q. (Note that Q may be non-convex.)

[*] This work was partially supported by the Grant-in-aid (MEXT/JSPS KAKENHI 23500024) for Scientific Research from Japan Society for the Promotion of Science and by National Natural Science Foundation of China under grant 61173034.

P. Widmayer, Y. Xu, and B. Zhu (Eds.): COCOA 2013, LNCS 8287, pp. 250–259, 2013.
© Springer International Publishing Switzerland 2013

In this paper, we restrict our attention to convex bodies. Clearly, $\mathcal{FW}_Q \in Q$ if Q is convex. Let c denote the supermum and c' denote the infimum of $\mu_Q^*/\Delta(Q)$ over all convex bodies Q in the plane, where $\Delta(Q)$ denotes the diameter of Q. Carmi, Har-Peled and Katz conjectured that $c = \frac{1}{3}$ and $c' = \frac{1}{6}$ [4]. Note that the supermum c is attained for a circular disk D since $\mu_D^* = \frac{\Delta(D)}{3}$, and the infimum c' can be achieved by constructing a flat rhombus P_ϵ such that $\mu_{P_\epsilon}^*$ tends to $\frac{\Delta(P_\epsilon)}{6}$ [4,6].

Carmi, Har-Peled and Katz were the first to show that $\frac{1}{6} \leq c' \leq \frac{1}{7}$ [4], and the lower bound for c' was later improved by Abu-Affash and Katz from $\frac{1}{7}$ to $\frac{5}{24}$ [1]. Very recently, Dumitrescu, Jiang and Tóth have proved that $c' = \frac{1}{6}$ [6]. For the other conjecture of $c = \frac{1}{3}$, Abu-Affash and Katz were the first to show that $c \leq \frac{2}{3\sqrt{3}}$ [1]. Their method is to transform P into several circular sectors, *without decreasing* the distance of any point to the considered center. Dumitrescu, Jiang and Tóth further refined the analysis of Abu-Affash and Katz, and showed that $c \leq \frac{2(4-\sqrt{3})}{13}$ [6]. The previously known methods have a limit in proving the conjectured value of $c = \frac{1}{3}$, as they do not handle the situation in which the distances of the transformed points to the considered center are decreased [1,6].

The main contribution of this paper is to prove that $\mu_Q^* < \frac{99-50\sqrt{3}}{36} \cdot \Delta(Q) < 0.3444 \cdot \Delta(Q)$. Our upper bound is rather close to the conjectured value of $\Delta(Q)/3$, as the difference between them is less than 0.0111. Moreover, our proof is simple and straightforward, since it is based on elementary geometric transformations.

As in the previous work [1,6], we actually give an upper bound on the average distance to the points in Q from the center of the smallest enclosing circle of Q. A new idea to evaluate the average distance from the Fermat-Weber center of Q is to transform Q into a circular sector of radius $\Delta(Q)/2$. Some points of Q may decrease their distances to the Fermat-Weber center in Q after the transformation, but the total amount of varied distances can be well controlled.

2   Preliminaries

We first review some known results related to our work. A *curve of constant width* is a convex planar shape whose width, defined as the perpendicular distance between two distinct parallel lines each intersecting its boundary at a single point, is the same regardless the direction of those two parallel lines. For instance, the width of a circle as well as the width of the Reuleaux triangle is constant: its diameter.

A basic result on curves of constant width is Barbier's theorem, which states that the perimeter of any curve of constant width is equal to the width multiplied by π.

The *isoperimetric inequality* is a geometric inequality involving the square of the circumference of a closed curve in the plane and the area of a plane region it encloses, as well as its various generalizations [3,5,9]. *Isoperimetric* literally means "having the same perimeter". Specifically, the isoperimetric inequality states, for the length L of a closed curve and the area A of the planar region that L encloses, that $4\pi A \le L^2$, and that the equality holds if and only if the curve is a circle.

By the isoperimetric inequality and Barbier's theorem, one can obtain the "isodiametric" theorem: In the class of all plane convex sets of diameter at most one, the circle of unit diameter has the largest area (see Section 11.3 of [3].) In our terminology, we have $area(Q) \le (\frac{\Delta(Q)}{2})^2\pi$, where $\Delta(Q)$ denotes the diameter of a convex body Q.

From the definition of the Fermat-Weber center, we can simply make the following observations.

Observation 1. *Let T be a circular sector of radius r and center angle α in the plane, and let o be the center of the sector T. Then,*

$$\mu_T(o) = \frac{\int_0^r \alpha x^2\, dx}{\int_0^r \alpha x\, dx} = \frac{\alpha r^3/3}{\alpha r^2/2} = \frac{2r}{3}.$$

Observation 2. *Let X, Y be two (not necessarily convex) bodies of the same area, and let p be a point in the plane. If the distance of a point of $X - (X \cap Y)$ from p is no more than that of any point of $Y - (X \cap Y)$, then $\mu_X(p) \le \mu_Y(p)$.*[1]

The following result is also needed for our proof of $c \le (99 - 50\sqrt{3})/36$.

Theorem 1. *[8]. Let S be a set of diameter $\Delta(S)$ in the plane. Then, S is contained in a circle of radius $\Delta(S)/\sqrt{3}$.*

3 $1/3 \le c \le (99 - 50\sqrt{3})/36$

Let P be a convex body in the plane. Denote by o and R the center and the radius of the smallest enclosing circle of P, respectively. From the convexity of P, we have $o \in P$. It has been shown by Abu-Affash and Katz that P can be transformed into a circular sector T of center angle α ($\le 2\pi$) and radius at most R, such that $area(P) = area(T)$ and $\mu_P(o) \le \mu_T(o) \le 2R/3$ [1]. The following result immediately follows.

Lemma 1. *(See [1]) In the special case that the radius R of the smallest enclosing circle of P is equal to $\Delta(P)/2$, $\mu_P^* \le \Delta(P)/3$.*

---

[1] This result is used several times in the proof of Theorem 3.3 of [1].

For ease of presentation, assume below that $\Delta(P) = 1$ and $1/2 < R \leq 1/\sqrt{3}$. Denote by D_r the disk of radius r, centered at o, and $area(D_r)$ the area of D_r. Also, we denote by A_r the set of the points of P, which are outside of D_r, and $area(A_r)$ the total area of the regions formed by the points of A_r.

We will first describe a method to transform P into a disk D_t, $t < 1/2$. Next, we describe a transformation from P into a circular sector of radius $\frac{1}{2}$. We show that $area(A_{\frac{1}{2}})$ is only a small portion (near a seventh) of $area(P)$. Since the decreased distance for any point of $A_{\frac{1}{2}}$ in our transformation is no more than $(1/\sqrt{3} - 1/2)$, a careful calculation gives $c \leq \frac{99-50\sqrt{3}}{36} < 0.3444$.

3.1   Transforming P into a Disk

From the "isodiametric" theorem [3], we can assume that $area(P1) = t^2\pi$ holds for some $t < 1/2$. Thus, P can be transformed into the disk D_t without changing its area. Since $R > 1/2$, the points of A_t clearly form at least three disjoint regions. See Fig. 1.

Lemma 2. *Suppose that K is a connected region of the point set A_t, and $Arc(K, C_x)$ is the intersection of K with the circle C_x of radius x, centered at o, $t \leq x \leq R$. Then, $Arc(K, C_x)$ is a single, connected arc of radius x. Moreover, the arc length of $Arc(K, C_x)$ is strictly larger than that of $Arc(K, C_y)$, if $t \leq x < y \leq R$.*

Proof. First, if $Arc(K, C_x)$ contains two disjoint arcs, then we can simply find two points, one per arc, such that the line segment connecting them is not completely contained in K. Since K is a region outside the disk D_t, the found segment is not completely contained in P either, contradicting the convexity of P. Analogously, if $x < y$ but the arc length of $Arc(G, C_x)$ is strictly smaller than that of $Arc(K, C_y)$, we can find, say, two extreme points of $Arc(K, C_x)$ and $Arc(K, C_y)$, such that the line segment connecting them is not completely contained in K and thus P, a contradiction again. Finally, if $x < y$ and the arc length of $Arc(G, C_x)$ is equal to that of $Arc(K, C_y)$, then all arcs between $Arc(G, C_x)$ and $Arc(K, C_y)$ are of the same length. From the convexity of P, all points of A_t can form at most two disjoint regions, and thus, R is equal to $1/2$, contradicting our assumption that $R > 1/2$. $\square$

Remark. The above result can be generalized a little more. Let us shrink the circle of radius t into that of radius c such that an arc of the circle of radius c outside of P degenerates into a point for the first time, see Fig. 1. Then, Lemma 3 also holds for any connected region of the point set A_c, because a line segment outside of the disk D_c is outside of the convex body P.

Lemma 3. *Suppose that $R > \Delta(P)/2$. Then, P can be transformed into a disk D_t, $t < 1/2$, such that $\mu_P(o) \leq \mu_{D_t}(o) + \frac{area(A_t)}{area(P)} \cdot (R - t)$.*

Proof. We first describe a transformation from P into D_t. Again, denote by K a connected region of A_t. For all circular arcs $\widehat{ab}$ in K, with two endpoints a and b on the boundary of K, we transform them into the arcs $\widehat{a'b'}$ of radius $\|oa'\|$, with two endpoints a' and b' inside D_t, such that $\|oa\| - t = t - \|oa'\|$ (i.e., $\|oa\| + \|oa'\| = 2t$), and the arc length of $\widehat{a'b'}$ is equal to that of $\widehat{ab}$. See Fig. 1. Let $u \in K$ be a point on the circle of radius R, and $v \in K$ the intersection point of the line segment ou with the circle of radius t. In particular, we assume that the point u is transformed into u' such that three points u, u' and v $(= v')$ are on the same line. See Fig. 1. (Note that for a point $x \in \widehat{ab}$ and its transformed point $x' \in \widehat{a'b'}$, three points o, x' and x are usually not on a line.) Denote by K' the region formed by the transformed points. Clearly, $area(K) = area(K')$.[2] Denote by A'_t the set of the transformed points. It immediately follows from Lemma 2 (as well as the remark after Lemma 2) that any two connected regions formed by the points of A'_t, cannot overlap each other, see Fig. 1.

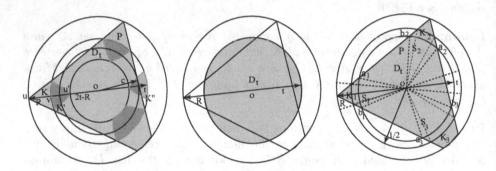

Fig. 1. Illustration for transforming P into the disk D_t, assuming that $\Delta(P) = 1$

Turn to the evaluation of distance changes occurred in the transformation. Let $\widehat{ab}$ and $\widehat{cd}$ denote the intersection arcs of a region K with two circles of radii x and y respectively, where $t + (R - t)/2 \leq x \leq t$ and $y = x + (R - t)/2$. Since $x < y$, the arc length of $\widehat{ab}$ is larger than that of $\widehat{cd}$ (Lemma 2). For every point $p \in \widehat{cd}$, we relate it to an unique point $q \in \widehat{ab}$ such that $\|op\| + \|oq\| = \|op'\| + \|oq'\| + 2(R - t)$, where p' and q' denote the points transformed from p and q, respectively. For the other points $z \in \widehat{ab}$, which are not related to any point of $\widehat{cd}$, we clearly have $\|oz\| \leq \|oz'\| + (R - t)$, where z' denotes the point transformed from z. Thus, $\mu_K(o) \leq \mu_{K'}(o) + (R - t)$. Therefore, $area(A_t) = area(A'_t)$ and $\mu_{A_t}(o) \leq \mu_{A'_t}(o) + (R - t)$.

---

[2] Note that if we stretch the arc of K on the circle of radius t into a line segment, without changing the distances of all points of K to the stretched segment, then K' is obtained by reflecting the stretched region using the stretched segment as a mirror.

Let $B = D_t - P \cap D_t$. Since $area(P) = area(D_t)$, we have $area(A_t') = area(B)$. Denote by A_t'' the set of the points, which is symmetric to A_t' about the center o. In Fig. 1, the region K'' is symmetric to K' about the center o. Clearly, A_t'' intersects with B, $area(A_t'') = area(A_t')$ and $\mu_{A_t'}(o) = \mu_{A_t''}(o)$. Let $A1 = A_t'' - A_t'' \cap B$ and $B1 = B - A_t'' \cap B$. Then, $area(A1) = area(B1)$. Note that the points of $A1$ and $B1$ are inside and outside of the convex body P, respectively. Denote by $A2$ $(\subseteq A1)$ the set of the points, whose distances to o are smaller than any point of $B1$, and $B2$ $(\subseteq B1)$ the set of the points, whose distances to o are larger than any point of $A1$. So, $area(A2) = area(B2)$, and $\mu_{A1-A2}(o) = \mu_{B1-B2}(o)$. Thus, the point set $(A1 - A2)$ is equivalently the same as $(B1 - B2)$, for the considered center o. Since the distance of any point of $A2$ to o is smaller than that of any point of $B2$, the point set $A2$ can then be transformed into $B2$ such that $\mu_{A2}(o) \le \mu_{B2}(o)$ (Observation 2). In summary, $\mu_{A_t'}(o) = \mu_{A1+A_t''\cap B}(o) = \mu_{A2+(A1-A2)+A_t''\cap B}(o) = \mu_{A2}(o)+\mu_{A1-A2}(o)+\mu_{A_t''\cap B}(o) \le \mu_{B2}(o)+\mu_{B1-B2}(o)+ \mu_{A_t''\cap B}(o) = \mu_{B1}(o) + \mu_{A_t''\cap B}(o) = \mu_B(o)$.

Finally, the lemma can be obtained by the following calculation:

$$
\begin{aligned}
\mu_P(o) &= \frac{\int_{x \in (P-A_t)} \|ox\| dx + \int_{x \in A_t} \|ox\| dx}{area(P)} \\
&= \frac{area(P - A_t) \cdot \mu_{(P-A_t)}(o) + area(A_t) \cdot \mu_{A_t}(o)}{area(P)} \\
&\le \frac{area(P - A_t) \cdot \mu_{(P-A_t)}(o) + area(A_t) \cdot (\mu_{A_t'}(o) + (R - t))}{area(P)} \\
&\le \frac{area(P - A_t) \cdot \mu_{(P-A_t)}(o) + area(B) \cdot \mu_B(o) + area(A_t) \cdot (R - t)}{area(P)} \\
&= \frac{\int_{x \in D_t} \|ox\| dx + area(A_t) \cdot (R - t)}{area(P)} \\
&= \mu_{D_t}(o) + \frac{area(A_t)}{area(P)} \cdot (R - t).
\end{aligned}
$$

□

3.2   Transforming P into a Circular Sector of Radius $\Delta(P)/2$

Following from Lemma 3, a key point in giving a good upper bound on the average distance from the Fermat-Weber center is to keep both $(R - t)$ and $area(A_t)$ to be small. Instead of a disk, P can analogously be transformed into a circular sector of radius $1/2$; in this case, the value $(R - t)$ is minimized. An important observation made in this paper is that the total area of the regions of P, outside of the circle of radius $1/2$, is roughly less than one seventh of the area of P.

Lemma 4. *Suppose that $\Delta(P) = 1$ and $R > 1/2$. Then, $area(A_{\frac{1}{2}}) < (1/2 - 2(1 - 1/\sqrt{3})^2) area(P)$.*

Proof. Let us reconsider the transformation of P into the disk D_t. Since we have assumed that $R > 1/2$, all the points of A_t form several connected regions. Denote by $K_1, K_2, \ldots, K_j$ the sequence of the regions, which are formed by the points of A_t. Denote by a_i, b_i two extreme points of the region K_i ($1 \leq i \leq j$) on the circle of radius t, and σ_i the center angles of the circular sector bounded by oa_i and ob_i. (The sector obtained by extending oa_i and ob_i in the disk D_R contains K_i.) See also Fig. 1.

Denote by S_i, $1 \leq i \leq j$, the circular sector of radius t and center angle σ_i, which is wholly contained in P. (So, S_i is adjacent to K_i.) The region of P contained in the congruent circular sector T_i, symmetric to S_i about the center o, is *not* a full circular sector; otherwise, P cannot be convex, a contradiction. From the convexity of P, the union of these circular sectors S_i and T_i cannot cover the disk D_t. Hence, the area of the union of all circular sectors S_i is strictly less than $t^2\pi/2$, or equally, $area(P)/2$.

Consider now a process of transforming P into a circular *sector* of radius s, in which s varies from t to $1/2$. From the assumption that $R > 1/2$ and the "isodiametric" theorem [3], we can assume that $area(P) = \alpha/8$ holds for some $\alpha < 2\pi$, i.e., P can be transformed into a sector of radius $1/2$ and center angle α. Denote by U_s the union of the (full) circular sectors of radius s ($t \leq s \leq 1/2$), contained in P. (See Fig. 1 for an example, where the circular sectors of radius $1/2$ are shown in thin, dotted line.) Denote by V_1 the union of the regions of P, which are outside of the circle of radius s, and V_2 the region $P - U_s - V_1$. In the process of changing s from t to $1/2$, some regions of V_1 are newly added into U_s, but some others drop out of U_s and are thus added into V_2. Since the total area (i.e., $area(P)$) is never changed, the decreased amount of $area(V_1)$ is equal to the increased amount of $area(V_2)$. That is, the amount of $area(U_s)$ is not changed at all! Hence, the area of the union of the circular sectors of radius $1/2$, which are contained in P, is strictly less than $area(P)/2$. All the regions $K_1, K_2, \ldots, K_j$ can then be gathered into a circular sector with the center angle less than $\alpha/2$.

Finally, let us analyze the size of $area(A_{\frac{1}{2}})$. Denote by K' the region which is transformed from a region K of the point set A_t, and β the center angle of the (longest) arc of K', which is on the circle of radius $1/2$. Let L be the circular region, between two circles of radii $1/2$ and $1 - R$, whose center angle γ is slightly larger than β. So, $area(L) = \gamma(1/4 - (1 - R)^2)/2$. Let the arc of K' on the circle of radius $1/2$ be completely contained in that of L. Since $\gamma > \beta$, the region L contains a circular arc, whose length is equal to the length of the arc of K' on the circle of radius $1/2$. From Lemma 2, the region K' is wholly contained in L. Hence, $area(K') < area(L)$. Since all the regions $K_1, K_2, \ldots, K_j$ can be gathered into a circular sector, whose center angle is less than $\alpha/2$, we can assume $\Sigma \gamma \leq \alpha/2$, and thus obtain $area(A'_{\frac{1}{2}}) < \alpha(1/4 - (1 - R)^2)/4$. Since $R \leq 1/\sqrt{3}$ and $area(P) = \alpha/8$, we have $area(A_{\frac{1}{2}}) < (1/2 - 2(1 - 1/\sqrt{3})^2) area(P)$. $\square$

By now, we can give the main result of this paper.

Theorem 2. *For any convex object P in the plane, we have*

$$\mu_P^* < \frac{99 - 50\sqrt{3}}{36} \cdot \Delta(P) < 0.3444 \cdot \Delta(P).$$

Proof. From Lemma 1, we discuss only the case $R > \Delta(P)/2$. Denote by I_x the set of the points of P, which are inside the circle of radius x. Let v be the value satisfying $\alpha(1/4 - v^2)/2 = 2area(A_{\frac{1}{2}})$. Then, $area(I_{\frac{1}{2}}) - area(I_v) = area(A_{\frac{1}{2}})$ and $area(I_v) = v^2\alpha/2$. Since $area(A_{\frac{1}{2}}) < \alpha(1/4 - (1 - R)^2)/4$, we have $v > 1 - R$.

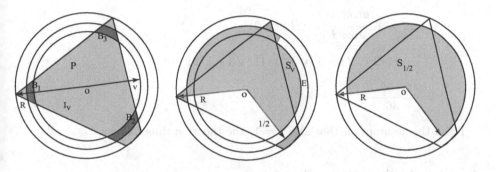

Fig. 2. Illustration for the transformation of P into $S_{\frac{1}{2}}$

Let us denote by S_x the circular sector of radius x and center angle α, centered at o. So, $area(S_{\frac{1}{2}}) = area(P)$. Assume also that $S_x \supseteq S_{x'}$ if $x' \leq x \leq 1/2$, as it can always be obtained by rotating $S_{x'}$ around the center o appropriately.

We describe below a transformation from P into $S_{\frac{1}{2}}$. Since I_v is a convex body and $area(I_v) = \alpha v^2/2$, all points of I_v can be transformed into the circular sector S_v, without decreasing the distance of any point to o after the transformation (see the proof of Theorem 3.3 of [1]). Hence, $\mu_{I_v}(o) \leq \mu_{S_v}(o)$. Denote by B_i a connected region of $I_{\frac{1}{2}} - I_v$. See Fig. 2. Then, B_i can be transformed (without changing the area) into a circular region B_i', which is the difference between two circular sectors of the same center angle α_i, and two radii $\frac{1}{2}$ and v. It follows from Observation 2 and Lemma 2 that $\mu_{B_i}(o) < \mu_{B_i'}(o)$. Since $area(I_{\frac{1}{2}}) - area(I_v) = \alpha(1/4 - v^2)/4$, we have $\Sigma \, \alpha_i = \alpha/2$ (see the middle of Fig. 2). Finally, denote by E the circular region, which is the difference between two circular sectors of the same center angle $\alpha/2$, and two radii $1/2$ and v. So, $area(E) = area(A_{\frac{1}{2}})$. Following from the proof of Lemma 4, the length of the arc of E on the circle of radius $1/2$ is larger than the total lengths of the arcs of $A_{\frac{1}{2}}$ on the circle of radius $1/2$. On the other hand, since $v > 1 - R$, we have $R - 1/2 > 1/2 - v$. As done in Section 3.1, the point set $A_{\frac{1}{2}}$ can then be transformed into E such that

$\mu_{A_{\frac{1}{2}}}(o) = \mu_E(o) + (R - 1/2)$. In this way, all points of P are transformed into $S_{\frac{1}{2}}$ (Fig. 2).

As in Section 3.1, we can then obtain the following result:

$$\mu_P(o) = \frac{\int_{x \in (P - A_{\frac{1}{2}})} \|ox\| dx + \int_{x \in A_{\frac{1}{2}}} \|ox\| dx}{area(P)}$$

$$\leq \frac{area(P - A_{\frac{1}{2}}) \cdot \mu_{(P - A_{\frac{1}{2}})}(o) + area(E) \cdot \mu_E(o) + area(A_{\frac{1}{2}}) \cdot (R - 1/2)}{area(P)}$$

$$= \frac{\int_{x \in S_{\frac{1}{2}}} \|ox\| dx + area(A_{\frac{1}{2}}) \cdot (R - 1/2)}{area(P)}$$

$$= \mu_{S_{\frac{1}{2}}}(o) + \frac{area(A_{\frac{1}{2}})}{area(P)} \cdot (R - 1/2)$$

$$\leq \frac{1}{3} + (1/2 - 2(1 - 1/\sqrt{3})^2) \cdot (1/\sqrt{3} - 1/2)$$

$$= \frac{99 - 50\sqrt{3}}{36}.$$

From the assumption that $\Delta(P) = 1$, the theorem thus follows. $\square$

4   Concluding Remarks

We have shown that for any convex object Q in the plane, $\mu_Q^* < 0.3444 \cdot \Delta(Q)$. This improves upon the previous upper bound of $0.3490 \cdot \Delta(Q)$ [6]. Our result is obtained by investigating several geometric transformations. As in [1,6], this new upper bound on the average distance from the Fermat-Weber center of a convex object can also be used to improve the known solutions of several geometric problems.

Finally, it is an interesting challenge to prove the conjecture of $c = 1/3$. Our method transforms P into a circular sector of radius $\Delta(P)/2$, which may always lead to a factor $c > 1/3$. To avoid this, one can make use of the transformation of P into the disk D_t. A good trade-off between $(R - t)$ and $area(A_t)$ may be helpful in completing the proof of $c = 1/3$.

References

1. Abu-Affash, A.K., Katz, M.J.: Improved bounds on the average distance to the Fermat-Weber center of a convex object. Inform. Process. Lett. 109, 329–333 (2009)
2. Aronov, B., Carmi, P., Katz, M.J.: Minimum-cost load-balancing partition. Algorithmica 54(3), 318–336 (2009)
3. Brass, P., Moser, W.O.J., Pach, J.: Research problems in discrete geometry. Springer (2005)

4. Carmi, P., Har-Peled, S., Katz, M.J.: On the Fermat-Weber center of a convex object. Comput. Geom. Theory and Appl. 32(3), 188–195 (2005)
5. Chavel, I.: Isoperimetric inequalities. Cambridge Univ. Press (2001)
6. Dumitrescu, A., Jiang, M., Tóth, C.T.: New bounds on the average distance from the Fermat-Weber center of a convex body. Discrete Optimization 8(3), 417–427 (2011)
7. Fekete, S.P., Mitchell, J.S.B., Weinbrecht, K.: On the continuous Weber and k-median problems. In: Proc. 16th ACM Symp. on Computational Geometry, pp. 70–79 (2000)
8. Jung, H.W.E.: Über der kleinsten Kreis, der eine ebene Figur einschließt. J. Angew. Math. 137, 310–313 (1910)
9. Osserman, R.: The isoperimetric inequality. Bull. Amer. Math. Soc. 84, 1182–1238 (1978)
10. Wesolowsky, G.: The Weber problem: History and perspectives. Location Sci. 1(1), 5–23 (1993)

Parameterized Complexity of Control and Bribery for d-Approval Elections[*]

Jianxin Wang[1], Min Yang[1], Jiong Guo[2], Qilong Feng[1], and Jianer Chen[1]

[1] School of Information Science and Engineering,
Central South University,
Changsha 410083, P.R. China
jxwang@mail.csu.edu.cn
[2] Universität des Saarlandes,
Campus E 1.7, D-66123 Saarbrücken, Germany
jguo@mmci.uni-saarland.de

Abstract. A d-Approval election consists of a set C of candidates and a set V of votes, where each vote v can be presented as a set of d candidates. For a vote $v \in V$, the protocol assigns one point to each candidate in v. The candidate getting the most points from all votes wins the election. An important aspect of studying election systems is the strategic behavior such as control and bribery problems. The control by deleting votes problem decides whether for a given election (C, V), a specific candidate c, and an integer k, it is possible to delete at most k votes such that c wins the resulting election. In the control by adding votes setting, one has two sets V and U of votes and asks for a subset $U' \subseteq U$ such that $|U'| \leq k$ and c becomes the winner in $V \cup U'$. The bribery problem has the same input as the vote deleting control problem and asks for changing at most k votes to make c win. All three problems have been shown NP-hard. We initialize the study of the parameterized complexity of these problems and present a collection of tractability and intractability results. In particular, we derive polynomial-size problem kernels for the standard parameterizations of the control by deleting votes and bribery problems, the seemingly first non-trivial problem kernels for the control and bribery problems of elections.

1 Introduction

An election consists of a set C of candidates, a set V of votes over the candidates, and an election protocol. A vote is normally a total linear order of the candidates. The winner of the election is determined by the election protocol. There are numerous different election protocols. The choice of election protocols may affect both the outcome of the election and the behavior of the voters. Most widely used protocols can be assigned to one of the two categories: *scoring protocols* and *protocols based on comparisons*. In a scoring protocol, a candidate is

[*] This work is supported by the National Natural Science Foundation of China under Grant (61232001, 61103033, 61173051, 61128006), the DFG Excellence Cluster on Multimodal Computing and Interaction (MMCI).

P. Widmayer, Y. Xu, and B. Zhu (Eds.): COCOA 2013, LNCS 8287, pp. 260–271, 2013.
© Springer International Publishing Switzerland 2013

assigned some points from a vote according to his position in this vote. Then, the candidates with the most points from all votes win the election. Most prominent scoring protocols include Plurality, Borda, and d-Approval. Protocols based on pairwise comparisons require that the winner should be "preferred" by most voters compared to each of other candidates, for instance, the Condorcet method. For more details of election protocols, we refer to [1,2].

Different aspects of elections have been explored, for instance, the strategic behavior. Some elections can be "controlled" by the authority conducting the election to achieve strategic results, for instance, to make a preferred candidate win the election. In this case, we have a so-called *constructive control* scenario [4]. The types of control can involve adding and deleting either candidates or votes. Two other strategic behaviors have also been particularly well studied, that is, *manipulation* and *bribery*. In the case of manipulation, voters may be better off revealing its preferences untruthfully. With bribery we refer to attacks where an outsider picks a group of votes and convinces them to vote in his interest.

In this work, we study strategic behavior of the d-Approval protocol, one of the most extensively studied scoring protocols [16,15]. It assigns one point to each of the first d candidates in a vote and the candidate getting the most points wins the election. We call that a vote v "approves" a candidate c, if c gets a point from v. In a d-Approval election, votes can be considered as size-d subsets of the candidates. If several candidates end up with the maximal amount of points, then they are called co-winners; otherwise, the election has a unique winner. We study the following control and bribery problems of d-Approval protocol.

Constructive Control by Deleting Votes for d-Approval (DV-d-Approval)
Input: A set C of candidates, a distinguished candidate c with $c \notin C$, a set V of size-d subsets of $C \cup \{c\}$, and an integer k
Question: Can we find a subset $V' \subseteq V$ such that $|V| - |V'| \leq k$ and c is the unique winner under the d-Approval protocol in $(C \cup \{c\}, V')$?

Constructive Control by Adding Votes for d-Approval (AV-d-Approval)
Input: A set C of candidates, a distinguished candidate c with $c \notin C$, two sets V and U of size-d subsets of $C \cup \{c\}$, and an integer k
Question: Can we find a subset $U' \subseteq U$ such that $|U'| \leq k$ and c is the unique winner under the d-Approval protocol in $(C \cup \{c\}, V \cup U')$?

Bribery for d-Approval (Bribery-d-Approval)
Input: A set C of candidates, a distinguished candidate c with $c \notin C$, a set V of size-d subsets of $C \cup \{c\}$, and an integer k
Question: Can we change at most k subsets in V such that c is the unique winner under the d-approval protocol in the new election?

Note that with "changing a vote v" in Bribery-d-Approval we mean to replace some candidates in v by others. However, the sets before and after the change should have the same cardinality d.

The computational study of election problems has been initialized by Bartholdi et al. [3]. The complexity of constructive control problems was studied first by

Bartholdi et al. [4] and later on by many others [10,12,16]. Faliszewski et al. gave the first classification of bribery problems concerning their computational complexity [9]. In particular, simple reductions from the NP-hard Exact Cover problem prove the NP-hardness of DV-d-Approval and AV-d-Approval, even d being a constant at least 3 and 4, respectively [15]. The NP-hardness of Bribery-d-Approval follows from a similar reduction and holds for $d \geq 3$ [9].

We focus here on the parameterized complexity of the above problems [8,17]. Many election problems turn out to be NP-hard. However, there are practical scenarios where some natural parameters of these problems take small values [7]. Hence, the analysis of their parameterized complexity with respect to various parameters might contribute new insight to the study of election protocols. We refer to [5] for a recent survey of parameterized complexity of election problems. A *parameterized problem* is a language $L \subseteq \Sigma^* \times N$, where Σ is an alphabet. Given an instance (I, k) of a parameterized problem L, we say L is *fixed-parameter tractable*, if there is an algorithm deciding in $f(k) \cdot |I|^{O(1)}$ time whether $(I, k) \in L$, where $f(k)$ is an arbitrary computable function depending only on k. The basic classes of fixed-parameter intractable problems are denoted by W[1] and W[2].

We consider two parameterizations of DV-d-Approval, that is, with k and $|V'|$ as parameters. Then, for AV-d-Approval and Bribery-d-Approval, we study the standard parameterization, namely, by the solution size k. It is easy to see that, if d is unbounded, then all four parameterized problems are not fixed-parameter tractable. However, in most practical settings of d-Approval elections, one assumes d is bounded by a constant. In these settings, we achieve fixed-parameter tractability results for all four parameterized problems. Our main contributions are two problem kernels of size $O(k^{d+2})$ for DV-d-Approval and Bribery-d-Approval. Given an instance (I, k) of a parameterized problem L, a *kernelization* computes an "equivalent" instance (I', k') in polynomial time, such that (1) $|I'| + k' \leq g(k)$ and (2) $(I, k) \in L$ iff $(I', k') \in L$. The new instance (I', k') is then called *problem kernel*. If the function $g(k)$ is polynomial, then (I', k') is a polynomial problem kernel. Refer to [6,11] for more details on kernelizations. In the last few years, several control and manipulation problems have been proven to be fixed-parameter tractable. However, almost all these results are based on some general classification methods, resulting in inpractically high running time [5]. Until now, we are not aware of any non-trivial kernelization results for strategic behavoir problems of elections.

2   Parameterized Complexity Results

In this section, we give a classification of parameterized complexity of the considered problems based on the value of d. Given a set C of candidates, a set V of votes over C, and a candidate c, we denote the set of votes which approve c with $V(c)$. Moreover, the *score* $s(c)$ of c achieved in (C, V) is defined as $|V(c)|$. For a vote v, let $C(v)$ denote the set of candidates approved by v. Clearly, in a d-approval election, $|C(v)| = d$ for all votes v.

2.1 Unbounded d

If the number d of approved candidates of the votes is unbounded, then all parameterizations considered here are not tractable.

Theorem 1. *DV-d-Approval with k or $|V'|$ as parameter, AV-d-Approval with k as parameter, and Bribery-d-Approval with k as parameter are W[2]-hard.*

Proof. We give here only reductions for DV-d-Approval; the other two problems can be shown in similar way.

For DV-d-Approval with k, we reduce from the W[2]-hard Hitting Set problem [8], where, given a ground set H, a collection $\mathcal{F}$ of subsets of H, and an integer l, we asks for a subset $H' \subseteq H$ with $|H'| \leq l$ such that $H' \cap F \neq \emptyset$ for every subset $F \in \mathcal{F}$. Given an instance $(H, \mathcal{F})$, we can safely assume that all subsets in $\mathcal{H}$ have the same cardinality t. We set $C = H$ and construct for every subset $H \in \mathcal{H}$ a vote $v = H$. Moreover, we add dummy candidates and votes such that all candidates in C have the same score in V. To this end, let $f(h_i)$ be the number of subsets in $\mathcal{F}$ that contain $h_i \in H$ and let $m := \max_i f(h_i)$. Then, for each h_i, add $m - f(h_i)$ many votes to V which approve h_i and $t - 1$ new dummy candidates. Note that each dummy candidate is approved only once. Finally, we add m many votes to V which approve c and $t - 1$ new dummy candidates. Finally, set $k := l$. It is easy to prove the equivalence between the instances. Every candidate in C has the same score in V as c. To make c win, we have to delete at least one vote for every candidate. Obviously, to delete the votes with dummy candidates is never better than to delete the votes according to the subsets in $\mathcal{H}$.

For the parameterization by $|V'|$, we reduce from the W[2]-hard Set Packing problem [8]. Here, given a ground set H, a collection $\mathcal{F}$ of subsets of H, and an integer l, one asks for a subcollection $\mathcal{F}' \subseteq \mathcal{F}$ such that $|\mathcal{F}'| \geq l$ and all subsets in $\mathcal{F}'$ are pairwisely disjoint. Again, we assume that all sets in $\mathcal{F}$ have the same cardinality t. The candidate set is set equal to H and each subset in $\mathcal{F}$ corresponds to a vote. Then, add two votes, each approving only c and $t - 1$ new dummy candidates. Finally, set $|V'| := l + 2$. Since the score of c is 2, the solution set V' of the election can contain for each other candidate c' at most one vote approving c'. Thus, the instances are clearly equivalent. □

2.2 Approved Candidates Bounded by a Constant

DV-d-Approval Parameterized by k. We prove that this parameterization is fixed-parameter tractable by giving an integer linear programming (ILP) formulation of this problem. Some of the following terms and definitions will be used again in the problem kernel section.

Given an instance (C, c, V, k) of DV-d-Approval, we divide the set C of candidates into two subsets with respect to their scores achieved in V. The first set E contains all "essential" candidates c' in C with $s(c') \geq s(c)$ and $I := C \setminus E$. Moreover, we partition the votes into three subsets. The first subset is set equal to $V(c)$, i.e., the set of all votes approving c. Then, the remaining votes are partitioned into two sets, $V(E)$ containing all votes approving at least one candidate in E but not c and $V(I) := V \setminus (V(c) \cup V(E))$. If a subset V' of V satisfies $|V| - |V'| \leq k$ and c is the unique winner in (C, V'), then V' is called a *solution set*.

The following observations are not hard to see and directly imply that DV-d-Approval is fixed-parameter tractable with respect to k.

Observation 1. *All votes in $V(I)$ can be safely removed, that is, (C, c, V, k) is a yes-instance if and only if $(C, c, V \setminus V(I), k)$ is a yes-instance.*

By this observation, we can from now on assume that $V = V(c) \cup V(E)$.

Observation 2. *If (C, c, V, k) is a yes-instance, then there exits a solution set V' satisfying $V(c) \subseteq V'$.*

Proof. Consider an arbitrary solution set V'. If V' satisfies the claim, then we are done; otherwise, set $V'' := V(c) \setminus V'$. The set $V' \cup V''$ is a solution set as well, since adding V'' to V' increases the score of c by $|V''|$. Thus, if c is the unique winner in (C, V'), then the same holds for $(C, V' \cup V'')$.   □

Observation 3. *If (C, c, V, k) is a yes-instance, then $|E| \leq d \cdot k$.*

Proof. By Observations 1 and 2, in order to construct a solution set V' from V, we need to delete votes from $V(E)$ such that for each candidate $c' \in E$, its score in (C, V') is upper-bounded by $s(c) - 1$. In other words, the deletion of these votes has to decrease the score of c' by at least $s(c') - s(c) + 1$ for each $c' \in E$. Moreover, by deleting one vote from V, we can decrease the scores of exact d candidates and decrease the score of each of these d candidates by one. If (C, c, V, k) is a yes-instance, then we delete at most k votes from V. This means that the scores of at most $d \cdot k$ candidates can be decreased, implying the claim.   □

We further partition the votes in $V(E)$ into disjoint classes $V(E')$, each class corresponding to a subset E' of E with $|E'| \leq d$ and containing all votes v in $V(E)$ with $C(v) \cap E = E'$. Recall that $C(v)$ is the set of the candidates that the vote v approves. With Observations 3, we have $O(k^d)$ many such classes. In the following, we give an integer linear programming formulation (ILP) for DV-d-Approval with the number of variables bound by a function of k. To this end, we define for each class $V(E')$ of $V(E)$ a variable $x_{E'}$, which takes integer values and represents the number of votes deleted from this class. The ILP consists of the following constraints:

$$\sum_{E' \subseteq E, |E'| \leq d} x_{E'} \leq k$$

$$0 \leq x_{E'} \leq |V(E')|, \quad \forall E' \subseteq E : |E'| \leq d$$

$$\sum_{E' \subseteq E, |E'| \leq d, c' \in E'} x_{E'} \geq s(c') - s(c) + 1, \quad \forall c' \in E$$

Clearly, this ILP has $O(k^d)$ many integer variables. With d being a constant, the Lenstra's results [14] directly imply the following theorem.

Theorem 2. *For constant d, DV-d-Approval is fixed-parameter tractable with k as parameter.*

Note that the results by Lenstra serve mainly for the classification purpose, resulting in an extremely high running time with respect to k. In Section 3, we improve Theorem 2 by presenting a problem kernel for DV-d-Approval.

DV-d-Approval Parameterized by $l := |V'|$ and AV-d-Approval. Note that Observation 2 clearly holds also for the parameterization with $l := |V'|$. That is, there is a solution set containing all votes approving c. Thus, if $l \leq s(c)$, then the given instance is trivial. Let $s_{V(c)}(c')$ be the score of an essential candidate c' achieved in $V(c)$. Recall that $V(c)$ is the set of votes approving c. For every essential candidate c', it must hold that $|V'(c') \setminus V(c)| < s(c) - s_{V(c)}(c')$, where $V'(c')$ is the set of votes in V' approving c'. Thus, DV-d-Approval can be reformulated as the problem of selecting $l - s(c)$ many votes from $V \setminus V(c)$ such that for every essential candidate c', the number of selected votes approving c' is less than $s(c) - s_{V(c)}(c')$. Therefore, the following problem can be considered as a generalization of DV-d-Approval.

Input: A set $C = \{c_1, c_2, \ldots, c_n\}$ associated with an integer vector $(b_1, b_2, \ldots, b_n)$, a set V of size-d subsets of C, an integer $l \geq 0$
Question: Is there a subset $V' \subseteq V$ such that $|V'| \geq l$ and V' has at most b_i subsets containing c_i for $1 \leq i \leq n$?

We call this problem Generalized d-Set Packing (d-GSP), since the well-known d-Set Packing problem is exactly the special case of d-GSP with $b_i = 1$ for all i's, that is, all subsets in V' are pairwisely disjoint. The parameterized algorithm for d-Set Packing in [13], that is based on the so-called "greedy localization" technique, can be easily modified for d-GSP: We only need to relax the disjoint condition in both greedy and localization phases of the algorithm to match b_i's. Thus, we arrive at the following theorem.

Theorem 3. *For constant d, DV-d-Approval is fixed-parameter tractable with respect to $|V'|$.*

In AV-d-Approval, we clearly add only those votes in U to V which approve c. Thus, we can assume that all votes in U approving c and the solution adds exactly k votes. The final score of c is $s(c) + k$. The problem is then to select k votes from U such that for every candidate c' in C the number of votes approving c' in the resulting election does not exceed $s(c) + k$. Since the score of c' in V is known, we can also reformulate AV-d-Approval as d-GSP.

Theorem 4. *For constant d, AV-d-Approval is fixed-parameter tractable with respect to k.*

Bribery-d-Approval. For the bribery problem, we can show similar observations as Observation 2. Here, we can assume that there is at least one vote in V not approving c, since, otherwise, it is trivial. Moreover, $|C| \geq d$, since, otherwise, the instance is clearly not solvable. Let $V(c)$ be the set of votes approving c.

Observation 4. *There exists a solution, where no vote in $V(c)$ is changed.*

Proof. Let V' be the set of votes that an optimal solution S changed. If $V' \cap V(c) = \emptyset$, then we are done; otherwise, let $v \in V' \cap V(c)$. Let $s'(c)$ and $s'(c')$ denote the scores of c and $c' \neq c$ in the election V'' resulting by changing V according to S but not v. Since v approves c, we have $s'(c) \geq s'(c')$ for all $c' \in C$. If there is a vote v' in V'' that does not approve c, then changing v' to approve c instead of changing v leads to another solution. If all votes in V'' approving c, then all candidates c' with $s'(c') = s'(c)$ must be approved by v, since, otherwise S could not be a solution. Moreover, these candidates c' have to be approved by all votes in V. However, since xthere is some vote $x \in V$ not approving c, the solution S has changed x. With c' being approved by x, we can modify S to change c' in x to some candidate not approved by v. □

The final score of c is clearly $s(c) + k$. We partition then the candidates into three sets: $C_1 := \{c' \mid s(c') \geq s(c) + k\}$, $C_2 := \{c' \mid s(c') \leq s(c)\}$, and $C_3 := C \setminus (C_1 \cup C_2)$. Since at most $d \cdot k$ candidates in C can have different scores in V and in the resulting election, the following observation is trivial.

Observation 5. *If $|C_1| \geq k \cdot d$, then the given instance is a no-instance.*

If $C_1 = \emptyset$, then it is trivial to solve: choose from $V \setminus V(c)$ arbitrarily k votes and in each of them, replace one of its approved candidates by c. Thus, the main task is to change the votes approving at least one candidate in C_1. With Observation 5, we can again partition the votes approving C_1 into $O(k^d)$ classes and give a similar ILP-formulation as for the control by deleting votes case.

Theorem 5. *For constant d, Bribery-d-Approval is fixed-parameter tractable with respect to k.*

3   Problem Kernels

In Section 2.2, for the DV-d-Approval problem, we already proved the bounds on the number of essential candidates and the number of the classes of votes in $V(E)$. However, in order to achieve a problem kernel, we need to bound the number of votes, that is, $|V(c)|$ and $|V(E)|$. Although the votes in $V(c)$ will never be deleted from V, we cannot simply ignore them and encode $s(c)$ with an integer. Such an integer is not available in an instance of DV-d-Approval. Moreover, the number of votes in one class of $V(E)$ could also be unbounded. The key for our kernelization is to partition the votes in $V(E)$ into two subsets. The first set D has a bounded size and contains all "deletable" votes that a solution set might delete from V. The votes in the second set R are irrelevant for the construction of solution sets and serve only to record the correct scores for the candidates in E. Based on these sets, the kernelization algorithm constructs a new, equivalent instance $(\mathcal{C}, c, \mathcal{V}, k)$ with $E \subseteq \mathcal{C}$ and $D \subseteq \mathcal{V}$. However, $V(c) \subseteq \mathcal{V}$ and $R \cap \mathcal{V} = \emptyset$. This new instance is then output as the problem kernel.

The set D of deletable votes is defined as follows. For each subset $E' \subseteq E$ with $|E'| \leq d$, if $|V(E')| \leq k$, then D contains all votes in $V(E')$. Otherwise, D contains arbitrary $k + 1$ many votes from $V(E')$. The set R is set to $V(E) \setminus D$. Clearly, both sets can be computed in $O(|V(E)|)$ time. The kernelization algorithm consists of two rules. We first apply the first rule exhaustively and then the second rule. Both rules create new votes approving exactly one candidate in $\mathcal{C}$ and some new, "dummy" candidates. Note that each dummy candidate is approved only once.

Rule 1. If there is a vote v in R approving more than one candidate in E, i.e., $|C(v) \cap E| > 1$, then, for each $c' \in C(v) \cap E$, add a new vote to R, which approves c' and $d - 1$ new, dummy candidates.

The second rule iteratively constructs the new instance. It needs as input the differences between the scores of the essential candidates and $s(c)$. Let $E = \{c_1, c_2, \ldots\}$ and set $t_i := s(c_i) - s(c)$.

Rule 2. _____

> **Input:** C, E, c, D, R, k, t_i's for all $c_i \in E$
> **Output:** $(\mathcal{C}, c, \mathcal{V}, k)$
> For each $c_i \in E$, compute its score s_i in $D \cup R$
> Set $\mathcal{C} := C$, $\mathcal{V} := D$, and $s_c := 0$
> For $i = 1$ to $|E|$, distinguish the following two cases:
>   Case 1. $s_i - s_c < t_i$
>   add $t_i - (s_i - s_c)$ many new votes to R,
>   each approving c_i and $d - 1$ new dummy candidates
>   set $s_i := s_c + t_i$
>   Case 2. $s_i - s_c > t_i$
>   Let $R(c_i)$ be the set of votes in R approving c_i
>   if $|R(c_i)| \geq s_i - s_c - t_i$, then
>   remove $s_i - s_c - t_i$ many votes in $R(c_i)$ from R
>   else
>   remove all votes in $R(c_i)$ from R
>   add $s_i - s_c - t_i - |R(c_i)|$ votes to $\mathcal{V}$,
>   each approving c and $d - 1$ new dummy candidates
>   set $s_c := s_i - t_i - |R(c_i)|$
>   For $j = 1$ to $i - 1$ do
>   add $t_j - (s_j - s_c)$ many new votes to R,
>   each approving c_j and $d - 1$ new dummy candidates.
>   set $s_j := s_c + t_j$
>   set $s_i := s_c + t_i$
> Add all dummy candidates to $\mathcal{C}$

Add all votes in R to $\mathcal{V}$
return $(\mathcal{C}, c, \mathcal{V}, k)$

Lemma 1. *Given two instances $I_1 = (C_1, c, V_1, k)$ and $I_2 = (C_2, c, V_2, k)$ of DV-d-Approval, let $s_1(c_1)$ and $s_2(c_2)$ denote the score of c_1 achieved in V_1 and the score of c_2 achieved in V_2, respectively. Moreover, I_1 and I_2 satisfy the following conditions:*

- *I_1 and I_2 have the same set of essential candidates, that is, $\{c' \in C_1 | s_1(c') \geq s_1(c)\} = \{c' \in C_2 | s_2(c') \geq s_2(c)\}$, and*
- *for every essential candidate c', we have $s_1(c') - s_1(c) = s_2(c') - s_2(c)$.*

Then, for a set S of votes with $S \subseteq (V_1 \cap V_2)$ and $|S| \leq k$, $V_1 \setminus S$ is a solution set for I_1 if and only if $V_2 \setminus S$ is a solution set for I_2.

Now we prove the correctness of both rules. Here, we say a rule is correct, if the instance $\mathcal{I}$ after one application of the rule is equivalent to the original instance I', that is, $\mathcal{I}$ is a yes-instance, if and only if I is a yes-instance. Note that Rule 2 applies only once.

Lemma 2. *Rule 1 is correct.*

Proof. Let $\mathcal{I} = (\mathcal{C}, c, \mathcal{V}, k)$ be the instance created by one application of Rule 1 to $I = (C, c, V, k)$. The rule "splits" one vote $v \in R$ into $|C(v) \cap E|$ votes, each approving one candidate in $C(v) \cap E$. Clearly, each candidate in E has the same score in both instances. If I is a yes-instance, then the score of c in V is at least 2, since the candidates in $C(v) \cap E$ have scores at least $k + 1$ in V. Therefore, the newly introduced dummy candidates in $\mathcal{C}$ are not essential. By Observations 1 and 2, we can only delete vote from $V(E)$. If there is a vote v in R is deleted by the solution set V' of I, then, by the precondition of Rule 1, we can replace v by a vote in D which approves the same set of essential candidates as v to derive a new solution set. Therefore, we can assume that V' deletes no vote from R. With Lemma 1, we can then conclude that V' represents a solution set for $\mathcal{I}$ as well. For the reversed direction, we can observe that there exists at least one solution set V' for $\mathcal{I}$ that does not delete the newly introduced votes. With Lemma 1, V' is a solution set for I. $\quad\square$

Lemma 3. *Rule 2 is correct.*

Proof. To show the equivalence of I and $\mathcal{I}$, we prove first that for every $c' \in E$, we have $s'(c') - s'(c) = s(c') - s(c)$, where $s'(c')$ and $s'(c)$ denote the scores of c' and c in $\mathcal{I}$, respectively. To this end, we need the following claim. Here, for $1 \leq i \leq |E|$, let $\mathcal{V}_i$ and R_i denote the current sets $\mathcal{V}$ and R after the i-th iteration of the outer For-loop of Rule 2.

Claim. After the i-th iteration of the outer For-loop of Rule 2, every candidate $c_j \in E$ with $1 \leq j \leq i$ satisfies $r_i(c_j) - r_i(c) = s(c_j) - s(c)$, where $r_i(c_j)$ and $r_i(c)$ are the scores of c_j and c achieved in $R_i \cup \mathcal{V}_i$, respectively.

Proof. We prove this claim by an induction on i. For $i = 1$, recall $t_1 = s(c_1) - s(c)$. Note that $R \cup D$ contains all votes approving c_1 but not c. By initializing s_1 as the score of c_1 in $R \cup D$, we have $s_1 \geq t_1$ before the first iteration. Since s_c is initialized as zero, Case 1 cannot apply to $i = 1$. If Case 2 applies, then votes are added to $\mathcal{V}$ or deleted from R such that, after this iteration, the difference between the scores of c_1 and c in $R_1 \cup \mathcal{V}_1$ is equal to t_1. Note that, due to Rule 1, all votes in $R(c_i)$ approve c_i but no other candidates in E.

For $i > 1$, assume the claim is true for $i - 1$, that is, for all c_j with $j \leq i - 1$, $r_{i-1}(c_j) - r_{i-1}(c) = s(c_j) - s(c) = t_j$. Consider now s_i, which is initialized as the score of c_i in $R \cup D$ at the beginning of Rule 2. Since in the iterations before, no vote approving c_i is deleted or added, $s_i = r_{i-1}(c_i)$ before the i-th iteration. Moreover, whenever the rule adds votes approving c to $\mathcal{V}$, the value of s_c is accordingly increased and always records the score of c in the new instance during the whole process of Rule 2. We have then $s_c = r_{i-1}(c)$ before the i-th iteration. If Case 1 is applied, then $t_i - (s_i - s_c)$ votes approving c_i are added to R and no other votes are deleted or added. This implies that $r_i(c_i) = r_{i-1}(c_i) + t_i - (s_i - s_c) = t_i + r_{i-1}(c) = t_i + r_i(c)$. This clearly holds for other candidates c_j with $j < i$.

The second case is more involved. For c_i, as in the first case, $r_i(c_i) = r_i(c) + t_i$ should hold at the end of this iteration. However, to achieve this, we added some votes approving c, which increase $r_i(c)$, compared to $r_{i-1}(c)$. To restore $r_i(c_j) - r_i(c) = r_{i-1}(c_j) - r_{i-1}(c) = t_j$ for $1 \leq j < i$, the second For-loop is applied. Note that at the end of the $(i-1)$-th iteration, s_j is set to $s_c + t_j$ with $s_c = r_{i-1}(c)$. Therefore, before the execution of the inner For-loop, there are s_j many votes approving c_j. With increased s_c, this For-loop adds then $s_c - (s_j - t_j)$ many votes approving c_j to restore $r_i(c_j) - r_i(c) = t_j$. Thus, the claim is true. □

By this claim, we have then that in the new instance $\mathcal{I}$ created by Rule 2, $s'(c_i) - s'(c) = t_i = s(c_i) - s(c)$ for all $1 \leq i \leq |E|$. By Observations 1 and 2, every solution set V' of I deletes only votes from $V(E)$. Moreover, by the definitions of D and R, we can conclude that, if V' deletes a vote v from R, then there is another solution set replacing v by a vote in D. Thus, by Lemma 1, if I is a yes-instance, then so is $\mathcal{I}$. The reversed direction is also easy to see. First, the dummy candidates added by Rule 2 are clearly not essential. By the above claim, the essential candidates in I remain essential in $\mathcal{I}$ and their score differences are not changed. If there is a solution set for $\mathcal{I}$ that does not delete any vote added by Rule 2, then Lemma 1 implies that I is a yes-instance. By Observation 2, the votes added by Rule 2, which approves c, are not deleted. Next, we prove that no solution set deletes the votes added by Rule 2, which approves only one essential candidate. Suppose this is not true. Assume there is such a vote v approving $c' \in E$ and deleted by a solution set. Then, we can assume that all votes in D approving c' are deleted by this solution set as well, since, otherwise, we could replace v by such a remaining vote. Let $D(c')$ be the set of votes in D approving c'. By the above assumption, $|D(c')| \leq k$. By the definition of D and R, there is no vote in $V(E) \setminus D$ approving c'. Thus, $s(c') - s(c) < |D(c')|$. Since $s'(c') - s'(c) = t_i = s(c') - s(c)$, deleting all votes in $D(c')$ results in that the new score of c' is less than $s(c)$, which implies that we do not have to delete v, a contradiction. □

Lemma 4. *Both rules can be executed in polynomial time.*

Proof. Clearly, Rule 1 can be applied $O(|V|)$ times and each application needs $O(d)$ time. Each iteration of the outer For-loop of Rule 2 creates at most $|V|$ votes, since the total score of all candidates is bounded by $d \cdot |V|$. Therefore, the total running time of Rule 2 is $O(|C| \cdot (|C| + d \cdot |V|))$. □

Theorem 6. *DV-d-Approval with d being a constant admits a problem kernel with $O(k^{d+2})$ votes and candidates.*

Proof. Since every vote can approve d candidates, it suffices to bound the number of votes. By Observation 3, there are at most $d \cdot k$ essential candidates and thus, $O(k^d)$ many classes in $V(E)$, each class corresponding to a subset $E' \subseteq E$ with $|E'| \le d$. By the definition of D, there are $O(k^{d+1})$ many votes in D. Next, we bound the number of votes in $V \setminus D$. According to Rules 1 and 2, these votes approve either c or one candidate in E. Therefore, we can partition them into $|E| + 1$ disjoint subsets, denoted by $\mathcal{V}(c), \mathcal{V}(c_1), \ldots,$ and $\mathcal{V}(c_{|E|})$, each containing the votes in $V \setminus D$ that approving c or $c_i \in E$. We prove first the size bound for $\mathcal{V}(c)$ and then the one for $\mathcal{V}(c_i)$.

We prove that $|\mathcal{V}(c)| \le k^{d+1}$ by analyzing the execution of Rule 2. Recall $t_i = s(c_i) - s(c)$ for $c_i \in E$. Clearly, $t_i \le k$ for all i's, since, otherwise we cannot make c the unique winner by deleting at most k votes. Before the first iteration of the outer For-loop of Rule 2, we have clearly $\mathcal{V}(c) = \emptyset$. In every iteration of this loop, Rule 2 adds some votes to $\mathcal{V}(c)$, only in Case 2. Moreover, these votes are added to $\mathcal{V}$, only when $R_i(c_i) = \emptyset$. Let $s_D(c_i)$ be the number of votes in D approving $c_i \in E$. Thus, $s_D(c_i) - |\mathcal{V}_i(c)| = t_j$ and $|\mathcal{V}_i(c)| \le s_D(c_i) \le k^{d+1}$.

By the claim proven in the proof of Lemma 3, after Rule 2, $s'(c_i) - s'(c) = t_i$ for each $c_i \in E$, where $s'(c_i)$ and $s'(c)$ are the scores of s' and s in $\mathcal{I}$, respectively. Since $s'(c) = |\mathcal{V}(c)| \le k^{d+1}$, we have $s'(c_i) - t_i \le k^{d+1}$. Since $s'(c_i) = s_D(c_i) + |R(c_i)|$, $|R(c_i)| \le k^{d+1}$ and thus, $|\mathcal{V}(c_i)| \le k^{d+1}$. □

By Observations 4 and 5, the same idea of essential candidates and the partition of votes into deletable and irrelevant votes can also apply to Bribery-d-Approval.

Corollary 1. *Bribery-d-Approval with d being a constant admits a problem kernel with $O(k^{d+2})$ votes and candidates.*

Corollary 2. *DV-d-Approval and Bribery-d-Approval can be solved in $O^*(k^{k \cdot (d+2)})$ time.*

4   Conclusion

We initialized a study of parameterized complexity of d-Approval election problems. In particular, we derived polynomial problem kernels for the control by deleting votes and bribery problems. It seems that similar idea could also be applied to solve other strategic behavior problems of d-Approval elections such as control by adding/deleting candidates or cloning candidates. Another future research direction would be to improve the running times of our algorithms, which are still inpractically high. Finally, the parameterized complexity study of strategic behavior of other election protocols remain mostly open [5].

References

1. Arrow, K., Sen, A., Suzumura, K.: Handbook of Social Choice and Welfare I. North-Holland (2002)
2. Arrow, K., Sen, A., Suzumura, K.: Handbook of Social Choice and Welfare II. North-Holland (2010)
3. Bartholdi III, J., Tovey, C.A., Trick, M.A.: Voting schemes for which it can be difficult to tell who won the election. Social Choice and Welfare 6, 157–165 (1989)
4. Bartholdi III, J., Tovey, C.A., Trick, M.A.: How hard is it to control an election? Mathematical and Computer Modelling 16(8/9), 27–40 (1992)
5. Betzler, N., Bredereck, R., Chen, J., Niedermeier, R.: Studies in Computational Aspects of Voting - A Parameterized Complexity Perspective. In: Bodlaender, H.L., Downey, R., Fomin, F.V., Marx, D. (eds.) Fellows Festschrift 2012. LNCS, vol. 7370, pp. 318–363. Springer, Heidelberg (2012)
6. Bodlaender, H.L.: Kernelization: New Upper and Lower Bound Techniques. In: Chen, J., Fomin, F.V. (eds.) IWPEC 2009. LNCS, vol. 5917, pp. 17–37. Springer, Heidelberg (2009)
7. Conitzer, V., Sandholm, T., Lang, J.: When are elections with few candidates hard to manipulate? Journal of the ACM 54(3), 1–33 (2007)
8. Downey, R.G., Fellows, M.R.: Parameterized Complexity. Springer (1999)
9. Faliszewski, P., Hemaspaandra, E., Hemaspaandra, L.A.: How Hard Is Bribery in Elections? J. Artif. Intell. Res (JAIR) 35, 485–532 (2009)
10. Faliszewski, P., Hemaspaandra, E., Hemaspaandra, L.A., Rothe, J.: Llull and Copeland Voting Computationally Resist Bribery and Constructive Control. J. Artif. Intell. Res (JAIR) 35, 275–341 (2009)
11. Guo, J., Niedermeier, R.: Invitation to data reduction and problem kernelization. ACM SIGACT News 38(1), 31–45 (2007)
12. Hemaspaandra, E., Hemaspaandra, L.A., Rothe, J.: Anyone but him: The complexity of precluding an alternative. Artificial Intelligence 171(5-6), 255–285 (2007)
13. Jia, W., Zhang, C., Chen, J.: An efficient parameterized algorithm for m-set packing. J. Algorithms 50(1), 106–117 (2004)
14. Lenstra, H.W.: Integer programming with a fixed number of variables. Mathematics of Operations Research 8, 538–548 (1983)
15. Lin, A.: The complexity of manipulating k-Approval elections. In: Proc. 3rd International Conference on Agents and Artificial Intelligence, vol. (2), pp. 212–218. SciTe Press (2011)
16. Meir, R., Procaccia, A., Rosenschein, J., Zohar, A.: Complexity of strategic behavior in multi-winner elections. Journal of Artificial Intelligence Research 33, 149–178 (2008)
17. Niedermeier, R.: Invitation to Fixed-Parameter Algorithms. Oxford University Press (2006)

Circular Convex Bipartite Graphs: Feedback Vertex Set

Zhao Lu[1], Min Lu[1], Tian Liu[1,*], and Ke Xu[2,*]

[1] Key Laboratory of High Confidence Software Technologies, Ministry of Education,
Institute of Software, School of Electronic Engineering and Computer Science,
Peking University, Beijing 100871, China
lt@pku.edu.cn
[2] National Lab of Software Development Environment,
Beihang University, Beijing 100191, China
kexu@nlsde.buaa.edu.cn

Abstract. A *feedback vertex set* is a subset of vertices, such that the removal of this subset renders the remaining graph cycle-free. The weight of a feedback vertex set is the sum of weights of its vertices. Finding a minimum weighted feedback vertex set is tractable for convex bipartite graphs, but $\mathcal{NP}$-complete even for unweighted bipartite graphs. In a *circular convex* (*convex*, respectively) bipartite graph, there is a circular (linear, respectively) ordering defined on one class of vertices, such that for every vertex in another class, the neighborhood of this vertex is a circular arc (an interval, respectively). The minimum weighted feedback vertex set problem is shown tractable for circular convex bipartite graphs in this paper, by making a Cook reduction (i.e. polynomial time Turing reduction) for this problem from circular convex bipartite graphs to convex bipartite graphs.

Keywords: Feedback vertex set, circular convex bipartite graph, convex bipartite graph, Cook reduction, tractability.

1 Introduction

A *feedback vertex set* is a subset of vertices, such that the removal of this subset renders the remaining graph cycle-free. The weight of a feedback vertex set is the sum of weights of its vertices. In a weighted graph, the weight of an FVS is the summation of weights over the vertices in the FVS, and the weight of each vertex is a positive integer. For unweighted graphs, each vertex has a unit weight. Finding a minimum weighed FVS (MFVS, in short) even in unweighted graphs is a classical $\mathcal{NP}$-complete problem [19,10] with many applications [8], and many algorithms have been developed for MFVS, such as approximate algorithms (e.g. [32]), randomized algorithms (e.g. [2]), parameterized algorithms (e.g. [4,21]),

---

* Corresponding authors. Partially supported by National 973 Program of China (Grant No. 2010CB328103) and Natural Science Foundation of China (Grant No. 61370052).

P. Widmayer, Y. Xu, and B. Zhu (Eds.): COCOA 2013, LNCS 8287, pp. 272–283, 2013.

exact algorithms (e.g. [9]), polynomial time algorithms for restricted graphs (e.g. [23,29,20,21]), algorithms based on statistical physics (e.g. [31]), algorithms to enumerate and count the number of MFVS (e.g. [7]), and also there are works to estimate the size of MFVS (e.g. [26,16]), and so on. A good survey on tractability of MFVS in various graph classes as well as on various kinds of MFVS algorithms is [8]. A recent good brief introduction is in [20]. See also [14].

It was known that MFVS is also $\mathcal{NP}$-complete on unweighed bipartite graphs [30], but tractable on weighted convex bipartite graphs [23] and on chordal bipartite graphs [20]. A natural question is

- *Where is the boundary between tractability and intractability of MFVS on bipartite graphs?*

In this paper, partial progress on this question has been made by investigating MFVS on circular convex bipartite graphs, which generalize convex bipartite graphs and constitute an interesting subclass of bipartite graphs.

In a *circular convex bipartite* [22] (*convex bipartite* [13], respectively) graph, there is a circular (linear, respectively) ordering defined on one class of vertices, such that for each vertex in another class, the neighborhood of this vertex is a circular arc (an interval, respectively). Circular convex bipartite graphs are natural models for scheduling problems. For example, the available working hours of a worker is usually a consecutive period of hours. A group of workers and their available hours can be modeled by a circular-convex bipartite graph [22].

Despite the apparent similarity in definitions of circular convex bipartite graphs and convex bipartite graphs, results for *circular convex bipartite* graphs are scarce for a long time, while results for convex bipartite graphs are plenty. Only maximum matching and Hamiltonian cycle and path were known linear time solvable for circular-convex bipartite graphs [22]. Recently, connected domination and independent domination were shown tractable for circular-convex bipartite graphs [25,24]. On the other hand, several $\mathcal{NP}$-complete problems for bipartite graphs were known *polynomial time* or even *linear time* solvable for *convex bipartite* graphs, such as feedback vertex set [30,23], variants of domination [6], Hamiltonian cycle and path [22], etc, see a brief survey in [15].

Here we show that MFVS is tractable for circular convex bipartite graphs, by making reduction from circular convex bipartite graphs to convex bipartite graphs. As in [25,24], the reduction is a Cook reduction (a polynomial time Turing reduction) [10]. The special properties of feedback vertex set makes this reduction possible. Before our works in [25,24] and here, only Karp reduction (polynomial many-one reduction) [10] from circular convex bipartite graphs to circular-arc graphs was used [22]. Our methods in this paper and in [25,24] may be of use to show more problems tractable for circular convex bipartite graphs.

To put our results more properly in the range of bipartite graphs, let us turn to other interesting superclasses of convex bipartite graphs, such as *chordal bipartite* graphs and *tree convex bipartite* graphs, see Figure 1. Chordal bipartite graphs were well known, see e.g. books [11,3]. In a *chordal bipartite* graph, every cycle of length at least *six* has a chord, where a *chord* of a cycle on a graph is

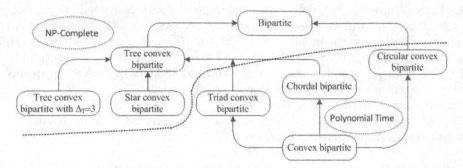

Fig. 1. (In)tractability of MFVS for various bipartite graphs

an edge between two vertices of the cycle but the edge itself is not a part of the cycle [12]. Feedback vertex set is tractable for chordal bipartite graphs [20].

Tree convex bipartite graphs were introduced recently as a natural extension to convex bipartite graphs [16,18,27,28,17,25,24]. In a *tree convex bipartite* graph, there is a tree defined on one class of vertices, such that for every vertex in another class, the neighborhood of this vertex is a subtree [16]. When the tree is a path (a star, a triad, of maximum degree three, respectively), the graph is called *convex bipartite* [13] (*star convex bipartite* [16], *triad convex bipartite* [18], *tree convex bipartite with* $\Delta_T = 3$ [28,17], respectively), where a *triad* is three pathes with a common end. Feedback vertex set is shown $\mathcal{NP}$-complete for star convex bipartite graphs and tree convex bipartite graphs with the maximum degree of the tree $\Delta_T = 3$ [16,28,17], but tractable for triad convex bipartite graphs [18,27,28,17,25,24]. Thus, the (in)tractability of feedback vertex set for various bipartite graph classes is well understood now.

We note in pass that the adjacent matrices of circular convex (convex, respectively) bipartite graphs have the so-called *circular* (*consecutive*, respectively) *ones property*, which is recognizable in linear time, see e.g. survey [5]. Tree convex bipartite graphs are also recognizable in linear time, see e.g. survey [1]. The associated circular orderings (linear orderings, trees, respectively) are all constructible in linear time, thus can safely be assumed as part of the inputs.

This paper is structured as follows. After introducing some necessary definitions and notations and some basic facts (Section 2), a polynomial time Turing reduction (i.e. a Cook reduction) is shown for feedback vertex set from circular convex bipartite graphs to convex bipartite graphs (Section 3), and finally are some concluding remarks and open problems (Section 4).

2   Preliminaries

The neighborhood of a vertex x in a graph G is denoted by $N_G(x) = \{y \mid y$ is adjacent to x in $G\}$. When G is clear from the context, we just write $N(x)$ instead of $N_G(x)$. In this paper, G will always denote an input graph, we just use $N(x)$ to denote the neighborhood of a vertex x in G. For any subgraph G' of G, we use $N_{G'}(x)$ to denote the neighborhood of a vertex x in G'.

For a circular-convex bipartite graph $G = (A, B, E)$, without loss of generality, we can always assume a canonical circular ordering $\prec$ on A,

$$a_1 \prec a_2 \prec \cdots \prec a_n \prec a_{n+1} = a_1,$$

such that for each vertex b in B, either $N(b) = \{a_i, a_{i+1}, \cdots, a_j\}$ or $N(b) = \{a_j, a_{j+1}, \cdots, a_n, a_1, \cdots, a_i\}$, for some i, j, where $1 \leq i \leq j \leq n+1$.

Recall that deleting a FVS makes the remaining graph a forest. Thus finding a MFVS in a graph is equivalent to finding a *maximum cycle-free set* (MCFS) in the same graph. We will show how to find a MCFS instead of a MFVS in circular convex bipartite graphs.

For a MCFS $F = (A', B', E')$ of a circular-convex bipartite graph $G = (A, B, E)$, we call two vertices a' and a'' *consecutive* in A', if either $a' = a_i$, $a'' = a_j$, $i < j$, and there is no a_x in A' for $i < x < j$, or $a' = a_j$, $a'' = a_i$, $i < j$, and there is no a_x in A' for $1 \leq x < i$ and $j < x \leq n$.

Notice that all graph properties we are interested in are invariant under the renaming of vertices, especially under the rotation of the circular ordering on A. By rotation on A we mean that changing the absolute positions of each a_i's on A, but not changing the relative ordering of all vertices on A. With these rotations in mind, without loss of generality, we may always assume that the neighborhood $N(b)$ of a vertex b is in the form $N(b) = \{a_i, a_{i+1}, \cdots, a_j\}$, where $1 \leq i \leq j \leq n$. This will greatly simplify our presentations. For example, in the above definition of consecutiveness on A', we only need to consider the following case, where $a' = a_i$, $a'' = a_j$, $i < j$, and there is no a_x in A' for $i < x < j$.

A very useful property of MCFS is listed as the following lemma, which is a key property for our reduction to work.

Lemma 1. *If G' is a subgraph of G and G' contains a MCFS of G, then any MCFS of G' is also a MCFS of G.*

Proof. Assume that F is a MCFS of G, F is contained in G', and G' is a subgraph of G. Now assume that U is any MCFS of G'. Since F is a MCFS of G, we know that F is cycle-free. Since F is contained in G' and U is a MCFS of G', we know that U is as large as F. Since U is cycle-free and also contained in G, we know that U is also a MCFS of G. □

We will use Lemma 1 in the following way. The graph G will be a *circular convex bipartite* graph whose MCFS F we want to find. The graph G' will be a *convex bipartite* graph which is a *subgraph* of G and also contains F. Then we can find a MCFS U of G' by a known efficient algorithm and U is *as good as F* for G. We will set different restrictions on F to get the proper G''s in different cases respectively, and these restrictions should cover all the possibilities. In each case, we are able to construct *not a single* G' satisfying the conditions in Lemma 1, but a *polynomial time constructible family* of G''s, with at least one of the G''s satisfying the conditions in Lemma 1. This constitutes the main frame of our reduction in this paper. Thus, our reduction will be a Cook reduction (polynomial time Turing reductions), rather than a Karp reductions (polynomial time many-one or mapping reductions). Notice that Lemma 1 holds for *weighted* graphs, so our reduction base on this lemma also works for weighted graphs.

3   Reduction

In this section, we will make a Cook reduction of the maximum cycle-free set (MCFS) from circular convex bipartite graphs to convex bipartite graphs.

A MCFS $F = (A', B', E')$ of $G = (A, B, E)$ must be in one of the following four cases.

1. There is only one vertex in A'.
2. There are exactly two vertices in A'.
3. There are at least three vertices in A' and there are two consecutive vertices in A' with no common neighbor in B'.
4. There are at least three vertices in A' and every two consecutive vertices in A' have a common neighbor in B'.

We will deal with the above four cases respectively as follows.

Case 1: There is only one vertex in A'

This is the easiest case to deal with among all cases. Notice that each MCFS must has at least one vertex from A. Indeed, for every vertex a in A, the subgraph induced by $\{a\} \cup B$ is still cycle-free. This is because that, to make a cycle $a'b'a''b'' \cdots a'$ in a bipartite graph with a bipartition of its vertices into two sets A and B, we need at least four vertices a', a'', b', b'', with at least two vertices a', a'' from A and at least two vertices b', b'' from B, respectively. If there is only one vertex in A', without loss of generality, we can assume that this vertex is a. Then $\{a\} \cup B$ is apparently cycle-free and is maximum with respect to the condition in case 1. Thus, in this case, the MCFS of G is $\{a\} \cup B$.

We define a set S_1 as follows.

$$S_1 = \{\{a\} \cup B \mid a \in A\}.$$

Then S_1 is computable in polynomial time $O(|A||B|)$, and S_1 contains a MCFS of G in case 1, as discussed above.

Case 2: There are exactly two vertices in A'

This is also an easy case to deal with among all cases. Notice that for each MCFS having two vertices in A', there will be at most one common neighbor of these two vertices in B'. Indeed, we have the following useful lemma, which is a key observation for our whole reduction to work in this case and in case 4 below.

Lemma 2. *Let $G = (A, B, E)$ be a circular-convex bipartite graph and $G' = (A', B', E')$ a cycle-free subgraph of G. Then for any two vertices a', a'' in A', $|B' \cap N(a') \cap N(a'')| \leq 1$.*

Proof. If there are two vertices a', a'' in A and two vertices b', b'' in $B' \cap N(a') \cap N(a'')$, then there is a cycle $a'b'a''b''a'$ in G', which is a contradiction to the assumption that G' is cycle-free. $\square$

Under the condition that there are exactly two vertices in A' and the intersection of their neighborhoods is nonempty, we know that there will be exact one common neighbor of these two vertices in B'. This is because that if the intersection of their neighborhoods is nonempty, then we can always put a common neighbor of them into B' still not making any cycle. On the other hand, if there are exactly two vertices in A' and the intersection of their neighborhoods is empty, then we can put all vertices in B into B' still not making any cycle.

More precisely, if there are exactly two vertices in A', without loss of generality, we may assume that these two vertices are a', a'', that is, $A' = \{a', a''\}$. If $N(a') \cap N(a'') \neq \emptyset$ and assume that $b \in N(a') \cap N(a'')$, then $\{a', a'', b\} \cup B \setminus (N(a') \cap N(a''))$ is apparently cycle-free, and also maximum with respect to the conditions that $A' = \{a', a''\}$ and $b \in N(a') \cap N(a'')$. Thus, in this case, the MCFS is $\{a', a'', b\} \cup B \setminus (N(a') \cap N(a''))$. If $N(a') \cap N(a'') = \emptyset$, then $\{a', a''\} \cup B$ is apparently cycle-free, and also maximum with respect to the conditions that $A' = \{a', a''\}$ and $N(a') \cap N(a'') = \emptyset$. Thus, in this case, the MCFS is $\{a', a''\} \cup B$. We define a set S_2 as follows.

$$S_2 = \{\{a', a'', b\} \cup B \setminus (N(a') \cap N(a'')) \mid a', a'' \in A, b \in N(a') \cap N(a'')\} \cup$$
$$\{\{a', a''\} \cup B \mid a', a'' \in A \text{ and } N(a') \cap N(a'') = \emptyset\}.$$

Then S_2 is computable in polynomial time $O(|A|^2 |B|^2)$, and S_2 contains a MCFS of G in case 2, as discussed above.

Case 3. There are at least three vertices in A' and there are two consecutive vertices in A' with no common neighbor in B'

This is a main case to deal with among all cases. In this case, we will make a *cut* on the circular ordering of G to get a convex bipartite subgraph G'.

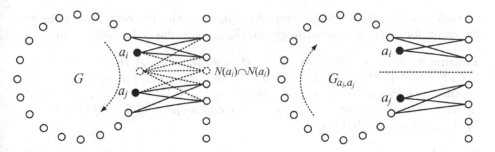

Fig. 2. Graph G and the graph G_{a_i, a_j}

Without loss of generality, we may assume that a_i, a_j are in A', but none of a_x's are in A' for $i < x < j$, and moreover, none of vertices in $N(a_i) \cap N(a_j)$ are in B', that is, $N(a_i) \cap N(a_j) \cap B' = \emptyset$. Then we can remove the vertices in $\{a_{i+1}, \cdots, a_{j-1}\} \cup (N(a_i) \cap N(a_j))$, and also remove all the edges incident with one of the removed vertices. Intuitively, in this way, we make a *cut* on the circular ordering of G between a_i and a_j to get a convex bipartite subgraph $G' = G_{a_i, a_j}$, whose MCFS is as large as a MCFS of G, see Figure 2.

More precisely, for any two vertices a_i, a_j in A, we define a graph G_{a_i, a_j} as follows.

$G_{a_i, a_j} = (A_{a_i, a_j}, B_{a_i, a_j}, E_{a_i, a_j})$, where
$A_{a_i, a_j} = A \setminus \{a_{i+1}, \cdots, a_{j-1}\}$,
$B_{a_i, a_j} = B \setminus (N(a_i) \cap N(a_j))$, and
$E_{a_i, a_j} = \{e \in E \mid e$ has its two ends in A_{a_i, a_j} and B_{a_i, a_j} respectively$\}$.

Figure 2 shows how we get the graph G_{a_i, a_j} from the original graph G for a pair of vertices a_i, a_j. The following lemmas justify the construction of G_{a_i, a_j}.

Lemma 3. *For any two vertices a_i, a_j in A, the graph G_{a_i, a_j} is convex bipartite.*

Proof. We prove by definition. After removing $\{a_{i+1}, ..., a_{j-1}\}$ from A, we define a new linear ordering on A_{a_i, a_j} as follows, see Figure 2 (right).

$$a_j \prec a_{j+1} \prec \cdots \prec a_{i-1} \prec a_i.$$

Then for each vertex b in B_{a_i, a_j}, its neighborhood $N_{G'}(b)$ is an interval on A_{a_i, a_j} under this new linear ordering. Indeed, assume that $N(b) = \{a_p, a_{p+1}, \cdots, a_q\}$ in G, where $1 \le p \le q \le n$. Then $N_{G'}(b) = N(b) \cap B_{a_i, a_j}$. In general, $N_{G'}(b)$ is an interval on A_{a_i, a_j}. More precisely, We have the following four cases to consider.

1. If $p \ge j$ and $q \le i$, then $N_{G'}(b) = N(b) = \{a_p, a_{p+1}, \cdots, a_{q-1}, a_q\}$.
2. If $p \le i \le q \le j$ or $j \le p \le i \le q$, then $N_{G'}(b) = \{a_p, a_{p+1}, \cdots, a_{i-1}, a_i\}$.
3. If $i \le p \le j \le q$ or $p \le j \le q \le i$, then $N_{G'}(b) = \{a_j, a_{j+1}, \cdots, a_{q-1}, a_q\}$.
4. If $i \le p \le q \le j$, then $N_{G'}(b) = \emptyset$.

In each case, $N_{G'}(b)$ is still an interval on A_{a_i, a_j} under the new linear ordering. By definition of convex bipartite graph, G_{a_i, a_j} is a convex bipartite graph.   □

Lemma 4. *For a circular convex bipartite graph $G = (A, B, E)$, if there is a MCFS $F = \{A', B', E'\}$ of G with two consecutive vertices a_i, a_j in A' and no common neighbor of a_i and a_j in B', then F is also a MCFS of G_{a_i, a_j}. Moreover, under the same assumption on the existence and property of F, any MCFS F_{a_i, a_j} of G_{a_i, a_j} is also a MCFS of G.*

Proof. Essentially the same as the proof of Lemma 1. For the first part of this lemma, we prove by contradiction. By the assumption on F, we know that during the construction of G_{a_i, a_j}, none of vertices or edges of F are removed, so F is still a CFS on G_{a_i, a_j}. If F is not a MCFS of G_{a_i, a_j}, then there is a MCFS F' of G_{a_i, a_j} which is strictly larger than F. Notice that any CFS of G_{a_i, a_j} is also a CFS of G, since G_{a_i, a_j} is a subgraph of G. Then F' is also a CFS of G and larger than F. This is a contradiction to the maximality of F in G.

For the second part of this lemma, notice that the MCFS F of G is also a MCFS of G_{a_i,a_j} by the first part of this lemma. Then by the assumption on F_{a_i,a_j}, we know that F_{a_i,a_j} is as large as F and F_{a_i,a_j} is also a CFS of G. Thus, F_{a_i,a_j} is also a MCFS of G.   □

We define a set S_3 as follows.

$$S_3 = \{F_{a_i,a_j} \mid a_i, a_j \in A \text{ and } F_{a_i,a_j} \text{ is a MCFS of } G_{a_i,a_j}\}.$$

Remark 1. For each pair of a_i, a_j, G_{a_i,a_j} is unique, but for each G_{a_i,a_j}, its MCFS F_{a_i,a_j} may not be unique. For our purpose, however, for each a_i, a_j, we only need one such MCFS F_{a_i,a_j} of G_{a_i,a_j} in S_3, see proof of Lemma 5 below.

Lemma 5. S_3 *is computable in polynomial time* $O(|A|^5 + |A|^4|E|)$.

Proof. By Lemma 3, for each pair of a_i, a_j, G_{a_i,a_j} is a convex bipartite graph, thus we can compute a MCFS of G_{a_i,a_j} by the known $O(|A|^3 + |A|^2|E|)$ time algorithm in [23]. As remarked in Remark 1, for each pair of a_i, a_j, we only need one such MCFS of G_{a_i,a_j} in S_3. Thus, by an enumeration of all $O(|A|^2)$ pairs (a_i, a_j) in $A \times A$, we can compute S_3 in polynomial time $O(|A|^5 + |A|^4|E|)$.   □

By lemma 5 and lemma 4, we know that S_3 is polynomial time computable and contains a MCFS of G under the conditions in case 3.

Case 4. There are at least three vertices in A' and every pair of connective vertices in A' has a common neighbor in B'

This is the last and also a main case to deal with among all cases. In this case, we can not directly make a *cut* on G, as in case 3, to get a convex bipartite subgraph. However, we can show that, in this case, A' is actually contained in a neighborhood of a *single* vertex in B'. This containment will enable us to deal with this case similarly as in case 3, when A' is restricted into the neighborhood of a single vertex in B'. Thus, the following lemma is a key observation for our reduction to work in this case.

Lemma 6. *Under the conditions of case 4, there is a vertex b in B' and a positive integer $k \geq 3$, such that $A' = \{a^{(1)}, a^{(2)}, \cdots, a^{(k)}\} \subseteq N(b)$.*

Proof. Under the conditions of case 4, without loss of generality, we can assume that $A' = \{a^{(1)}, a^{(2)}, \cdots, a^{(k)}\}$ for $k \geq 3$, $a^{(i)}$ and $a^{(i+1)}$ are consecutive in A' with $a^{(k+1)} = a^{(1)}$ for $1 \leq i \leq k$, and there is a common neighbor $b^{(i)}$ in B' to $a^{(i)}$ and $a^{(i+1)}$ for $1 \leq i \leq k$.

Now, consider the following vertex sequence $a^{(1)}b^{(1)}a^{(2)}b^{(2)} \cdots a^{(k)}b^{(k)}a^{(1)}$. Each two consecutive vertices in this sequence are adjacent in F. Below, we will show by induction on k that, if there are two *distinct* vertices among $b^{(i)}$'s, then there is a *cycle* in F, whose vertices are all among this sequence.

The base step $k = 3$. If $b^{(1)}, b^{(2)}, b^{(3)}$ are all distinct, then the sequence $a^{(1)}b^{(1)}a^{(2)}b^{(2)}a^{(3)}b^{(3)}a^{(1)}$ itself is a cycle of length six in F. If two of $b^{(1)}, b^{(2)}, b^{(3)}$ are equal, say $b^{(1)} = b^{(2)}$, then $a^{(1)}b^{(1)}a^{(3)}b^{(3)}a^{(1)}$ is a cycle of length four in F.

The induction step. Assume that for sequence $a^{(1)}b^{(1)} \cdots a^{(k')}b^{(k')}a^{(1)}$ with $3 \leq k' \leq k$, if there are two distinct vertices among $b^{(i)}$'s, then there is a cycle of length at most $2k'$ in F. Let us consider $k+1$ vertices $b^{(1)}, \cdots, b^{(k+1)}$. If all $b^{(i)}$'s are distinct, then the sequence $a^{(1)}b^{(1)}a^{(2)}b^{(2)} \cdots a^{(k+1)}b^{(k+1)}a^{(1)}$ itself is a cycle of length $2k+2$ in F. Assume that at least two of $b^{(i)}$'s are equal, say $b^{(i)} = b^{(j)}$ with $1 \leq i < j \leq k+1$. If $j = i+1$, then by induction on sequence $a^{(1)}b^{(1)} \cdots a^{(i)}b^{(j)}a^{(j+1)}b^{(j+1)} \cdots a^{(k+1)}b^{(k+1)}a^{(1)}$, there is a cycle of length at most $2k$ in F. If $j > i+1$, then by induction on two sequence $a^{(i)}b^{(i)} \cdots a^{(j)}b^{(j)}a^{(i)}$ and $a^{(j)}b^{(j)}a^{(j+1)}b^{(j+1)} \cdots a^{(k+1)}b^{(k+1)}a^{(1)}b^{(1)} \cdots a^{(i)}b^{(i)}a^{(j)}$ respectively, there is a cycle of length at most $\max(2(j-i), 2(k+1-j+i)) \leq 2k$ in F.

The existence of such a cycle in F always contradicts to the cycle-freeness of F. Thus, we can conclude that for $1 \leq i \leq k$, all $b^{(i)}$'s are equal to a single vertex, say b in B'.   □

Based on this lemma, for any two consecutive vertices $a^{(i)}$ and $a^{(i+1)}$ in A', there is no vertex in $N(a^{(i)}) \cap N(a^{(i+1)}) \cap B'$ other than the vertex b. That is, for any $1 \leq i \leq k$, $N(a^{(i)}) \cap N(a^{(i+1)}) \cap B' = \{b\}$. Thus, we can make a *cut* on the circularly ordered set $N(b)$ between $a^{(i)}$ and $a^{(i+1)}$, to get a convex bipartite graph, see Figure 3.

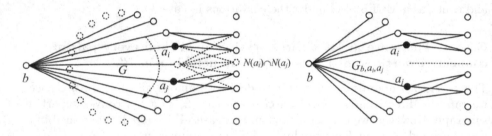

Fig. 3. The graph G and its subgraph G_{b,a_i,a_j}

More precisely, for each vertex b in B, without loss of generality, we may assume that $N(b) = \{a_p, a_{p+1}, \cdots, a_q\}$, where $1 \leq p < q \leq n$. Then for any i, j, where $1 \leq p \leq i < j \leq q \leq n$, we define a graph G_{b,a_i,a_j} as follows.

$$G_{b,a_i,a_j} = (A_{b,a_i,a_j}, B_{b,a_i,a_j}, E_{b,a_i,a_j}), \text{ where}$$
$$A_{b,a_i,a_j} = N(b) \setminus \{a_{i+1}, a_{i+2} \cdots, a_{j-1}\},$$
$$B_{b,a_i,a_j} = B \setminus (N(a_i) \cap N(a_j)) \cup \{b\}, \text{ and}$$
$$E_{a_i,a_j} = \{e \in E \mid e \text{ has its two ends in } A_{a_i,a_j} \text{ and } B_{a_i,a_j} \text{ respectively}\}.$$

Figure 3 shows how we get the graph G_{b,a_i,a_j} from G for each triple (b, a_i, a_j). The following lemmas justify the construction of G_{b,a_i,a_j}.

Lemma 7. *For each vertex b in B and any two of its neighbors a_i, a_j, the graph G_{b,a_i,a_j} is a convex bipartite graph.*

Proof. Almost the same as the proof of Lemma 3. Notice that $N(b) = A_{b,a_i,a_j}$ is an interval on A_{b,a_i,a_j}. □

Lemma 8. *Under the conditions of case 4, any MCFS F of G containing two consecutive vertices a_i and a_j in A and their common neighbor b in B is also a MCFS of G_{b,a_i,a_j}, and any MCFS F_{b,a_i,a_j} of G_{b,a_i,a_j} is also a MCFS of G.*

Proof. Essentially the same proofs as Lemma 1 and Lemma 4. For the first part, notice that in construction of G_{b,a_i,a_j}, no vertices of F is removed. Indeed, since a_i and a_j are consecutive in F, we know that $a_{i+1}, \cdots, a_{j-1}$ all are not in F. For the second part, notice that G_{b,a_i,a_j} is a subgraph of G and G_{b,a_i,a_j} contains a MCFS of G under the conditions of case 4. □

Now we define a set $\mathcal{S}_4$ as follows.

$$\mathcal{S}_4 = \{F_{b,a_i,a_j} \mid b \in B, a_i, a_j \in N(b) \text{ and } F_{b,a_i,a_j} \text{ is a MCFS of } G_{b,a_i,a_j}\}.$$

Remark 2. For each tuple of (b, a_i, a_j), G_{b,a_i,a_j} is unique, but for each G_{b,a_i,a_j}, its MCFS may not be unique. For our purpose, however, for each triple (b, a_i, a_j), we only need one such MCFS of G_{b,a_i,a_j} in $\mathcal{S}_4$, see proof of Lemma 9 below.

Lemma 9. $\mathcal{S}_4$ *is computable in polynomial time* $O(|B|(|A|^5 + |A|^4|E|))$.

Proof. By Lemma 7, for each vertex b in B and each pair of a_i, a_j in $N(b)$, G_{b,a_i,a_j} is a convex bipartite graph, thus we can compute a MCFS of G_{b,a_i,a_j} by the known $O(|A|^3 + |A|^2|E|)$ time algorithm in [23]. As remarked in Remark 2, for each triple (b, a_i, a_j), we only need one such MCFS of G_{b,a_i,a_j} in $\mathcal{S}_4$. Thus, by an enumeration of all $O(|B||A|^2)$ triples (b, a_i, a_j) in $B \times A \times A$, we can compute $\mathcal{S}_4$ in polynomial time $O(|B|(|A|^5 + |A|^4|E|))$. □

By lemma 9 and lemma 8, we know that $\mathcal{S}_4$ is polynomial time computable and contains a MCFS of G under the conditions in case 4.

The overall reduction

Finally, we can define a set $\mathcal{S} = \mathcal{S}_1 \cup \mathcal{S}_2 \cup \mathcal{S}_3 \cup \mathcal{S}_4$. By above discussions in from case 1 to case 4, we know that $\mathcal{S}$ is polynomial time computable and $\mathcal{S}$ contains a MCFS of G in all cases. Now our overall reduction works as follows.

– Compute the set $\mathcal{S}$ and pick out a cycle-free set with the maximum weight.

Since cycle-freeness is verifiable in polynomial time, this reduction is a polynomial time Turing reduction (Cook reduction) of MCFS (or MFVS) from circular convex bipartite graphs to convex bipartite graphs. Thus, by the known tractability result of MCFS (or MFVS) in [23] and the complementary relationship between MCFS and MFVS, we have the following main theorem.

Theorem 1. *The minimum weighthed feedback vertex set problem (MFVS) is polynomial time solvable for circular convex bipartite graphs.*

4   Conclusions

We have shown that the minimum weighted feedback vertex set (MFVS) is tractable for circular convex bipartite graphs, by making a Cook reduction (polynomial time Turing reduction) of this problem from circular convex bipartite graphs to convex bipartite graphs. We still have the following open problems.

First, the existence of Karp reduction rather than Cook reduction of this problem for these graphs is unknown. Second, our main concern here is on tractability of MFVS rather than on practical efficiency of algorithms of MFVS for these graphs. The existence of linear time algorithms of MFVS for these graphs is unknown. Third, any problem which is $\mathcal{NP}$-complete for circular convex bipartite graphs but tractable for convex bipartite graphs is unknown. Finally, can we show more problems tractable for circular convex bipartite graphs?

Acknowledgments. We thank Kaile Su for encouragement and support, Francis Y.L. Chin for rising the question on tractability of MFVS for circular convex bipartite graphs during FAW-AAIM 2011, and anonymous reviewers for helpful comments.

References

1. Bao, F.S., Zhang, Y.: A review of tree convex sets test. Computational Intelligence 28(3), 358–372 (2012); Previous version: A survey of tree convex sets test. arXiv.0906.0205 (2009)
2. Becker, A., Bar-Yehuda, R., Geiger, D.: Randomized Algorithms for the Loop Cutset Problem. J. Artif. Intell. Res. 12, 219–234 (2000)
3. Brandstad, A., Le, V.B., Spinrad, J.P.: Graph Classes - A Survey. Society for Industrial and Applied Mathematics, Philadelphia (1999)
4. Cao, Y., Chen, J., Liu, Y.: On Feedback Vertex Set: New Measure and New Structures. In: Kaplan, H. (ed.) SWAT 2010. LNCS, vol. 6139, pp. 93–104. Springer, Heidelberg (2010)
5. Dom, M.: Algorithmic aspects of the consecutive ones property. Bulletin of the EATCS 98, 27–59 (2009)
6. Damaschke, P., Muller, H., Kratsch, D.: Domination in Convex and Chordal Bipartite Graphs. Inform. Proc. Lett. 36, 231–236 (1990)
7. Fomin, F.V., Gaspers, S., Pyatkin, A., Razgon, I.: On the Minimum Feedback Vertex Set Problem: Exact and Enumeration Algorithms. Algorithmica 52(2), 293–307 (2008)
8. Festa, P., Pardalos, P.M., Resende, M.G.C.: Feedback set problems. In: Handbook of Combinatorial Optimization, (suppl. vol. A), pp. 209–258. Kluwer Academic Publishers (1999)
9. Fomin, F.V., Villanger, Y.: Finding Induced Subgraphs via Minimal Triangulations. In: Proc. of STACS, pp. 383–394 (2010)
10. Garey, M.R., Johnson, D.S.: Computers and Intractability, A Guide to the Theory of NP-Completeness. W.H. Freeman and Company (1979)
11. Golumbic, M.C.: Algorithmic Graph Theory and Perfect Graphs. Academic Press, New York (1980)

12. Golumbic, M.C., Goss, C.F.: Perfect elimination and chordal bipartite graphs. J. Graph Theory. 2, 155–163 (1978)
13. Grover, F.: Maximum matching in a convex bipartite graph. Nav. Res. Logist. Q. 14, 313–316 (1967)
14. Guo, J.: Undirected feedback vertex set. Encyclopedia of Algorithms, 995–996 (2008)
15. Hung, R.-W.: Linear-time algorithm for the paired-domination problem in convex bipartite graphs. Theory Comput. Syst. 50, 721–738 (2012)
16. Jiang, W., Liu, T., Ren, T., Xu, K.: Two hardness results on feedback vertex sets. In: Atallah, M., Li, X.-Y., Zhu, B. (eds.) FAW-AAIM 2011. LNCS, vol. 6681, pp. 233–243. Springer, Heidelberg (2011)
17. Jiang, W., Liu, T., Wang, C., Xu, K.: Feedback vertex sets on restricted bipartite graphs. Theor. Comput. Sci. (in press, 2013), doi: 10.1016/j.tcs.2012.12.021
18. Jiang, W., Liu, T., Xu, K.: Tractable feedback vertex sets in restricted bipartite graphs. In: Wang, W., Zhu, X., Du, D.-Z. (eds.) COCOA 2011. LNCS, vol. 6831, pp. 424–434. Springer, Heidelberg (2011)
19. Karp, R.: Reducibility among combinatorial problems. In: Complexity of Computer Computations, pp. 85–103. Plenum Press, New York (1972)
20. Kloks, T., Liu, C.H., Pon, S.H.: Feedback vertex set on chordal bipartite graphs. arXiv:1104.3915 (2011)
21. Kloks, T., Wang, Y.L.: Advances in graph algorithms. Manuscipt of a book (2013)
22. Liang, Y.D., Blum, N.: Circular convex bipartite graphs: maximum matching and Hamiltonian circuits. Inf. Process. Lett. 56, 215–219 (1995)
23. Liang, Y.D., Chang, M.S.: Minimum feedback vertex sets in cocomparability graphs and convex bipartite graphs. Acta Informatica 34, 337–346 (1997)
24. Lu, M., Liu, T., Xu, K.: Independent Domination: Reductions from Circular- and Triad-Convex Bipartite Graphs to Convex Bipartite Graphs. In: Fellows, M., Tan, X., Zhu, B. (eds.) FAW-AAIM 2013. LNCS, vol. 7924, pp. 142–152. Springer, Heidelberg (2013)
25. Lu, Z., Liu, T., Xu, K.: Tractable Connected Domination for Restricted Bipartite Graphs (Extended Abstract). In: Du, D.-Z., Zhang, G. (eds.) COCOON 2013. LNCS, vol. 7936, pp. 721–728. Springer, Heidelberg (2013)
26. Madelaine, F.R., Stewart, I.A.: Improved upper and lower bounds on the feedback vertex numbers of grids and butterflies. Discrete Math. 308, 4144–4164 (2008)
27. Song, Y., Liu, T., Xu, K.: Independent domination on tree convex bipartite graphs. In: Snoeyink, J., Lu, P., Su, K., Wang, L. (eds.) AAIM 2012 and FAW 2012. LNCS, vol. 7285, pp. 129–138. Springer, Heidelberg (2012)
28. Wang, C., Liu, T., Jiang, W., Xu, K.: Feedback vertex sets on tree convex bipartite graphs. In: Lin, G. (ed.) COCOA 2012. LNCS, vol. 7402, pp. 95–102. Springer, Heidelberg (2012)
29. Wang, F.H., Wang, Y.L., Chang, J.M.: Feedback vertex sets in star graphs. Inform. Process. Lett. 89(4), 203–208 (2004)
30. Yannakakis, M.: Node-deletion problem on bipartite graphs. SIAM J. Comput. 10, 310–327 (1981)
31. Zhou, H.: The feedback vertex set problem: a spin glass approach. arXiv:1307.6948 (2013)
32. Van Zuylen, A.: Linear programming based approximation algorithms for feedback set problems in bipartite tournaments. Theor. Comput. Sci. (in press)

The Multi-parameterized Cluster
Editing Problem

Faisal N. Abu-Khzam

Department of Computer Science and Mathematics
Lebanese American University
Beirut, Lebanon
faisal.abukhzam@lau.edu.lb

Abstract. The Cluster Editing problem seeks a transformation of a
given undirected graph into a transitive graph via a minimum number of
edge-edit operations. Existing algorithms often exhibit slow performance
and could deliver clusters of no practical significance, such as singletons.
A constrained version of Cluster Editing is introduced, featuring more
input parameters that set a lower bound on the size of a clique-cluster
as well as upper bounds on the amount of both edge-additions and deletions per vertex. The new formulation allows us to solve Cluster Editing
(exactly) in polynomial time when edge-edit operations per vertex is
smaller than half the minimum cluster size. Moreover, we address the
case where the new edge addition and deletion bounds (per vertex) are
small constants. We show that Cluster Editing has a linear-size kernel in
this case.

1 Introduction

Given an undirected loopless graph $G = (V, E)$ and an integer $k > 0$, the Cluster
Editing Problem asks whether k or less edge additions or deletions can transform
G into a graph whose connected components are cliques. Cluster Editing is NP-
Complete [13, 16], but can be solved in polynomial-time when the parameter k
is fixed [4, 10]. In other words, the problem is fixed-parameter tractable when
parameterized by the total number of edge-edit operations[1].

The Cluster Editing problem received considerable attention recently. This
can be seen from a long sequence of continuous algorithmic improvements (see
[1–3, 5, 6, 10, 11]). The current asymptotically fastest fixed-parameter algorithm
runs in $O^*(1.618^k)$ [2]. Moreover, a kernel of order $2k$ was achieved recently in [6].
In other words, an arbitrary Cluster Edit instance can be reduced in polynomial-
time into an equivalent instance where the number of vertices is at most $2k$.

In application domains, Cluster Editing is also known under the name "Cor-
relation Clustering" where it can be viewed as a model for accurate unsupervised
clustering. In such context, edges to be deleted/added from a given instance are

---

[1] We assume familiarity with the notion of fixed-parameter tractability and kernelization algorithms (see [7, 8, 15]).

P. Widmayer, Y. Xu, and B. Zhu (Eds.): COCOA 2013, LNCS 8287, pp. 284–294, 2013.

considered false positives/negatives. Such errors could be small in some practical applications, and they tend to be even smaller per input object, or vertex, unless the said object is an outlier. Moreover, in some applications clusters are not expected to be too small or differ significantly in size.

Motivated by the above, we consider a parameterized version of Cluster Editing, dubbed Constrained Cluster Editing, which assumes a cluster is bounded below by some application-specific threshold and both the number of edges that can be deleted and the number of edges that can be added, per vertex, are also bounded. We refer to these two bounds by *error parameters*. We introduced these parameters in [9] where an experimental study was conducted on their effect on the running time of branching algorithms. Similar work appeared recently in [12] where the <u>total</u> number of edge-edit operations, per vertex, and the number of clusters in the target solution are used as additional parameters. We note here that setting separate bounds on the two parameters could affect the complexity as well as algorithmic solutions of the problem. Moreover, according to computational biologists [14], the expected frequency of false positives and false negatives differ in many applications.

We present a polynomial time algorithm that solves the optimization version of Cluster Edit whenever the sum of the two error parameters is smaller than half the expected minimum cluster size in the target solution. This is a rather common case whenever the clusters are of significant minimum size and error parameters are not expected to be high, being often due to noise.

We shall discuss the complexity of Constrained Cluster Editing when the error-parameters are small constants. In particular, we show that Minimum Cluster Editing is solvable in polynomial-time when at most one edge can be deleted and at most one edge can be added per vertex. Moreover, in the case where error parameters are small constants, we show that our simple reduction procedure leads to a problem kernel whose <u>size</u> is linear in the parameter k.

The paper is organized as follows: section 2 presents some preliminaries; section 3 is devoted to the main reduction procedure; our main complexity result is presented in section 4; in section 5 we study the optimization version of Cluster Editing when the error parameters are small constants, and section 6 concludes with a summary.

2   Preliminaries

We adopt common graph theoretic terminologies, such as neighborhood, degrees, adjacency, path, etc. The term non-edge is used to designate a pair of non-adjacent vertices. Given a graph $G = (V, E)$, and a set $S \subset V$, the subgraph induced by S is denoted by $G[S]$. A clique in a graph is a subgraph induced by a set of pair-wise adjacent vertices. An edge-edit operation is either a deletion or an addition of an edge. We shall use the term *cluster graph* to denote a transitive graph, which consists of a disjoint union of cliques, as connected components.

For a given graph G and parameter k, the Parameterized Cluster Editing problem asks whether G can be transformed into a cluster graph via k or less edge-edit operations. In this paper, we consider a parameterized version of this problem, called *Constrained Cluster Editing*, which can be formally defined as follows.

Input: A graph G, parameters k, s, a, d, and two functions $a : V(G) \rightarrow \{0, 1, \cdots, a\}$, $d : V(G) \rightarrow \{0, 1, \cdots, d\}$.
Question: Can G be transformed into a disjoint union of cliques by at most k edge-edit operations such that each clique is of size s or more and, for each vertex $v \in V(G)$, the number of added (deleted) edges incident on v is at most $a(v)$ ($d(v)$ respectively)?

We shall use some special terminology to help us present our algorithm. The expression *solution graph* may be used instead of cluster graph, when dealing with a specific input instance. Edges that are not allowed to be in the cluster graph are called *forbidden* edges, while edges that are (decided to be) in the solution graph are *permanent*. An induced path of length two is called a *conflict triple*, which is so named because it can never be part of a solution graph. To *cliquify* a set S of vertices is to transform $G[S]$ into a clique by adding edges.

A clique is permanent if each of its edges are permanent. To *join* a vertex v to clique C is to add all edges between v and vertices of C that are not in $N(v)$. This operation makes sense only when C is permanent or when turning C to a permanent clique. If v already contains C in its neighborhood, then *joining* v to C is equivalent to making $C \cup \{v\}$ a permanent clique. To *detach* v from C is the opposite operation (of deleting all edges between v and the vertices of C).

The first, and simplest, algorithm for Cluster Editing finds a conflict triple in the input graph and "resolves" it by exploring two cases: either delete one of the two edges in the path or insert the missing edge, In both cases, the algorithm proceeds recursively. As such, the said algorithm runs in $O(3^k)$ (3 cases per conflict triple). The same idea has been used in almost all subsequent algorithms, which added more sophisticated "branching rules."

A kernelization algorithm, with respect to an input parameter k, is a polynomial-time reduction procedure that yields an equivalent problem instance whose size, or order, is bounded by a function of the input parameter. Known kernelization algorithms for Cluster Editing have so far obtained kernels whose order (number of vertices only) is bounded by a linear function of k [5, 6, 11]. The most recent order-bound is $2k$ [6]. We shall present a simple reduction procedure that delivers kernels whose number of edges is bounded by $2(a+3d)^2k$. The same reduction procedure is used to obtain our main result, which states that Minimum Cluster Editing is solved exactly in polynomial time when $s > 2(a+d)$.

When $s = 1$, the bound on a cluster size is not important. The corresponding version of the problem can be denoted by (a, d)-Cluster Editing. This version is different from the one introduced in [12] where a bound c is placed on the total number of edge-edit operations per vertex. We refer to this problem as c-Cluster Editing. When $c > 3$, c-Cluster Editing is NP-hard (shown also in [12]).

Note that 4-Cluster Editing is not equivalent to (a, d)-Cluster Editing where $a + d = 4$. According to our definition, $a + d = 4$ means that that (a, d) is one of the five elements of $\{(i, j) : i + j = 4\}$. Therefore, the NP-hardness of c-Cluster Editing for $c \geq 4$ does not imply that (a, d)-Cluster Editing is NP-hard when $a + d = 4$. Note that, for any a, $(a, 0)$-Cluster Editing is solvable in polynomial time: any solution must add all edges to get rid of all conflict-triples, if possible.

We note here that if (a, d)-Cluster Editing is NP-hard then so is (a', d')-Cluster Edit for all $a' \geq a$ and $d' \geq d$. This follows immediately from the definition since every instance of the first is an instance of the second. We conjecture that $(a, 1)$-Cluster Editing is NP-hard for $a > 1$, and we prove in section 5 that $(1, 1)$-Cluster Editing is solvable in polynomial-time when the cluster size is at least four. We pose as open problem whether $(2, 1)$-Cluster Editing is NP-hard.

Finally, as an additional practical objective, our reduction procedure assumes that no edge can be added between two different connected components. This can be motivated by the common practice in social networks to (try to) connect people who have mutual friends/connections. This constraint may be made stronger by requiring that two vertices with disjoint neighborhoods must belong to different clusters. The latter constraint is not studied in this paper.

3 A Reduction Procedure

In general, a problem-reduction procedure is based on reduction rules, each of the form $\langle condition, action \rangle$, where *action* is an operation that can be performed to obtain an equivalent instance of the problem whenever *condition* holds. If a reduction is not possible, or the *action* violates a problem-specific constraint, then we have a no instance. Moreover, a reduction rule is sound if its action results in an equivalent instance.

Let (G, k) be an instance of (a, d, s, k)-Cluster Editing. In what follows, a set is cliquifiable if it can be made a clique by adding all pairs of missing edges without violating any of the constraints.

The main reduction rules are given below. They are assumed to be applied successively in such a way that a rule is not applied, or re-applied, until all the previous rules have been applied exhaustively. We shall prove the soundness of non-obvious reduction rules only.

3.1 Base-Case Reductions

Reduction Rule 1. *The reduction algorithm terminates and reports a no instance, whenever any of $k, a(v)$, or $d(v)$ becomes negative for some vertex $v \in V(G)$.*

Reduction Rule 2. *For any vertex v, if $d(v) = 0$ (or becomes zero), then $N(v)$ is cliquified.*

Note that applying Rule 2 may yield negative parameters, which triggers Rule 1 and causes the algorithm to terminate with a negative answer.

Reduction Rule 3. *If $a(u) = 0$, then set every non-edge of u to forbidden.*

Reduction Rule 4. *If $a(u) = d(u) = 0$, then delete $N[u]$. This results in deleting all edges connecting $N(u)$ to $V(G) \setminus N(u)$ and decrementing all the corresponding parameters.*

Soundness: In this case, $N[u]$ is cliquified by Rule 2. If the algorithm does not terminate, and since no new neighbors can be added to u, $N[u]$ is a cluster.

3.2   Reductions Based on Conflict-Triples

Reduction Rule 5. *If uv and uw are permanent edges and vw is a non-edge, then add vw and decrement each of k, $a(v)$ and $a(w)$ by one. If vw is a non-permanent edge, then set vw as permanent.*

Reduction Rule 6. *If uv is a permanent edge and uw is a forbidden edge, then set vw as forbidden. If vw exists, delete it and decrement k, $d(v)$ and $d(w)$ by one.*

3.3   Reductions Based on Common Neighbors

Reduction Rule 7. *If two non-adjacent vertices u and v have more than $d(u) + d(v)$ common neighbors (or just $> 2d$ common neighbors), then add edge uv and decrement each of k, $a(u)$ and $a(v)$ by one.*

Soundness: If u and v are not in the same clique in the solution graph, then at least one of them has to lose more than d edges, which is not possible.

Reduction Rule 8. *If two adjacent vertices, u and v, have at least $2d$ common neighbors then set uv as permanent edge.*

Soundness: Same argument as in Rule 7 above.

Reduction Rule 9. *If two vertices u and v are such that $|N(u) \setminus N(v)| > a + d$ then set edge uv as forbidden. If u and v are adjacent, then delete uv and decrement each of k, $d(u)$ and $d(v)$ by one.*

Soundness: For u and v to be in the same cluster, at most d neighbors may be deleted from $N(u)$ and at most a neighbors can be added to $N(v)$.

3.4   Reductions Based on Cluster-Size

Reduction Rule 10. *If $s > 1$ and there is a vertex v satisfying: $0 \le degree(v) < s - a(v) - 1$, then return No.*

Soundness: Obviously, v needs more than $a(v)$ edges to be a member of a cluster in a solution graph.

Reduction Rule 11. *If $s > 1$ and there is a vertex v satisfying: $0 \leq degree(v) - d(v) < s - a(v) - 1$, then set $d(v) = degree(v) - s + a + 1$.*

Soundness: If $d(v)$ edges incident on v are deleted, we get a no-instance by Rule 10.

Reduction Rule 12. *If $s > 2$ and two non-adjacent vertices u and v have less than $s - 2a$ common neighbors, then set edge uv as forbidden.*

Soundness: For u and v to belong to the same cluster, they must have at least $s - 2$ common neighbors. After adding uv, the maximum number of common neighbors we can add is $2a - 2$ ($a - 1$ edges between u and $N(v)$ and vice versa). The total number of common neighbors after adding all possible edges remains less than $s - 2 (= s - 2a + 2a - 2)$.

Reduction Rule 13. *If $s > 2$ and two adjacent vertices u and v have $< s - 2a - 2$ common neighbors, then delete edge uv.*

Soundness: The argument is similar to the previous case, except that each vertex must add at least a neighbors of the other to obtain $s - 2$ common neighbors.

3.5   Permanent and Isolated Cliques

Deleting all isolated cliques is a sound reduction rule for the general Cluster Editing problem. In our formulation we do not allow a cluster to be of size $< s$. Therefore, and since we do not allow the addition of edges between different connected components, we shall assume that an isolated clique of size $< s$ yields a no instance.

Reduction Rule 14. *If a clique C is such that $N(C) \setminus C = \emptyset$ and $|C| < s$ then we have a no-instance.*

Reduction Rule 15. *If a clique C is such that $N(C) \setminus C = \emptyset$, and $|C| \geq s$, then delete C.*

Soundness: This follows from our assumption that no edges are to be added between different connected components.

Finally, the presence of permanent cliques can yield problem reductions that are not obtained by exhaustive applications of the above reduction rules. Note that a permanent edge is a special case of a permanent clique.

Reduction Rule 16. *If a vertex v has more than d neighbors in a permanent clique C, then v is joined to C.*

Reduction Rule 17. *Let C be a permanent clique of size $> a$. If a vertex v has less than $|C| - a$ neighbors in C, then v is detached from C.*

4   Complexity of Constrained Cluster Editing

An instance (G, a, d, s, k) of Constrained Cluster Editing is said to be reduced if the above reduction rules have been exhaustively applied to the input graph G. The following key Lemma follows from the reduction procedure.

Lemma 1. *Let (G, a, d, s, k) be a reduced yes-instance of Constrained Cluster Editing. Then every vertex of G has at most $a + 3d$ neighbors.*

Proof. Assume there is a vertex v such that $|N(v)| > a + 3d$. By Rule 7, any vertex u is either a neighbor of v or has at most $2d$ common neighbors with v. In the latter case, v has more than $a + d$ vertices that are not common with u. Edge uv would then be forbidden by Rule 9. By Rules 8 and 9, every edge incident on v is either deleted or becomes permanent. Applying Rules 5 and 6 (exhaustively) leads to cliquifying and isolating $N[v]$, which then results in deleting $N[v]$.

We now consider the optimization version of Cluster Editing, which seeks a minimum number of edge-edit operations. So k is not a parameter in this case, but we keep the three constraints a, d and s.

Theorem 1. *When $s > 2(a + d)$, and $ad > 0$, the Minimum Cluster Editing problem is solvable in polynomial time.*

Proof. By Rules 7 and 12, any two vertices u, v of a reduced instance that are at distance two from each other satisfy: $s - 2a \leq |N(u) \cap N(v)| \leq 2d$. When $s - 2a > 2d$, and provided the reduction procedure does not detect a no instance, the reduction rules will automatically lead to a P_2-free graph since no two non-adjacent vertices can have common neighbors.

In practice, the total number of errors per data element is expected to be small and should be much smaller than a cluster size. In the seemingly common case where the cluster size is greater than two times such error, our theorem asserts that optimum clustering is solvable in polynomial time.

When $s \leq 2(a + d)$, the Minimum Constrained Cluster Editing problem remains NP-hard. In this case, the reduction procedure may still help to obtain faster parameterized algorithms. Moreover, when the error-parameters a and d are small constants and the size of a cluster is not important (i.e., $s = 1$), applying the above reduction rules yields equivalent instances whose size is bounded by a linear function of the main parameter k.

Theorem 2. *There is a polynomial-time reduction algorithm that takes an arbitrary instance of Constrained Cluster Editing and either determines that no solution exists or produces an equivalent instance whose order is bounded by $2k(a + 3d + 1)$.*

Proof. Let (G, a, d, k) be a reduced instance of Cluster Editing. By Lemma 1, each remaining vertex in a reduced instance has degree $\leq a + 3d$. Let A be the set of vertices of G that are incident to an edge that must be deleted or a non-edge that must be added to obtain a minimum solution. If (G, a, d, k) is a yes instance, then $|A| \leq 2k$. Let $B = N(A)$ and $C = V(G) \setminus (A \cup B)$. Observe that any member of a conflict triple is either in A or in B. It follows that C must be empty since any isolated clique is deleted in a reduced instance. The proof follows from the fact that $|A| \leq 2k$ and every vertex of A has at most $a + 3d$ neighbors in B.

The following Corollary follows easily from Theorem 2 and Lemma 1.

Corollary 1. *There is a polynomial-time reduction algorithm that takes an arbitrary instance of Constrained Cluster Editing and either determines that no solution exists or produces an equivalent instance whose size is bounded by $2k(a + 3d)^2$.*

5   The (a, d, s)-Cluster Editing Problem

In [9], a Cluster Editing algorithm was modified so that it solves $(a, d, 1, k)$-Cluster Editing. Experiments show that using the add and delete parameters improves the running time on yes instances, despite the fact that a few combinations of values of a and d were tried on each instance before finding a solution. It was observed that solutions were always found with small values of a and d. Motivated by these experiments, we discuss the complexity of the optimization version of Cluster Editing when a, d and s are small constants. We shall refer to this optimization version of the problem as (a, d, s)-Cluster Editing.

It was shown in [12] that Cluster Editing is NP-hard when the total number of edge-edit operations per vertex is four or more. Unfortunately, the result (and proof) of [12] cannot be used in studying the complexity of each of the separate cases where $a + d = 4$ ($(a, d) \in \{(0, 4), (1, 3), (2, 2), (3, 1)\}$). It was also shown in [12] that Cluster Editing is NP-hard when $a = 0$ and $d = 2$. This implies the NP-hardness of $(a, 2, s)$-Cluster Editing for $a \geq 0$ and $s \geq 1$. Motivated by these results, and by the need for non-trivial clusters, we show that our reduction procedure solves $(1, 1, s)$-Cluster Editing in polynomial time. We shall address the case $s \geq 4$ in the rest of this paper [2].

Lemma 2. *A reduced instance of $(1, 1, s)$-Cluster Editing is triangle free.*

Proof. If the reduced instance has a triangle $T = \{u, v, w\}$ that is not contained in (or does not form) a clique-cluster, then we distinguish two cases:

(i) Some vertex x of G has exactly two neighbors, say v and w, in T. This triggers Rule 8, which forces edge vw to become permanent. Then Rule 16 applies to both u and x, forcing $\{u, v, w, x\}$ to become a clique in the solution graph (if possible).

(ii) Some vertex of G has exactly one neighbor in T, which triggers Rule 17 and deletes the edge between the said vertex and T.

---

[2] It can be shown that the problem is solvable in polynomial-time for any $s > 0$, but the proof is too long to include in this paper.

Lemma 3. *In a reduced yes instance of* $(1, 1, s)$*-Cluster Editing, the maximum degree is bounded above by three.*

Proof. By Lemma 1 every vertex is of degree ≤ 4. Let v be a vertex of degree 4. Since at most one incident edge can be deleted, v must belong to a cluster that contains at least three of its neighbors. Since G is triangle-free, two edges would be needed to connect each such neighbor to the other two, which is impossible.

If $s > 4$, and by Theorem 1, the problem is completely solved by the reduction procedure since $s > 4 = 2(a + d)$. When $s = 4$, any yes-instance cannot contain a vertex of degree one. Moreover, edges incident on a degree-two vertex cannot be deleted. Therefore neighbors of a degree-two vertex must be in the same cluster. This is guaranteed by Rule 11, which sets $d(v)$ to sero for any degree-two vertex. This results in introducing triangles and triggering other rules (as described above) that yield automatic isolation (and removal) of clusters whenever possible. The remaining graph must be 3-regular.

The following reduction rule is specific to reduced (triangle-free) instances of $(1, 1, s)$-Cluster Editing.

Reduction Rule 18. *If u and v are adjacent degree-three vertices of a reduced $(1, 1, s)$-Cluster Editing instance, and none of the neighbors of u is adjacent to a neighbor of v (i.e., $N(N(u)) \cap N(v) = \emptyset$), then delete edge uv.*

Soundness: If uv is permanent, and since $d = 1$, at least one neighbor of each is in the same cluster of u and v. This requires adding two edges to a neighbor of each, which is impossible.

Lemma 4. *In a reduced yes instance of* $(1, 1, 4)$*-Cluster Editing, if two vertices have a common neighbor then they must belong to a cycle of length four.*

Proof. Every vertex of the reduced instance is of degree-three. If two adjacent vertices are not part of a C_4, then Rule 18 applies.

Since G is triangle-free and 3-regular, exactly one edge must be deleted per vertex. Moreover, every edge of G is part of a cycle of length four. It follows that edges that are not deleted by an optimum solution, if one exists, form a disjoint union of cycles of length four each (the remaining subgraph, before performing any edge addition, is 2-regular). To find an optimum solution, we transform the problem into an instance of *Independent Edge Cover*, which seeks a minimum set of edges S such that every vertex is incident on exactly one element of S. In fact, we just treat a reduced instance as an instance of Edge Cover since each and every vertex has a d value of one. We now obtain the following.

Theorem 3. $(1, 1, 4)$*-Cluster Editing is solvable in polynomial-time.*

Proof. This follows from the above reduction to Independent Edge Cover, which can be solved in polynomial-time [17].

When $s = 3$, the reduction procedure would also result in a 3-regular graph and the reduction to Independent Edge Cover can also be used. To see this note that deleting a single edge of a connected component results in subsequent reductions that turn it into a cluster graph, unless a no instance is detected. It follows that $(1, 1, 3)$-Cluster Editing is solvable in polynomial-time. We note that the case $s \geq 1$ is also solvable in polynomial-time, via a reduction to Weighted Edge Cover.

6   Conclusion

We studied a multi-parameterized version of the Cluster Editing problem and showed that a simple reduction procedure solves Minimum Cluster Editing in polynomial-time when the smallest acceptable cluster size exceeds twice the allowable edge operations per vertex. Such operations are viewed as errors or false positives/negatives in the input.

On the other hand, when the bound on the edge-edit operations per vertex is constant, we showed that a linear-size kernel is obtained. Previously known kernels have a linear bound on the number of vertices only, and they are not easily applicable to the constrained version.

We observed that (a, d, s, k)-Cluster Editing is NP-hard when $d \geq 2$, and conjectured that $(a, 1, s, k)$-Cluster Editing is NP-hard, in general, for $a \geq 2$. To prove this claim, it would be enough to show the NP-hardness for the case $a = 2$ and $d = 1$.

References

1. Böcker, S., Briesemeister, S., Bui, Q.B.A., Truss, A.: Going weighted: Parameterized algorithms for cluster editing. Theor. Comput. Sci. 410(52), 5467–5480 (2009)
2. Böcker, S.: A golden ratio parameterized algorithm for cluster editing. In: Iliopoulos, C.S., Smyth, W.F. (eds.) IWOCA 2011. LNCS, vol. 7056, pp. 85–95. Springer, Heidelberg (2011)
3. Böcker, S., Briesemeister, S., Klau, G.W.: Exact algorithms for cluster editing: Evaluation and experiments. Algorithmica 60(2), 316–334 (2011)
4. Cai, L.: Fixed-parameter tractability of graph modification problems for hereditary properties. Inf. Process. Lett. 58(4), 171–176 (1996)
5. Cao, Y., Chen, J.: Cluster editing: Kernelization based on edge cuts. Algorithmica 64(1), 152–169 (2012)
6. Chen, J., Meng, J.: A $2k$ kernel for the cluster editing problem. In: Thai, M.T., Sahni, S. (eds.) COCOON 2010. LNCS, vol. 6196, pp. 459–468. Springer, Heidelberg (2010)
7. Downey, R.G., Fellows, M.R.: Parameterized Complexity. Springer (1999)
8. Flum, J., Grohe, M.: Parameterized Complexity Theory. Springer (2006)
9. Ghrayeb, A.: Improved search-tree algorithms for the cluster edit problem. MS thesis, Lebanese American University (2011)
10. Gramm, J., Guo, J., Hüffner, F., Niedermeier, R.: Graph-modeled data clustering: Exact algorithms for clique generation. Theor. Comp. Sys. 38(4), 373–392 (2005)

11. Guo, J.: A more effective linear kernelization for cluster editing. Theor. Comput. Sci. 410(8-10), 718–726 (2009)
12. Komusiewicz, C., Uhlmann, J.: Cluster editing with locally bounded modifications. Discrete Applied Mathematics 160(15), 2259–2270 (2012)
13. Krivánek, M., Morávck, J.: NP -hard problems in hierarchical-tree clustering. Acta Inf. 23(3), 311–323 (1986)
14. Langston, M.A.: Private communication (2012)
15. Niedermeier, R.: Invitation to fixed-parameter algorithms. Oxford University Press (2006)
16. Shamir, R., Sharan, R., Tsur, D.: Cluster graph modification problems. Discrete Applied Mathematics 144(1-2), 173–182 (2004)
17. van Rooij, J.M.M., Bodlaender, H.L.: Exact algorithms for edge domination. Algorithmica 64(4), 535–563 (2012)

Fast Order-Preserving Pattern Matching

Sukhyeun Cho[1], Joong Chae Na[2], Kunsoo Park[3], and Jeong Seop Sim[1,*]

[1] School of Computer and Information Engineering, Inha University, Korea
csukhyeun@inha.edu, jssim@inha.ac.kr
[2] Department of Computer Science and Engineering, Sejong University, Korea
jcna@sejong.ac.kr
[3] School of Computer Science and Engineering, Seoul National University, Korea
kpark@theory.snu.ac.kr

Abstract. Given a text T and a pattern P, the order-preserving pattern matching (OPPM) problem is to find all substrings in T which have the same relative orders as P. The OPPM has been studied in the fields of finding some patterns affected by relative orders, not by their absolute values. For example, it can be applied to time series analysis like share prices on stock markets and to musical melody matching of two musical scores. In this paper, we present a new method of deciding the order-isomorphism between two strings even when there are same characters. Then, we show that the bad character rule of the Horspool algorithm for generic pattern matching problems can be applied to the OPPM problem. Finally, we present a fast algorithm for the OPPM problem and give experimental results to show that our algorithm is about 2 to 5 times faster than the KMP-based algorithm in reasonable cases.

Keywords: order-preserving pattern matching, order-isomorphism, Horspool algorithm.

1 Introduction

Given a text T and a pattern P, the order-preserving pattern matching (OPPM for short) problem is to find all substrings in T which have the same relative orders as P. For example, when $P = (35, 40, 23, 40, 40, 28, 30)$ and $T = (10, 20, 15, 28, 32, 12, 32, 32, 20, 25, 15, 25)$ are given, P has the same relative orders as the substring $T' = (28, 32, 12, 32, 32, 20, 25)$ of T. In T' (resp. P), the first character 28 (resp. 35) is the 4-th smallest, the second character 32 (resp. 40) is the 5-th smallest, the third character 12 (resp. 23) is the smallest, and so on. See Figure 1. The OPPM has been studied in the fields of finding some patterns affected by relative orders, not by their absolute values. For example, it can be applied to time series analysis like share prices on stock markets and to musical melody matching of two musical scores [1].

Recently, several results were presented on the OPPM problem. For the OPPM problem, the order-isomorphism must be defined. Kim et al. [1] defined the order-isomorphism as the equivalence of permutations converted from strings with an

---

* Corresponding author.

P. Widmayer, Y. Xu, and B. Zhu (Eds.): COCOA 2013, LNCS 8287, pp. 295–305, 2013.

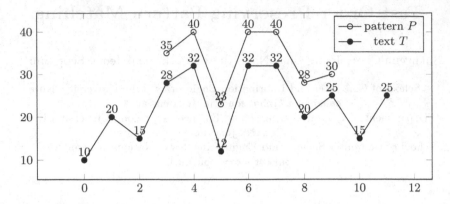

Fig. 1. An OPPM example for pattern $P = (35, 40, 23, 40, 40, 28, 30)$ and text $T = (10, 20, 15, 28, 32, 12, 32, 32, 20, 25, 15, 25)$

assumption that all the characters in a string are distinct. Given T ($|T| = n$) and P ($|P| = m$), they proposed an algorithm for the OPPM problem running in $O(n + m \log m)$ time based on the Knuth-Morris-Pratt (KMP) algorithm [4]. Meanwhile, Kubica et al. [2] defined the order-isomorphism as the equivalence of all relative orders between two strings, and presented a method of deciding the order-isomorphism of two strings even when there are same characters. They independently proposed an algorithm for the OPPM problem based on the KMP algorithm running in $O(n + m \log m)$ time for a general alphabet and $O(n + m)$ time for an integer alphabet. More recently, Crochemore et al. [3] introduced order-preserving suffix trees, and they suggested an algorithm finding all occurrences of P in T running in $O((m \log n) / \log \log n + z)$ time where z is the number of occurrences.

In this paper, we propose a fast algorithm for the OPPM problem based on the Horspool algorithm [6–8]. Experimental results show that our algorithm is about 2 to 5 times faster than the KMP-based algorithm in reasonable cases. Our contributions are as follows.

- We present a new method of deciding the order-isomorphism between two strings even when there are same characters. We show that Kubica et al.'s method [2] may decide incorrectly when there are same characters.
- We show that the bad character rule can be applied to the OPPM problem by defining groups of characters as one character. Kim et al. [1] mentioned the hardness of applying the Boyer-Moore algorithm [5] to the OPPM problem. The good suffix rule could be well-defined but the bad character rule could not be directly applied to the OPPM problem.
- We present a space-efficient algorithm computing the shift table for text search based on a factorial number system. Let q be the size of the group of characters and $|\Sigma|$ be the size of the alphabet. Then, our algorithm uses $O(q!)$ space for the shift table while the algorithms of [6, 7] for the generic pattern matching problem use $O(|\Sigma|^q)$ space for the shift table.

Table 1. $LMax_P$, $LMin_P$, $\mu(P)$ for $P = (35, 40, 23, 40, 40, 28, 30)$

i	0	1	2	3	4	5	6
$P[i]$	35	40	23	40	40	28	30
$LMax_P[i]$	-1	0	-1	1	3	2	5
$LMin_P[i]$	-1	-1	0	1	3	0	0
$\mu(P)[i]$	0	1	0	3	4	1	2

This paper is organized as follows. In Section 2, we describe the previous works related to the OPPM problem. In Section 3, we present a new method of deciding the order-isomorphism between two strings. In Section 4, we present an algorithm for the OPPM problem. In Section 5, we show experimental results comparing our algorithm with the KMP-based algorithm.

2 Preliminaries

Let Σ denote an alphabet and $\sigma = |\Sigma|$. Let $|x|$ denote the length of a string x. A string x is described by a sequence of characters $(x[0], x[1], \ldots, x[|x| - 1])$. For a string x, let a substring $x[i..j]$ be $(x[i], x[i + 1], \ldots, x[j])$.

Now, we formally define the order-isomorphism and the order-preserving pattern matching problem. Two strings x and y of the same length over Σ are called *order-isomorphic*, written $x \approx y$, if

$$x[i] \leq x[j] \Leftrightarrow y[i] \leq y[j] \text{ for } 0 \leq i, j < |x|.$$

If two strings x and y are not order-isomorphic, we write $x \not\approx y$. Given a text $T[0..n - 1]$ and a pattern $P[0..m - 1]$, we say that T *matches* P at position i if $T[i - m + 1..i] \approx P$. In the previous example shown in Figure 1, T matches P at position 9 because $T[3..9] \approx P$. The *order-preserving pattern matching problem* is to find all positions of T matched with P.

Let us define a *prefix table* $\mu(x)$ of string x:

$$\mu(x)[i] = |\{j : x[j] \leq x[i] \text{ for } 0 \leq j < i\}|.$$

For the previous example, the prefix table of P is $\mu(P)[i] = (0, 1, 0, 3, 4, 1, 2)$. See Table 1.

Lemma 1. *For two strings x and y, if $x \approx y$, then $\mu(x) = \mu(y)$.*

Proof. By the assumption that $x \approx y$, $x[i] \leq x[j] \Leftrightarrow y[i] \leq y[j]$ for $0 \leq i < j < |x|$. Hence, $\mu(x) = \mu(y)$.

Lemma 2. *Assume that $x[0..t] \approx y[0..t]$. For all $0 \leq i, j \leq t$, if $x[i] < x[j]$, then $y[i] < y[j]$, and if $x[i] = x[j]$, then $y[i] = y[j]$.*

Proof. We first prove by contradiction the first proposition (when $x[i] < x[j]$). Suppose that $y[i] \geq y[j]$. Then, by the definition of order-isomorphism, $x[i] \geq x[j]$, which contradicts the assumption that $x[i] < x[j]$.

Next, consider the case when $x[i] = x[j]$. Then, since $x[i] \leq x[j]$, $y[i] \leq y[j]$ by the definition of order-isomorphism. Moreover, since $x[j] \leq x[i]$, $y[j] \leq y[i]$. Since $y[i] \leq y[j]$ and $y[j] \leq y[i]$, $y[i] = y[j]$.

Kubica et al. [2] used location tables called $LMax$ and $LMin$ for the order information of prefixes of P:
Given a string x, for $i = 0, \dots, |x| - 1$,

$$LMax_x[i] = j \text{ if } x[j] = \max\{x[k] : k \in [0, i-1], x[k] \leq x[i]\};$$

if there is no such j then $LMax_x[i] = -1$. Similarly

$$LMin_x[i] = j \text{ if } x[j] = \min\{x[k] : k \in [0, i-1], x[k] \geq x[i]\},$$

and $LMin_x[i] = -1$ if no such j exists. If more than one such j exist, we select the rightmost one among them. Intuitively, $LMax_x[i]$ indicates the position of the largest character which is not larger than $x[i]$ in $x[0..i-1]$, and $LMin_x[i]$ indicates the position of the smallest character which is not smaller than $x[i]$ in $x[0..i-1]$. For the previous example, the location tables of P are $LMax_P[i] = (-1, 0, -1, 1, 3, 2, 5)$ and $LMin_P[i] = (-1, -1, 0, 1, 3, 0, 0)$. See Table 1. Notice the location tables of x can be computed in $O(|x|)$ time for an integer alphabet and in $O(|x| \log |x|)$ time for a general alphabet [2].

3   New Decision of Order-Isomorphism

In this section, we show that Kubica et al.'s method [2] for deciding the order-isomorphism of two strings may be incorrect when there are same characters and present a new method which works correctly even when there are same characters.

Kubica et al. [2] claimed that the order-isomorphism of two strings x and y could be decided using the location tables as follows.

Lemma 3 (see [2]). *Assume that $x[0..t] \approx y[0..t]$, $t < |x| - 1, |y| - 1$ and $a = LMax_x[t+1]$, $b = LMin_x[t+1]$. Then, $x[0..t+1] \approx y[0..t+1] \Leftrightarrow y[a] \leq y[t+1] \leq y[b]$. In case a or b is equal to -1, we omit the respective inequality in the condition.*

For example, assume two strings $x = (1, 3, 2)$, $y = (2, 5, 4)$, and the location tables $LMax_x = (-1, 0, 0)$ and $LMin_x = (-1, -1, 1)$ are given. Then, $y \approx x$ since $y[LMax_x[i]] \leq y[i] \leq y[LMin_x[i]]$ for all $0 \leq i < 3$.

However, this method may decide incorrectly when there are same characters. For example, consider two strings $x = (1, 3, 2)$ and $y = (1, 2, 2)$.

Then, $y[LMax_x[i]] \leq y[i] \leq y[LMin_x[i]]$ for all $0 \leq i < 3$. But, by the definition of order-isomorphism, $y \not\approx x$ because $x[1] \not\leq x[2]$ and $y[1] \leq y[2]$. The reasons why Lemma 3 may not hold when there are same characters in the given strings are as follows. In the proof of the necessary condition of Lemma 3, to show $x[0..t+1] \approx y[0..t+1]$ (when $y[a] \leq y[t+1] \leq y[b]$), they tried to prove that $x[i] \leq x[t+1] \Leftrightarrow y[i] \leq y[t+1]$ for $i \leq t$. For this, they proved that $x[i] \leq x[t+1] \Rightarrow y[i] \leq y[t+1]$ and $x[i] \geq x[t+1] \Rightarrow y[i] \geq y[t+1]$. But, it is not equivalent to $x[i] \leq x[t+1] \Leftrightarrow y[i] \leq y[t+1]$. Instead of the latter $x[i] \geq x[t+1] \Rightarrow y[i] \geq y[t+1]$, it should be proven that $x[i] > x[t+1] \Rightarrow y[i] > y[t+1]$. As seen in our example, however, $x[1] > x[2] \not\Rightarrow y[1] > y[2]$.

We show a new lemma for deciding whether two strings are order-isomorphic or not even when there are same characters.

Lemma 4. *Assume that $x[0..t] \approx y[0..t]$, $t < |x|-1, |y|-1$ and $a = LMax_x[t+1]$, $b = LMin_x[t+1]$. Let p be the condition $y[a] < y[t+1]$ and q be the condition $y[t+1] < y[b]$. Then, $x[0..t+1] \approx y[0..t+1] \Leftrightarrow (p \wedge q)$ or $(\neg p \wedge \neg q)$. In case a or b is equal to -1, we assume the respective condition p or q is true.*

Proof. Without loss of generality, we assume that $a \neq -1$ and $b \neq -1$. Since $x[a] \leq x[b]$ by definitions of $LMax$ and $LMin$, $y[a] \leq y[b]$ by definition of the order-isomorphism. Hence, $(\neg p \wedge \neg q)$, i.e., $y[a] \geq y[t+1] \geq y[b]$ is equal to $y[a] = y[t+1] = y[b]$.

($\Rightarrow$) By definitions of $LMax$ and $LMin$, $x[a] \leq x[t+1] \leq x[b]$. We have two cases according to whether $x[a] = x[b]$ or not.

- Case when $x[a] = x[b]$: In this case, $x[a] = x[t+1] = x[b]$. Since $x[0..t+1] \approx y[0..t+1]$, $y[a] = y[t+1] = y[b]$ by Lemma 2.

- Case when $x[a] < x[b]$: First we prove that $x[a] \neq x[t+1] \neq x[b]$. Without loss of generality, suppose $x[t+1] = x[a]$. Then, $x[a]$ is the smallest character which is not smaller than $x[t+1]$ in $x[0..t+1]$. That is, $x[a] = x[b]$, which contradicts the condition that $x[a] < x[b]$. Since $x[a] \neq x[t+1] \neq x[b]$, $x[a] < x[t+1] < x[b]$ and thus $y[a] < y[t+1] < y[b]$ by Lemma 2.

Therefore, $x[0..t+1] \approx y[0..t+1] \Rightarrow (y[a] < y[t+1] < y[b])$ or $(y[a] = y[t+1] = y[b])$.

($\Leftarrow$) Since we have already $x[0..t] \approx y[0..t]$ (assumption), to show $x[0..t+1] \approx y[0..t+1]$, we only need to prove that for all $i \leq t$,

$$x[i] \leq x[t+1] \Leftrightarrow y[i] \leq y[t+1] \text{ and } x[t+1] \leq x[i] \Leftrightarrow y[t+1] \leq y[i].$$

We only consider the former, i.e., $x[i] \leq x[t+1] \Leftrightarrow y[i] \leq y[t+1]$. (The latter can be proven in a similar way.) First, we show that $x[i] \leq x[t+1] \Rightarrow y[i] \leq y[t+1]$ when $(p \wedge q)$ or $(\neg p \wedge \neg q)$. By the definition of $LMax$, $x[i] \leq x[a]$. Since $x[0..t] \approx y[0..t]$, $y[i] \leq y[a]$. Finally, by the hypothesis $(p \wedge q)$ or $(\neg p \wedge \neg q)$, $y[a] \leq y[t+1]$. Hence, we get $y[i] \leq y[t+1]$.

Next, we show that $y[i] \leq y[t+1] \Rightarrow x[i] \leq x[t+1]$ when $(p \wedge q)$ or $(\neg p \wedge \neg q)$. We have two cases according to the hypothesis $(p \wedge q)$ or $(\neg p \wedge \neg q)$.

- Case when $y[a] = y[t + 1] = y[b]$ ($\neg p \wedge \neg q$): Since $y[i] \leq y[t + 1] = y[a]$ and $x[0..t] \approx y[0..t]$, $x[i] \leq x[a]$. Moreover, since $y[a] = y[b]$, $x[a] = x[b]$ by Lemma 2, and then $x[t + 1] = x[a] = x[b]$. Hence, $x[i] \leq x[a] = x[t + 1]$.
- Case when $y[a] < y[t + 1] < y[b]$ ($p \wedge q$): We prove it by contradiction. Suppose $x[i] > x[t + 1]$. Then, $x[i] \geq x[b]$ by the definition of $LMin$, and thus $y[i] \geq y[b]$ due to $x[0..t] \approx y[0..t]$. Moreover, since $y[b] > y[t + 1]$, we have $y[i] > y[t + 1]$. It contradicts the condition that $y[i] \leq y[t + 1]$.

Therefore, $(p \wedge q)$ or $(\neg p \wedge \neg q) \Rightarrow x[0..t + 1] \approx y[0..t + 1]$.

$\square$

For example, let us consider again the two strings $x = (1, 3, 2)$, $y = (1, 2, 2)$ and the location tables $LMax_x = (-1, 0, 0)$, $LMin_x = (-1, -1, 1)$ shown as the counter-example. Obviously, $x[0..1] \approx y[0..1]$ by the definition of the order-isomorphism. Then, $y \not\approx x$ because $y[LMax_x[2]] < y[2] = y[LMin_x[2]]$.

4 Fast Order-Preserving Pattern Matching Algorithm

4.1 Basic Idea

Basically, our algorithm for the OPPM problem is based on the Horspool algorithm widely used for generic pattern matching problems. The Horspool algorithm for generic pattern matching problems uses the shift table for filtering mismatched positions to expect sublinear behavior. (This method is well known as the bad character rule.) That is, when a mismatch occurs, the generic Horspool algorithm shifts the pattern using the shift table by setting the character of T compared with $P[m - 1]$ as the bad character.

However, as mentioned in [1], it is not easy to apply the bad character rule to the OPPM problem since the order-isomorphism is defined using the orders of characters, not just the character itself. Consider the previous example again, i.e., $T = (10, 20, 15, 28, 32, 12, 32, 32, 20, 25, 15, 25)$ and $P = (35, 40, 23, 40, 40, 28, 30)$. If we apply the generic Horspool algorithm to this case, we should compare $T[6]$ with $P[6]$ first. $T[6] \approx P[6]$ by the definition of order-isomorphism but as we can see, $T[0..6] \not\approx P$. If we set $T[6]$ as the bad character as the generic Horspool algorithm, the shift table for $T[6]$ is hard to be defined since $T[6]$, no matter what character it is, will match every character in P by the definition of the order-isomorphism.

There exist some variants of the Horspool algorithm using the notion of q-grams which consider q consecutive characters as one character [6, 7]. When a mismatch occurs, the q-gram based algorithms shift the pattern farther than the original Horspool algorithm by some modification of the shift table. Given a q-gram x and a pattern P of length m over Σ, the shift table D in [6, 7] is defined as follows:

Let $k = \max\{i \mid P[i - q + 1..i] = x$ for $q - 1 \leq i < m - 1\}$. Then,

$$D[f(x)] = \min(m - q + 1, m - k - 1). \tag{1}$$

In (1), k means the last position of P matching a q-gram x. To index the shift table D, they defined a fingerprint $f(x)$ which maps a q-gram x to an integer. Intuitively, using $f(x)$, a q-gram x is mapped to a character over an alphabet whose size is σ^q. For a q-gram x, the fingerprint $f(x)$ is defined as follows.

$$f(x) = \sum_{k=0}^{q-1} x[k] \cdot \sigma^k$$

We use q-grams to solve the hardness of defining bad characters in the OPPM. For this, we should modify the shift table and the fingerprint. Given a q-gram x and a pattern P of length m, we define the shift table D indexed by the fingerprint $f(x)$ as follows:

Let $k = \max\{i \mid \mu(P[i-q+1..i]) = \mu(x)$ for $q-1 \le i < m-1\}$. Then,

$$D[f(x)] = \min(m - q + 1, m - k - 1). \tag{2}$$

In (2), the meaning of k is the same as in (1), but we find the position of P matching a q-gram using the prefix table and a new fingerprint for space-efficiency of the shift table. Note that even if we use the prefix table instead of the location tables, we do not miss any position of P that matches the q-gram x by Lemma 1. We use a factorial number system [9] for our new fingerprint. Note that we can use the factorial number system since there are $i+1$ possible values for the i-th element of the prefix table. Refer to [9–11] for more details. For a q-gram x, we define a fingerprint $f(x)$ as follows.

$$f(x) = \sum_{k=0}^{q-1} \mu(x)[k] \cdot k! \tag{3}$$

Since the fingerprint $f(x)$ in (3) has the factorial number system, the prefix tables are uniquely mapped to integers from 0 to $q! - 1$ [9–11]. Thus, our shift table D needs $O(q!)$ space.

4.2 Search Algorithm

Our algorithm consists of two steps. In the first step, we compute the location tables $LMax_P$, $LMin_P$ and the shift table D of pattern P. As mentioned above, the location tables can be computed in $O(m \log m)$ time for a general alphabet and can be computed in $O(m)$ time for an integer alphabet [2]. To compute D, all the fingerprints of q-grams of P must be computed. For all the q-grams of P, prefix tables can be computed in $O(m q \log q)$ time using dynamic order-statistics trees [1] for a general alphabet and can be computed in $O(m q)$ time using word-encoded sets [11] for an integer alphabet where $\sigma = 2^{\lfloor w/q \rfloor - 1}$ and w is the word size. Then, after computing all the prefix tables, all the fingerprints can be computed in $O(m q)$ time by Horner's rule [4]. Finally, D can be computed in $O(q! + m q \log q)$ time [6,7]. Note that we need $O(q!)$ time for initialization of D. The first step takes $O(q! + m q \log q + m \log m)$ for a general alphabet.

Algorithm 1 shows a pseudo-code of the second step, where we search for P in T using the shift table D. Suppose we check if P matches $T[i - m + 1..i]$. We first compare the last q-grams of P and $T[i - m + 1..i]$ using their fingerprints, i.e., $f(P[m - q..m - 1])$ and $f(T[i - q + 1..i])$. If they are the same, we check the order-isomorphism of P and $T[i - m + 1..i]$ character by character using $LMax_P$ and $LMin_P$ (Lemma 4). Otherwise, we do not compare P and $T[i - m + 1..i]$ because $T[i - m + 1..i]$ cannot be order-isomorphic to P by Lemma 1. Then, we shift P forward by $D[f(T[i - q + 1..i])]$. We repeat this until P reaches the rightmost of T. Figure 2 shows a part of process of Algorithm 1 on the previous example shown in Figure 1. We first compare the fingerprints $f(T[4..6]) = 4$ and $f(P[4..6]) = 2$. Since they are distinct, we shift P by $D[f(T[4..6])] = D[4] = 3$. Next, since $f(T[7..9])$ and $f(P[4..6])$ are the same, we compare P and $T[3..9]$ using Lemma 4. Since $P \approx T[3..9]$, Algorithm 1 reports the position 9 as an occurrence. Since the second step takes $O(nm + nq \log q)$ time for a general alphabet, Algorithm 1 takes $O(nm + nq \log q + q!)$ time overall. For an integer alphabet of size $\sigma = 2^{\lfloor w/q \rfloor - 1}$ where w is the word size, Algorithm 1 takes $O(nm + nq + q!)$ time.

Algorithm 1. Text Search

1: Preprocess $D, LMax_P, LMin_P$
2: $m \leftarrow |P|, n \leftarrow |T|$
3: $t \leftarrow f(P[m - q..m - 1])$
4: $i \leftarrow m - 1$
5: **while** $i < n$ **do**
6:     $c \leftarrow f(T[i - q + 1..i])$
7:     **if** $c = t$ **then**   ▷ Compare the last q-grams
8:        **if** $T[i - m + 1..i] \approx P$ **then**
9:           print "pattern occurs at position" i
10:    $i \leftarrow i + D[c]$ ▷ Shift P by $D[c]$

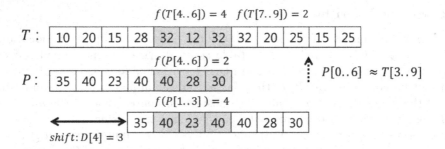

Fig. 2. Performing search on $T = (10, 20, 15, 28, 32, 12, 32, 32, 20, 25, 15, 25)$ with $P = (35, 40, 23, 40, 40, 28, 30)$ using Algorithm 1

Algorithm 2. Fingerprint Computation

$q \leftarrow |x|, c \leftarrow 0$
for $i \leftarrow q - 1$ **downto** 1 **do**
 $t \leftarrow 0$
 for $j \leftarrow 0$ **to** $i - 1$ **do**
  if $x[j] \leq x[i]$ **then** $t \leftarrow t + 1$ $\triangleright$ Compute $\mu(x)[i]$
 $c \leftarrow (c + t) \cdot i$ $\triangleright$ Horner's rule
return c

5 Experimental Results

Table 2. Search times (in seconds) for 1,000 random patterns in a random text of length 5,000,000

σ	m	5			10			15		
	q	3	4	5	3	4	5	3	4	5
2^{30}	OKMP		41.76			41.78			41.84	
	OHq	28.81	39.31	82.17	17.22	13.17	14.79	15.49	8.86	8.71
10	OKMP		41.17			41.28			41.22	
	OHq	28.75	39.50	82.57	17.39	13.26	14.82	15.79	8.99	8.75
4	OKMP		41.43			41.28			41.29	
	OHq	30.92	40.89	83.18	18.55	14.20	15.24	16.86	9.86	9.11
2	OKMP		40.46			41.10			40.90	
	OHq	37.99	47.08	86.56	24.55	19.41	18.60	21.72	14.21	11.67

We conducted experiments to compare the practical performance of our algorithm (OHq) and the KMP-based algorithm (OKMP). The KMP-based algorithm was implemented based on the algorithms of [1,2]. We checked the order-isomorphism using Lemma 4 in both algorithms. We used a naive approach (Algorithm 2) to compute the fingerprints instead of using dynamic order-statistics trees or word-encoded sets because they are less practical when implemented. Algorithm 2 runs in $O(q^2)$ time.

The experimental environments and parameters are as follows. Both algorithms were implemented in C++ and compiled with Microsoft's C/C++ compiler (x86) version 17.00.50727.1, and O2 (maximizing speed) and Oi (generating intrinsic functions) options were used as optimization options. The experiments were performed on a Windows 7 PC (64bit) with 32 GB RAM and Intel Core i7 3820 processor. We tested for a random text T of length $n = 5,000,000$ from an integer alphabet and searched for 1,000 random patterns of length $m = 5, 10, 15$, respectively. We performed experiments with varying q from 3 to 5 and $\sigma = 2^{30}, 10, 4, 2$.

Table 2 shows search times. As the pattern length m becomes longer, OHq runs faster compared to OKMP. Especially, for example, when $\sigma = 2^{30}$, $m = 15$, and $q = 5$, OHq is about 5 times faster than OKMP. Whereas when $m = 5$, OHq does not work well compared to OKMP. The reason why OHq is relatively slower in this case is because it is based on the Horspool algorithm which works better as patterns are longer and σ is larger. When $m = 5$ and $q = 5$, OKMP beats OHq for all cases because $q = m$ and q-gram technique has no effect on speedup. From our experiment, it seems that setting $q = 4$ is adequate for short patterns ($m \leq 15$). Also, it is worthy of remark that the search times for each algorithm are almost the same regardless of the alphabet size. That is, the alphabet size hardly affects the search time in the order-preserving pattern matching.

Acknowledgments. This work was supported by the National Research Foundation of Korea (NRF) grant funded by the Korea government (MSIP) (No. 2012R1A2A2A01014892). This work was supported by the IT R&D program of MSIP/KEIT [10038768, The Development of Supercomputing System for the Genome Analysis]. This work was supported by the Industrial Strategic technology development program (10041971, Development of Power Efficient High-Performance Multimedia Contents Service Technology using Context-Adapting Distributed Transcoding) funded by the Ministry of Knowledge Economy (MKE, Korea). This research was supported by Basic Science Research Program through the National Research Foundation of Korea (NRF) funded by the Ministry of Science, ICT & Future Planning (2011-0007860). This research was supported by Next-Generation Information Computing Development Program through the National Research Foundation of Korea (NRF) funded by the Ministry of Science, ICT & Future Planning (2011-0029924).

References

1. Kim, J., Eades, P., Fleischer, R., Hong, S., Iliopoulos, C.S., Park, K., Puglisi, S.J., Tokuyama, T.: Order preserving matching. CoRR, abs/1302.4064 (2013); Submitted to Theor. Comput. Sci.
2. Kubica, M., Kulczynski, T., Radoszewski, J., Rytter, W., Walen, T.: A linear time algorithm for consecutive permutation pattern matching. Information Processing Letters 113(12), 430–433 (2013)
3. Crochemore, M., Iliopoulos, C.S., Kociumaka, T., Kubica, M., Langiu, A., Pissis, S.P., Radoszewski, J., Rytter, W., Walen, T.: Order-preserving suffx trees and their algorithmic applications. CoRR, abs/1303.6872 (2013)
4. Cormen, T.H., Leiserson, C.E., Rivest, R.L., Stein, C.: Introduction to Algorithms, 3rd edn. The MIT Press (2009)
5. Boyer, R.S., Moore, J.S.: A fast string searching algorithm. Comm. ACM 20(10), 762–772 (1977)
6. Baeza-Yates, R.: Improved string searching. Software: Practice and Experience 19(3), 257–271 (1989)

7. Tarhio, J., Peltola, H.: String matching in the DNA alphabet. Software: Practice and Experience 27(7), 851–861 (1997)
8. Horspool, R.N.: Practical fast searching in strings. Software: Practice and Experience 10(6), 501–506 (1980)
9. Knuth, D.E.: The Art of Computer Programming, 3rd edn. Seminumerical Algorithms, vol. 2. Addison-Wesley (1997)
10. Myrvold, W., Ruskey, F.: Ranking and unranking permutations in linear time. Information Processing Letters 79(6), 281–284 (2001)
11. Mares, M., Straka, M.: Linear-time ranking of permutations. Algorithms-ESA, 187–193 (2007)

Scheduling for Electricity Cost in Smart Grid

Mihai Burcea[1,*], Wing-Kai Hon[2], Hsiang-Hsuan Liu[2],
Prudence W.H. Wong[1], and David K.Y. Yau[3]

[1] Department of Computer Science, University of Liverpool, UK
{m.burcea,pwong}@liverpool.ac.uk
[2] Department of Computer Science, National Tsing Hua University, Taiwan
{wkhon,hhliu}@cs.nthu.edu.tw
[3] Information Systems Technology and Design,
Singapore University of Technology and Design, Singapore
david_yau@sutd.edu.sg

Abstract. We study an offline scheduling problem arising in demand response management in smart grid. Consumers send in power requests with a flexible set of timeslots during which their requests can be served. For example, a consumer may request the dishwasher to operate for one hour during the periods 8am to 11am or 2pm to 4pm. The grid controller, upon receiving power requests, schedules each request within the specified duration. The electricity cost is measured by a convex function of the load in each timeslot. The objective of the problem is to schedule all requests with the minimum total electricity cost. As a first attempt, we consider a special case in which the power requirement and the duration a request needs service are both unit-size. For this problem, we present a polynomial time offline algorithm that gives an optimal solution and show that the time complexity can be further improved if the given set of timeslots is a contiguous interval.

1 Introduction

We study an offline scheduling problem arising in "demand response management" in smart grid [7, 9, 18]. The electrical smart grid is one of the major challenges in the 21st century [6, 28, 29]. The smart grid uses information and communication technologies in an automated fashion to improve the efficiency and reliability of production and distribution of electricity. Peak demand hours happen only for a short duration, yet makes existing electrical grid less efficient. It has been noted in [4] that in the US power grid, 10% of all generation assets and 25% of distribution infrastructure are required for less than 400 hours per year, roughly 5% of the time [29]. *Demand response management* attempts to overcome this problem by shifting users' demand to off-peak hours in order to reduce peak load [3, 12, 17, 20, 23, 25]. This is enabled technologically by the advances in smart meters [13] and integrated communication. Research initiatives in the area include GridWise [10], the SeeLoad^TM system [16], EnviroGrid^TM [24], peak demand [27], etc.

---

* Supported by EPSRC Studentship.

P. Widmayer, Y. Xu, and B. Zhu (Eds.): COCOA 2013, LNCS 8287, pp. 306–317, 2013.
© Springer International Publishing Switzerland 2013

The smart grid operator and consumers communicate through smart metering devices. We assume that time is divided into integral timeslots. A consumer sends in a power request with the power requirement, required duration of service, and the time intervals that this request can be served (giving some flexibility). For example, a consumer may want the dishwasher to operate for one hour during the periods from 8am to 11am or 2pm to 4pm. The grid operator upon receiving all requests has to schedule them in their respective time intervals using the minimum energy cost. The *load* of the grid at each timeslot is the sum of the power requirements of all requests allocated to that timeslot. The *energy cost* is modeled by a convex function on the load. As a first attempt to the problem, we consider in this paper the case that the power requirement and the duration of service requested are both unit-size, a request can specify several intervals during which the request can be served, and the power cost function is any convex function.

Previous Work. Koutsopoulos and Tassiulas [12] has formulated a similar problem to our problem where the cost function is piecewise linear. They show that the problem is NP-hard, and their proof can be adapted to show the NP-hardness of the general problem studied in this paper for which jobs have arbitrary duration or arbitrary power requirement (see elaboration in Section 6). They also presented a fractional solution and some online algorithms. Salinas et al. [25] considered a multi-objective problem to minimize energy consumption cost and maximize some utility. A closely related problem is to manage the load by changing the price of electricity over time, which has been considered in a game theoretic manner [3, 20, 23]. Heuristics have also been developed for demand side management [17]. Other aspects of smart grid have also been considered, e.g., communication [4, 14, 15], security [19]. Reviews of smart grid can be found in [7, 9, 18].

The combinatorial problem we defined in this paper has analogy to the traditional load balancing problem [2] in which the machines are like our timeslots and the jobs are like our power requests. The main difference is that the aim of load balancing is usually to minimize the maximum load of the machines. Another related problem is deadline scheduling with speed scaling [1, 31] in which the cost function is also a convex function, nevertheless a job can be served using varying speed of the processor. Two problems that are more closely related are the minimum cost maximum flow problem [5] with convex functions [21, 26] when we have unit power requirement and unit duration for each job; and the maximum-cardinality minimum-weight matching on a bipartite graph. Yet, existing algorithms for the problem cater for more general input [8, 11, 22, 30]. They are more powerful and have higher time complexity than necessary to solve our problem.

Our Contributions. In this paper we study an optimization problem in demand response management in which requests have unit power requirement, unit duration, arbitrary timeslots that the jobs can be served, and the cost function is a general convex function. We propose a polynomial time offline algorithm that gives an optimal solution. We show that the time complexity of the algorithm is $\mathcal{O}(n^2\tau)$, where n is the number of jobs and τ is the number of timeslots. We further show that if the feasible timeslots for each job to be served forms a contiguous interval, we can improve the time complexity to $\mathcal{O}(n\tau \log n)$.

Technically speaking, we use a notion of "feasible graph" to represent alternative assignments. After scheduling a job, we can look for improvement via this feasible graph. We show that we can maintain optimality each time a job is scheduled. For the analysis, we compare our schedule with an optimal schedule via the notion of "agreement graph", which captures the difference of our schedule and an optimal schedule. We then show that we can transform our schedule stepwise to improve the agreement with the optimal schedule, without increasing the cost, thus proving the optimality of our algorithm.

Organization of the Paper. Section 2 gives the definition of the problem and notions required. Section 3 describes our algorithm and its properties. In Section 4, we prove that our algorithm gives an optimal solution, while in Section 5 we prove its time complexity. We give some concluding remarks in Section 6.

2   Preliminaries

We consider an offline scheduling problem where the input consists of a set of unit-sized jobs $\mathcal{J} = \{J_1, J_2, \ldots, J_n\}$. The time is divided into integral timeslots $T = \{1, 2, 3, \ldots, \tau\}$ and each job $J_i \in \mathcal{J}$ is associated with a set of feasible timeslots $I_i \subseteq T$, in which it can be scheduled. In this model, each job J_i must be assigned to exactly one feasible timeslot from I_i. The *load* $\ell(t)$ of a timeslot t represents the total number of jobs assigned to the timeslot. We consider a general convex cost function f that measures the cost used in each timeslot t based on the load at t. The total cost used is the sum of cost over time. Over all timeslots this is $\sum_{t \in T} f(\ell(t))$. The objective is to find an assignment of all jobs in $\mathcal{J}$ to feasible timeslots such that the total cost is minimized. We first describe the notions required for discussion.

Feasible Graph. Given a particular job assignment A, we define a *feasible graph* G which is a directed multi-graph that shows the potential allocation of each job in alternative assignments. In G each timeslot is represented by a vertex. If job J_i is assigned to timeslot r in A, then for all $w \in I_i \backslash \{r\}$ we add a directed edge (r, w) with J_i as its label.

Legal-Path in a Feasible Graph. A path (t, t') in a feasible graph G is a *legal-path* if and only if the load of the starting point t is at least 2 more than the load of the ending point t', i.e., $\ell(t) - \ell(t') \geq 2$. Note that if there is a legal-path in the feasible graph G, the corresponding job assignment is not optimal.

Agreement Graph. We define an *agreement graph* $G_a(A, A^*)$ which is a directed multi-graph that measures the difference between a job assignment solution A and an optimal assignment A^*. In $G_a(A, A^*)$ each timeslot is represented by a vertex. For each job J_i such that J_i is assigned to different timeslots in A and A^*, we add an arc from t to t', where t and t' are the timeslots that J_i is assigned to by A and A^*, respectively. The arc (t, t') is labelled by the tuple $(J_i, +/-/=)$. The second value in the tuple is "+" or "−" if moving job J_i from timeslot t to timeslot t' causes the total cost of assignment A to increase or decrease, respectively. The value is "=" if moving the job does not cause any change in the total cost of assignment A.

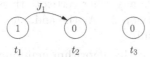

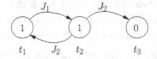

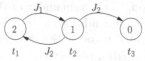

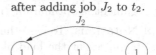

(a) The feasible graph G
after adding job J_1 to t_1.

(b) The feasible graph G
after adding job J_2 to t_2.

(c) Left: The feasible graph immediately after J_3 is added to slot t_1. The path (t_1, t_3) is a legal-path and we shift by moving J_1 to t_2 and J_2 to t_3. Right: The feasible graph after the shift, with no more legal-paths.

Fig. 1. Let $\mathcal{J} = \{J_1, J_2, J_3\}$, $T = \{t_1, t_2, t_3\}$, $I_1 = \{t_1, t_2\}$, $I_2 = \{t_1, t_2, t_3\}$, and $I_3 = \{t_1\}$. The number inside the vertices denotes their load. Suppose the algorithm schedules the jobs in order of their indices. (a) and (b) Jobs J_1 and J_2 are arbitrarily assigned their feasible minimum load slots. (c) A legal-path and the corresponding shift after assigning J_3.

Observation 1. *By moving J_i from t_1 to t_2 the overall energy cost* **(i)** *decreases if $\ell(t_1) > \ell(t_2) + 1$,* **(ii)** *remains the same if $\ell(t_1) = \ell(t_2) + 1$, and* **(iii)** *increases if $\ell(t_1) < \ell(t_2) + 1$.*

Shifting. By Observation 1, existence of a legal-path implies that the assignment is not optimal and we can execute a *shift* and decrease the total cost of the assignment. Given a legal-path P, a shift moves each job corresponding to an arc e along P from the original assigned timeslot to the timeslot determined by e. More precisely, if the path contains an arc (r, w) with J as its label, then job J is moved from r to w. It is easy to see from Observation 1 that such a shift decreases the cost, implying that the original assignment is not optimal.

On the other hand, when there is no legal-path, it is not as straightforward to show that the assignment is optimal. Nevertheless, we will prove this is the case in Lemma 6.

3   Our Algorithm

The Algorithm. We propose a polynomial time offline algorithm that minimizes the total cost (Figure 1 shows an illustration). The algorithm arranges the jobs in $\mathcal{J}$ in arbitrary order, and runs in stages. At any Stage i, we have three steps:
(1) Assign J_i to a feasible timeslot with minimum load, breaking ties arbitrarily;
(2) Suppose J_i is assigned to timeslot t. We update the feasible graph G to reflect this assignment in the following way. If applicable, we add arcs from t labelled by J_i to any other feasible timeslots (vertices) of J_i;

(3) If there exists any legal-path in G from t to any other vertex t', the algorithm executes a shift along the legal-path (see Section 2). At the end, the algorithm updates the feasible graph G to reflect this shift.

Invariants. In the next section, we show that the algorithm maintains the following two invariants. At the end of each stage:

(I1) There is no legal-path in the resulting feasible graph;

(I2) The assignment is optimal for the jobs considered so far.

Additional Notations. To ease the discussion, in the remainder of the paper, we use $\ell'_i(t)$ to represent the load of timeslot t after adding J_i (but before the shift), $\ell_i(t)$ to represent the load of timeslot t at the end of Stage i, and $\ell'_i(s,t)$ and $\ell_i(s,t)$ to represent $\ell'_i(s) - \ell'_i(t)$ and $\ell_i(s) - \ell_i(t)$, respectively.

4   Correctness

Theorem 1. *Our algorithm finds an optimal assignment.*

Framework. Consider any stage. After Step (2), there may be a legal-path in the resulting feasible graph G. In Lemma 1, we show that if a legal-path exists in G after adding J_i to timeslot r, there is at least one legal-path starting from r. Suppose the algorithm chooses the legal-path (r, t) and executes the shift along this path in Step (3). In Lemma 3, we show that if there is no legal-path in the feasible graph G before adding a job, then after adding a job and executing the corresponding shift by the algorithm, the resulting feasible graph has no legal-paths. Therefore, Step (3) of the algorithm needs to be applied only once and there will be no legal-path left, implying that Invariant (I1) holds. In Lemma 6, we show that if there is no legal-path in a feasible graph G, the corresponding assignment is optimal and hence Invariant (I2) holds.

Proof of Invariant (I1)

Lemma 1. *Suppose that before adding job J_i to timeslot r the feasible graph G has no legal-path. If there is any legal-path after adding J_i, there is at least one legal-path starting from r.*

Proof. Assume that there is a legal-path (s, t) after assigning J_i to timeslot r, so that $\ell'_i(s,t) \geq 2$. If $r = s$, we have obtained a desired legal-path. Otherwise, $r \neq s$, there are two cases:

Case 1. G contains an (s, t) path before adding J_i. Since $r \neq s$, $\ell_{i-1}(s) = \ell'_i(s)$ and $\ell_{i-1}(t) \leq \ell'_i(t)$ (the latter inequality comes from the fact that r may be equal to t). This implies $\ell_{i-1}(s,t) \geq \ell'_i(s,t) \geq 2$, which contradicts the precondition that there is no legal-path before adding J_i. Thus, Case 1 cannot occur.

Case 2. G does not contain any (s, t) path before adding J_i. Since (s, t) becomes a legal-path after adding J_i, it must be the case that assigning J_i to timeslot r adds some new edge (r, w) (with J_i as its label) to G, which connects an existing (s, r) path and an existing (w, t) path. We know that $\ell_{i-1}(s) - \ell_{i-1}(r) \leq 1$ because there is no legal-path before adding J_i. Also, $\ell'_i(s) = \ell_{i-1}(s)$ and

$\ell'_i(r) = \ell_{i-1}(r) + 1$ because the new job J_i is assigned to r, with $r \neq s$. Hence, $\ell'_i(r,t) \geq \ell'_i(s,t)$, so that the (r,t) subpath is also a legal-path. ▫

Lemma 2. *If before adding a job the feasible graph G does not have a legal-path, then after adding one more job there will be no legal-paths where the load of the starting point is at least 3 more than the load of the ending point. In other words, the load difference corresponding to any new legal-path, if it exists, is exactly 2.*

Lemma 2 will be proved in the full paper and we proceed with Invariant (I1).

Lemma 3. *Suppose that G is a feasible graph with no legal-paths. Then after adding a job and executing the corresponding shift by the algorithm, the resulting feasible graph has no legal-paths.*

Proof. Suppose that there were no legal-paths in G after Stage $i-1$, but there is a new legal-path in G after assigning J_i. By Lemma 1, there must be one such legal-path (s,t) where s is the timeslot assigned to J_i, and without loss of generality, let it be the one that is selected by our algorithm to perform the corresponding shift. Let the ordering of the vertices in the path be $[s, v_1, v_2, \ldots, v_k, t]$, and P denote the set of these vertices.

We define $In(r)$ to be the set of vertices w such that a (w,r) path exists before adding J_i, and $Out(r)$ to be the set of vertices w such that an (r,w) path exists before adding J_i. We assume that $r \in In(r)$ and $r \in Out(r)$ for the ease of later discussion. Similarly, we define $In''(r)$ to be the set of vertices w such that a (w,r) path exists after shifting, and we define $Out''(r)$ analogously. Given a set R of vertices, let $IN(R) = \bigcup_{r \in R} In(r)$ and $OUT(R) = \bigcup_{r \in R} Out(r)$. The notation $IN''(R)$ and $OUT''(R)$ are defined analogously.

Briefly speaking, we upper bound the load of a vertex in $IN''(P)$, and lower bound the load of a vertex in $OUT''(P)$, as any legal-path that may exist after the shift must start from a vertex in $IN''(P)$ and end at a vertex in $OUT''(P)$. Based on the bounds, we shall argue that there are no legal-paths as the load difference of any path after the shift will be at most 1. Note that after the shift, only the load of t is increased by one, whereas the load of any other vertex remains unchanged. Now, concerning the legal-path (s,t), there are two cases:

Case 1. There was an arc from s to v_1 in the feasible graph G before adding J_i. In this case, it is easy to check that $IN''(P) \subseteq IN(P)$,[1] and $OUT''(P) \subseteq OUT(P) \cup OUT(I_i)$.[2]

---

[1] Otherwise, let z be a vertex in $IN''(P)$ but not in $IN(P)$. Take the shortest path from z to some vertex in P after the shift. Then all the intermediate vertices of such a path are not from P. However, the jobs assigned to those intermediate vertices are unchanged, so that such a path also exists before the shift, and z is in $IN(P)$. A contradiction occurs.

[2] Otherwise, let z be a vertex in $OUT''(P)$ but not in $OUT(P) \cup OUT(I_i)$. Take the shortest path that goes to z starting from some vertex in P after the shift. Then all the intermediate vertices of such a path are not from P. If such a path does not involve vertices from I_i, then this path must exist before the shift, so that z is in $OUT(P)$. Else, z is in $OUT(I_i)$. A contradiction occurs.

Suppose that $\ell_{i-1}(s) = x$. Then, $\ell_{i-1}(t) = x - 1$ because there is no legal-path before adding J_i but there is one after adding J_i. This implies $\ell_{i-1}(v_h) \leq x$ for any $h \in [1, k]$, or there was a legal-path (v_h, t) before adding J_i. The load of any vertex in $IN(P)$ is at most x or there was a legal-path entering t before adding J_i. The load of any vertex in $OUT(P)$ is at least $x - 1$ or there was a legal-path leaving s before adding J_i. For any vertex r in I_i, $\ell_{i-1}(r) \geq x$, since $s \in I_i$ has the minimum load. This implies that the load for any vertex in $OUT(I_i)$ is at least $x - 1$, or there was a legal-path leaving a vertex in I_i before adding J_i. Thus, after the shift, the load of any vertex in $IN''(P)$ is at most x, and the load of any vertex in $OUT''(P)$ is at least $x - 1$, so no legal-paths will exist.

Case 2. There were no arcs from s to v_1 in the feasible graph G before adding J_i. In this case, J_i must be involved in the shift, so that the jobs assigned to s after the shift will be the same as if J_i was not added. Consequently, if there is still a legal-path after the shift, the starting vertex must be from $IN''(P\backslash\{s\})$, while the ending vertex must be from $OUT''(P\backslash\{s\})$. Similar to Case 1, it is easy to check that $IN''(P\backslash\{s\}) \subseteq IN(P\backslash\{s\})$ and $OUT''(P\backslash\{s\}) \subseteq OUT(P\backslash\{s\}) \cup OUT(I_i)$. Suppose that $\ell_{i-1}(s) = x$, so that $\ell'_i(s) = x + 1$. Because adding J_i creates a new legal-path (s, t), by Lemma 2, $\ell'_i(t) = \ell_{i-1}(t) = x - 1$. Thus, the load of any vertex in $IN(P\backslash\{s\})$ is at most x, since there was no legal-path entering t before adding J_i. On the other hand, $\ell_{i-1}(v_1) \geq x$ otherwise job J_i would be assigned to v_1. However, $\ell_{i-1}(v_1) \leq x$ or there is a legal-path (v_1, t). Hence, $\ell_{i-1}(v_1) = x$. This implies that the load of any vertex in $OUT(P\backslash\{s\})$ is at least $x - 1$, since there was no legal-path leaving v_1 before adding J_i. As for the vertices in $OUT(I_i)$, we can use a similar argument as in Case 1 to show that their load is at least $x - 1$. Thus, after the shift, the load of any vertex in $IN''(P\backslash\{s\})$ is at most x, and the load of any vertex in $OUT''(P\backslash\{s\})$ is at least $x - 1$, so no legal-path will exist. $\qquad\square$

Proof of Invariant (I2)

We now prove in Lemma 6 (the other key lemma for the correctness) that non-existence of legal-paths implies the assignment is optimal. The rough ideas are as follows. Consider an optimal assignment A^* (satisfying some constraints as to be defined). In Lemma 5, we show that there is a sequence of agreement graphs $G_a(A_1, A^*)$, $G_a(A_2, A^*)$, $\ldots$, $G_a(A_k, A^*)$ where the cost is non-increasing every step, A_1 is the original assignment of jobs given by our algorithm, and A_k is an optimal assignment. We prove Lemma 6 by contradiction, assuming there is no legal-path in the feasible graph G but the assignment A is not optimal. We then consider the sequence of agreement graphs given in Lemma 5 and show that either there is no agreement graph in the sequence involving strict decrease of overall cost (which means A is already optimal) or that there is a legal-path in the feasible graph G, leading to a contradiction.

Note that Lemma 5 considers an optimal assignment A^* such that $G_a(A, A^*)$ is acyclic. The existence of such A^* is stated here and proved in the full paper.

Lemma 4. *There exists an optimal assignment A^* such that $G_a(A, A^*)$ is acyclic.*

Lemma 5. *Suppose A is not optimal and A^* is an optimal assignment such that $G_a(A, A^*)$ is acyclic. Then we can have a sequence of agreement graphs $G_a(A_1, A^*), G_a(A_2, A^*), \ldots, G_a(A_k, A^*)$ such that $A_1 = A$, $A_k = A^*$, and the cost is non-increasing every step.*

Proof. Consider the agreement graph $G_a(A_i, A^*)$, for $i \geq 1$, starting from $A_1 = A$. In each step, from $G_a(A_i, A^*)$ to $G_a(A_{i+1}, A^*)$, one arc is removed. For $i \geq 1$, we consider in $G_a(A_i, A^*)$ any arc labelled with either a "$-$" or an "$=$" and we execute the move corresponding to this arc. Through this move, we remove one arc, and thus we do not introduce any new arcs. However, the $+/-/=$ label of other arcs may change. If the resulting graph $G_a(A_{i+1}, A^*)$ does not contain any more "$-$" or "$=$" arcs, we stop. Otherwise, we repeat the process.

Note that the cost is non-increasing in every step. By the time we stop, if the resulting graph, say, $G_a(A_h, A^*)$, does not contain any more arcs, we have obtained the desired sequence of agreement graphs. Otherwise, we are left only with "$+$" labelled arcs in $G_a(A_h, A^*)$; however, in the following, we shall show that such a case cannot happen, thus completing the proof of the lemma.

Firstly, $cost(A_h) \geq cost(A^*)$ since A^* is an optimal assignment. Next, by Lemma 4, $G_a(A_1, A^*)$ is acyclic and the resulting graph $G_a(A_h, A^*)$ by removing all "$-$" and "$=$" labelled arcs is also acyclic. Thus, in $G_a(A_h, A^*)$, there must exist at least one vertex with in-degree 0 and one vertex with out-degree 0. We look at all such (v_1, v_i) paths in $G_a(A_h, A^*)$, where v_1 has in-degree 0 and v_i has out-degree 0. For any such (v_1, v_i) path, we show that by executing all moves of the path (i) the overall cost is increasing, and (ii) the labels of all arcs not contained in the (v_1, v_i) path remain "$+$". After executing all moves of the path, all arcs of the (v_1, v_i) path are removed.

(i) Suppose the vertices of the path are $[v_1, v_2, \ldots, v_i]$ and $\ell(v_1) = x$. As all arcs in (v_1, v_i) are labelled with "$+$" (i.e., the cost is increasing), $\ell(v_j) \geq x$, for $j > 1$. By executing all moves in the path, $\ell(v_1) = x - 1$, $\ell(v_j)$ is unchanged, for $1 < j < i$, and $\ell(v_i)$ is increased by one. Thus, the overall cost is increasing.

(ii) We show that the labels of all arcs not contained in the (v_1, v_i) path remain "$+$". There may be out-going arcs from v_1 to other vertices not in the (v_1, v_i) path initially labelled by "$+$". Before executing all the moves in the (v_1, v_i) path, the load of all other vertices is at least x as we assume $\ell(v_1) = x$. After the move, $\ell(v_1) = x - 1$ and out-going arcs from v_1 point to vertices with load at least x. Thus, an arc from v_1 to any other vertex denotes a further increase in the cost and the labels of the arcs do not change. For vertices v_j, for $1 < j < i$, the load of v_j remains unchanged and thus the labels of the arcs incoming to or outgoing from v_j remain the same. For v_i, there may be incoming arcs. Suppose $\ell(v_i) = y$ before executing all the moves in the (v_1, v_i) path. Then the load of all other vertices pointing to v_i is at most y and the arcs are labelled by "$+$". After executing all the moves in the (v_1, v_i) path, $\ell(v_i) = y + 1$, and thus any subsequent moves from vertices pointing to v_i cause further increases in the cost, i.e., the labels do not change.

Thus, the overall cost is increasing. We repeat this process until there are no more such (v_1, v_i) paths. We end up with $cost(A_k) > cost(A^*)$, which contradicts the fact that $cost(A_k) = cost(A^*)$ as $A_k = A^*$. Thus, the case where we are left only with "+" labelled arcs in $G_a(A_h, A^*)$ cannot happen, and the lemma follows.   □

Lemma 6. *If there is no legal-path in the feasible graph G, the corresponding assignment is optimal.*

Proof. Suppose by contradiction there is no legal-path in the feasible graph G, but the corresponding assignment A is not optimal. Let $A^*, A_1 = A, A_2, \ldots, A_k = A^*$ be the assignments as defined in Lemma 5. Note that each arc in the agreement graph $G_a(A_1, A^*)$ corresponds to an arc in the feasible graph G (since G captures all possible moves). Because the sequence of agreement graphs in Lemma 5 only involves removing arcs, each arc in all of $G_a(A_i, A^*)$ corresponds to an arc in G.

Suppose $G_a(A_j, A^*)$ is the first agreement graph in which a "−" labelled arc is considered between some timeslots t_a and t_b. If there is no such arc, then A is already an optimal solution (since the sequence will be both non-increasing by Lemma 5 and non-decreasing as no "−" labelled arc is involved). Otherwise, if there is such an arc in $G_a(A_j, A^*)$, we show that there must have existed a legal-path in the feasible graph G, leading to a contradiction. We denote by $\ell(A_i, t)$ the load of timeslot t in the agreement graph $G_a(A_i, A^*)$. Suppose $\ell(A_j, t_a) = x$, then $\ell(A_j, t_b) \leq x - 2$ as the overall energy cost would be decreasing by moving a job from t_a to t_b. If $\ell(A_1, t_a) = x$ and $\ell(A_1, t_b) \leq x-2$ in the original assignment, then there is a legal-path in G, which is a contradiction. Otherwise, we claim that there are some timeslots u_{i_y} and v_{k_z} such that $\ell(A_1, u_{i_y}) \geq x$ and $\ell(A_1, v_{k_z}) \leq x - 2$, and there is a path from u_{i_y} to v_{k_z} in G. This forms a legal-path in G, leading to a contradiction.

To prove the claim, we first consider finding u_{i_y}. We first set $i_0 = j$ and $u_{i_0} = t_a$. If $\ell(A_1, u_{i_0}) \geq x$, we are done. Else, since $\ell(A_j, u_{i_0}) = x$ and $\ell(A_1, u_{i_0}) < x$, there must be some job that is moved to u_{i_0} before A_j. Let $i_1 < i_0$ be the latest step such that a job is added to u_{i_0} and the job is moved from u_{i_1}. Note that since this move corresponds to an arc with label "=", $\ell(A_{i_1}, u_{i_1}) = x$ and $\ell(A_{i_1}, u_{i_0}) = x - 1$. If $\ell(A_1, u_{i_1}) \geq x$, we are done. Otherwise, we can repeat the above argument to find u_{i_2} and so on. The process must stop at some step $i_y < i_0$ where $\ell(A_1, u_{i_y}) \geq x$. Similarly, we set $k_0 = j$ and $v_{k_0} = t_b$, so that we can find a step $k_z < k_0$ such that $\ell(A_1, v_{k_z}) \leq x - 2$. Recall that since each arc in $G_a(A_1, A^*)$ corresponds to an arc in the feasible graph G and in all subsequent agreement graphs we only remove arcs, there is a path from u_{i_y} and v_{k_z} in G. Therefore, we have found a legal-path from u_{i_y} to v_{k_z} in G.   □

5   Time Complexity

We prove the time complexity of our algorithm in Theorem 2 and show that this can be improved for the case where the feasible timeslots associated with each job are contiguous.

Theorem 2. *We can find the optimal schedule in $\mathcal{O}(n^2\tau)$ time.*

Proof. We add jobs one by one. Each round when we assign the job J_i to timeslot t, we add arcs (t, w) labelled by J_i for all vertices w that $w \in I_i$ in the feasible graph. By Lemma 1, there is a legal-path starting from t if there is a legal-path after assigning J_i to timeslot t. When J_i is assigned to t, we start breadth-first search (BFS) at t. By Lemma 2, if there is a node w which can be reached by the search and the number of jobs assigned to w is two less than the number of jobs assigned to t, it means that there is a legal-path (t, w). Then we shift the jobs according to the (t, w) legal-path. After shifting there will be no legal-paths anymore by Lemma 3. Finally we update the edges of the vertices on the legal-path in the feasible graph.

Adding J_i to the feasible graph needs $\mathcal{O}(|I_i|)$ time. Because $|I_i|$ is at most the total number of timeslots in T, $|I_i| = \mathcal{O}(\tau)$ where τ is the number of timeslots. The BFS takes $\mathcal{O}(\tau + n\tau)$ time because there are at most $n\tau$ edges in the feasible graph. If a legal-path exists after adding J_i and its length is l, the shifting needs $\mathcal{O}(l)$ time, which is $\mathcal{O}(\tau)$ because there are at most τ vertices in the legal-path. After the shift, at most $n\tau$ edges are updated in the feasible graph, taking $\mathcal{O}(n\tau)$ time. The total time for adding n jobs is thus bounded by $\mathcal{O}(n^2\tau)$. $\qquad\square$

We now consider the special case where each job $J_i \in \mathcal{J}$ is associated with an interval of contiguous timeslots $I_i = [\rho_i, \delta_i]$, for positive integers $\rho_i \leq \delta_i$, and each job J_i must be assigned to exactly one feasible timeslot s_i, for $\rho_i \leq s_i \leq \delta_i$. We give a sketch here, while the full proof can be found in the full paper.

Theorem 3. *We can find the optimal schedule in $\mathcal{O}(n\tau \log n)$ time for the case where the feasible timeslots associated with each job are contiguous.*

Proof (Sketch). For the special case, we use data structure techniques for the speed up. For each timeslot $t_i \in T$, we use two balanced binary search trees that contain the feasible intervals for all jobs assigned to t_i. For each job J_j with $I_j = [\rho_j, \delta_j]$ assigned to t_i, the first binary tree keeps the value of ρ_j, while the second binary tree keeps the value of δ_j. The binary trees are updated whenever a job is moved to and from t_i accordingly, and each such update takes $\mathcal{O}(\log n)$ time. We can query a minimum and a maximum value of the two trees, respectively, in order to establish the *directly reachable interval* of timeslot t_i, i.e., the other timeslots that jobs from t_i can be moved to. Because of the contiguous property of the feasible intervals, the set of timeslots is contiguous. We denote this interval of timeslots by $[\alpha_i, \beta_i]$ and we have that $\alpha_i \leq t_i \leq \beta_i$.

We further find the set of the ending vertices of all the paths of length at most $\tau - 1$ that start from t_i, which we call *reachable interval*. Note that the ending vertices of paths of length 2 from t_i can be found by checking the binary search trees of each timeslot in $[\alpha_i, \beta_i]$, which can then be used to find vertices at distance 3 from t_i and so on. Finding the reachable interval requires $\mathcal{O}(\tau)$ time. We can then identify any legal path in $\mathcal{O}(\tau)$ time.

In summary, adding a job to the feasible graph takes $\mathcal{O}(\log n)$ time. Finding the reachable interval and legal path takes $\mathcal{O}(\tau)$ time. Shifting of jobs along the legal path found takes $\mathcal{O}(\tau \log n)$ time. Thus the time taken to add one job is bounded by $\mathcal{O}(\tau \log n)$. The overall time for adding all n jobs is thus bounded by $\mathcal{O}(n\tau \log n)$. □

6 Conclusion

In this paper we study an offline scheduling problem arising in demand response management in smart grid. We focus on the particular case where requests have unit power requirement and unit duration. We give a polynomial time offline algorithm that gives an optimal solution. Natural generalization extends to arbitrary power requirement and arbitrary duration. The problem where requests have unit power requirement and arbitrary duration has been shown to be NP-hard [12] by a reduction from the bin packing problem. Using a similar idea, it can be shown that the problem where requests have arbitrary power requirement and unit duration is also NP-hard. An obvious research direction is to develop approximation algorithms for the general problem. It would be also interesting to consider online algorithms for the problem.

Acknowledgement. We would like to thank the reviewers for very helpful comments leading to improvement in the time complexities of our algorithms.

References

1. Albers, S.: Energy-efficient algorithms. Communication ACM 53(5), 86–96 (2010)
2. Azar, Y.: On-line load balancing. In: Fiat, A., Woeginger, G.J. (eds.) Online Algorithms 1996. LNCS, vol. 1442, pp. 178–195. Springer, Heidelberg (1998)
3. Caron, S., Kesidis, G.: Incentive-based energy consumption scheduling algorithms for the smart grid. In: IEEE Smart Grid Comm., pp. 391–396 (2010)
4. Chen, C., Nagananda, K.G., Xiong, G., Kishore, S., Snyder, L.V.: A communication-based appliance scheduling scheme for consumer-premise energy management systems. IEEE Trans. Smart Grid 4(1), 56–65 (2013)
5. Edmonds, J., Karp, R.M.: Theoretical improvements in algorithmic efficiency for network flow problems. J. ACM 19(2), 248–264 (1972)
6. European Commission. Europen smartgrids technology platform (2006), ftp://ftp.cordis.europa.eu/pub/fp7/energy/docs/smartgrids_en.pdf
7. Hamilton, K., Gulhar, N.: Taking demand response to the next level. IEEE Power and Energy Magazine 8(3), 60–65 (2010)
8. Hochbaum, D.S., Shanthikumar, J.G.: Convex separable optimization is not much harder than linear optimization. J. ACM 37(4), 843–862 (1990)
9. Ipakchi, A., Albuyeh, F.: Grid of the future. IEEE Power and Energy Magazine 7(2), 52–62 (2009)
10. Kannberg, L.D., Chassin, D.P., DeSteese, J.G., Hauser, S.G., Kintner-Meyer, M.C., Pratt, R.G., Schienbein, L.A., Warwick, W.M.: GridWiseTM: The benefits of a transformed energy system. CoRR, nlin/0409035 (September 2004)
11. Karzanov, A.V., McCormick, S.T.: Polynomial methods for separable convex optimization in unimodular linear spaces with applications. SIAM J. Comput. 26(4), 1245–1275 (1997)

12. Koutsopoulos, I., Tassiulas, L.: Control and optimization meet the smart power grid: Scheduling of power demands for optimal energy management. In: Proc. e-Energy, pp. 41–50 (2011)
13. Krishnan, R.: Meters of tomorrow (in my view). IEEE Power and Energy Magazine 6(2), 96–94 (2008)
14. Li, H., Qiu, R.C.: Need-based communication for smart grid: When to inquire power price? CoRR, abs/1003.2138 (2010)
15. Li, Z., Liang, Q.: Performance analysis of multiuser selection scheme in dynamic home area networks for smart grid communications. IEEE Trans. Smart Grid 4(1), 13–20 (2013)
16. Martin, L.: SEELoad^{TM} Solution,
    http://www.lockheedmartin.co.uk/us/
    products/energy-solutions/seesuite/seeload.html
17. Logenthiran, T., Srinivasan, D., Shun, T.Z.: Demand side management in smart grid using heuristic optimization. IEEE Trans. Smart Grid 3(3), 1244–1252 (2012)
18. Lui, T., Stirling, W., Marcy, H.: Get smart. IEEE Power and Energy Magazine 8(3), 66–78 (2010)
19. Ma, C.Y.T., Yau, D.K.Y., Rao, N.S.V.: Scalable solutions of markov games for smart-grid infrastructure protection. IEEE Trans. Smart Grid 4(1), 47–55 (2013)
20. Maharjan, S., Zhu, Q., Zhang, Y., Gjessing, S., Basar, T.: Dependable demand response management in the smart grid: A stackelberg game approach. IEEE Trans. Smart Grid 4(1), 120–132 (2013)
21. Minoux, M.: A polynomial algorithm for minimum quadratic cost flow problems. European Journal of Operational Research 18(3), 377–387 (1984)
22. Minoux, M.: Solving integer minimum cost flows with separable convex cost objective polynomially. In: Gallo, G., Sandi, C. (eds.) Netflow at Pisa. Mathematical Programming Studies, vol. 26, pp. 237–239. Springer, Heidelberg (1986)
23. Mohsenian-Rad, A.-H., Wong, V., Jatskevich, J., Schober, R.: Optimal and autonomous incentive-based energy consumption scheduling algorithm for smart grid. In: Innovative Smart Grid Technologies (ISGT) (2010)
24. REGEN Energy Inc. ENVIROGRID^{TM} SMART GRID BUNDLE.,
    http://www.regenenergy.com/press/
    announcing-the-envirogrid-smart-grid-bundle/
25. Salinas, S., Li, M., Li, P.: Multi-objective optimal energy consumption scheduling in smart grids. IEEE Trans. Smart Grid 4(1), 341–348 (2013)
26. Sokkalingam, P.T., Ahuja, R.K., Orlin, J.B.: New polynomial-time cycle-canceling algorithms for minimum-cost flows. Networks 36(1), 53–63 (2000)
27. Toronto Hydro Corporation. Peaksaver Program,
    http://www.peaksaver.com/peaksaver_THESL.html
28. UK Department of Energy & Climate Change. Smart grid: A more energy-efficient electricity supply for the UK (2013),
    https://www.gov.uk/smart-grid-a-more-energy-
    efficient-electricity-supply-for-the-uk
29. US Department of Energy. The Smart Grid: An Introduction (2009),
    http://www.oe.energy.gov/SmartGridIntroduction.htm
30. Végh, L.A.: Strongly polynomial algorithm for a class of minimum-cost flow problems with separable convex objectives. In: Proceedings of the 44th Symposium on Theory of Computing, STOC 2012, pp. 27–40. ACM, New York (2012)
31. Yao, F., Demers, A., Shenker, S.: A scheduling model for reduced CPU energy. In: Proceedings of IEEE Symposium on Foundations of Computer Science (FOCS), pp. 374–382 (1995)

Uniform-Circuit and Logarithmic-Space Approximations of Refined Combinatorial Optimization Problems

Tomoyuki Yamakami

Department of Information Science, University of Fukui
3-9-1 Bunkyo, Fukui, 910-8507 Japan

Abstract. We lay out a refined framework to discuss various approximation algorithms for combinatorial optimization problems residing inside the optimization class PO. We are focused on optimization problems characterized by computation models of uniform NC^1-circuits, uniform-AC^0, and logarithmic-space Turing machines. We present concrete optimization problems and prove that they are indeed complete under reasonably weak reductions. We also show collapses and separations among refined optimization classes.

Keywords: optimization problem, approximation-preserving reduction, approximation algorithm, NC^1 circuit, AC^0 circuit, logarithmic space.

1 Introduction

A *combinatorial optimization problem* asks to find an "optimal" solution among all feasible solutions associated with each admissible instance, where the optimality usually takes a form of either maximization or minimization according to a certain fixed ordering over all solutions. A significant progress was made in a field of fundamental research during 1990s and its trend has continued promoting our understandings of the approximability of optimization problems. In particular, *NP optimization problems* (or *NPO problems*, in short) have been a centerfold of our interests in a direct connection to NP decision problems. Let NPO express the collection of such optimization problems. NPO problems that can be exactly solved in polynomial time form a "tractable" optimization class PO, whereas APX (which is denoted in this paper by APXP for technicality) consists of NPO problems whose optimum solutions are relatively approximated within constant factors in polynomial time. A large number of NPO problems that have been studied are classified into those complexity classes.

Those classifications of optimization problems are all described from a viewpoint of polynomial-time computability and any systematic discussion on optimization problems inside PO has been vastly neglected except for [7], in which *logarithmic-space optimization problems* (or *NLO problems*) were discussed. Note that Àlvarez and Jenner [1] also studied from a slightly different viewpoint a class OptL of functions computing optimal solutions using only logarithmic space.

P. Widmayer, Y. Xu, and B. Zhu (Eds.): COCOA 2013, LNCS 8287, pp. 318–329, 2013.

A number of intriguing optimization problems have been already known to reside within PO. As a typical example, the problem, MIN ST-CUT, of finding a minimal s-t cut of a given directed graph is well-known to belong to PO. Another example is the *minimum path weight problem* (MIN PATH-WEIGHT), which is to find a path $\mathcal{S} = (v_1, v_2, \ldots, v_k)$ with $k \geq 2$ from given source v_1 to sink v_k on a given directed graph such that the path weight $\sum_{i=1}^{k} w(v_i)$ is minimum, where $w(v)$ is a given weight (expressed in binary) of vertex v. This problem is also in PO. Although MIN ST-CUT is a "complete" problem for PO, MIN PATH-WEIGHT, which is actually an NLO problem, seems to have low complexity within PO.

Here, we wish to raise a question of whether it is possible to obtain a finer classification inside PO. To achieve our goal, we seek to develop a *new, finer framework*—a low-complexity world of optimization problems—and reexamine the computational complexity of optimization problems within this new framework. In particular, we look into a world of optimization problems that can be approximated by *logarithmic-space* (or *log-space*) Turing machines and by uniform families of NC^1-*circuits*. For this purpose, we need to reshape the existing framework of expressing optimization complexity classes by clarifying the scope and complexity of verification processes for solutions and objective functions. For instance, to expand the approximation class APX into lower complexity classes, we intend to use a new notation $APXP_{NPO}$ to emphasize the polynomial-time approximability of NP optimization problems. Similarly, we describe a collection of NP optimization problems that are P-solvable as PO_{NPO}.

As logarithmic-space computation has often exhibited intriguing features in the past decades, significant differences also exist between NPO and NLO. For example, unlike NPO problems, weak computation models do not seem to support a typical reduction between minimization problems and maximization problems within NLO unless they are polynomially bounded (see Section 3.2).

The optimization class LO_{NLO} was introduced in [7] as a collection of NL optimization problems that are L-solvable (i.e., solvable by multi-tape deterministic Turing machines using logarithmic space). If we replace directed graphs of MIN PATH-WEIGHT by undirected forests, then the resulted problem, called MIN FOREST-PATH-WEIGHT, belongs to LO_{NLO}. In a similar way, using log-space uniform families of NC^1-circuits and AC^0-circuits in place of logarithmic-space Turing machines, we can define NC^1O_{NLO} and AC^0O_{NLO}, respectively.

We will present a number of concrete optimization problems that are "complete" for the aforementioned refined classes of optimization problems under weak reductions. We need such weak reductions among low-complexity optimization problems because strong reductions tend to obscure essential characteristics of "complete" problems. We will also prove relationships among those classes.

2   Optimization and Approximation Preliminaries

We will refine an existing framework for studying combinatorial optimization problems of, in particular, low computational complexity. Throughout this paper, the notation $\mathbb{N}$ denotes the set of all *natural numbers* (i.e., nonnegative integers) and $\mathbb{Q}$ indicates the set of all *rational numbers*. Two special notations $\mathbb{Q}^{>1}$ and $\mathbb{Q}^{\geq 1}$ respectively express the sets $\{q \in \mathbb{Q} \mid q > 1\}$ and $\{q \in \mathbb{Q} \mid q \geq 1\}$. Given two numbers $m, n \in \mathbb{N}$ with $m \leq n$, an *integer interval* $[m, n]_{\mathbb{Z}}$ is a set $\{m, m + 1, m + 2, \ldots, n\}$. A *string* (or a *word*) over alphabet Σ is a finite series of symbols taken from Σ. The *empty string* is denoted λ. Given a binary string w, $rep(w)$ denotes the positive integer represented by w in binary.

2.1   Models of Computation

As a model of computation, we will use the following basic form of *Turing machine*, which is equipped with a *random-access* input tape, an input-index tape, multiple work tapes, and possibly an output tape. A tape is called *read-once* if it is a read-only tape and its tape head either stays at the same cell without reading any information (whose move is called an λ-move or ε-move) or moves to the right cell to scan another symbol. Similarly, a *write-only* tape indicates that, whenever its tape head writes a nonempty symbol in a tape cell, the head should move immediately to its right tape cell. In this paper, "output tapes" are always assumed to be write-only tapes.

An *auxiliary Turing machine* is the above-mentioned deterministic Turing machine equipped with an extra read-once *auxiliary tape* on which a sequence of symbols is provided as an extra input. This machine can therefore read off two symbols at once from an input tape and an auxiliary tape to make a deterministic move. Let auxL denotes the collection of all sets A for which there exist a polynomial p and a log-space auxiliary Turing machine M such that, for every x and y, (i) $(x, y) \in A$ implies $|y| \leq p(|x|)$ and (ii) M accepts (x, y) iff $(x, y) \in A$, where y is given on an auxiliary tape. Its functional version (with polynomially-bounded output symbols) is denoted by auxFL.

We assume that the reader is familiar with four complexity classes, P, NP, L, and NL, and two function classes, FP and FL. For circuit-based complexity classes AC^0 and NC^1 (and their functional versions FAC^0 and FNC^1), we use a standard notion of *Boolean circuits*, which are composed only of three basic gates AND, OR, and NOT. A family of NC^1-circuits requires *log-space uniformity*, whereas a family of AC^0-circuits requires *DLOGTIME-uniformity*.

It is important to note that, on an output tape of a machine, a natural number is represented in binary, where the least significant bit is always placed at the right end of the output bits.

2.2   Refined Optimization Classes

Combinatorial optimization problems that we will extensively discuss in this paper can be formulated in the following manner. Since our purpose is to examine lower-complexity problems, it is better to reformulate an existing framework of *NP optimization problems* or *NPO problems* (see, e.g., [2]) in terms of auxiliary Turing machines.

An NPO problem $P = (I, SOL, m, goal)$:

○ I is a finite set of *admissible instances*. There must be a deterministic Turing machine (DTM) that recognizes I in polynomial time.

○ SOL is a function mapping I to a collection of certain finite sets, where $SOL(x)$ is a set of *feasible solutions* of input instance x. There must be a polynomial q such that (i) for every $x \in I$ and every $y \in SOL(x)$, it holds that $|y| \leq q(|x|)$ and (ii) the set $I \circ SOL = \{(x, y) \mid x \in I, y \in SOL(x)\}$ is recognized in time polynomial in $|x|$ by a certain auxiliary Turing machine stating with x on an input tape and y on an auxiliary tape.

○ *goal* is either MAX or MIN. When *goal* = MAX, P is called a *maximization problem*; when *goal* = MIN, it is a *minimization problem*.

○ m is a *measure function* (or an *objective function*) from $I \circ SOL$ to $\mathbb{N}$ whose value $m(x, y)$ is computed in time polynomial in $|x|$ by a certain auxiliary Turing machine starting with x on an input tape and y on an auxiliary tape. For any instance $x \in I$, $m^*(x)$ denotes the "optimal" value $goal\{m(x, y) \mid y \in SOL(x)\}$. Moreover, $SOL^*(x)$ expresses the "optimal" set $\{y \in SOL(x) \mid m(x, y) = m^*(x)\}$ of x.

Since a polynomial-time Turing machine can copy y into its work tape and manipulate it freely, the above use of auxiliary Turing machines does not alter the existing notion of NPO problems. Let the notation NPO express the class of all NPO problems. We say that an NPO problem P is P-*solvable* if there exists a polynomial-time deterministic algorithm M such that, for every instance $x \in I$, M returns an optimal solution y in $SOL(x)$ (possibly together with its optimal value $m^*(x)$). To analyze log-space optimization problems, Tantau [7] considered *NL optimization problems* (or *NLO problems*, in short), which are obtained simply by replacing the term "polynomial time" in the above definition of NPO problems with "logarithmic space." For NLO problems, the use of auxiliary Turing machine is essential and it may not be replaced by any Turing machine having no read-once tapes. To express the class of all NLO problems, we use the succinct notation of NLO. Moreover, MinNL (MaxNL, resp.) denotes the class of all minimization (maximization, resp.) problems in NLO. Given a class C of optimization problems, the notation PO_C expresses the class of all optimization problems in C that are P-solvable. Similarly, we can define the notions of LO_C, NC^1O_C, and AC^0O_C by replacing the term "P-solvable" with "L-solvable," "NC^1-solvable," and "AC^0-solvable," respectively. Conventionally, PO_{NPO} is written as PO, and LO_{NLO} is noted briefly as LO in [7].

A measure function m is called *polynomially bounded* if there exists a polynomial p such that $m(x, y) \leq p(|x|, |y|)$ holds for all pairs $(x, y) \in I \circ SOL$. Moreover, an optimization problem is said to be *polynomially bounded* if its measure function is polynomially bounded. We use a succinct notation PBO to denote the collection of all optimization problems that are polynomially bounded.

Next, we will define approximation classes using a notion of γ-approximation. Given an optimization problem $P = (I, SOL, m, goal)$, the *performance ratio* of solution y with respect to instance x is defined as $R(x, y) = \max\{|m(x, y)/m^*(x)|, |m^*(x)/m(x, y)|\}$, provided that neither $m(x, y)$ nor $m^*(x)$ is zero. Notice that $R(x, y) = 1$ iff $y \in SOL^*(x)$. Let $\gamma > 1$ be a constant indicating an upper bound of performance ratio. We say that P is *polynomial-time γ-approximable* if there exists a polynomial-time deterministic Turing machine M such that, for any instance x, $R(x, M(x)) \leq \gamma$. Such a machine is also called a *γ-approximate algorithm*. The γ-approximability implies that the set $\{x \in I \mid SOL(x) \neq \varnothing\}$ is in P. We also define three extra notions of "log-space γ-approximation" [7], "NC^1 γ-approximation," and "AC^0 γ-approximation" by replacing "polynomial-time Turing machine" in the above definition with "logarithmic-space (auxiliary) Turing machine," "uniform family of NC^1-circuits," and "uniform family of AC^0-circuits," respectively.

The notation $APXP_\mathcal{C}$ denotes a class consisting of problems P in class $\mathcal{C}$ of optimization problems such that, for a certain fixed constant $\gamma > 1$, P is polynomial-time γ-approximable. Similarly, we introduce the notations of $APXL_\mathcal{C}$, $APXNC_\mathcal{C}^1$, and $APXAC_\mathcal{C}^0$ using "log-space γ-approximation," "NC^1 γ-approximation," and "AC^0 γ-approximation," respectively. Notice that $APXP_{NPO}$ is conventionally expressed as APX.

2.3 Approximation-Preserving Reductions

We will use three types of reductions between two optimization problems. Given two optimization problems $P = (I_1, SOL_1, m_1, goal)$ and $Q = (I_2, SOL_2, m_2, goal)$, P is *polynomial-time AP-reducible* (or *APP-reducible*, in short) to Q, denoted $P \leq_{AP}^P Q$, if there are two functions f, g and a constant $c \geq 1$ such that the following *APP-condition* is satisfied:

- for any instance $x \in I_1$ and any $r \in \mathbb{Q}^{>1}$, it holds that $f(x, r) \in I_2$,
- for any $x \in I_1$ and any $r \in \mathbb{Q}^{>1}$, if $SOL_1(x) \neq \varnothing$ then $SOL_2(f(x, r)) \neq \varnothing$,
- for any $x \in I_1$, any $r \in \mathbb{Q}^{>1}$, and any $y \in SOL_2(f(x, r))$, it holds that $g(x, y, r) \in SOL_1(x)$,
- $f(x, r)$ and $g(x, y, r)$ are computed by two deterministic auxiliary Turing machines that run in time polynomial in $(|x|, |y|)$ for any fixed $r \in \mathbb{Q}^{>1}$, and
- for any $x \in I_1$, any $r \in \mathbb{Q}^{>1}$, and any $y \in SOL_2(f(x, r))$, $R_2(f(x, r), y) \leq r$ implies $R_1(x, g(x, y, r)) \leq 1 + c(r - 1)$.

When the above APP-condition holds, we also say that P *APP-reduces* to Q. The triplet (f, g, c) is called a *polynomial-time AP-reduction* (or an *APP-reduction*) from P to Q.

Notice that the above definition excludes the case of $r = 1$. As a result, PO_{NPO} is not closed under polynomial-time AP-reductions. Since our main target is problems inside PO_{NPO}, we further need to introduce another type of reduction (f, g), in which g "exactly" transforms in polynomial time an optimal solution for Q to another optimal solution for P. We write $P \leq_{EX}^P Q$ when the following *EX-condition* holds:

- for any instance $x \in I_1$, it holds that $f(x) \in I_2$,
- for any $x \in I_1$, if $SOL_1(x) \neq \varnothing$ then $SOL_2(f(x)) \neq \varnothing$,
- for any $x \in I_1$ and any $y \in SOL_2(f(x))$, it holds that $g(x, y) \in SOL_1(x)$,
- $f(x)$ and $g(x, y)$ are computed by two deterministic auxiliary Turing machines that run in time polynomial in $(|x|, |y|)$, and

○ for any $x \in I_1$ and $y \in SOL_2(f(x))$, $R_2(f(x), y) = 1$ implies $R_1(x, g(x, y)) = 1$, where R_1 and R_2 respectively express performance ratios for P_1 and P_2.

The above pair (f, g) is called a *polynomial-time EX-reduction* from P to Q.

By combining $\leq_{AP}^P$ and $\leq_{EX}^P$, we define the third notion of *polynomial-time strong AP-reduction* (or strong APP-reduction), denoted $\leq_{sAP}^P$, obtained from $\leq_{AP}^P$ by allowing r to be chosen from $\mathbb{Q}^{\geq 1}$ (instead of $\mathbb{Q}^{>1}$).

By replacing the requirement of "polynomial time" in the above (strong) APP-condition with "logarithmic-space," "uniform family of NC^1-circuits," and "uniform family of AC^0-circuits," we obtain *(strong) APL-reduction* ($\leq_{AP}^L$, $\leq_{sAP}^L$), *(strong) APNC^1-reduction* ($\leq_{AP}^{NC^1}$, $\leq_{sAP}^{NC^1}$), and *(strong) APAC^0-reduction* ($\leq_{AP}^{AC^0}$, $\leq_{sAP}^{AC^0}$), respectively. The following lemma is immediate.

Lemma 1. *For any reduction type* $c \in \{P, L, NC^1, AC^0\}$, $P_1 \leq_{sAP}^c P_2$ *implies both* $P_1 \leq_{AP}^c P_2$ *and* $P_1 \leq_{EX}^c P_2$.

Given a type of reduction, say, $\leq$ discussed above as well as a class $\mathcal{C}$ of optimization problems, an optimization problem P is called $\leq$-*hard* for $\mathcal{C}$ if $Q \leq P$ holds for every problem Q in $\mathcal{C}$. Moreover, P is said to be $\leq$-*complete* for $\mathcal{C}$ if P is in $\mathcal{C}$ and it is $\leq$-hard for $\mathcal{C}$.

3   Complete Problems

In Section 2, we have introduced basic classes of low-complexity optimization problems. Note that, for any given class $\mathcal{C}$ of optimization problems, $NC^1O_{\mathcal{C}} \subseteq LO_{\mathcal{C}} \subseteq PO_{\mathcal{C}}$ and $APXNC_{\mathcal{C}}^1 \subseteq APXL_{\mathcal{C}} \subseteq APXP_{\mathcal{C}}$. Moreover, it holds that $NC^1O_{\mathcal{C}} \subseteq APXNC_{\mathcal{C}}^1$, $LO_{\mathcal{C}} \subseteq APXL_{\mathcal{C}}$, and $PO_{\mathcal{C}} \subseteq APXP_{\mathcal{C}}$.

3.1   General Complete Problems

Hereafter, we will discuss complete problems for refined optimization classes. We first note that the type of reduction is often crucial. The $\leq_{AP}^L$- and $\leq_{EX}^L$-reductions are quite powerful so that all problems in $APXL_{NLO}$ and LO_{NLO} become reducible to problems even in $APXAC_{NLO}^0$ and AC^0O_{NLO}, respectively.

Proposition 1. *1.* $APXL_{NLO} = \{P \in NLO \mid \exists Q \in APXAC_{NLO}^0 \, [P \leq_{AP}^L Q]\}$.
2. $LO_{NLO} = \{P \in NLO \mid \exists Q \in AC^0O_{NLO} \, [P \leq_{EX}^L Q]\}$.

In a given graph, a *path* of G is a sequence $(v_1, v_2, \ldots, v_k)$ of vertices satisfying that (v_i, v_{i+1}) is an edge for every index $i \in [k-1]$. A path is called *simple* if there are no repeated vertices in it. The *maximum vertex weight problem* (MAX VERTEX) takes a directed graph, a source $s \in V$, and a weight function $w : V \to \mathbb{N}^+$ and finds a path from s to a certain vertex $t \in V$ so that the weight of t is maximum. It follows from [7] that MAX VERTEX is $\leq_{AP}^L$-complete for $APXL_{MaxNL}$.

Proof Sketch of Proposition 1.   We will show only (1). ($\subseteq$) Since MAX VERTEX is in $APXL_{NLO}$, take a constant $\gamma > 1$ and a log-space deterministic

Turing machine M that produces γ-approximate solutions for MAX VERTEX. Letting MAX VERTEX $= (I, SOL, m, \text{MAX})$, we modify it as follows and obtain a new problem, say, P_{max}. Instances of P_{max} are of the form (x, t_0), where $x \in I$ and $t_0 \in V$, satisfying the condition that (*) for every $v \in V$, $w(t_0) \leq w(v) \leq \gamma w(t_0)$. Consider an AC^0-circuit that outputs t_0 on input (x, t_0). Since $w(t_0) \leq m^*(x, t_0) \leq \gamma w(t_0)$, P_{max} must belong to $APXAC^0_{NLO}$.

Let $r \geq 1$ and define $f(x, r) = (x, t_0)$ and $g(x, y, r) = y$. Since t_0 can be obtained by running M on x, f is in FL. Note that the performance ratio $R_2(f(x, r), y)$ equals $R_1(x, g(x, y, r))$. Thus, MAX VERTEX APL-reduces to P. Moreover, because MAX VERTEX is $\leq^L_{AP}$-complete for $APXL_{MaxNL}$, we conclude that every maximization problem in $APXL_{NLO}$ is $\leq^L_{AP}$-reducible to P_{max}. The case of minimization is similar.

($\supseteq$) Let $P \in NLO$ and $Q \in APXAC^0_{NLO}$ satisfying $P \leq^L_{AP} Q$. It is not difficult to prove that $Q \in APXAC^0_{NLO}$ implies $P \in APXL_{NLO}$. $\square$

The complete problems presented in the proof of Proposition 1 does not seem to capture the essence of problems in $APXL_{NLO}$ as well as NLO. Therefore, in what follows, we intend to look into weaker notions of reducibilities. In particular, we want to limit our attention within $\leq^{NC^1}_{AP}$-complete and $\leq^{NC^1}_{EX}$-complete problems.

Let DSTCON denote the well-known *s-t connectivity problem* on directed graphs. Let us recall the *minimum path weight problem* (MIN PATH-WEIGHT) introduced in Section 1. Notice that, if we set a weight of every vertex of a given input graph to be 1, then MIN PATH-WEIGHT is equivalent to a problem of finding the shortest s-t path in the graph. We will prove that MIN PATH-WEIGHT is $\leq^{NC^1}_{sAP}$-complete for MinNL.

Proposition 2. MIN PATH-WEIGHT *is* $\leq^{NC^1}_{sAP}$-*complete for* MinNL.

Proof Sketch. For notational convenience, let MIN PATH-WEIGHT $= (I_0, SOL_0, m_0, \text{MIN})$. It is not difficult to show that MIN PATH-WEIGHT belongs to NLO. Next, we will show that every minimization problem in NLO is $\leq^{NC^1}_{sAP}$-reducible to MIN PATH-WEIGHT. Let $P = (I, SOL, m, \text{MIN})$ be any minimization problem in NLO. For m, we choose an appropriate log-space auxiliary Turing machine M with three tapes computing m. Recall that any solution candidate is written on M's auxiliary read-once tape. We define a *partial configuration* of M as a $\langle a, \sigma, b, \tau, c, u, d, \xi \rangle$, where an input tape-head scans σ at cell a, an auxiliary-tape head scans τ at cell b, u indicates the entire content of an $O(\log n)$-space work tape with its head scanning at cell c, and an output tape-head writes ξ in cell d, where all cell numbers are expressed in binary. The weight of this vertex is defined as ξ (expressed in binary). For convenience, we call this graph a *configuration graph* of M on input x. Let us define an instance of MIN PATH-WEIGHT as follows. Let $f(x, r)$ denote the configuration graph of M on input x. Let y be any path of the graph $f(x, r)$. Let $g(x, y, r)$ denote the content of the auxiliary tape that is reconstructed from labels attached to vertices along the path y. Clearly, f and g are in FNC^1. It is not difficult to show that $m(f(x,r),y) = m_0(x,g(x,y,r))$. Therefore, MIN PATH-WEIGHT is $\leq_{sAP}^{NC^1}$-complete for MinNL.   □

Under the assumption that $L = NL$, we can prove that the optimization problem MIN PATH-WEIGHT is $\leq_{sAP}^{NC^1}$-complete for $NLO = MaxNL \cup MinNL$.

Lemma 2. *If* $L = NL$, *then* MIN PATH-WEIGHT *is* $\leq_{sAP}^{NC^1}$-*complete for NLO.*

Proof Sketch. By Proposition 2, it suffices to show that every maximization problem P_1 in NLO is sAPAC^0-reducible to a certain minimization problem P_2 in NLO (since $AC^0 \subseteq NC^1$). Let $P_1 = (I_1, SOL_1, m_1, MAX)$ in NLO. We construct a minimization problem $P_2 = (I_2, SOL_2, m_2, MIN)$ in NLO as follows. Take an appropriate polynomial p satisfying that $b(x) = 2^{p(n)} \geq m_1^*(x)$ for every $x \in I$. Let $I_2 = I_1$ and $SOL_2 = SOL_1$. Moreover, for every $(x,y) \in I_2 \circ SOL_2$, let $m_2(x,y) = \lceil \frac{b(x)^2}{m_1(x,y)} \rceil$ if $m_1(x,y) > 0$; $b(x)^2$ otherwise. Here, we define $f(x,r) = x$ and $g(x,y,r) = y$. If $R_2(f(x,r),y) \leq r$ with $r \geq 1$, then $R_1(x,g(x,y,r))$ equals $\frac{m_1^*(x)}{m_1(x,y)}$, which is at most $\frac{b(x)^2}{m_2^*(x)+1} / \frac{b(x)^2}{m_2(x,y)} = \frac{m_2(x,y)}{m_2^*(x)} \leq \frac{rm_2^*(x)}{m_2^*(x)+1} \leq 1 + c(r-1)$, where $c = 1$, since $m_2(x,y) \leq rm_2^*(x)$.

To complete the proof, assuming that $L = NL$, we still need to prove that the measure function m_2 is in auxFL. Consider the following procedure: on input $x \in I$, guess a number e and a series of carry-on integers, check bit by bit whether $em_1(x,y) \leq b(x)^2$ and $(e+1)m_1(x,y) > b(x)^2$, check that all carry-on numbers are correct, and output e. Under the assumption of $L = NL$, this procedure can be implemented on a log-space auxiliary Turing machine.   □

When we consider an undirected-graph version of MIN PATH-WEIGHT, denoted MIN UPATH-WEIGHT, it is log-space $2^{n^{O(1)}}$-approximable because, by the result of [6], using log space, we not only determine whether there exists a feasible solution for MIN UPATH-WEIGHT but also find at least one feasible solution if any. When all admissible input graphs of MIN UPATH-WEIGHT are restricted to be forests, we call the corresponding problem MIN FOREST-PATH-WEIGHT, where a *forest* is an acyclic undirected graph.

Proposition 3. MIN FOREST-PATH-WEIGHT *is* $\leq_{EX}^{NC^1}$-*complete for* LO$_{NLO}$.

Different from standard terminology, we will define a *mixed graph* $G = (V, E)$ to be induced from a directed graph (V_1, E_1) and an undirected graph (V_2, E_2) as $V = V_1 \times V_2$ and $E = \{((v_1, v_2), (v_1', v_2')) \mid (v_1, v_1') \in E_1, (v_2, v_2') \in E_2\}$.

The *minimum mixed path weight problem* (MIN MIX-PATH-WEIGHT) takes an instance of a mixed graph $G = (V, E)$ induced from (V_1, E_1) and (V_2, E_2), a source pair $(s_1, s_2) \in V$, and a weight function $w : V \to \mathbb{N} \times \mathbb{N}$ with two extra conditions: (i) (V_2, E_2) is a forest and (ii) $w_2(v_2) \leq w_1(v_1) \leq 2w_2(v_2)$ for every $(v_1, v_2) \in V$, where $w(v_1, v_2) = (w_1(v_1), w_2(v_2))$. The problem is to find a (mixed) path $\mathcal{S}$ of G starting at (s_1, s_2) and ending at (t_1, t_2) for which the partial path weight $\sum_{(v_1,v_2) \in \mathcal{S}} w_1(v_1)$ is minimum.

Proposition 4. MIN MIX-PATH-WEIGHT *is* $\leq_{sAP}^{NC^1}$-*complete for* APXL$_{MinNL}$.

Proof Sketch. It is not difficult to show that MIN MIX-PATH-WEIGHT is in APXL$_{NLO}$ (and thus APXL$_{MinNL}$). For simplicity, we set MIN MIX-PATH-WEIGHT $= (I_0, SOL_0, m_0, MIN)$ with measure function $m_0(x, \mathcal{S}) = \sum_{(v_1,v_2)\in\mathcal{S}} w_1(y_1)$.

Next, let $P = (I, SOL, m, MIN)$ be any minimization problem in APXL$_{NLO}$. Our goal is to show that P is $\leq_{AP}^{NC^1}$-reducible to MIN MIX-PATH-WEIGHT. Let M_1 be a log-space auxiliary Turing machine computing m and let M_2 be a log-space γ-approximation algorithm for P, where $\gamma > 1$ is a constant. This implies that (*) $m^*(x)/\gamma \leq m(x, M_2(x)) \leq m^*(x)$. Let M_3 compute $m(x, M_2(x))$ using log space. As in the proof of Proposition 2, we consider a pair of partial configurations of M_1 and M_3. Those pairs constitute a mixed graph. For each $i \in [3]$, let s_i be the initial configuration of M_i and let y_i be the final and accepting configuration of M_i. The weight of a path corresponds to the value of m. Given auxiliary input pair (z_1, z_3) of M_1 and M_3, let $h(z_1, z_3)$ denote the associated series of pairs of partial configurations of M_1 and M_3, respectively. Let us define $f(x, r)$ to be $\langle G, (s_1, s_2), (t_1, t_2), w\rangle$ and let $g(x, (y_1, y_3), r) = h^{-1}(y_1, y_3)$. The desired weight function $w(v_1, v_2) = (w_1(v_1), w_2(v_2))$ is defined as follows. Note that $m(x, y) = m_0(f(x, r), (y_1, y_3))$. If $\gamma \leq 2$, then Condition (*) implies $m_0^*(f(x, r))/2 \leq m_0(f(x, r), (y_1, y_3)) \leq m_0^*(f(x, r))$. Next, we assume that $\gamma > 2$. Fix $x \in I$. Let us define $\Delta = (\gamma - 2)m(x, M_2(x))$. We define $w_2(v_2)$ to be Δ plus the output value produced by M_1 that appears inside partial configuration v_2. Similarly, let $w_1(v_1)$ be Δ plus the value outputted by M_3 inside v_1. Note that w is computed from x using log space. Let $b(x) = m(x, M_2(x))$. The ratio $\frac{m_0(f(x,r),(y_1,y_3))}{m_0^*(f(x,r))}$ equals $\frac{b(x)+\Delta}{m^*(x)+\Delta} \leq \frac{b(x)+\Delta}{\gamma b(x)+\Delta} = \frac{1}{2}$ by (*), as requested. Therefore, $P \leq_{AP}^{NC^1}$ MIN-PATH-WEIGHT holds. $\square$

3.2   Polynomially-Bounded Problems

For low-complexity optimization classes, polynomially-bounded optimization problems play a quite special role. Hereafter, we are focused on those problems.

Lemma 3. *1. Let P be a minimization (maximization, resp.) problem in* APXL$_{NLO} \cap$ PBO. *There exists a maximization (minimization, resp.) problem Q in* APXL$_{NLO} \cap$ PBO *such that P is $\leq_{sAP}^{AC^0}$-reducible to Q.*
2. *For any minimization (maximization, resp.) problem P in* NLO$\cap$PBO, *there exists a maximization (minimization, resp.) problem Q in* NLO $\cap$ PBO *such that P is $\leq_{sAP}^{AC^0}$-reducible to Q.*
3. *For any minimization (maximization, resp.) problem P in* LO$_{NLO} \cap$ PBO, *there exists a maximization (minimization, resp.) problem Q in* LO$_{NLO} \cap$ PBO *such that P is $\leq_{sAP}^{AC^0}$-reducible to Q.*

The *maximum bounded vertex weight problem* (MAX B-VERTEX) takes an undirected graph $G = (V, E)$, a source $s \in V$, and a weight function $w : V \to \mathbb{N}$ satisfying $w(v) \leq |V|$ for every $v \in V$, and finds a path of G starting at s and ending at a certain vertex t of the maximum non-zero weight.

Proposition 5. MAX B-VERTEX *is* $\leq_{\text{EX}}^{\text{NC}^1}$*-complete for* LO$_{\text{NLO}}$ ∩ PBO.

The *maximum Boolean formula value problem* (MAX BFVP) takes a set of Boolean formulas and a Boolean assignment σ for variables in the formulas and finds a maximal set of satisfied formulas by σ. Note that MAX BFVP is known to be NC^1-complete.

Lemma 4. MAX BFVP *is* $\leq_{\text{EX}}^{\text{NC}^1}$*-complete for* NC^{1}O$_{\text{NLO}}$ ∩ PBO.

If we use $\leq_{\text{AP}}^{\text{L}}$-reductions instead of $\leq_{\text{sAP}}^{\text{NC}^1}$-reductions, then it is possible to prove that APXL$_{\text{NLO}}$ contains polynomially-bounded $\leq_{\text{AP}}^{\text{L}}$-complete problems.

Lemma 5. *There exists a polynomially-bounded optimization problem that is* $\leq_{\text{AP}}^{\text{L}}$*-complete for* APXL$_{\text{NLO}}$.

Proof Sketch. It is shown in [7] that APXL$_{\text{NLO}}$ ∩ PBO has an $\leq_{\text{AP}}^{\text{L}}$-complete maximization problem. To prove the lemma, we want to show that every minimization problem $P_1 = (I_1, SOL_1, m_1, \text{MIN})$ in APXL$_{\text{NLO}}$ is $\leq_{\text{AP}}^{\text{L}}$-reducible to a certain maximization problem $P_2 = (I_2, SOL_2, m_2, \text{MAX})$ in APXL$_{\text{NLO}}$ ∩ PBO. Assume that, for an appropriate constant $\gamma > 1$, P_1 is γ-approximable by a log-space deterministic Turing machine M_1. Let $b(x) = m_1(x, M_1(x))$ for every $x \in I$. Note that $b(x)/\gamma \leq m_1^*(x) \leq b(x)$. For convenience, set $c = \gamma \log \gamma + \gamma - 1$.

Since the case where $1 + c(r-1) \geq \gamma$ is easy, we consider the other case where $1 + c(r-1) < \gamma$. For brevity, we set $\delta = 1 + c(r-1)$. Define $k = \lceil \log \gamma / \log \delta \rceil$. Note that $\delta^{k-1} \leq \gamma \leq \delta^k$. For convenience, we define $I_0 = \{((x,r,i) \mid x \in I, r \geq 1, 0 \leq i \leq k\}$ and $SOL_0(x,r,i) = \{y \in SOL_1(x) \mid m_1(x,y) \in (b(x)/\delta^{i+1}, b(x)/\delta^i]\}$. Let i_0 be the maximum integer i satisfying that $0 \leq i \leq k$ and $SOL_0(x,r,i) \neq \emptyset$. Note that $m_1^*(x) \in (b(x)/\delta^{i_0+1}, b(x)/\delta^{i_0}]$.

Let $I_2 = \{(x,r) \mid x \in I_1, r \geq 1\}$ and $SOL_2(x,r) = \{\langle y_0, y_1, \ldots, y_k\rangle \mid \exists i \in [0,k]_{\mathbb{Z}} \forall j \in [i+1,k]_{\mathbb{Z}} [y_i \in SOL_0(x,r,i) \wedge y_j \notin SOL_0(x,r,j)]\}$. Note that $I_2 \in$ L and $I_2 \circ SOL_2 \in$ auxL. We set $m_2((x,r),y) = i+1$ if $y = \langle y_0, \ldots, y_k\rangle$ and i is the maximum integer satisfying $y_i \in SOL_0(x,r,i)$. Note that $m_2 \in$ auxFL. Define $f(x,r) = (x,r)$ and $g((x,r),y,r) = y_i$ where $i = m_2((x,r),y)$. Take any $y \in SOL_2(x,r)$ for which $R_2(x,y) \leq r$. Since $m_2^*(x,r) = i_0 + 1$, it follows that $\frac{i_0+1}{r} \leq m_2((x,r),y) \leq i_0 + 1$. We then obtain $b(x)/\delta^{i_0+1} \leq m_1(x,y_i) \leq b(x)/\delta^{(i_0+1)/r}$. Thus, $R_1(x, g((x,r),y,r)) = R_1(x,y_i) = \frac{m_1(x,y_i)}{m_1^*(x)} \leq \frac{b(x)/\delta^{(i_0+1)/r}}{b(x)/\delta^{i_0+1}} = \delta^{(i_0+1)(1-1/r)}$. Since $\log z \geq \frac{z-1}{z}$ for all real numbers $z \in [1,2]$, it follows that $k \leq \frac{\log \gamma}{\log \delta} + 1 \leq \frac{\delta \log \delta}{\delta - 1} + 1 = \frac{c}{\delta - 1}$. Hence, $r = \frac{\delta - 1}{c} + 1 \leq \frac{1}{k} + 1$. From this inequality, we obtain $(i_0 + 1)(1 - 1/r) \leq (k+1)(1 - 1/r) = 1$. This implies $R_1(x, g((x,r),y,r)) \leq \delta = 1 + c(r-1)$. Therefore, $P_1 \leq_{\text{AP}}^{\text{L}} P_2$ holds. □

4 Relations among Refined Optimization Classes

We will turn our attention to relationships among basic optimization problems introduced in Section 2. We start with claiming that two classes APXP$_{\text{NLO}}$ and PO$_{\text{NLO}}$ coincide with NLO.

Lemma 6. $\mathrm{APXP_{NLO}} = \mathrm{PO_{NLO}} = \mathrm{NLO}$.

Proof Sketch. Note that $\mathrm{PO_{NLO}} \subseteq \mathrm{APXP_{NLO}}$. First, we claim that $\mathrm{APXP_{NLO}} \subseteq \mathrm{NLO}$. By the definition of $\mathrm{APXP_{NLO}}$, all problems in $\mathrm{APXP_{NLO}}$ must be NLO problems, and hence they are in NLO. Next, we show that $\mathrm{NLO} \subseteq \mathrm{PO_{NLO}}$. Let $P = (I, SOL, m, goal)$ be any problem in NLO. We consider only the case of $goal = \mathrm{MAX}$. We want to show that P also belongs to $\mathrm{PO_{NLO}}$. Let x be any instance in I. Consider the following algorithm on x. Define $D = \{(x, y) \in I \circ SOL \mid \exists z \in SOL(x)\,[z \geq y \wedge m(x, z) \geq m(x, y)]\}$, where $\geq$ is the lexicographic ordering. Note that $D \in \mathrm{NL} \subseteq P$. By a binary search technique using D, we can find a maximal solution $y \in SOL^*(x)$ in polynomial time. Therefore, $\mathrm{NLO} \subseteq \mathrm{PO_{NLO}} \subseteq \mathrm{APXP_{NLO}} \subseteq \mathrm{NLO}$. $\qquad\square$

Proposition 6. *1. [7]* $\mathrm{L} = \mathrm{NL}$ *iff* $\mathrm{LO_{NLO}} \cap \mathrm{PBO} = \mathrm{NLO} \cap \mathrm{PBO}$.
2. $\mathrm{NC^1} = \mathrm{L}$ *iff* $\mathrm{NC^1O_{NLO}} \cap \mathrm{PBO} = \mathrm{LO_{NLO}} \cap \mathrm{PBO}$.
3. $\mathrm{L} = \mathrm{P}$ *iff* $\mathrm{LO_{NPO}} \cap \mathrm{PBO} = \mathrm{PO_{NPO}} \cap \mathrm{PBO}$.

Proposition 7. *1. [7] If* $\mathrm{L} \neq \mathrm{NL}$, *then* $\mathrm{LO_{NLO}} \neq \mathrm{APXL_{NLO}} \neq \mathrm{NLO}$.
2. If $\mathrm{L} \neq \mathrm{P}$, *then* $\mathrm{PO_{NPO}} \not\subseteq \mathrm{APXL_{NPO}}$.
3. If $\mathrm{NC^1} \neq \mathrm{NL}$, *then* $\mathrm{NC^1O_{NLO}} \neq \mathrm{APXNC^1_{NLO}}$.

Proof Sketch. We will show only (3). Assume that $\mathrm{NC^1O_{NLO}} = \mathrm{APXNC^1_{NLO}}$. Consider DSTCON (on unweighted directed graphs), which is $\leq^{\mathrm{NC^1}}_m$-complete for NL. Taking a constant $\gamma > 1$, let us define a restricted version of MIN PATH-WEIGHT, called MIN REST-PATH$(\gamma) = (I, SOL, m, \mathrm{MIN})$, as follows. An instance of MIN REST-PATH is $w = \langle G, s, t, p_0 \rangle$, where $G = (V, E)$ is a directed graph, and s, t are distinct vertices in G, p_0 is a special path from s to t. Let $len(p)$ denote the *length* of a path p. A solution of w is a path p from s to t satisfying $len(p_0)/\gamma \leq len(p) \leq len(p_0)$. We use the length of a path as the measure. It is not difficult to show that MIN REST-PATH(γ) is in $\mathrm{APXAC^0_{NLO}}$, which is included in $\mathrm{APXNC^1_{NLO}}$.

By our assumption, MIN REST-PATH$(\gamma) \in \mathrm{NC^1O_{NLO}}$ for any $\gamma > 1$. Now, we take $\gamma = 2$. Given an instance $\langle G, s, t \rangle$ of DSTCON, define $\langle G', s', t, p_0 \rangle$ as follows. Let $n = |V|$ and let $G' = (V', E')$, where $V' = V \cup \{v_1, v_2, \ldots, v_n, s'\}$ and $E' = E \cup \{(s', v_1), (v_n, s), (s, w_1), (w_n, t)\} \cup \{(v_i, v_{i+1}), (w_i, w_{i+1}) \mid i \in [n-1]\}$. Moreover, let $p_0 = (s', v_1, v_2, \ldots, v_n, s, w_1, w_2, \ldots, w_n, t)$. If p is a path from s to t, let $p' = (s', v_1, \ldots, v_n, s) * p$, which is a concatenation of two paths. Since $len(p') = n + 1 + len(p)$, it follows that $n \leq len(p') \leq 2n$. Thus, an appropriate $\mathrm{NC^1}$-circuit computes the minimal path p by our assumption. If $len(p) < len(p_0)$, then we accept the input; otherwise, we reject the input.

Notice that $\langle G', s', t, p_0 \rangle$ may be quite larger than $\langle G, s, t \rangle$ and, in general, we cannot produce $\langle G, s, t, p_0 \rangle$ on a log-space work tape. However, we can avoid this pitfall as follows. Whenever a circuit needs information on vertices $\{v_1, \ldots, v_n, w_1, \ldots, w_n, s'\}$, the circuit automatically answer the question. This implies that DSTCON $\in \mathrm{NC^1}$. Since DSTCON is $\leq^{\mathrm{NC^1}}_m$-complete for NL, DSTCON $\in \mathrm{NC^1}$ implies $\mathrm{NC^1} = \mathrm{NL}$. $\qquad\square$

Proposition 8. *1.* $\mathrm{NC^1 O_{NLO}} \nsubseteq \mathrm{APXAC^0_{NLO}}$.
2. $\mathrm{AC^0 O_{NLO}} \neq \mathrm{APXAC^0_{NLO}}$.

The *parity function* π is defined as $\pi(x_1, x_2, \ldots, x_n) = x_1 \oplus x_2 \oplus \cdots \oplus x_n$, where each $x_i \in \{0, 1\}$. Let $\pi^*(x_{11}, \ldots, x_{1n}, x_{21}, \ldots, x_{2n}, \ldots, x_{n1}, \ldots, x_{nn})$ be the n-bit string $\pi(x_{11}, \ldots, x_{1n}) \pi(x_{21}, \ldots, x_{2n}) \cdots \pi(x_{n1}, \ldots, x_{nn})$. It is not difficult to show that π^* is in $\mathrm{FNC^1}$ but not in $\mathrm{FAC^0}$ because π resides in $\mathrm{NC^1} - \mathrm{AC^0}$. Given $y \in \{0, 1\}^+$, $rep(y)$ expresses *one plus* the natural number represented in binary as y.

Proof Sketch of Proposition 8. We will show only (1). Here, we consider the minimization problem $\textsc{Min M-Parity} = (I, SOL, m, \textsc{min})$ defined as follows. Let $I = \bigcup_{n \in \mathbb{N}^+} \{0, 1\}^{n^2}$ and $SOL(x) = \{y \in \{0, 1\}^n \mid rep(y) \geq rep(\pi^*(x))\}$ for each $x \in I$ with $|x| = n^2$. Let $m(x, y) = rep(y)$. Clearly, I is in L, $I \circ SOL$ is in auxL, and m is in $\mathrm{FNC^1}$. Hence, $\textsc{Min M-Parity} \in \mathrm{NLO}$. Since $SOL^*(x) = \{\pi^*(x)\}$ for every $x \in I$, it follows from $\pi^* \in \mathrm{FNC^1}$ that $\textsc{Min M-Parity}$ is $\mathrm{NC^1}$-solvable.

Next, we will prove that $\textsc{Min M-Parity} \notin \mathrm{APXAC^0_{NLO}}$. Assume otherwise. There exists a uniform family $\{C_n\}_{n \in \mathbb{N}^+}$ of $\mathrm{AC^0}$-circuits such that, for every $x \in I$, $C_{|x|}(x)$ computes a string y in $SOL(x)$ such that $(1/\gamma) rep(y) \leq rep(\pi^*(x)) \leq rep(y)$. Take any number n satisfying $2^n > \gamma$ and any string $x \in \{0, 1\}^n$. Let us consider $\pi^*(x^n)$. Define $C_{n^2}(x^n) = y$. If $\pi(x) = 1$, then we have $2^{n+1} \leq rep(y) \leq \gamma \cdot 2^{n+1}$ because of $rep(\pi^*(x^n)) = 2^{n+1}$. Since $|y| = n$, it must hold that $rep(y) = 2^{n+1}$; that is, $y = 1^n$. However, if $\pi(x) = 0$, then we obtain $1 \leq rep(y) \leq \gamma$ since $rep(\pi^*(x^n)) = 1$. Since $\gamma < 2^n$, y has the form $0z$. Hence, $\pi(x)$ equals the first bit of y. This gives an $\mathrm{AC^0}$-circuit that computes π. This is a contradiction against the fact that π is not in $\mathrm{AC^0}$. Therefore, $\textsc{Min M-Parity}$ does not belong to $\mathrm{APXAC^0_{NLO}}$. $\square$

References

1. Àlvarez, C., Jenner, B.: A very hard log-space counting class. Theoret. Comput. Sci. 107, 3–30 (1993)
2. Ausiello, G., Crescenzi, P., Gambosi, G., Kann, V., Marchetti-Spaccamela, A., Protasi, M.: Complexity and Approximation: Combinatorial Optimization Problems and Their Approximability Properties. Springer (2003)
3. Gabow, H.N.: A matroid approach to finding edge connectivity and packing arborescences. In: STOC 1991, pp. 112–122. ACM Press (1991)
4. Goldschlager, L.M., Shaw, R.A., Staples, J.: The maximum flow problem is log space complete for P. Theoret. Comput. Sci. 21, 105–111 (1982)
5. Karger, D.R.: Global min-cuts in RNC, and other ramifications of a simple min-cut algorithm. In: SODA 1993, pp. 21–30 (1993)
6. Reingold, O.: Undirected connectivity in log-space. J. ACM 55, article 17 (2008)
7. Tantau, T.: Logspace optimisation problems and their approximation properties. Theory Comput. Syst. 41, 327–350 (2007)

An Optimal Single-Machine Scheduling with Linear Deterioration Rate and Rate-Modifying Activities

Sheng Yu

School of Business Administration, Zhongnan University of Economics and Law, Wuhan, 430073, P.R. China
yusheng@znufe.edu.cn

Abstract. This paper considers a single-machine scheduling with linear deterioration rate of processing speed and multiple rate-modifying activities simultaneously. A rate-modifying activity can change the processing rate of machine under consideration, which means after each rate-modifying activity the speed of the machine is fully recovered. The integration of these two concept is motivated by human operators and semi-automatic systems that experience performance degradation over time and require rate-modifying activities for recovery. The objective is to minimize the makespan. We need to decide the sequence of jobs and when to schedule the rate-modifying activities. An optimal schedule is proposed, which can solve the problem in $O(n \log n)$ time where n is the number of jobs.

Keywords: Scheduling,Rate-modifying activity, Makespan, Linear deterioration.

1 Introduction

In many scheduling environments with human operators or semi-automatic systems, such as goods loading and unloading operations in a warehouse, the speeds of machines usually deteriorate during processing due to human fatigue and other factors, and thus a job started later consumes a longer time for satisfaction. In such environments, machine maintenance and rest time are necessary to recover the speeds or efficiencies of the machines.

Eilon [1] is the first to introduce a mathematical model, aiming to maximize the total revenue, and to determine the optimal placement and duration of a single rest period with the assumption of a linear rate of recovery. Gentzler et al. [2] further extended the model with multiple uniform length rest periods each of which ensures full recovery of the speed, and the speed of machine declines at a linear or exponential rate. Bechtold et al. [3] considered the problem to determine the optimal number, placement, and duration of rest periods during a fixed length time horizon with the assumption of linear decay of work rate and linear recovery rate.

P. Widmayer, Y. Xu, and B. Zhu (Eds.): COCOA 2013, LNCS 8287, pp. 330–339, 2013.

Lee and Leon [4] first adopted the concept of *rate-modifying activity* for machine maintenance or rest period. They study a single-machine scheduling with one rate-modifying activity, which changes the speed of the machine, inducing the processing time of $\alpha_i p_i$ for job J_i where p_i is the initial processing time of J_i before the rate-modifying activity and α_i is the activity effect for job J_i. They present polynomial algorithms for both problems of minimizing the makespan and minimizing total completion time. Lee and Lin [5] considered a single-machine scheduling such that the speed of machine changes via rate-modifying activity (or machine maintenance) and randomized machine breakdowns cause one repair operation consuming a longer time than in rate-modifying activity. Both resumable and non-resumable cases considering several objectives are studied. Lodree and Geiger [6] investigated a single-machine scheduling problem with time-dependent processing times and one rate-modifying activity which fully recovers job processing times, aiming to minimize the makespan. With the assumption that the first job is with unit length and specific definition of processing times of other jobs, they prove that the optimal policy is to schedule the activity in the middle of the job sequence. Lodree et al. [7] made a comprehensive review on scheduling with human factors, and call such scenario *the human task-sequencing problem*, which includes scheduling with rate-modifying activities.

One quite related line is that the speed of machine is constant while jobs deteriorate during waiting. Browne and Yechiali [8] first introduced the problem of scheduling deteriorating jobs on a single machine such that the *actual processing time* p_j^A of a job J_j is equal to its *initial processing time* p_j plus α_j times of the starting time s_j of the job where α_j is named the deterioration rate of job J_j, i.e., $p_j^A = p_j + \alpha_j s_j$. They prove that the schedule in non-decreasing order of p_i/α_i minimizes the makespan or total completion time. In Mosheiov [9], the actual processing time of job J_j is defined as $p_j^A = p_j + bs_j$ where b is called the *job-independent deterioration rate*. For more details on scheduling with deteriorating effect, please refer to Alidaee and Womer [10], and Cheng et al. [11]. The difference between deteriorating jobs and deteriorating machine speed lies in that the actual processing time of a job is much dependent on its starting time in the former scenario, while in the latter one, the actual processing time of a job is dependent on how long the job waits since the last rate-modifying activity.

In this paper, we investigate a single-machine scheduling with multiple rate-modifying activities and linear deterioration rate of processing speed. Similar to Gentzler et al. [2], and Lodree and Geiger [6], we assume that each rate-modifying activity fully recovers the speed of the machine. Moreover, we define a more reasonable linear deterioration function than that in Lodree and Geiger [6]. The objective is to minimize the makespan. Since there are multiple activities, the problem under consideration becomes more difficult in deciding the number of jobs to be processed between any two rate-modifying activities. We provide a polynomial time algorithm for the problem.

2   Problem Statement

There are n jobs $J_1, J_2, \ldots, J_n$ to be processed on a single machine. All the jobs are available at time zero and preemption is not allowed. The machine is with a linear job-independent deterioration rate, that is, its speed decreases linearly during processing and thus the actual processing time of a job becomes longer if it is started later. However, the speed of the machine can be fully recovered at any time via one rate-modifying activity.

There are totally $l \geq 1$ rate-modifying activities each of which consumes t units of time. No job is allowed to be processed during each rate-modifying activity, and then the l rate-modifying activities partition the whole (non-delay) processing sequence into $l+1$ subsequences π_i $(1 \leq i \leq l+1)$. Between π_i and π_{i+1} $(1 \leq i \leq l)$ is the ith rate-modifying activity with length of t. Let n_i denote the number of jobs in π_i, $J_{[i,j]}$ the jth job in π_i, and $C_{[i,j]}$ the completion time of $J_{[i,j]}$. Then $\sum_{i=1}^{l+1} n_i = n$.

Combining the idea of rate-modifying activity in Lodree and Geiger [6] and linear job-independent deterioration rate for the model without rate-modifying activities in Zhao and Tang [12], we define the *actual processing time* of job $J_{[i,j]}$ by $p_{[i,j]}^A = p_{[i,j]} + bS_{[i,j]}$, which assumes that the deterioration rate of this single-machine scheduling problem with time-dependent processing time is constant, and comparing with the position-dependent processing times set in Lodree and Geiger [6] this is more general. Here, $S_{[i,j]}$ is the sum of the actual processing times of jobs preceding $J_{[i,j]}$ in Π_i. We call $S_{[i,j]}$ the *modified starting time* and give its formulaic definition as follows: for all i $(1 \leq i \leq l+1)$,

$$S_{[i,j]} = \begin{cases} 0, & j = 1; \\ \sum_{u=1}^{j-1} p_{[i,u]}^A = (1+b)S_{[i,j-1]} + p_{[i,j-1]}, & j = 2, \ldots, n_i. \end{cases}$$

By the above definition of $p_{[i,j]}^A$, it is a linear function of $S_{[i,j]}$ but not of the starting time of job $J_{[i,j]}$, and $p_{[i,1]}^A = p_{[i,1]}$ tells that the speed of the machine is fully recovered after the $(i-1)$th rate-modifying activity.

Consider the completion time of each job. For the first job of each π_i, since $S_{[i,1]} = 0$, $C_{[1,1]} = p_{[1,1]}$ and $C_{[i,1]} = C_{[i-1,n_{i-1}]} + t + p_{[i,1]}^A = C_{[i-1,n_{i-1}]} + t + p_{[i,1]}$ $(2 \leq i \leq l+1)$. For $j = 2, 3, \ldots, n_i$ $(1 \leq i \leq l+1)$, we have $C_{[i,j]} = C_{[i,j-1]} + p_{[i,j]}^A = C_{[i,j-1]} + p_{[i,j]} + bS_{[i,j]}$. The objective of the problem is to minimize the makespan $C_{max} = C_{[l+1,n_{l+1}]}$.

We denote the problem as $1|p_{[j]}^A = p_{[j]} + bS_{[j]}, l - rms|C_{max}$, where the parameter $l - rms$ denotes that there are l rate-modifying activities. For a given processing sequences $\pi_1, \pi_2, \ldots, \pi_{l+1}$, with simple algebra, the makespan is equal to

$$C_{max} = \sum_{i=1}^{l+1} \sum_{j=1}^{n_i} p_{[i,j]}(1+b)^{n_i-j} + lt.$$

By the above formula of C_{max}, we observe that the contribution of each subsequence π_i depends on the number n_i of jobs, normal processing times of the jobs, and their processing sequence within π_i. Thus, the rest problem is to decide how to allocate each job to some suitable π_i and find optimal processing rule for the jobs within each π_i.

3   Optimal Policy for $1|p^A_{[j]} = p_{[j]} + bS_{[j]}, l - rms|C_{\max}$

We first consider the optimal processing rule for jobs in each subsequence π_i and give the following lemma.

Lemma 1. *In an optimal solution of problem* $1|p^A_{[j]} = p_{[j]} + bS_{[j]}, l - rms|C_{\max}$, *the jobs in each* π_i *are processed in the shortest processing time (SPT) order.*

Proof. We prove the lemma by adjacent pairwise interchange argument. Given a schedule $\pi = (\pi_1, \ldots, \pi_i, \ldots, \pi_{l+1})$. If π_i contains exactly one job, then the theorem is trivial. We assume that π_i contains at least two jobs, i.e., $n_i \geq 2$. Let $\pi_i = (J_{[i,1]}, \ldots, J_{[i,r]}, J_{[i,r+1]}, \ldots, J_{[i,n_i]})$ where $p_{[i,r]} \leq p_{[i,r+1]}$. We construct a corresponding schedule $\pi'_i = (\ldots, J_{[i,r-1]}, J_{[i,r+1]}, J_{[i,r]}, J_{[i,r+2]}, \ldots)$. The only difference between π'_i and π_i is the position exchange between jobs $J_{[i,r]}$ and $J_{[i,r+1]}$. To show that π_i dominates π'_i, it suffices to show that $C_{\max}(\pi_1, \ldots, \pi_i, \ldots, \pi_{l+1}) \leq C_{\max}(\pi_1, \ldots, \pi'_i, \ldots, \pi_{l+1})$. The makespans of schedules $(\pi_1, \ldots, \pi_i, \ldots, \pi_{l+1})$ and $(\pi_1, \ldots, \pi'_i, \ldots, \pi_{l+1})$ are given respectively by,

$$
\begin{aligned}
&C_{\max}(\pi_1, \ldots, \pi_i, \ldots, \pi_{l+1}) \\
&= \sum_{h=1}^{i-1}\sum_{q=1}^{n_h} p_{[h,q]}(1+b)^{n_h-q} + \sum_{h=i+1}^{l+1}\sum_{q=1}^{n_h} p_{[h,q]}(1+b)^{n_h-q} + lt \\
&\quad + \sum_{q=1}^{r-1} p_{[i,q]}(1+b)^{n_i-q} + p_{[i,r]}(1+b)^{n_i-r} + p_{[i,r+1]}(1+b)^{n_i-(r+1)} \\
&\quad + \sum_{q=r+2}^{n_i} p_{[i,q]}(1+b)^{n_i-q},
\end{aligned}
\tag{1}
$$

and

$$
\begin{aligned}
&C_{\max}(\pi_1, \ldots, \pi'_i, \ldots, \pi_{l+1}) \\
&= \sum_{h=1}^{i-1}\sum_{q=1}^{n_h} p_{[h,q]}(1+b)^{n_h-q} + \sum_{h=i+1}^{l+1}\sum_{q=1}^{n_h} p_{[h,q]}(1+b)^{n_h-q} + lt \\
&\quad + \sum_{q=1}^{r-1} p_{[i,q]}(1+b)^{n_i-q} + p_{[i,r+1]}(1+b)^{n_i-r} + p_{[i,r]}(1+b)^{n_i-(r+1)} \\
&\quad + \sum_{q=r+2}^{n_i} p_{[i,q]}(1+b)^{n_i-q}.
\end{aligned}
\tag{2}
$$

Taking the difference between (1) and (2), we have

$$C_{\max}(\pi_1, \ldots, \pi_i', \ldots, \pi_{l+1}) - C_{\max}(\pi_1, \ldots, \pi_i, \ldots, \pi_{l+1})$$
$$= (p_{[i,r+1]} - p_{[i,r]})((1+b)^{n_i - r} - (1+b)^{n_i - (r+1)})$$
$$\geq 0.$$

where the inequality is due to $p_{[i,r]} \leq p_{[i,r+1]}$ and $0 < b$. Thus, π_i dominates π_i'. Repeating this interchange argument for any two adjacent jobs in each π_i yields the theorem. □

We are now ready to deal with the optimal partition of the n jobs into $l+1$ subsequences. By arranging the n jobs in the non-decreasing order of p_j, let σ denote the sequence $(J_1, J_2, \ldots, J_n)$ where $p_1 \leq p_2 \leq \ldots \leq p_n$. Define $j^{\mathrm{mod}} = j \bmod (l+1)$. For each job J_j $(1 \leq j \leq n)$, we assign it to subsequence Π_i such that

$$i = \begin{cases} l+1, & j^{\mathrm{mod}} = 0; \\ j^{\mathrm{mod}}, & \text{otherwise.} \end{cases}$$

By Lemma 1, we schedule the jobs in each Π_i in SPT order, and assign one rate-modifying activity between any two adjacent subsequences. We call this schedule $\Pi(l) = (\Pi_1, \Pi_2, \ldots, \Pi_{l+1})$ the *coresidual schedule*. We will prove this coresidual schedule is an optimal schedule for this problem.

Assume $n = (l+1)x + y$, where $x = \lfloor n/(l+1) \rfloor$ is the quotient of n divided by $(l+1)$ and $y = n \bmod (l+1)$ is the residue of the division. According to the coresidual schedule, the qth job in Π_h is exactly the $((q-1)(l+1)+h)$th job in sequence σ, i.e., $J_{[h,q]} = J_{(q-1)(l+1)+h}$.

Theorem 1. *For problem $1|p_{[j]}^A = p_{[j]} + bS_{[j]}, l - rms|C_{\max}$, an optimal policy is as follows: process jobs in the coresidual schedule $\Pi(l)$ and assign one rate-modifying activity to the end of each Π_i $(1 \leq i \leq l)$.*

Proof. For $\Pi(l)$, if $y = 0$, then there are uniformly x jobs in each Π_u $(1 \leq u \leq l+1)$; otherwise if $y \neq 0$, then there are $x+1$ jobs in each Π_u $(1 \leq u \leq y)$ and x jobs in each Π_v $(y+1 \leq v \leq l+1)$, respectively. We consider the latter case where $y \neq 0$. For the former case it can be similarly discussed. With $y \neq 0$, the makespan of $\Pi(l)$ is given by

$$C_{\max}(\Pi(l)) = \sum_{h=1}^{y} \sum_{q=1}^{x+1} p_{(q-1)(l+1)+h}(1+b)^{x+1-q} + lt$$

$$+ \sum_{h=y+1}^{l+1} \sum_{q=1}^{x} p_{(q-1)(l+1)+h}(1+b)^{x-q}. \tag{3}$$

From (3), we observe that for any two subsequences Π_u and Π_w both containing x jobs, the qth job in Π_u and that in Π_w make a total contribution of

$[p_{(q-1)(l+1)+u} + p_{(q-1)(l+1)+w}](1+b)^{x-q}$ to the makespan, which keeps the same if the two jobs are exchanged between the two subsequences. For the case that both Π_u and Π_w contain $x+1$ jobs, the same conclusion holds with the similar reasoning. Thus, to prove the optimality of $C_{\max}(\Pi(l))$, it suffices to show that: (i) By interchanging the jth job in Π_u $(1 \leq u \leq y)$ with the jth job in Π_v $(y+1 \leq v \leq l+1)$ which produces schedule $\widetilde{\Pi}$, $C_{\max}(\widetilde{\Pi}) \geq C_{\max}(\Pi(l))$; (ii) By moving one job $J_{i(l+1)+s}$ from Π_s to the corresponding position (by SPT order) in Π_m $(s \neq m)$ which produces schedule $\hat{\Pi}$, $C_{\max}(\hat{\Pi}) \geq C_{\max}(\Pi(l))$; (iii) By interchanging the ith job in Π_s with the jth job in Π_m $(j \neq i, s \neq m)$ which produces schedule $\check{\Pi}$ by SPT order, $C_{\max}(\check{\Pi}) \geq C_{\max}(\Pi(l))$.

(a) The proof of part (i).

After the interchange, job $J_{(j-1)(l+1)+u}$ is scheduled in the jth position of Π_v and job $J_{(j-1)(l+1)+v}$ is scheduled in the jth position of Π_u, while all the other jobs in $\Pi(l)$ are processed in the same positions.

$$
\begin{aligned}
C_{\max}(\widetilde{\Pi}) = & \sum_{1 \leq h \leq y, h \neq u} \sum_{q=1}^{x+1} p_{(q-1)(l+1)+h}(1+b)^{x+1-q} \\
& + \sum_{q=1}^{j-1} p_{(q-1)(l+1)+u}(1+b)^{x+1-q} + p_{(j-1)(l+1)+v}(1+b)^{x+1-j} \\
& + \sum_{q=j+1}^{x+1} p_{(q-1)(l+1)+u}(1+b)^{x+1-q} \\
& + \sum_{y+1 \leq h \leq l+1, h \neq v} \sum_{q=1}^{x} p_{(q-1)(l+1)+h}(1+b)^{x-q} \\
& + \sum_{q=1}^{j-1} p_{(q-1)(l+1)+v}(1+b)^{x-q} + p_{(j-1)(l+1)+u}(1+b)^{x-j} \\
& + \sum_{q=j+1}^{x} p_{(q-1)(l+1)+u}(1+b)^{x-q} + lt.
\end{aligned} \tag{4}
$$

Taking the difference between (3) and (4), we have that

$$
\begin{aligned}
& C_{\max}(\widetilde{\Pi}) - C_{\max}(\Pi(l)) \\
& = \left(p_{(j-1)(l+1)+v}(1+b)^{x+1-j} + p_{(j-1)(l+1)+u}(1+b)^{x-j}\right) \\
& \quad - \left(p_{(j-1)(l+1)+u}(1+b)^{x+1-j} + p_{(j-1)(l+1)+v}(1+b)^{x-j}\right) \\
& = \left(p_{(j-1)(l+1)+v} - p_{(j-1)(l+1)+u}\right)\left((1+b)^{x+1-j} - (1+b)^{x-j}\right) \\
& > 0.
\end{aligned}
$$

where the inequality holds since $b > 0$ and $v > u$, implying $p_{(j-1)(l+1)+v} \geq p_{(j-1)(l+1)+u}$. The proof of part (i) is completed.

(b) The proof of part (ii).

There are totally six cases including $1 \leq s < m \leq y$, $1 \leq m < s \leq y$, $1 \leq s \leq y < m \leq l+1$, $1 \leq m \leq y < s \leq l+1$, $y < s < m \leq l+1$, and $y < m < s \leq l+1$. We consider the last case where $y < s < m \leq l+1$. For the other five cases, their proofs are similar to that of this case. By case condition and Lemma 1,

$$
C_{\max}(\hat{\Pi}) = \sum_{h=1}^{y} \sum_{q=1}^{x+1} p_{(q-1)(l+1)+h}(1+b)^{x+1-q}
$$

$$
+ \sum_{y+1 \leq h \leq l+1,\, h \neq s, m} \sum_{q=1}^{x} p_{(q-1)(l+1)+h}(1+b)^{x-q} + lt
$$

$$
+ \sum_{q=1}^{i-1} p_{(q-1)(l+1)+s}(1+b)^{x-1-q} + \sum_{q=i+1}^{x} p_{(q-1)(l+1)+s}(1+b)^{x-q}
$$

$$
+ \sum_{q=1}^{i-1} p_{(q-1)(l+1)+m}(1+b)^{x+1-q} + p_{(i-1)(l+1)+s}(1+b)^{x+1-i}
$$

$$
+ \sum_{q=i}^{x} p_{(q-1)(l+1)+m}(1+b)^{x-q}. \tag{5}
$$

Taking the difference between (3) and (5), we have

$$
C_{\max}(\hat{\Pi}) - C_{\max}(\Pi(l))
$$

$$
= \sum_{q=1}^{i-1} p_{(q-1)(l+1)+s}(1+b)^{x-1-q} + p_{(i-1)(l+1)+s}(1+b)^{x+1-i}
$$

$$
+ \sum_{q=1}^{i-1} p_{(q-1)(l+1)+m}(1+b)^{x+1-q}
$$

$$
- \left(\sum_{q=1}^{i} p_{(q-1)(l+1)+s}(1+b)^{x-q} + \sum_{q=1}^{i-1} p_{(q-1)(l+1)+m}(1+b)^{x-q} \right)
$$

$$
= b \sum_{(q=1}^{i-1} \left(p_{(q-1)(l+1)+m}(1+b)^{x-q} - p_{(q-1)(l+1)+s}(1+b)^{x-q-1} \right)
$$

$$
+ b p_{(i-1)(l+1)+s}(1+b)^{x-i}.
$$

Since $p_i \leq p_j$ for $i < j$ and $b > 0$, $C_{\max}(\hat{\Pi}) > C_{\max}(\Pi(l))$ which completes the proof of part (ii).

(c) The proof of part (iii).

Without loss of generality, assume that $i < j$. There are six cases which are the same as those in part (ii), and we only consider one of these cases where $y < s < m \leq l+1$. Again, the proofs of the other five cases are similar to that of this case and we omit them. For the case where $y < s < m \leq l+1$, Note that after the interchange, the ith job in Π_s is still in the ith position in the new subsequence by SPT order, and the same conclusion holds for the jth job in Π_m. By case condition and Lemma 1,

$$C_{\max}(\breve{\Pi})$$

$$= \sum_{h=1}^{y} \sum_{q=1}^{x+1} p_{(q-1)(l+1)+h}(1+b)^{x+1-q}$$

$$+ \sum_{y+1 \le h \le l+1, h \ne s, m} \sum_{q=1}^{x} p_{(q-1)(l+1)+h}(1+b)^{x-q} + lt$$

$$+ \sum_{q=1}^{i-1} p_{(q-1)(l+1)+s}(1+b)^{x-q} + \sum_{q=i}^{j-1} p_{q(l+1)+s}(1+b)^{x-q+1}$$

$$+ p_{(j-1)(l+1)+m}(1+b)^{x-j} + \sum_{q=j+1}^{x} p_{(q-1)(l+1)+s}(1+b)^{x-q}$$

$$+ \sum_{q=1}^{i-1} p_{(q-1)(l+1)+m}(1+b)^{x-q} + p_{(i-1)(l+1)+s}(1+b)^{x-i}$$

$$+ \sum_{q=i+1}^{j} p_{(q-2)(l+1)+m}(1+b)^{x-q-1}$$

$$+ \sum_{q=j+1}^{x} p_{(q-1)(l+1)+m}(1+b)^{x-q}. \tag{6}$$

Taking the difference between (3) and (6), we have

$$C_{\max}(\breve{\Pi}) - C_{\max}(\Pi(l))$$

$$= \sum_{q=i}^{j-1} p_{q(l+1)+s}(1+b)^{x-q} + p_{(j-1)(l+1)+m}(1+b)^{x-j}$$

$$+ p_{(i-1)(l+1)+s}(1+b)^{x-i} + \sum_{q=i+1}^{j} p_{(q-2)(l+1)+m}(1+b)^{x-q}$$

$$- \sum_{q=i}^{j} p_{(q-1)(l+1)+s}(1+b)^{x-q} - \sum_{q=i}^{j} p_{(q-1)(l+1)+m}(1+b)^{x-q}$$

$$= \sum_{q=i+1}^{j-1} \left((p_{q(l+1)+s} - p_{(q-1)(l+1)+m}) - (p_{(q-1)(l+1)+s} - p_{(q-2)(l+1)+m}) \right)(1+b)^{x-q}$$

$$- (p_{(j-1)(l+1)+s} - p_{(j-2)(l+1)+m})(1+b)^{x-j}$$

$$+ (p_{i(l+1)+s} - p_{(i-1)(l+1)+m})(1+b)^{x-i}$$

$$= \sum_{q=i}^{j-1} (p_{q(l+1)+s} - p_{(q-1)(l+1)+m})((1+b)^{x-q} - (1+b)^{x-q-1}).$$

Since $p_i \leq p_j$ for $i < j$ and $b > 0$, we obtain $C_{\max}(\breve{\Pi}) \geq C_{\max}(\Pi(l))$ which completes the proof of part (iii). Thus, the schedule $\Pi(l)$ is with the minimum makespan. The theorem follows. $\Box$

For time complexity of solving the optimal makespan, the jobs need to be renumbered in non-decreasing order of p_j, taking $O(n \log n)$ time. Thus problem $1|p_{[j]}^A = p_{[j]} + bS_{[j]}, l - rms|C_{\max}$ can be solved in $O(n \log n)$ time.

Corollary 1. *For problem* $1|p_{[j]}^A = p_{[j]} + bS_{[j]}, l - rms|C_{\max}$, *the number N of optimal schedules is*

$$N = \begin{cases} (l + 1)! \, (y!)^x ((l + 1 - y)!)^{x-1} & \text{if } y \neq 0; \\ ((l + 1)!)^x & \text{if } y = 0. \end{cases}$$

where $x = \lfloor n/(l+1) \rfloor$ *and* $y = n \mod (l + 1)$.

Proof. Assume that $n \geq l+1$. We first consider the case $y \neq 0$. From the proof of Theorem 1, we see that for the first y or last $(l-1-y)$ subsequences with the same number of jobs, jobs ranked in the same position can be arbitrarily interchanged without changing the objective value. By (3), each of the first y subsequences has $x + 1$ jobs, and each of the last $(l - 1 - y)$ subsequences has x jobs. So, the permutation of the jobs in the same position of the first y subsequences is $(y!)^{x+1}$, and that of the last $(l-1-y)$ subsequences is $((l+1-y)!)^x$. At this time, the first y subsequences has already been permutated, so as the last $(l-1-y)$ subsequences. Therefore, the permutation of the $(l + 1)$ subsequences is $\frac{P(l+1,l+1)}{P(y,y)P(l+1-y,l+1-y)}$. Then, the number of optimal schedules is $(l + 1)! \, (y!)^x ((l + 1 - y)!)^{x-1}$ when $y \neq 0$. It is analogous to prove $N = ((l + 1)!)^x$ when $y = 0$. $\Box$

For one variation where the machine may arbitrarily adopt $k \in [0, l]$ rate-modifying activities during job processing, we denote the problem by $1|p_{[j]}^A = p_{[j]} + bS_{[j]}, rms|C_{\max}$. This variation is reasonable for the case when the value of t is not that small compared to p_j, and it may be optimal to adopt none of the rate-modifying activities in some case. Hence, we shall be more careful to select rate-modifying activities, that is, the makespan may be less by using some of the l rate-modifying activities than using all of them. In this variation, it is necessary to figure out what is the optimal number of rate-modifying activities adopted. Hence, we get the following theorem:

Theorem 2. *For problem* $1|p_{[j]}^A = p_{[j]} + bS_{[j]}, rms|C_{\max}$, *there exists an optimal schedule $\Pi(L)$ such that* $\Pi(L) = \min\{ \min_{1 \leq i \leq l} \Pi(i), \Pi(0) \}$ *and* $\Pi(0) = (J_1 J_2 \ldots J_n)$ *without rate-modifying activities.*

4   Conclusion

This paper investigated one single-machine scheduling problem to minimize the makespan. We considered the scenario such that the machine is with linear deterioration rate of processing speed and rate-modifying activities. Each rate-modifying activity, consuming a fixed length of time, can fully recover the speed of the machine. We presented an optimal schedule with time complexity of $O(n \log n)$ where n is the number of jobs, and gave one result on an extended problem where an optimal schedule may not adopt all the rate-modifying activities. Future research topics include extending our results to multi-machine scheduling problem as well as problems with different objective functions.

Acknowledgements. This work was supported by NSFC(No. 71301168) and the Scientific Research Foundation of ZNUFE (Zhongnan University of Economics and Law, No. 31541310813).

References

[1] Eilon, S.: On a mechanistic approach to fatigue and rest periods. Int. J. Prod. Res. 3, 327–332 (1964)

[2] Gentzler, G.L., Khalil, T.M., Sivazlian, B.B.: Quantitative models for optimal rest period scheduling. Omega-Int. J. Manag. Sci. 5, 215–220 (1977)

[3] Bechtold, S.E., Janaro, R.E., Sumners, D.L.: Maximization of labor productivity through multi-rest break scheduling. Manag. Sci. 30, 1442–1458 (1984)

[4] Lee, C.Y., Leon, V.J.: Machine scheduling with a rate-modifying activity. Eur. J. Oper. Res. 128, 119–128 (2001)

[5] Lee, C.Y., Lin, C.S.: Single-machine scheduling with maintenance and repair rate-modifying activities. Eur. J. Oper. Res. 135, 493–513 (2001)

[6] Lodree, E.J., Geiger, C.D.: A note on the optimal sequence position for a rate-modifying activity under simple linear deterioration. Eur. J. Oper. Res. 201, 644–648 (2010)

[7] Lodree, E.J., Geiger, C.D., Jiang, X.: Taxonomy for integration scheduling theory and human factors: Review and research opportunities. Int. J. Ind. Ergonom. 39, 39–51 (2009)

[8] Browne, S., Yechiali, U.: Scheduling deteriorating job on a single processor. Oper. Res. 38, 495–498 (1990)

[9] Mosheiov, G.: Λ-shaped policies of schedule deteriorating jobs. J. Oper. Res. Soc. 47, 1184–1191 (1996)

[10] Alidaee, B., Womer, N.K.: Scheduling with time dependent processing times: Review and extensions. J. Oper. Res. Soc. 50, 711–720 (1999)

[11] Cheng, T.C.E., Ding, Q., Lin, B.M.T.: A concise survey of scheduling with time-dependent processing times. Eur. J. Oper. Res. 152, 1–13 (2004)

[12] Zhao, C.L., Tang, H.Y.: A note to due-window assignment and single machine scheduling with deteriorating jobs and a rate-modifying activity. Comput. Oper. Res. 39, 1300–1303 (2012)

A Loopless Algorithm for Generating Multiple Binary Tree Sequences Simultaneously*

Ro-Yu Wu[1], Jou-Ming Chang[2], Hung-Chang Chan[3], and Kung-Jui Pai[4]

[1] Department of Industrial Management,
Lunghwa University of Science and Technology, Taoyuan, Taiwan
[2] Institute of Information and Decision Sciences,
National Taipei College of Business, Taipei, Taiwan
[3] Department of Computer Science and Information Engineering,
Yuanpei University, Hsinchu, Taiwan
[4] Department of Industrial Engineering and Management,
Ming Chi University of Technology, New Taipei City, Taiwan

Abstract. Pallo and Wu et al. respectively introduced the left-weight sequences (LW-sequences) and right-weight sequences (RW-sequences) for representing binary trees. In this paper, we introduce two new types of binary tree sequences called the left-child sequences (LC-sequences) and right-child sequences (RC-sequences). Next, we propose a loopless algorithm associated with rotations of binary trees for generating LW-, RW-, LC-, and RC-sequences simultaneously. Moreover, we show that LW- and RW-sequences are generated in Gray-code order, and LC- and RC-sequences are generated so that each sequence can be obtained from its predecessor by changing at most two digits. Our algorithm is shown to be more efficient in both space and time than the existing known algorithms.

Keywords: Binary trees, Loopless generating algorithms, Gray-code order, Constant amortized time.

1 Introduction

Combinatorial objects are usually encoded by integer sequences and are generated in a particular order. Since a large amount of objects need to be generated, the design of efficient generating schemes are necessary. Ehrlich [1] defined that an algorithm for generating objects is a *loopless generating algorithm* provided the computation of changing one object into the next one only takes a constant time. In particular, if the change between two successive objects is limited by only one integer, the generation results in a Gray-code order. Loopless algorithms for generating binary trees [6, 8, 11–13, 17] and their generalization called k-ary trees [2–4, 9, 18] have attracted a great deal of attention. In these literature, some of these generating scheme are based on tree rotations [2–4, 6, 8, 12].

---

* This work was partially supported by the National Science Council of Taiwan under contracts NSC102-2221-E-262-013 and NSC102-2221-E-141-001-MY3.

P. Widmayer, Y. Xu, and B. Zhu (Eds.): COCOA 2013, LNCS 8287, pp. 340–350, 2013.

Although many types of integer sequences have been introduced to encode binary trees, the so-called left-weight sequences (LW-sequences) are the most common type of adoption [7]. For generating LW-sequences of binary trees, Pallo [7] has proposed a constant amortized time (CAT for short) algorithm that generates sequences in lexicographic order. Shortly afterwards, van Baronaigien and Ruskey [10] designed a recursive CAT algorithm that generates LW-sequences in a Gray-code order. Roughly a decade later, Vajnovszki [12] developed a loopless algorithm associated with rotations of binary trees for generating LW-sequences by way of Williamson's idea [14]. At a later time, based on the traversal in a rotation graph constructed by a special type of rotation on the left-arm (or right-arm) of a tree, Wu et al. [15] provided a recursive CAT algorithm to generate LW-sequences of binary trees in lexicographic order. In that paper, to efficiently transform binary tree sequences, they made use of the concept of mirror image of LW-sequences to define the right-weight sequences (RW-sequences) for binary trees.

Because the usual binary trees are implemented so that nodes are allocated by structures and children of a node are accessed through pointers (i.e., the structure and pointer representation), this motivates us to define integer sequences connoting child's information to represent binary trees. Two new types of binary tree sequences called the left-child sequences (LC-sequences) and the right-child sequences (RC-sequences) are introduced in the next section. In this paper, we present a loopless algorithm associated with rotations of binary trees for generating four types of binary tree sequences simultaneously, including LW-, RW-, LC-, and RC-sequences. In particular, LW- and RW-sequences are generated in Gray-code order, and LC- and RC-sequences are generated so that each sequence can be obtained from its predecessor by changing at most two digits. To the best of our knowledge, this is the first time that a loopless generating algorithm can simultaneously generate multiple types of binary tree sequences. Moreover, we show that the proposed algorithm has a high efficiency in the space and time requirements. In fact, to generate sequences of binary trees with n internal nodes, our loopless algorithm requires totally $6n + \mathcal{O}(1)$ memory space, and each generation needs no more than 4 comparisons and takes an amortized cost with at most 8 assignments when n is larger.

2 Preliminaries

A binary tree T considered here is a rooted, ordered tree with n internal nodes numbered by $1, 2, \ldots, n$ in inorder (i.e., visit recursively the left subtree, root and then the right subtree of T) such that every internal node has exactly two children called the *left child* and the *right child*. Such a binary tree is also referred to an *extended binary tree* [5]. For a node $i \in T$, the subtree rooted at i is denoted by T_i. Also, the subtree rooted at the left child (respectively, right child) of i is called the *left subtree* (respectively, *right subtree*) of i and is denoted by L_i (respectively, R_i).

2.1   Left-Weight and Right-Weight Sequences

The *weight* of a binary tree T, denoted by $w(T)$, is defined to be the number of leaves in T. Clearly, $w(T) = n + 1$ and T_i contains $w(T_i) - 1$ internal nodes for $1 \leqslant i \leqslant n$. The *left weight* of a node $i \in T$, denoted by $w_\ell(T, i)$, is defined to be the number of leaves in L_i, i.e., $w_\ell(T, i) = w(L_i)$. Pallo [7] further defined the integer sequence $w_\ell(T) = (w_\ell(T, 1), w_\ell(T, 2), \ldots, w_\ell(T, n))$ to be the *left-weight sequence* (LW-sequence for short) of T. Similarly, the *right weight* of a node $i \in T$ is $w_r(T, i) = w(R_i)$ and the integer sequence $w_r(T) = (w_r(T, 1), w_r(T, 2), \ldots, w_r(T, n))$ is called the *right-weight sequence* (RW-sequence for short) of T in [15]. For notational convenience, if the tree T is clear from the context, we simply write $w_\ell(i)$ and $w_r(i)$ instead of $w_\ell(T, i)$ and $w_r(T, i)$, respectively. For example, Fig. 1 shows a binary tree T with 9 internal nodes whose LW-sequence and RW-sequence are $w_\ell(T) = (1, 2, 1, 1, 3, 6, 1, 1, 3)$ and $w_r(T) = (1, 4, 2, 1, 1, 4, 2, 1, 1)$, respectively.

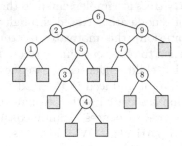

Fig. 1. A binary tree T of weight 10

For a binary tree T, there is a linear time transformation between $w_\ell(T)$ and $w_r(T)$ in [15]. In addition, Pallo [7] and Wu et al. [15] respectively characterized an integer sequence to be an LW-sequence or RW-sequence of a binary tree as follows.

Theorem 1.   [7,15] *Let $w = (w_1, w_2, \ldots, w_n)$ be an integer sequence. Then,*

(a) *w is the LW-sequence of a binary tree T with n internal nodes if and only if the following conditions are satisfied for all $i \in \{1, 2, \ldots, n\}$: (i) $1 \leqslant w_i \leqslant i$ and (ii) $i - w_i \leqslant j - w_j$ for all $j \in [i - w_i + 1, i]$.*

(b) *w is the RW-sequence of a binary tree T with n internal nodes if and only if the following conditions are satisfied for all $i \in \{1, 2, \ldots, n\}$: (i) $1 \leqslant w_i \leqslant n - i + 1$ and (ii) $i + w_i \geqslant j + w_j$ for all $j \in [i, i + w_i - 1]$.*

For a binary tree T, a path from the root to its leftmost (respectively, rightmost) leaf is called the *left arm* (respectively, *right arm*). The following lemma shows that a node on the left arm or right arm of a subtree is easy to be checked. In particular, if we imagine that T is the right subtree (respectively, left subtree) of a dummy node numbered by 0 (respectively, $n + 1$), it is easy to check from

Lemma 1 that a node x is in the left arm (respectively, right arm) of T if and only if $x = w_\ell(x)$ (respectively, $x + w_r(x) = n+1$), and x is the root of T if and only if $w_\ell(x) + w_r(x) = n + 1$.

Lemma 1. *[15] Let T be a binary tree and $i \in T$ an internal node. If x is a descendant of i, the following statements hold:*

(a) *The node x is contained in the left arm of $R_i \Leftrightarrow x - w_\ell(x) = i$.*
(b) *The node x is contained in the right arm of $L_i \Leftrightarrow x + w_r(x) = i$.*

Lemma 2. *[15] Let T be a binary tree and $i \in T$ an internal node. Suppose that i is not the root of T and let p be the parent of i. Then, the following statements hold:*

(a) *The node i is the right child of $p \Leftrightarrow p = i - w_\ell(i) \Leftrightarrow w_r(p) = w_\ell(i) + w_r(i)$.*
(b) *The node i is the left child of $p \Leftrightarrow p = i + w_r(i) \Leftrightarrow w_\ell(p) = w_\ell(i) + w_r(i)$.*

2.2 Left-Child and Right-Child Sequences

We now introduce new types of sequences to represent binary trees. For a binary T with n internal nodes $1, 2, \ldots, n$ numbered in inorder, the *left-child sequence* (LC-sequence for short) of T, denoted by $c_\ell(T) = (c_\ell(T, 1), c_\ell(T, 2), \ldots, c_\ell(T, n))$, is an integer sequence such that the term $c_\ell(T, i)$, $1 \leqslant i \leqslant n$, is defined as follows:

$$c_\ell(T, i) = \begin{cases} 0 & \text{if the left child of } i \text{ is a leaf;} \\ j & \text{if } j \text{ is the left child of } i \text{ in } T. \end{cases} \tag{1}$$

Similarly, we denote $c_r(T) = (c_r(T, 1), c_r(T, 2), \ldots, c_r(T, n))$ as the *right-child sequence* (RC-sequence for short) of T, where we use the right child instead of the left child in (1) to define the term $c_r(T, i)$. Obviously, $c_\ell(T, i) < i$ and either $c_r(T, i) = 0$ or $c_r(T, i) > i$. In particular, $c_\ell(T, 1) = c_r(T, n) = 0$. Customarily, we omit the parameter T in the terms $c_\ell(T, i)$ and $c_r(T, i)$ if it is clear from the context. For instance, the LC-sequence and RC-sequence of the binary tree T shown in Figure 1 are $c_\ell(T) = (0, 1, 0, 0, 3, 2, 0, 0, 7)$ and $c_r(T) = (0, 5, 4, 0, 0, 9, 8, 0, 0)$, respectively.

We can also reformulate $c_\ell(T, i)$ and $c_r(T, i)$ as follows.

Lemma 3. *Let T be a binary tree and $i \in T$ an internal node. Then*

(a) $c_\ell(i) = \begin{cases} 0 & \text{if the left child of } i \text{ is a leaf;} \\ \min\{x \in [1, i-1] \mid x + w_r(x) = i\} & \text{otherwise.} \end{cases}$

(b) $c_r(i) = \begin{cases} 0 & \text{if the right child of } i \text{ is a leaf;} \\ \max\{x \in [i+1, n] \mid x - w_\ell(x) = i\} & \text{otherwise.} \end{cases}$

Proof. By symmetry, we only prove (a). Suppose that the left child of i is not a leaf of T and let $c_\ell(i) = j$. By Lemma 1, $x + w_r(x) = i$ if and only if x is contained in the right arm of T_j (i.e., L_i). In particular, $j \leqslant x$ whenever x is contained in the right arm of T_j. $\square$

Lemma 4. *Let T be a binary tree and $i \in T$ an internal node. The following conditions hold:*

(a) $w_\ell(i) = \begin{cases} 1 & \text{if } c_\ell(i) = 0 \\ w_\ell(c_\ell(i)) + i - c_\ell(i) & \text{otherwise;} \end{cases}$

(b) $w_r(i) = \begin{cases} 1 & \text{if } c_r(i) = 0 \\ w_r(c_r(i)) + c_r(i) - i & \text{otherwise.} \end{cases}$

Proof. By symmetry, we only prove (a). Clearly, if $c_\ell(i) = 0$, then $w_\ell(i) = 1$. Otherwise, by Lemma 2 we have $i = c_\ell(i) + w_r(c_\ell(i))$. Thus, $w_\ell(i) = w(L_i) = w(T_{c_\ell(i)}) = w_\ell(c_\ell(i)) + w_r(c_\ell(i)) = w_\ell(c_\ell(i)) + i - c_\ell(i)$. □

2.3 Tree Rotations

A rotation is a simple operation that reconstructs a binary tree into another tree and preserves its inorder. For a binary tree T and two nodes $x, y \in T$ where x is the right child of y, a *left rotation* at x, denoted by $\rho_\ell(T, x)$, is an operation that raises x to the place of y, such that y becomes the new left child of x and the left subtree of x becomes the new right subtree of y, while the remaining parts of the tree are unchanged. A *right rotation* at a node x, denoted by $\rho_r(T, x)$, is the reverse operation of $\rho_\ell(T, x)$, i.e., if we apply the right rotation after a left rotation at a node, it reconstructs the original tree. See Fig. 2 for an illustration. Also, for notational convenience, we use the same notation $\rho_\ell(T, x)$ or $\rho_r(T, x)$ to denote the resulting tree that performs the rotation.

The following propositions directly follow from the definition of rotations and thus we omit the proof since its validity can easily be checked from Lemma 2.

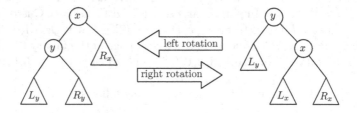

Fig. 2. Left rotation and right rotation

Proposition 1. *For a binary tree T and two nodes $x, y \in T$ where x is the right child of y. Let $L = \rho_\ell(T, x)$. Then*

(a) $w_r(L, y) = w_\ell(T, x)$ *and* $w_\ell(L, x) = w_\ell(T, x) + w_\ell(T, y)$;
(b) $c_r(L, y) = c_\ell(T, x)$ *and* $c_\ell(L, x) = y$;
(c) *If y is the left child of a node z, then $z = x + w_r(x)$ and $c_\ell(L, z) = x$;*
(d) *If y is the right child of a node z, then $z = y - w_\ell(y)$ and $c_r(L, z) = x$;*
(e) *Except for the above changes, the left child and right child (respectively, left weight and right weight) of every other node in L remain unchanged.*

Proposition 2. *For a binary tree T and two nodes $x, y \in T$ where y is the left child of x. Let $R = \rho_r(T, x)$. Then*

(a) $w_\ell(R, x) = w_r(T, y)$ *and* $w_r(R, y) = w_r(T, y) + w_r(T, x)$;
(b) $c_\ell(R, x) = c_r(T, y)$ *and* $c_r(R, y) = x$;
(c) *If x is the left child of a node z, then $z = x + w_r(x)$ and $c_\ell(R, z) = y$;*
(d) *If x is the right child of a node z, then $z = x - w_\ell(x)$ and $c_r(R, z) = y$;*
(e) *Except for the above changes, the left child and right child (respectively, left weight and right weight) of every other node in R remain unchanged.*

3 A Loopless Generation of Binary Tree Sequences

Let $\mathscr{T}_n$ be the set of binary trees with n internal nodes. It is well-known that $|\mathscr{T}_n| = \frac{1}{n+1}\binom{2n}{n}$. A systematic way to describe all binary tree sequences is the use of coding trees. A coding tree $\mathbb{T}_n$ is a rooted tree consisting of n levels such that every node is associated with a label and the full labels along a path from the root to a leaf in $\mathbb{T}_n$ represent the sequence of a binary tree $T \in \mathscr{T}_n$. To facilitate the description of $\mathbb{T}_n$, the following terms are used in [16]. We say that a non-leaf node $x \in \mathbb{T}_n$ has an *up-fragment* (respectively, a *down-fragment*) if the labels of x's children in $\mathbb{T}_n$ are arranged from left to right in increasing order (respectively, decreasing order). In particular, a coding tree is called a *flip-flap tree* if the following conditions hold: (1) every non-leaf node has either an up-fragment or a down-fragment; (2) if a node has an up-fragment, then its adjacency siblings (if exist) must have a down-fragment, and vice versa.

Accordingly, we construct a coding tree such that the root has label 1 and, for each level $i \geq 2$, the two particular labels 1 and i always appear in any fragment of level i. Moreover, if we restrict that the two nodes with labels 1 and i are on the boundary in every fragment and the two types of fragments (i.e., up-fragments and down-fragments) alternately appear in each level of $\mathbb{T}_n$, then the resulting coding tree is a flip-flap tree. In this case, if the full labels along the path in the left arm of $\mathbb{T}_n$ are determined, then so is the arrangement of $\mathbb{T}_n$. Since we have two choices to label a node (except the root) in the left arm of $\mathbb{T}_n$, there are at least 2^{n-1} different ways to construct a flip-flap tree. For instance, Fig. 3 shows a flip-flap tree $\mathbb{T}_5$ begins with the sequence $(1, 2, 3, 4, 5)$ in its left arm. In this case, the initial fragment in each level of $\mathbb{T}_5$ is given by a down-fragment.

In the above arrangement of $\mathbb{T}_n$, if u and v are two adjacent leaves, the sequences from the root to u and to v are said to be *consecutive*. The following lemma establishes the base of our loopless algorithm such that an enumeration of LW-sequences in Gray-code order can be achieved.

Lemma 5. *In a flip-flap tree, any two consecutive LW-sequences differ in exactly one digit.*

Proof. Let $w_\ell(T_1)$ and $w_\ell(T_2)$ be consecutive sequences with respect to two adjacent leaves u and v in $\mathbb{T}_n$, and let w be the lowest common ancestor of u

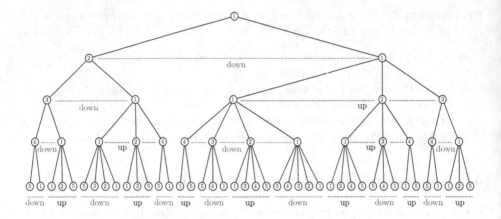

Fig. 3. A flip-flap tree $\mathbb{T}_5$ for LW-sequences

and v. It is clear that labels collected from the root to node w are the same for leaves u and v, and are different for w's children. By the flip-flap arrangement, the remaining labels for leaves u and v must also be the same. Thus, there is only one position having different digits for $w_\ell(T_1)$ and $w_\ell(T_2)$.   □

A procedure called NextTree() is developed to perform either a left rotation or a right rotation at node i for generating the next tree sequences. Arrays $w_\ell(1..n)$, $w_r(1..n)$, $c_\ell(1..n)$ and $c_r(1..n)$ are used to store the current LW-, RW-, LC- and RC-sequences, respectively. We initially set $w_\ell(j) = j$, $w_r(j) = 1$, $c_\ell(j) = j - 1$ and $c_r(j) = 0$ for $j = 1, \ldots, n$. That is, the first generated tree is the left-skew tree. In addition, two auxiliary arrays $\rho(1..n)$ and $F(1..n)$ are employed to record important information for nodes $1, 2, \ldots, n$. The setting $\rho(i) = 1$ (respectively, $\rho(i) = -1$) indicates that the current operation performed at node i is a left rotation (respectively, a right rotation), and in this moment the fragment at level i of $\mathbb{T}_n$ will be an up-fragment (respectively, a down-fragment). Also, we use $F(i)$ to keep the track of positions to be processed for node i. In the beginning, let $i = n$ and, for $1 \leqslant j \leqslant n$, we set up $\rho(j) = -1$ and $F(j) = j$. When NextTree() is invoked, it performs the following steps:

1. Determine the case to perform a left rotation or a right rotation at node i.
    1.1 The case of a left rotation (i.e., an up-fragment at level i of $\mathbb{T}_n$):
        – Let p be the parent of i (in this case, i is the right child of p). If p is not the root of the current tree T, let q be the parent of p.
        – Perform the rotation $T' = \rho_\ell(T, i)$.
        – According to Proposition 1(c) and 1(d), if p is the left child of q in T, update $c_\ell(T', q) = i$; otherwise, update $c_r(T', q) = i$.
        – According to Proposition 1(a) and 1(b), update $w_r(T', p)$, $w_\ell(T', i)$, $c_r(T', p)$ and $c_\ell(T', i)$.
        – If i is not on the left arm of T', perform $\rho_\ell(T', n)$ in the next iteration.

1.2 The case of a right rotation (i.e., a down-fragment at level i of $\mathbb{T}_n$):

- Let q be the left child of i. If i is not the root of the current tree T, let p be the parent of i.
- Perform the rotation $T' = \rho_r(T, i)$.
- According to Proposition 2(c) and 2(d), if i is the left child of p in T, update $c_\ell(T', p) = q$; otherwise, update $c_r(T', p) = q$.
- According to Proposition 2(a) and 2(b), update $w_\ell(T', i)$, $w_r(T', q)$, $c_\ell(T', i)$ and $c_r(T', q)$.
- If i still has a left child in T', perform $\rho_r(T', n)$ in the next iteration.

2. If the fragment is in face of a boundary (i.e., i is on the left arm of T' for Step 1.1 or i has no left child in T' for Step 1.2), then we perform the following:

- Change the type of rotation in the next iteration(i.e., $\rho(i) = -\rho(i)$).
- Propagate the information of F from an upper level to a lower level in $\mathbb{T}_n$ (i.e., $F(i) = F(i - 1)$).
- Recover the information of F for the upper level of $\mathbb{T}_n$ (i.e., $F(i - 1) = i - 1$).
- Set $F(n)$ to be the node that will be rotated in the next iteration (i.e., $i = F(n)$).
- Retrieve the initial value of $F(n)$ (i.e., $F(n) = n$).

3. When $i = 1$, all tree sequences have been generated.

Fig. 4 gives the implementation in detail. Table 1 shows the list of binary tree sequences generated by NextTree() when $n = 5$.

Table 1. The lists of binary tree sequences generated by NextTree() when $n = 5$

Tree	LW-seq.	RW-seq.	LC-seq.	RC-seq.	Tree	LW-seq.	RW-seq.	LC-seq.	RC-seq.
1	(1,2,3,4,5)	(1,1,1,1,1)	(0,1,2,3,4)	(0,0,0,0,0)	22	(1,1,1,2,4)	(5,3,1,1,1)	(0,0,0,3,2)	(5,4,0,0,0)
2	(1,2,3,4,1)	(1,1,1,2,1)	(0,1,2,3,0)	(0,0,0,5,0)	23	(1,1,1,2,5)	(4,3,1,1,1)	(0,0,0,3,1)	(2,4,0,0,0)
3	(1,2,3,1,1)	(1,1,3,2,1)	(0,1,2,0,0)	(0,0,4,5,0)	24	(1,1,1,1,5)	(4,3,2,1,1)	(0,0,0,0,1)	(2,3,4,0,0)
4	(1,2,3,1,2)	(1,1,3,1,1)	(0,1,2,0,4)	(0,0,5,0,0)	25	(1,1,1,1,4)	(5,3,2,1,1)	(0,0,0,0,2)	(5,3,4,0,0)
5	(1,2,3,1,5)	(1,1,2,1,1)	(0,1,2,0,3)	(0,0,4,0,0)	26	(1,1,1,1,3)	(5,4,2,1,1)	(0,0,0,0,3)	(2,5,4,0,0)
6	(1,2,1,1,5)	(1,3,2,1,1)	(0,1,0,0,2)	(0,3,4,0,0)	27	(1,1,1,1,2)	(5,4,3,1,1)	(0,0,0,0,4)	(2,3,5,0,0)
7	(1,2,1,1,3)	(1,4,2,1,1)	(0,1,0,0,3)	(0,5,4,0,0)	28	(1,1,1,1,1)	(5,4,3,2,1)	(0,0,0,0,0)	(2,3,4,5,0)
8	(1,2,1,1,2)	(1,4,3,1,1)	(0,1,0,0,4)	(0,3,5,0,0)	29	(1,1,2,1,1)	(5,1,3,2,1)	(0,0,2,0,0)	(3,0,4,5,0)
9	(1,2,1,1,1)	(1,4,3,2,1)	(0,1,0,0,0)	(0,3,4,5,0)	30	(1,1,2,1,2)	(5,1,3,1,1)	(0,0,2,0,4)	(3,0,5,0,0)
10	(1,2,1,2,1)	(1,4,1,2,1)	(0,1,0,3,0)	(0,4,0,5,0)	31	(1,1,2,1,4)	(5,1,2,1,1)	(0,0,2,0,3)	(5,0,4,0,0)
11	(1,2,1,2,3)	(1,4,1,1,1)	(0,1,0,3,4)	(0,5,0,0,0)	32	(1,1,2,1,5)	(4,1,2,1,1)	(0,0,2,0,1)	(3,0,4,0,0)
12	(1,2,1,2,5)	(1,3,1,1,1)	(0,1,0,3,2)	(0,4,0,0,0)	33	(1,1,2,3,5)	(4,1,1,1,1)	(0,0,2,3,1)	(4,0,0,0,0)
13	(1,2,1,4,5)	(1,2,1,1,1)	(0,1,0,2,4)	(0,3,0,0,0)	34	(1,1,2,3,4)	(5,1,1,1,1)	(0,0,2,3,4)	(5,0,0,0,0)
14	(1,2,1,4,1)	(1,2,1,2,1)	(0,1,0,2,0)	(0,3,0,5,0)	35	(1,1,2,3,1)	(5,1,1,2,1)	(0,0,2,3,0)	(4,0,0,5,0)
15	(1,1,1,4,1)	(3,2,1,2,1)	(0,0,0,1,0)	(2,3,0,5,0)	36	(1,1,2,4,1)	(3,1,1,2,1)	(0,0,2,1,0)	(3,0,0,5,0)
16	(1,1,1,4,5)	(3,2,1,1,1)	(0,0,0,1,4)	(2,3,0,0,0)	37	(1,1,2,4,5)	(3,1,1,1,1)	(0,0,2,1,4)	(3,0,0,0,0)
17	(1,1,1,3,5)	(4,2,1,1,1)	(0,0,0,2,1)	(4,3,0,0,0)	38	(1,1,3,4,5)	(2,1,1,1,1)	(0,0,1,3,4)	(2,0,0,0,0)
18	(1,1,1,3,4)	(5,2,1,1,1)	(0,0,0,2,4)	(5,3,0,0,0)	39	(1,1,3,4,1)	(2,1,1,2,1)	(0,0,1,3,0)	(2,0,0,5,0)
19	(1,1,1,3,1)	(5,2,1,2,1)	(0,0,0,2,0)	(4,3,0,5,0)	40	(1,1,3,1,1)	(2,1,3,2,1)	(0,0,1,0,0)	(2,0,4,5,0)
20	(1,1,1,2,1)	(5,4,1,2,1)	(0,0,0,3,0)	(2,4,0,5,0)	41	(1,1,3,1,2)	(2,1,3,1,1)	(0,0,1,0,4)	(2,0,5,0,0)
21	(1,1,1,2,3)	(5,4,1,1,1)	(0,0,0,3,4)	(2,5,0,0,0)	42	(1,1,3,1,5)	(2,1,2,1,1)	(0,0,1,0,3)	(2,0,4,0,0)

	Procedure NextTree( )
	begin
1	if $\rho(i) = 1$ then // Step 1.1
2	$p \leftarrow i - w_\ell(i)$; // Set i to be the right child of p; see Lemma 2(a);
3	if $w_\ell(p) + w_r(p) \neq n+1$ then // Test, if p is not the root of T;
4	if $w_\ell(p + w_r(p)) = w_\ell(p) + w_r(p)$ then
	// Test, if p is the left child of its parent; see Lemma 2(b);
5	$c_\ell(p + w_r(p)) \leftarrow i$; // $p + w_r(p)$ is the parent of p; see Proposition 1(c);
6	else
7	$c_r(p - w_\ell(p)) \leftarrow i$; // $p - w_\ell(p)$ is the parent of p; see Proposition 1(d);
8	$w_r(p) \leftarrow w_\ell(i)$; // See Proposition 1(a);
9	$w_\ell(i) \leftarrow w_\ell(i) + w_\ell(p)$;
10	$c_r(p) \leftarrow c_\ell(i)$; // See Proposition 1(b);
11	$c_\ell(i) \leftarrow p$;
12	if $w_\ell(i) \neq i$ then // Test, if i is not on the left arm of T';
13	$i \leftarrow n$; return; // Perform $\rho_\ell(T', n)$ in the next iteration;
14	else // Step 1.2
15	$q \leftarrow c_\ell(i)$; // Set q to be the left child of i;
16	if $w_\ell(i) + w_r(i) \neq n+1$ then // Test, if i is not the root of T;
17	if $w_\ell(i + w_r(i)) = w_\ell(i) + w_r(i)$ then
	// Test, if i is the left child of its parent; see Lemma 2(b);
18	$c_\ell(i + w_r(i)) \leftarrow q$; // $i + w_r(i)$ is the parent of i, see Proposition 2(c);
19	else
20	$c_r(i - w_\ell(i)) \leftarrow q$; // $i - w_\ell(i)$ is the parent of i, see Proposition 2(d);
21	$w_\ell(i) \leftarrow w_r(q)$; // See Proposition 2(a);
22	$w_r(q) \leftarrow w_r(q) + w_r(i)$;
23	$c_\ell(i) \leftarrow c_r(q)$; // See Proposition 2(b);
24	$c_r(q) = i$;
25	if $w_\ell(i) \neq 1$ then // Test, if i still has a left child in T';
26	$i \leftarrow n$; return; // Perform $\rho_r(T', n)$ in the next iteration;
27	$\rho(i) \leftarrow -\rho(i)$; // Step 2;
28	$F(i) \leftarrow F(i-1)$;
29	$F(i-1) \leftarrow i-1$;
30	$i \leftarrow F(n)$;
31	$F(n) \leftarrow n$;

Fig. 4. The loopless procedure NextTree()

Lemma 6. *NextTree() takes an amortized cost with at most 4 comparisons and* $8 + \frac{3}{2n-1}$ *assignments.*

Proof. Let N_c and N_a denote the expected number of comparisons and the expected number of assignments used in the procedure. We can check from NextTree() that there are 3 comparisons in the best case and 4 comparisons in the worst case. It follows that $N_c \leqslant 4$. For computing the upper bound of N_a, it is easy to see that the procedure requires at most 7 assignments when the rotation is performed at node n. Otherwise, it requires at most 11 assignments. Moreover, from the observation of a flip-flap tree, we can see that the number of rotations at node n is equal to the difference between the number of leaves and the number of their parents, i.e., $|\mathscr{T}_n| - |\mathscr{T}_{n-1}|$ (e.g., we have $42 - 14 = 28$ for $\mathbb{T}_5$). Thus,

$$N_a \leqslant \frac{7(|\mathcal{T}_n| - |\mathcal{T}_{n-1}|) + 11|\mathcal{T}_{n-1}|}{|\mathcal{T}_n|} = 7 + 4 \cdot \frac{n+1}{n} \cdot \frac{\binom{2n-2}{n-1}}{\binom{2n}{n}} = 8 + \frac{3}{2n-1}.$$

$\square$

We summarize our main result as the following theorem.

Theorem 2. *All binary trees with n internal nodes encoded by LW-, RW-, LC-, and RC-sequences can be generated simultaneously in $\mathcal{O}(|\mathcal{T}_n|)$ time using $6n + \mathcal{O}(1)$ memory space and each generation requires only constant time. In particular, LW-sequences and RW-sequences are generated in Gray-code order, and discriminatingly, LC-sequences and RC-sequences are generated so that each sequence can be obtained from its predecessor by changing at most two digits.*

Proof. For NextTree(), the memory requirement is obvious as those stated in the procedure (see Fig. 4) and the time complexity directly follows from Lemma 6. Since the procedure NextTree() performs either a left rotation or a right rotation, by Propositions 1 and 2, the four types of binary tree sequences can be correctly generated. Accurately, by Lemma 5, we have known that LW-sequences is generated in a Gray-code order. From another point of view, we can check the procedure to see that a digit in the current LW-sequence (respectively, RW-sequence) will be changed either in Line 9 (respectively, Line 8) for a right rotation or in Line 21 (respectively, Line 22) for a left rotation. Thus, for these two types of sequences, there only one digit is changed for each generation. By contrast, for LC-sequences (respectively, RC-sequences), the change of the two digits occurs either in the case that we perform a right rotation at node i whose parent is a left child (respectively, right child) of a node or in the case that we perform a left rotation at node i who is a left cild (respectively, right child) of its parent (see Lines 5 and 11 or Lines 18 and 23 for LC-sequences, and Lines 7 and 10 or Lines 20 and 24 for RC-sequences). For all other cases in each of the generation of LC-sequences or RC-sequences, there is only one digit to be changed. $\square$

4   Concluding Remarks

In this paper, we design a loopless algorithm associated with rotations of binary trees for generating LW-, RW-, LC-, and RC-sequences simultaneously. Also, we prove that our algorithm is possessed of a high efficiency in space and time requirements. Compared with the loopless algorithm that associates with rotations to generate LW-sequences of binary trees in [12], ours is more efficient in both space and time. Our algorithm needs $6n + \mathcal{O}(1)$ space and each generation requires 3/4 comparisons and 7/11 assignments in the best/worst case between two successive sequences. In fact, a result pointed in [18] shows that Vajnovszki's algorithm needs $8n + \mathcal{O}(1)$ space and each generation requires 3/5 comparisons and 9/13 assignments in its best/worst case between two successive sequences.

We know that almost all combinatorial objects have integer sequence representation. Since different representations are derived from some structural properties of these objects, there must be a natural correspondence between the different representations. As to the future research, we believe that the idea of this paper could easily be imitated to generate multiple types of sequences for other combinatorial objects, such as k-ary trees. Applying this technique to generate other combinatorial objects is still open.

References

1. Ehrlich, G.: Loopless algorithms for generating permutations, combinations, and other combinatorial configurations. J. ACM 20, 500–513 (1973)
2. Korsh, J.F.: Loopless generation of k-ary tree sequences. Inform. Process. Lett. 52, 243–247 (1994)
3. Korsh, J.F., LaFollette, P.: Loopless generation of Gray codes for k-ary trees. Inform. Process. Lett. 70, 7–11 (1999)
4. Korsh, J.F., Lipschutz, S.: Shift and loopless generation of k-ary trees. Inform. Process. Lett. 65, 235–240 (1998)
5. Knuth, D.E.: The Art of Computer Programming. Fascicle 4A — Generating All Trees, vol. 4. Addison-Wesley (2005)
6. Lucas, J.M., Roelants van Baronaigien, D., Ruskey, F.: On rotations and the generation of binary trees. J. Algorithms 15, 343–366 (1993)
7. Pallo, J.: Enumerating, ranking and unranking binary trees. Comput. J. 29, 171–175 (1986)
8. Roelants van Baronaigien, D.: A loopless algorithm for generating binary tree sequences. Inform. Process. Lett. 39, 189–194 (1991)
9. Roelants van Baronaigien, D.: A loopless Gray-code algorithm for listing k-ary trees. J. Algorithms 35, 100–107 (2000)
10. Roelants van Baronaigien, D., Ruskey, F.: A Hamiltonian path in the rotation lattice of binary trees. Congr. Numer. 59, 313–318 (1987)
11. Takaoka, T.: O(1) time algorithms for combinatorial generation by tree traversal. Comput. J. 42, 400–408 (1999)
12. Vajnovszki, V.: On the loopless generation of binary tree sequences. Inform. Process. Lett. 68, 113–117 (1998)
13. Vajnovszki, V.: Generating a Gray code for P-sequences. J. Math. Model. Algorithms 1, 31–41 (2002)
14. Williamson, S.G.: Combinatorics for Computer Science. Computer Science Press, Rockville (1985)
15. Wu, R.-Y., Chang, J.-M., Wang, Y.-L.: A linear time algorithm for binary tree sequences transformation using left-arm and right-arm rotations. Theor. Comput. Sci. 355, 303–314 (2006)
16. Wu, R.-Y., Chang, J.-M., Wang, Y.-L.: Loopless Generation of non-regular trees with a prescribed branching sequence. Comput. J. 53, 661–666 (2010)
17. Xiang, L., Ushijima, K.: On O(1) time algorithms for combinatorial generation. Comput. J. 44, 292–302 (2001)
18. Xiang, L., Ushijima, K., Tang, C.: Efficient loopless generation of Gray codes for k-ary trees. Inform. Process. Lett. 76, 169–174 (2000)

Touring Disjoint Polygons Problem Is NP-Hard

Arash Ahadi, Amirhossein Mozafari, and Alireza Zarei

Department of Mathematical Sciences
Sharif University of Technology

Abstract. In the Touring Polygons Problem (TPP) there is a start point s, a sequence of simple polygons $\mathcal{P} = (P_1, \ldots, P_k)$ and a target point t in the plane. The goal is to obtain a path of minimum possible length that starts from s, visits in order each of the polygons in $\mathcal{P}$ and ends at t. This problem has a polynomial time algorithm when the polygons in $\mathcal{P}$ are convex and is NP-hard in general case. But, it has been open whether the problem is NP-hard when the polygons are pairwise disjoint. In this paper, we prove that TPP is also NP-hard when the polygons are pairwise disjoint in any L_p norm even if each polygon consists of at most two line segments. This result solves an open problem from *STOC '03* and complements recent approximation results.

1 Introduction

A natural and well studied problem in computational geometry is to find a *shortest path* from a start point s to a target point t, having some properties in the plane. In some applications, the shortest path must visit a set of regions according to a given order. Zoo-keeper [5], Safari [9] and Watchman route [2,8] problems are famous examples of such applications. In the fixed source version of the Safari and Zoo-keeper problems, we are given a simple polygon P, a start point s inside it and a set of disjoint convex polygons (cages) $\{P_1, \ldots, P_k\}$ inside P each of which sharing exactly one edge with P. In the Zoo-keeper problem, we seek a tour of minimum possible length that visits the cages only at their boundaries but never enters any of them while in the Safari problem the tour can enter the cages. In the Watchman route problem (fixed-source version) we have a simple polygon P and a start point s inside it and the goal is to find a shortest tour from s inside P such that every point in P can be seen from at least one point of the tour. It is not difficult to show that in the Zoo-keeper and Safari problems, the shortest tour must visit the cages in the same order as they lie on the boundary of P and in the Watchman route problem, the shortest tour must visit its *essential pockets* in their order around P [2]. What all these problems have in common is that we should find a shortest path visiting some polygons in a given order. The Touring Polygons Problem (TPP) is the general problem having this visiting property. In TPP, we are given a start point s, a sequence $\mathcal{P} = (P_1, \ldots, P_k)$ of simple polygons and a target point t in the plane and the goal is to find a shortest path that starts from s, visits all polygons according to their order and ends at t. This problem was introduced by Dror *et al.* [3] in 2003.

P. Widmayer, Y. Xu, and B. Zhu (Eds.): COCOA 2013, LNCS 8287, pp. 351–360, 2013.

They also discussed the constrained version of TPP in which an ordered sequence $\mathcal{F} = (F_0, \dots, F_k)$ of simple polygons called fences are also given and the portion of the desired path between P_i and P_{i+1} must lie inside F_i (consider s as P_0 and t as P_{k+1}). They proved that TPP is NP-hard in general case and proposed an $O(kn \log(n/k))$ time algorithm for the case that the polygons are convex and pairwise disjoint and an $O(nk^2 \log n)$ time algorithm for the constrained case with convex polygons where n is the total number of vertices of the polygons. But the complexity of TPP (constrained and unconstrained versions) when the polygons are disjoint and allowed to be non-convex has remained open so far even for the L_1 norm. We call this case Touring pairwise Disjoint Polygons Problem (TDPP). Fig 1. shows an example of TDPP.

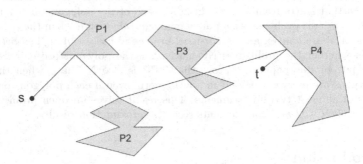

Fig. 1. TDPP with sequence $(s, P_1, P_2, P_3, P_4, t)$ and its solution

Recently, several approximation algorithms have been proposed for TPP and TDPP [6,7,10]. In 2010, Pan *et al.* [10] gave a linear time approximation algorithm for TDPP without answering what the complexity of this problem is. In this paper, we prove that TDPP is NP-hard for any L_p norm even if each polygon is composed of at most two joint line segments whose angles with the x-axis are in $\{0, \pm\pi/4, \pi/2\}$. Our proof complements the Dror *et. al* proof of NP-hardness of TPP [3] which is based on the Canny-Reif proof of NP-hardness of three-dimensional shortest path problem [1].

2   Proof of NP-Hardness

To prove the NP-hardness, we use a reduction from the 3-SAT problem. Let $\Phi(X, C)$ be an instance of the 3-SAT problem where $X = \{x_1, \dots, x_n\}$ is its variable set and $C = \{C_1, \dots, C_m\}$ is its clause set such that each variable appears at most once in each clause. We build an instance $\mathcal{P}_\Phi = (s, P_1, \dots, P_k, t)$ of TDPP with total complexity of $O(n+m)$ and a real number L_Φ in polynomial time such that the solution of TDPP on $\mathcal{P}_\Phi$ is no longer than L_Φ if and only if Φ has a satisfying assignment. To construct $\mathcal{P}_\Phi$, we consider the octal Cartesian coordinate system in the plane that is similar to the ordinary Cartesian coordinate system but the x-axis and the y-axis are numbered by octal numbers and the coordinate of each point in the plane is an ordered pair of two octal numbers.

We define a *proper path* as a path that starts from s, intersects the polygons of $\mathcal{P}_\Phi$ according to their order and ends at t. An *optimal path* is a proper path of least possible length. $\mathcal{P}_\Phi$ contains the start point s and the target point t as the first and the last polygons respectively. Between s and t in $\mathcal{P}_\Phi$, there is a sequence (G_i) of *gadgets* each of which consists of a sequence of polygons which are either a line segment or joined two line segments whose angles with the x-axis are in $\{0, \pm\pi/4, \pi/2\}$. So, a proper path should start from s and traverse the gadgets according to their order and end at t. A *semi-optimal path* of G_i is a path that starts from s and optimally traverses G_j gadgets $(1 \le j \le i)$ according to their order and an *incoming semi-optimal path* of G_i is a semi-optimal path of G_{i-1} that optimally enters into the G_i. Note that there may be many semi-optimal paths for one gadget. We arrange the gadgets in the plane in such a way that all optimal and semi-optimal paths can only be horizontal from one gadget to its next gadget. So the height of any incoming semi-optimal path of G_i is its vertical distance from the x-axis when it traverses from G_{i-1} to G_i. This height determines how it has traversed the previous gadgets.

To construct $\mathcal{P}_\Phi$, we assign two sequences of numbers namely $\{\alpha_i\}$ and $\{\beta_i\}$ to $\Phi(X, C)$. We define numbers α_i and β_i $(1 \le i \le n)$ as follows :

$$\alpha_i = \sum_{x_i \in C_j} 8^{j-1} \quad , \quad \beta_i = \sum_{\bar{x}_i \in C_j} 8^{j-1}$$

In other words, for $\alpha_i = (a_{m-1} \dots a_0)_8$, $a_j = 1$ if $x_i \in C_{j+1}$ and otherwise $a_j = 0$. Similarly, for $\beta_i = (b_{m-1} \dots b_0)_8$, $b_j = 1$ if $\bar{x}_i \in C_{j+1}$ and otherwise $b_j = 0$. Let Γ be the following set:

$$\Gamma = \{ \sum_{i=1}^{n} \gamma_i \mid \gamma_i = \alpha_i \ or \ \gamma_i = \beta_i \ \}.$$

Note that the elements of Γ are octal numbers whose digits are in $\{0, 1, 2, 3\}$.

Lemma 1. Φ has a satisfying assignment if and only if Γ contains a number with no zero digit in its m digit octal representation.

Proof. Let l_{ij} be the j'th literal of the i'th clause. If $\Phi(X, C)$ has a satisfying assignment, each clause C_i has a true literal l_{ij}. Let $\gamma = \sum \gamma_w$ in which $\gamma_w = \alpha_w$ if $x_w \in L$ or $\gamma_w = \beta_w$ if $\bar{x}_w \in L$ where $L = \{l_{ij}\}$ is the set of true literals in the satisfying assignment of $\Phi(X, C)$. Trivially, $\gamma \in \Gamma$ and has no zero digit in its octal representation. On the other hand, if such a γ number exists, the w'th digit of α_i or β_i is 1 for all w. So, we can build a satisfying assignment by assigning *true* for the first case and *false* for the second case to the variable x_i. $\square$

Based on the above discussion, the sketch of the reduction is generating 2^n equivalent semi-optimal paths whose heights are all numbers in Γ and then stretch the paths whose height has a zero digit in its m digit octal representation. So if Φ doesn't have a satisfying assignment, we have a longer solution for TDPP. To this end, we use three kinds of gadgets : *Splitter gadget*, *Filter gadget* and *Eliminator gadget*. We assign a *legal region* to the left side of each gadget such that each gadget performs a special operation on any incoming semi-optimal path that enters horizontally to its legal region.

The Splitter gadget $S(r)$ has one positive real parameter r. This gadget generates two equivalent semi-optimal paths from any incoming semi-optimal path whose heights differ by r. Precisely, if an incoming semi-optimal path with height h_0 enters horizontally into the legal region of this gadget, it can leave this gadget at one of two equivalent possible heights $h_0 + c_{S(r)}$ and $h_0 + c_{S(r)} + r$ while it remains semi-optimal where $c_{S(r)}$ is constant for this gadget. Therefore, it doubles the number of incoming semi-optimal paths. Note that some of these paths may be coincident after leaving the gadget.

The Filter gadget $F(d)$ with non-negative integer parameter d, increases the length of any incoming semi-optimal path of height $(x_d...x_0)_8$ with $x_d = 0$ by a sufficiently small $\epsilon > 0$ compare to any incoming semi-optimal path with $x_d \neq 0$. In fact, this increment guarantees that the length of any *extension* of paths with $x_d = 0$ is greater than L_Φ, where extension, we meant following this path to traverse next gadgets to reach t (Note that some of these extensions can still be an optimal path).

An Eliminator gadget $E(d)$ with the non-negative integer parameter d, eliminates the leftmost digit of height of any incoming semi-optimal path whose height is a $d+1$ digit octal number. Strictly speaking, if an incoming semi-optimal path with height $(x_d x_{d-1}...x_0)_8$ enters to the legal region of $E(d)$, it leaves this gadget with height $(x_{d-1}...x_0)_8$ while it remains semi-optimal. Also $E(0)$ reduces the height of any incoming semi-optimal path with height $(x_0)_8$ to 0.

Now, we can describe the structure of $\mathcal{P}_\Phi$ as follows : By locating s at an appropriate height and putting n consecutive splitter gadgets $(S_1, \ldots, S_n)$ in front of s, we generate 2^n equivalent semi-optimal paths whose heights are exactly all numbers of Γ. Then, we stretch all semi-optimal paths whose heights have a zero digit in their octal representation. This can be done by checking all m digits of heights of these paths by putting m consecutive Filter-Eliminator gadgets $(F_{m-1}E_{m-1}, \ldots, F_0E_0)$ and putting t after the last Eliminator. We set L_Φ as the length of an optimal path if we remove all filter gadgets. So, $\Phi(X, C)$ has a satisfying assignment if and only if the length of a solution of TDPP on $\mathcal{P}_\Phi$ is no longer than L_Φ. Fig. 2 shows the total structure of $\mathcal{P}_\Phi$.

Now, we go through the details of the gadgets. We inductively arrange these gadgets in the plane in such a way that all semi-optimal paths must traverse horizontally from s to the first gadget, between the gadgets and from the last gadget to t. In addition, each gadget must be located in the plane in such a way that all leaving semi-optimal paths from its previous gadget horizontally enter to the legal region.

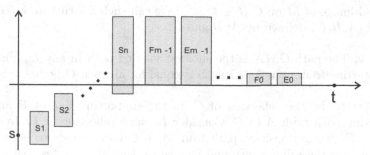

Fig. 2. Total configuration of $\mathcal{P}_\Phi$

Fig. 3 shows the structure of the Splitter gadget $S(r)$. This gadget consists of 8 disjoint polygons $(P_1, \ldots, P_8)$ and generates two equivalent semi-optimal paths from each incoming semi-optimal path. Let p be such a path and O be the last intersection of p and the last gadget before entering into the Splitter. After p intersects P_1, it has two equivalent choices to intersect P_2. Let p_1 and p_2 be such paths as indicated in the figure. Because the angle between P_3 and horizontal line is $-\pi/4$, these paths traverse equal distances to reach P_4.

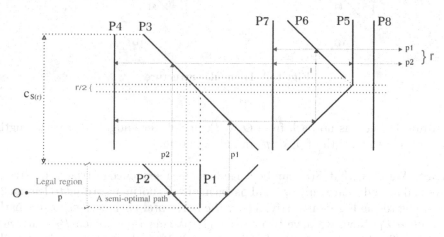

Fig. 3. Splitter gadget S(r)

On the other hand, the distance that p_1 traverses from P_5 to P_6 is equal to the distance that p_2 traverses from the point I to P_5 plus the distance it traverses from P_5 to the point I to reach P_7. So, the distances that p_1 and p_2 traverse from P_4 to P_8 are equal. So, the distances that p_1 and p_2 traverse from O to P_8 are equal.

Let (A, B, C) be a sequence of non-intersecting line segments that completely lie on the isosceles triangle of their supporting lines. Also, the supporting line of A is either vertical or horizontal and is perpendicular to the supporting line of C. Let O be a point on A such that M is its C directed image on B and N is

A directed image of *M* on *C* (Fig 4. (a)). We call such a structure a *triangular structure* (A, B, C) defined by *A*, *B* and *C*.

Lemma 2. The path OMN is the shortest visiting path in any L_p norm from *A* to *C* starting from *O* and its length is equal for all such *O* points.

Proof. Let C' be the reflection of *C* on the supporting line of *B* and OIJ be a visiting path from *A* to *C*. Consider IJ' as a reflection of IJ by the extension of *B*. So, any visiting path from *A* to *C* corresponds to a path from *A* to C' having the same length and vice versa. But, the shortest path from *A* to C' in any L_p norm is the straight line that corresponds to the path OMN (Fig. 4(b)).   □

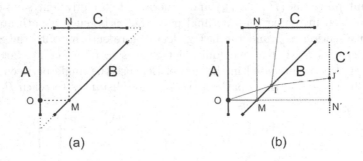

Fig. 4. A Triangular Structure

Lemma 3. There is no path from *O* to P_8 in Splitter gadget $S(r)$ with length shorter than the length of p_1 or p_2 in any L_p norm.

Proof. We show that $S(r)$ can be considered as a sequence of triangular structures and p_1 and p_2 are only optimal paths in any L_p norm according to lemma 2. According to the Fig. 5, it is trivial that the horizontal path is the shortest path from *O* to P_1. Now, we have two cases : (1) passing P_1 to reach P_2 (path p_1) and (2) reflecting at P_1 to reach P_2 (path p_2). In each case an optimal path should intersect DF to reach P_3. So for the case (1), we have the triangular structure (P_1, BC, EF) and for the case (2) we have (P_1, AB, DE) and then, we have (DF, P_3, P_4). It is obvious that there is no path for case (2) from P_4 to P_8 with length shorter than the length that p_2 traverses from P_4 to P_8. In the case (1), an optimal path should intersect P_4 in HG and we have (HG, IJ, LK) and then (LK, P_6, NM) to reach P_7 and a straight line to P_8. Therefore, there is no path from P_1 to P_8 with the length shorter than the length of p_1 and p_2 in any L_p norm and the result of this lemma follows.   □.

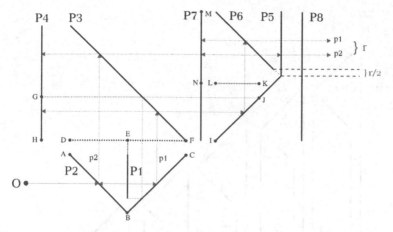

Fig. 5. Splitter proof

Fig. 6 shows the structure of the Filter gadget $F(d)$. Let p be an incoming semi-optimal path for this gadget with height $(x_d, \ldots, x_0)_8$. This gadget, increases the length of p by a fixed $\epsilon > 0$ if $x_d = 0$ and let p pass the gadget without bending if $x_d \neq 0$. It is trivial that in both cases, p can't traverse the gadget by a shorter distance.

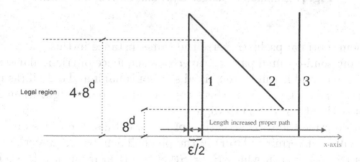

Fig. 6. Filter Gadget $F(d)$

Fig. 7 shows the structure of the Eliminator gadget $E(d)$ that decreases the height of any incoming semi-optimal path p with height $(x_d, \ldots, x_0)_8$ by removing the $(d+1)$'th digit (which is the highest digit) of the octal representation of its height.

The legal region of this gadget is all heights $(x_d, \ldots, x_0)_8$ such that $x_i \in \{0, 1, 2, 3\}$ $(0 \le i \le d)$. In this gadget, for all $0 \le k \le 5$ the upper endpoint $3k + 2$ polygons is sufficiently close to the intersection point of segments of $3k + 1$ polygons. Since all digits of heights of all incoming semi-optimal paths are in $\{0, 1, 2, 3\}$, these paths are not close to the joint point of $3k + 1$ polygons and they enter to the legal region of this gadget. It is easy to check that all

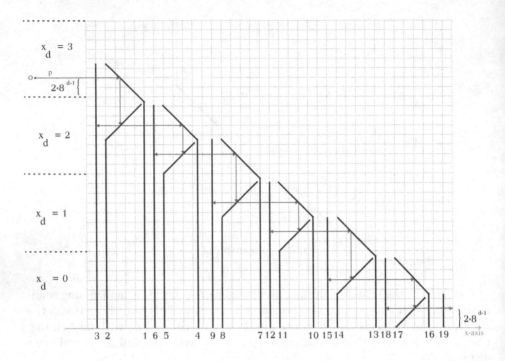

Fig. 7. Eliminator Gadget $E(d)$ and an example path

incoming semi-optimal paths traverse the same distance in this gadget. In fact, each incoming semi-optimal path p traverses a sequence of triangular structures and straight lines and it will be remained semi-optimal if and only if (as it shown in Fig. 7) its height when leaving this gadget is equal to its height when entering this gadget without its $(d + 1)$'th digit.

Now, we describe the total structure of $\mathcal{P}_\Phi$ in details. We put s on the point $(0, y_0)$ (y_0 will be determined later) in the plane. Then, we arrange the sequence of splitters $(S_1, \ldots, S_n)$ in which $S_i = S(|\alpha_i - \beta_i|)$ in front of s in such a way that a semi-optimal path from s goes directly and horizontally to the legal region of S_1 and for $2 \leq i \leq n$, we put a sufficiently large S_i in front of S_{i-1} such that all possible semi-optimal paths of S_{i-1} go directly and horizontally to the legal region of S_i. By choosing y_0 in such a way that the height of the lowest semi-optimal path of S_n be equal to the minimum element of Γ, the heights of all possible semi-optimal paths of S_n exactly correspond to all the numbers in Γ. Note that all semi-optimal paths of S_n have non-negative integer heights less than 8^m. Then, we use the sequence $(F(m-1), E(m-1), ..., F(0), E(0))$ of Filter-Eliminator gadgets to force semi-optimal paths with zero digit in the octal representation of their heights traverse longer distances. Finally, all semi-optimal paths must leave $E(0)$ with height zero. We end $\mathcal{P}_\Phi$ by putting the target point t in front of $E(0)$ along the x-axis. So we have $\mathcal{P}_\Phi = (s, S_1, \ldots, S_n, F(m-1), E(m-1), \ldots, F(0)E(0), t)$. Let G be any Splitter or Eliminator gadget.

We let $L(G)$ be the distance that any optimal path traverses in G. Let G_1 and G_2 be gadgets in $\mathcal{P}_\Phi$ such that G_1 appears earlier than G_2 in the sequence and let $D(G_1, G_2)$ be the horizontal distance between the last segment of G_1 and the first segment of G_2. So, we define L_Φ as

$$L_\Phi = D(s, S_1) + \sum_{i=1}^{n-1}(L(S_i) + D(S_i, S_{i+1})) + L(S_n) + D(S_n, E(m-1))$$

$$+ L(E_{m-1}) + \sum_{i=0}^{m-2}(L(E(i)) + D(E(i+1), E(i))) + D(E(0), t).$$

So L_Φ is polynomially computable and according to the reduction $\Phi(X, C)$ has a satisfying assignment if and only if the length of the solution of the TDPP on $\mathcal{P}_\Phi$ is not greater than L_Φ.

3   Conclusion

In this paper, we proved that the touring polygons problem is NP-hard for disjoint polygons. This result complements the proof of Dror *et.al* [3] given only for the intersecting polygons. In our reduction we use segments with the exponential length ratio. What is the complexity of TDPP for polygons such that the ratio between each pair of them are constant or polynomial ? We leave this question as an open problem. Also there is no hardness result for TPP and TDPP. Finding such hardness results can be considered as future works.

References

1. Canny, J., Reif, J.H.: New lower bound techniques for robot motion planning problems. In: Proc. 28th Annu. IEEE Sympos. Found. Comput. Sci., pp. 49–60 (1987)
2. Chin, W., Ntafos, S.: Shortest Watchman Routes in Simple Polygons. Discrete and Computational Geometry 6(1), 9–31 (1991)
3. Dror, M., Efrat, A., Lubiw, A., Mitchell, J.: Touring a sequence of polygons. In: STOC 2003, pp. 473–482 (2003)
4. Dror, M.: Polygon plate-cutting with a given order. IIE Transactions 31, 271–274 (1999)
5. Hershberger, J., Snoeyink, J.: An efficient solution to the zookeeper's problem. In: Proc. 6th Canadian Conf. on Comp. Geometry, pp. 104–109 (1994)
6. Li, F., Klette, R.: Rubberband algorithms for solving various 2D or 3D shortest path problems. In: Proc. Computing: Theory and Applications, The Indian Statistical Institute, The Indian Statistical Institute, Kolkata, pp. 9–18. IEEE (2007)
7. Li, F., Klette, R.: Approximate Algorithms for Touring a Sequence of Polygons. MI-tech TR-24, The University of Auckland, Auckland (2008),
   http://www.mi.auckland.ac.nz/tech-reports/MItech-TR-24.pdf

8. Tan, X., Hirata, T.: Constructing Shortest Watchman Routes by Divide and Conquer. In: Ng, K.W., Balasubramanian, N.V., Raghavan, P., Chin, F.Y.L. (eds.) ISAAC 1993. LNCS, vol. 762, pp. 68–77. Springer, Heidelberg (1993)
9. Tan, X., Hirata, T.: Shortest safari routes in simple polygons. In: Du, D.-Z., Zhang, X.-S. (eds.) ISAAC 1994. LNCS, vol. 834, pp. 523–531. Springer, Heidelberg (1994)
10. Pan, X., Li, F., Klette, R.: Approximate shortest path algorithms for sequences of pairwise disjoint simple polygons. In: Proc. Canadian Conf. Computational Geometry, Winnipeg, Canada, pp. 175–178 (2010)

Walking in Streets with Minimal Sensing

Azadeh Tabatabaei[1] and Mohammad Ghodsi[2,*]

[1] Department of Computer Engineering, Sharif University of Technology
atabatabaei@ce.sharif.edu
[2] Sharif University of Technology and School of Computer Science,
Institute for Research in Fundamental Sciences (IPM)
ghodsi@sharif.edu

Abstract. We consider the problem of walking in an unknown street, starting from a point s, to reach a target t by a robot which has a minimal sensing capability. The goal is to decrease the traversed path as short as possible. The robot cannot infer any geometric properties of the environment such as coordinates, angles or distances. The robot is equipped with a sensor that can only detect the discontinuities in the depth information (gaps) and can locate the target point as soon as it enters in its visibility region. In addition, a pebble as an identifiable point is available to the robot to mark some position of the street. We offer a data structure similar to Gap Navigation Tree to maintain the essential sensed data to explore the street. We present an *online* strategy that guides such a robot to navigate the scene to reach the target, based only on what is sensed at each point and is saved in the data structure. Although the robot has a limited capability, we show that the detour from the shortest path can be restricted such that generated path by our strategy is at most 11 times as long as the shortest path to target.

1 Introduction

Path planning is one of the basic problems in computational geometry, online algorithms, and robotics [6, 8, 12]. Specifically, path planning appears in many applications where the environment is unknown and no geometric map of the scene is available [3]. In robot path planning, the robot's sensor is the only tool to collect information from the scene, and the volume of the information gathered from the environment depends on the capability of the sensor. A robot with a simple sensing model has many advantage such as: it is low cost, less sensitive to failure, robust against sensing uncertainty and noise, and applicable to many situations [3].

The robot that we use in this research, has a limited ability. It has an abstract sensor that can only detect the order of discontinuities in the depth information (or gaps) in its visibility region. Each discontinuity corresponds to a portion of the environment that is not visible to the robot, (Fig. 1). The robot assigns to every gap g a label L or R depending on which side of the gap the hidden

---

* This research was in part supported by a grant from IPM. (No. CS1390-2-01).

P. Widmayer, Y. Xu, and B. Zhu (Eds.): COCOA 2013, LNCS 8287, pp. 361–372, 2013.

region is. Also, the robot recognizes a target point t when it is in the robot's omnidirectional and unbounded field of view. In order to cover the hidden region behind each gap, the robot moves towards the gap in an arbitrary steps. Note that the robot cannot measure any angles or distances to the walls of the scene or infer its position. In addition, we assume that the robot has access to a single pebble which is a detectable object that can be put anyplace and can be lifted again.

Throughout this paper, the workspace is assumed to be a restricted simple polygon called a street. A simple polygon P with two vertices s and t is called a street if the counter-clockwise polygonal chain R_{chain} from s to t and the clockwise chain L_{chain} from s to t are mutually weakly visible [7]. This means that each point on the left chain L_{chain} can see at least one point on the right chain R_{chain} and vice versa, (Fig. 1.a). In some literatures, a street is also known as L-R visible polygon [2]. A point robot that is equipped with the gap sensor starts navigating this environment from s to reach its target t. The robot has no geometric map of the scene and only based on the information gathered through the sensor has to make decisions to achieve the target.

Klein proposed the first competitive *online* strategy for searching a target point in a street [7]; called *walking in streets*. The robot employed in [7] is equipped with a 360 degree vision system. Also, it can measure each angle or distance to the walls of the street. As the robot moves, a partial map is constructed from what has been seen so far. Klein proved an upper bound of 5.72 for the competitive ratio (the ratio of the length of the traversed path to the shortest path from s to t) of this problem. Also, it was proved later that there is no strategy with the competitive ratio less than $\sqrt{2}$ for this problem. A strategy similar to Klein's with the competitive ratio of $\pi + 1$ has been introduced in [9, 10] which is robust under small navigation errors. Other researchers have presented several algorithms with the competitive ratios between $\sqrt{2}$ and the upper bound of 5.72 [8, 10]. Icking *et al.* presented an optimal strategy with the competitive ratio of $\sqrt{2}$ [6].

The limited sensing model that we use in this paper was first introduced by Lavalle *et al.* [16]. Gap Navigation Tree (GNT) has been proposed to maintain and update the gaps seen along the navigating path. This tree is built by detecting the discontinuities in the depth information and updated by the topological changes of the information. The topological changes are: appearances, disappearances, merges, and splits of gaps. Once the GNT is completed, it can encode the shortest path from its root (start point of the navigation) to any place in a simply connected environment. It is shown in [15] that, using this data structure, the globally optimal navigation is impossible in multiply connected environments, but locally optimal exploration can be achieved. Guilamo *et al.* [5, 13] presented an online algorithm for the well-known visibility problem pursuit-evasion in an unknown simply connected environment using GNT. As mentioned in [15], GNT is well suited for solving other visibility problems. An optimal search strategy using GNT is presented for a disc robot to find a target point t, starting from s in a simply connected environment [11].

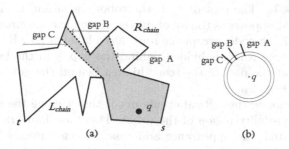

Fig. 1. (a) A street in which L_{chain} is the left chain and R_{chain} is the right chain. The colored region is the visibility polygon of the point robot q in the street. (b) The position of discontinuities in the depth information detected by the sensor.

Another minimal sensing model introduced by Suri and Vicari [14] for a simple robot. They assume that the robot can only sense the combinatorial (non-metric) properties of their surroundings. The sensor can detect vertices of the polygon in its visibility region, and can report if there is a polygon edge between consecutive vertices. The information maintain in two combinatorial vectors, called the combinatorial visibility vector (cvv) and the point identification vector (piv). Despite of minimal capability, they shown the robot can obtain many geometric reasoning and can accomplish many non-trivial tasks.

In this paper we propose an *online* search strategy for a point robot equipped with the gap sensor and the single pebble to reach the target point t in a street environment, starting from s. The minimal sensing model that we use here is in contrast with the strong sensing model that Klein used for walking in streets problem. A data structure that is maintained and updated similar to GNT is introduced for designing the robot search path. We show that the search path which is generated by our strategy is at most 11 times as long as the shortest path. Also, we show that if the robot has access to many pebbles, this ratio reduces to 9. To our knowledge, this is the first result providing some competitive ratio for walking in streets with the minimal sensing model.

2   GNT Data Structure and the Sensing Model

2.1   Gap Sensor

Gap sensor is a naive visual sensing model. At any position q of the environment, a cyclically ordered location of the depth discontinuities in the visibility region of the point $(V(q))$ is what the robot's sensor detects, as shown in Fig. 1. When the robot reports the discontinuities counterclockwise from a visibility region, it assigns a left label to a transition from far to near and assigns a right label to a transition from near to far [15]. The robot can only walk towards the gaps.

GNT data structure has been introduced as a mean to navigate in an unknown scene for the robot system. Here, we briefly explain the data structure from [15], and refer to it as T_g. The root of T_g is the robot's location. Each child of the root is a gap g that appears as the robot moves; these gaps are circularly ordered around the root. Each node, except the root, has a label of L or R. L means that the part of the scene which is hidden behind the gap is in the left side of the gap. R means that the part of the scene hidden behind the gap is in the right side of the gap, (Fig. 2).

As the robot moves, the critical events occur that change the combinatorial structure of the visibility region of the robot. There are four critical events in which T_g is updated: the appearance and disappearance events happen when the robot crosses the inflection rays, the merge and split events occur when the robot crosses the bitangent complements. In the disappearance event in which a gap g disappears, the node g will be eliminated from T_g. When a gap appears, a child is augmented to the root of T_g in a location that the circular ordering of the gaps is maintained. Each of these added nodes shows a portion of the environment that was so far visible, and now is invisible. These new nodes are specified as primitive (others are non-primitive). If a gap g splits into g_1 and g_2, then it will be replaced by the new nodes g_1 and g_2, (Fig. 2). If two gaps g_1 and g_2 merge into g , then g_1 and g_2, the adjacent children of the root, will be the children of a new node g which is added to the root.

The robot follows the non-primitive gaps until it reaches a point at which all leaf nodes are primitive. At this point, the robot has observed the entire environment. This data structure, after completion, can encode the shortest paths from the start point to any point of the environment.

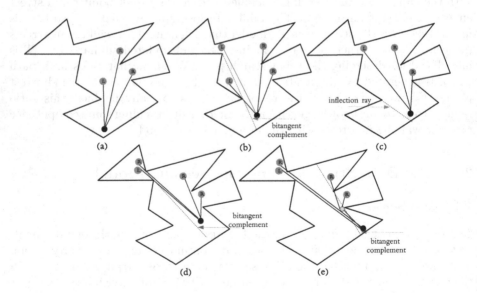

Fig. 2. The dark circle denotes the location of the robot. (a) Existing gaps at the beginning. (b) A split event. (c) A disappearance event. (d) Another split event. (e) A merge event.

2.2   The Sensor and Motion Primitive

All times, our robot's sensor reports the gaps, with their labels, in their counterclockwise cyclic order as they appear in its visibility region. The robot carries a pebble which is a marker device and is distinguishable for it. The robot can orient its heading with the gaps, and walks towards them in an arbitrary number of steps, for example: 2 steps towards a gap g_x, or 4 steps towards a gap g_y. Each step is a constant distance which is already specified for the robot by its manufacturer, for example it may be 1 meter, 2 meters and etcetera. When the robot moves towards a gap, it comes close to the gap, but the robot cannot report its distance to gaps and walls, size of gaps, and angles. Whenever a new event in the sensing of the environment happens, the robot can stop to make a reliable decision to reach the target. Also, the robot can move towards the pebble and the target, as soon as they enter in its visibility region, until it touches them.

3   Preliminarily Results

At each point p of the robot's search path, the gap sensor either sees the target, or achieves a set of gaps with the label of L or R (l-gap and r-gap for abbreviation). If the target is seen, the robot moves towards the goal and reaches it. In the other case in which the robot reports the position of the gaps (nonprimitive gaps), the robot should move towards the gaps to achieve the target.

Definition 1. *In the set of l-gaps, the gap which is in the right side of the others is called the most advanced left gap and is denoted by g_l. Analogously, in the set of r-gaps, the gap which is in the left side of the others is called the most advanced right gap and is denoted by g_r, (Fig. 3.a).*

Each gap is adjacent to a reflex vertex. The corresponding reflex vertices of g_l and g_r are denoted by v_l and v_r, (Fig. 3.a). The two gaps have the following property.

Lemma 1. *On any point of the robot search path, if the target is not visible, then it is behind one of the most advanced gaps.*

Proof. Let the target be behind of another gap, except g_l or g_r. Without loss of generality, it is behind an r-gap, so the points that are immediately behind g_r are not visible by any point on the opposite chain, this contradicts the definition of the street.

Above attribute of the two gaps is similar to the main feature of top most left packet and top most right pocket in [7].

As the robot moves in the environment, g_l and g_r may dynamically change. The critical events in which the structure of the robot's visibility region changes, can also change g_l and g_r. In the next section, we show how the critical events change the left most advanced gaps such that a sequence of the left most advanced gaps, $[g_{l1}, g_{l2}, ..., g_{lm}]$, appears in the robot's visibility region, while

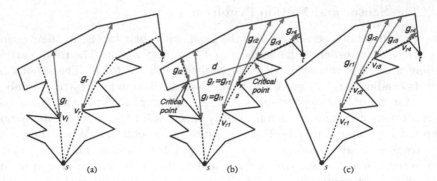

Fig. 3. (a) g_r and g_l are the most advanced gaps at the start point s. v_r and v_l are the corresponding reflex vertices. (b) Sequences of the most advanced gaps may occur, as the robot moves. The funnel situation which ends as soon as the robot crosses over the segment d. Dotted chains, starting from s, are the two convex chains of the funnel. (c) In this case there is only one most advanced gap, at start point s.

exploring the street. Similarly the sequence of the right most advanced gaps, $[g_{r1}, g_{r2}, ..., g_{rn}]$, may occur, (Fig. 3.b).

At each point, if there is exactly one of the two gaps (g_r or g_l), then the goal is hidden behind that gap. Thus, there is no ambiguity and the robot moves towards the gap, (Fig. 3.c). If both of g_r and g_l exist, then the target is hidden behind one of these gaps. This case is called a *funnel*, (Fig. 3.b). As soon as the robot enters a point in which both of g_r and g_l exist, a funnel situation starts. This case continues until one of g_r or g_l disappears, (Fig. 3.b), or they become collinear, (point 2 in Fig. 4.a). When the robot enters a point in which there is a funnel situation, the only non-trivial case in this navigation occurs.

In a funnel situation, previous strategies proposed by Kelin *et.al* [6–10] were based on choosing a walking direction within the angle between v_r and v_l. In other words, in this case, their robots select a point to move towards which is in equal distance with v_r and v_l and repeat this process until the funnel case ends. But, the robot that we use in this research cannot compute the point between v_r and v_l. So, applying their strategy for this robot is impossible. Before describing our strategy, we state some features of a street and the gaps that are applied in the algorithm.

When the robot enters in a funnel situation, there are two convex chains in front of it: *the left convex chain* that lies on the left chain (L_{chain}) of the street, and *the right convex chain* that lies on the right chain (R_{chain}) of the street, (Fig. 3.b). The two chains have the following main property.

Lemma 2. *When a funnel situation starts, shortest path from s to t lies completely on the left convex chain, or on the right convex chain of the funnel.*

Proof. Obviously, the point in which the funnel situation starts, belongs to shortest path from s to t. So, this claim is a straight result of the lemma 1 and the theorem below.

Theorem 1. *[4] For any vertex $v_j \in L_{chain}(or, v_j \in R_{chain})$, shortest path from s to v_j makes a left turn (respectively, a right turn) at every vertex of L_{chain} (respectively, R_{chain}) in the path.*

Definition 2. *Between the two convex chains, in a funnel situation, the one which is a part of the shortest path is called exact chain of the funnel.*

Lemma 3. *Each of the two convex chains, in a funnel situation, contains a point in which the funnel situation ends or a new funnel situation starts.*

Proof. While the robot explores the street in a funnel case, the situation ends in two conditions: (1) When the robot enters a point in which one of the most advanced gaps (g_l or g_r) disappears. The inflection ray of the gap intersects the two convex chains. The intersection points are the points which are claimed, (Fig. 3.b). (2) When the robot enters a point in which the two most advanced gaps are collinear. The bitangent of the corresponding reflex vertices of the current most advanced gaps intersects the two convex chains. So, the claimed points exist, (points 2 and 3 in Fig. 4.a).

We refer to the points, which is mentioned in the above lemma as *funnel critical points*. Clearly, one of the two points belongs to shortest path from s to t.

Lemma 4. *Assume the robot is walking along one of the convex chain of a funnel. The exact chain of the funnel can be specified as soon as the robot touches the critical point that belongs to the chain.*

Proof. There are two situations in which the robot touches a critical point: (1) The robot reaches a point in which one of the most advanced gaps disappears, obviously the convex chain which contains the existing gap is the exact chain, (Fig. 3.b). (2) The robot reaches a point in which g_l and g_r are collinear. If the point is the corresponding reflex vertex of the gap that the robot was moving towards, the chain that the robot was walking on it is the exact chain, (point 3 in Fig. 4.a). Otherwise the other chain is the exact chain, (point 2 in Fig. 4.a).

4   Main Strategy

Now, we explain our strategy for the robot to move in the street from s to t such that the generated path is at most a constant times as long as the shortest path. In the situation in which only one of the most advanced gaps exists each reasonable strategy directs the robot towards the gap.

The robot, based on the information gathered through its sensor and the pebble which it is equipped with, searches the scene. In the funnel situation, we lead the robot to reach the critical point. Our idea for directing the robot in this case is inspired by the algorithm for searching a point on a line, called doubling. In the doubling strategy, the robot moves back and forth on the line, such that at each stage i, it walks 2^i steps in one direction, comes back to the origin, walks 2^{i+1} steps in the opposite direction until the target is reached.

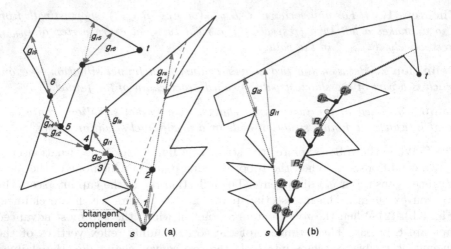

Fig. 4. (a)There is a funnel situation at start point s. Points 2 and 3 are the critical points of the funnel. g_{rs} and g_{ls} are the most advanced gaps at the start point. g_{ri} and g_{li} are the most advanced gaps at point i, for $i = 1, 2, ..., 6$. (b) Illustration of constructing and updating the data structure, as the robot walks along the right convex chain. Dark circle denote the robot's location and it is the root of the data structure. The path leads to R_g is the return path.

Theorem 2. *[1] The doubling strategy for searching a point on a line has a competitive factor of 9, and this is optimal.*

If we assume the two convex chains as a line then, by applying the doubling strategy on this line, we can find the critical point. Therefore, directing the robot along these two chains avoiding any detour to other places of the environment is what is important in this exploration.

Lemma 5. *The robot traces the left/ right convex chain and detects its critical point if and only it walks towards the left/ right most advanced gap maintaining the dynamically changes of the two most advanced gaps.*

Proof. When a funnel situation starts, the first vertex of the left/ right convex chain coincides with the corresponding reflex vertex of the currently left/ right most advanced gap. This most advanced gap doesn't change until the robot touches the reflex vertex. Then the first segment of the convex chain and robot's path are the same. Other segments are similarly coincident. As soon as the robot reaches a point in which one of the most advanced gaps disappears, or they are collinear, the critical point is achieved.

In the following subsection, we describe the process of constructing and updating the required data structure for leading the robot along the two convex chains and coming back to the origin (the point in which funnel case starts).

4.1   Data Structure

In the funnel situation, the robot puts a pebble on the point to mark this point as origin. In order to follow each of the two convex chains to reach the critical point, it must dynamically maintain the changes of g_r and g_l. Furthermore, the required information to come back to the origin must be preserved. This data is saved in a tree which we called S-GNT (street GNT).

The root of this tree is the start point of the funnel (current location of the robot). g_r and g_l are the only leaf of the tree, at the point. As the robot moves, the critical events, appearance, disappearance, merge and split, may dynamically change g_r and g_l. Moreover, these critical events generate the comeback path to origin as follows: (Assume the robot follows the right convex chain, in other words it moves towards g_r. The situation in which it moves towards g_l is symmetric and the S-GNT is constructed analogously.)

– When the robot crosses a bitangent complement of g_l and another l-gap, then g_l splits and will be replaced by the l-gap, (point 1 in Fig. 4.a).
– When the robot crosses a bitangent complement of g_l and an r-gap, then g_l splits into two gaps. g_r will be replaced by the r-gap. At this point g_l and g_r are collinear and the funnel situation ends, (point 2 in Fig. 4.a).
– When the robot crosses a bitangent of g_r and another r-gap, at the point in which g_r disappears, g_r will be replaced by the r-gap, in the tree. (disappearance and split events occur simultaneously.) In this situation, if there are more than one gap similar to the r-gap, g_r will be replaced by the one which is in the left side of the others, (point v_{r1} in Fig. 3.b).
– When the robot crosses a bitangent of g_r and another l-gap, at the point in which g_r disappears, g_l will be replaced by the l-gap, in the tree. (disappearance and split events occur simultaneously.) In this situation, if there are more than one gap similar to the l-gap, g_l will be replaced by the one which is in the right side of the others, (point 5 in Fig. 4.a).
– When the robot crosses over an inflection ray, each of g_l or g_r which is adjacent to the ray, disappears and is eliminated from the data structure. each of the critical point of the funnel in Fig. 3.b is an example for this event.
– When the robot crosses over an inflection ray, a gap may appear. If this gap hides the pebble that was so far visible, a child is augmented to the root of S-GNT in a location that the circular ordering of the gap and g_r and g_l is maintained. We refer to this gap as comeback gap. This gap is maintained in the tree for generating the comeback path to the origin, (point v_{r1} in Fig. 3.b). Other appearance events don't change the data structure.
– When the robot crosses a bitangent of reflex vertex of g_r and reflex vertex of the comeback gap, these two gaps merge. So, the comeback gap will be a child of a new node which is added to the root, (point v_{r2} in Fig. 3.b).

Note, the last two events update the data structure such that the return path to the origin is generated. In Fig. 4.b, the process of constructing the data structure, as the robot traces the right convex chain in the funnel situation in Fig. 3.b, is illustrated.

4.2   Algorithm

The robot starts navigating the environment based on the information gathered about the most advanced gaps until reaches a point in which the target is visible.

At each point, if there is exactly one of the two gaps (g_r or g_l), then the goal is hidden behind that gap. Thus, there is no ambiguity and the robot moves towards the gap.

In the funnel case in which both of g_r and g_l exist, the robot is not sure that the target is hidden behind which of these gaps. The robot put a pebble at this point, and saves the location of the two most advanced gaps at this point as g_{rf} and g_{lf}. At each stage i, the robot while constructing S-GNT, walks 2^i steps along the right convex chain, and returns to origin by following the return path, then walks 2^{i+1} steps along the opposite convex chain until a critical point of the funnel is achieved. As soon as the robot touches a critical point of current funnel, from lemma 4, the exact chain of the funnel is determined. So, at the critical point, the robot returns to the origin to pick up the pebble and walks along the exact chain to reach the target while constructing the S-GNT. The robot continues walking along the convex chain until the target is achieved or a new funnel case starts again. In the later case, the procedure for a funnel case (the doubling procedure) is repeated.

Note that at each stage i in a funnel case that the robot start going forth along one of the two convex chains, g_r and g_l in S-GNT are set to g_{rf} and g_{lf}, and as the robot moves S-GNT dynamically is constructed again, as explained in the previous section. Also, When no pebble is put on the environment, no return path generates in the S-GNT.

5   Correctness and Analysis

In this section, we show that the robot by following the path generated with our strategy achieves the target t starting from s. Also, we compare the length of the generated path with the shortest path and prove a constant competitive ratio for our strategy.

Theorem 3. *By executing our strategy, the robot rightly reaches target t, starting from start point s.*

Proof. In the walking in streets problem, the target constantly lies behind g_l or g_r. Thus, the events in which g_r and g_l are updated must be considered as critical events in the problem. Now, we show that these critical events are only the two types of the critical events: Split and disappearance. Each appearance event creates a primitive gap which was once visible by the robot. Obviously, the target is not behind this gap. A critical event which merges g_r and g_l occurs when the robot crosses the bitangent complement of the corresponding reflex vertices of the two gaps. As shown in Fig. 4.a, the bitangent complement either is in the left side of the line which connects the current position of the robot to g_l or is in the right side of the line which connects the current position of the robot to g_r. According to the algorithm, the robot cannot cross over the bitangent complement. Also, a most advanced gap merges with another gap at the point in which the most advanced gap will disappear. Hence, just the split and the disappearance are the critical events which change g_r and g_l. The merge and appearance are handled for constructing the return path to origin in data structure S-GNT.

We now compare the length of the path constructed by our *online* search strategy, and the length of shortest path. Each *online* walking strategy for a robot with the minimal sensing capability (the gap sensor) can significantly detour from the shortest path. Here, we prove a competitive ratio for the length of the generated search path by the algorithm.

Lemma 6. *In each funnel case, if we eliminate the robot movement to reach the critical point of the funnel, and comeback path to the origin of the funnel from the generated path by our strategy, the remained path and shortest path are coincide.*

Proof. When a funnel situation stars, the robot isn't sure which chain is the exact convex chain. From lemma 4, if the robot achieves the critical point, the exact chain is specified. So, the only detour from the shortest path is the movement to reach the critical point and comeback path to origin, in each funnel case.

Theorem 4. *By executing our strategy the robot can search a goal in an unknown street with a competitive ratio of at most 11. If the robot was allowed carrying many pebbles, it can search a goal with a competitive ratio of at most 9.*

Proof. By lemma 6, if an algorithm achieves a competitive factor in each funnel case, then it achieves the same ratio in every streets. So, we compare these two paths in one funnel in order to find the detour from the shortest path. The robot, using the information gather through its sensor about the most advanced gaps, searches the convex chains of the funnel, by executing the doubling strategy, until it reaches the critical point of the funnel. The robot traverses at most 9 times as long as the shortest path to reach the critical point. At this point, the robot comes back to the origin to pick up the pebble then walks along the exact chain. So, in each funnel situation the robot traverses at most 11 times as long as the length of shortest path to reach the critical point of the funnel which is on the shortest path.

6   Conclusions

In this study, we proposed an *online* strategy for the walking in streets problem for a point robot that has a minimal sensing capability. The robot can only detect the gaps and the target in the street. Also, it carry a pebble to mark some locations of the environment. Our strategy generates a path with a bounded detour from the shortest. We proved that our strategy has a competitive ratio of 11. Improving this upper bound can be considered as an opportunity for future research. Introducing more general classes of polygons which admit competitive searching with minimal sensing is an interesting open problem.

References

1. Baezayates, R.A., Culberson, J.C., Rawlins, G.J.: Searching in the plane. Information and Computation 106(2), 234–252 (1993)
2. Das, G., Heffernan, P.J., Narasimhan, G.: LR-visibility in polygons. Computational Geometry 7(1), 37–57 (1997)
3. Gfeller, B., Mihalák, M., Suri, S., Vicari, E., Widmayer, P.: Counting targets with mobile sensors in an unknown environment. In: Kutyłowski, M., Cichoń, J., Kubiak, P. (eds.) ALGOSENSORS 2007. LNCS, vol. 4837, pp. 32–45. Springer, Heidelberg (2008)
4. Ghosh, S.K.: Visibility algorithms in the plane. Cambridge University Press (2007)
5. Guilamo, L., Tovar, B., LaValle, S.M.: Pursuit-evasion in an unknown environment using gap navigation trees. In: Proceedings of the 2004 IEEE/RSJ International Conference on Intelligent Robots and Systems (IROS 2004), vol. 4, pp. 3456–3462. IEEE (September 2004)
6. Icking, C., Klein, R., Langetepe, E.: An optimal competitive strategy for walking in streets. In: Meinel, C., Tison, S. (eds.) STACS 1999. LNCS, vol. 1563, pp. 110–120. Springer, Heidelberg (1999)
7. Klein, R.: Walking an unknown street with bounded detour. Computational Geometry 1(6), 325–351 (1992)
8. Kleinberg, J.M.: On-line search in a simple polygon. In: Proceedings of the Fifth Annual ACM-SIAM Symposium on Discrete Algorithms, pp. 8–15. Society for Industrial and Applied Mathematics (January 1994)
9. Lopez-Ortiz, A., Adviser-Ragde, P.: On-line target searching in bounded and unbounded domains. University of Waterloo (1996)
10. Lopez-Ortiz, A., Schuierer, S.: Simple, efficient and robust strategies to traverse streets. In: Proc. 7th Canad. Conf. on Computational Geometry (1995)
11. Lopez-Padilla, R., Murrieta-Cid, R., LaValle, S.M.: Optimal Gap Navigation for a Disc Robot. In: Frazzoli, E., Lozano-Perez, T., Roy, N., Rus, D. (eds.) Algorithmic Foundations of Robotics X. STAR, vol. 86, pp. 123–138. Springer, Heidelberg (2013)
12. Mitchell, J.S.: Geometric shortest paths and network optimization. In: Handbook of computational geometry. Elsevier Science Publishers B.V. North-Holland, Amsterdam (1998)
13. Sachs, S., LaValle, S.M., Rajko, S.: Visibility-based pursuit-evasion in an unknown planar environment. The International Journal of Robotics Research 23(1), 3–26 (2004)
14. Suri, S., Vicari, E., Widmayer, P.: Simple robots with minimal sensing: From local visibility to global geometry. The International Journal of Robotics Research 27(9), 1055–1067 (2008)
15. Tovar, B., Murrieta-Cid, R., LaValle, S.M.: Distance-optimal navigation in an unknown environment without sensing distances. IEEE Transactions on Robotics 23(3), 506–518 (2007)
16. Tovar, B., La Valle, S.M., Murrieta, R.: Optimal navigation and object finding without geometric maps or localization. In: Proceedings of the IEEE International Conference on Robotics and Automation, ICRA 2003, vol. 1, pp. 464–470. IEEE (September 2003)

Robust Optimization for the Hazardous Materials Transportation Network Design Problem

Chunlin Xin[1,2], Qingge Letu[1,2], and Yin Bai[3]

[1] School of Economics and Management, Beijing University of Chemical Technology, 100029, China
[2] Research Center for Operations Management and Strategy Decision
[3] Data Mining Group, NEC Laboratories, 100084, China
xinchl@mail.buct.edu.cn

Abstract. We consider the problem of designing a transportation network for hazardous materials (HTNDP). For HTNDP, it was shown that deciding whether there exists an optimal path of risk 0 is NP-hard. A natural way to handle NP-hard problems is approximation solutions or FPT algorithms. We prove that HTNDP does not admit any approximation, neither any FPT algorithm, unless P=NP. The hazmat network design problem faces significant uncertainty in conflicting numbers of edge risk reported by different researchers and many factors affecting edge risk could induce different results since the edge risk is often difficult to characterize. In this paper, we use maximum regret criterion robust optimization to model the problem as a bi-level integer programming problem under edge risk uncertainty where an interval of possible risk values is associated with each arc. We present a heuristic approach that always finds a robust and stable hazmat network. At the end, we test our method on a random instance on a network of Guangdong province in China to illustrate the efficiency of our model and algorithm. Our experimental tests illustrate that the robust interval risk scenario network performs very well, and can handle the risk change better compared with the deterministic scenario network. Overall, the numerical analysis reveals that the maximum regret criterion robust optimization used in HTNDP is more conservative but has the merit of robustness.

Keywords: Hazmat network design, Computational Complexity, Robust optimization, Heuristic method.

1 Introduction

The production and transportation of hazardous materials (hazmats) plays an important role in industrial development. Although the probability of hazmat transportation accidents is low, the vast majority of transporting hazmats and inadequate supervision by the government causes frequent hazmat transportation accidents, which arouse close national attention in China. The total annual hazmat transport volume reached about 400 million tons in China. Among them,

P. Widmayer, Y. Xu, and B. Zhu (Eds.): COCOA 2013, LNCS 8287, pp. 373–386, 2013.

about 95% of hazmats need to be transported from one place to another and road transportation accounts for approximately 82%. What differentiates hazmat transport from the transportation of other materials is the risk associated with an accidental release of hazmats during transportation. Due to the nature of the hazmats, in case accidents happen, it may cause unpredictable harm to people, property and the environment. Thus, mitigation of hazmat transportation risk is an increasingly significant challenge and concern for the government of China. One way to mitigate hazmat transportation risk for government is to close certain roads to hazmat vehicles since it cannot force specific routes for individual shipments.

1.1   The Hazmat Transportation Network Design Problem

At present, most of the literature on hazmat transportation focused on risk assessment, routing and scheduling, and facility location. The Hazmat Transportation Network Design Problem (HTNDP) was first proposed by Kara and Verter in 2004 [2] and received more attention of researchers recently. They studied the problem of hazardous material transportation network design where the government can decide which road segments have to be closed to hazmats so as to minimize the overall risk of the shipments; and the carriers choose the routes on the designated network to minimize their route costs. HTNDP consists of finding a sub-graph of the existing road network so that the total risk resulting from the carriers' route choices is minimized. They provided a bi-level programming formulation for this network design problem, but such a bi-level model may fail to find a stable solution when there exist multiple minimum cost routes with different risk selected by the carriers in the follower model, which may result in a much higher risk than the government expected. Erkut and Alp [3] modeled HTNDP as a Steiner tree problem, and formulated the problem as a single level integer programming with the objective of minimizing the total risk. Erkut and Gzara [4] considered a similar problem to generalize their model to the undirected case. In order to protect the government from the worst case when the problem becomes unstable, they proposed a heuristic algorithm to handle the bi-level model stability. Verter and Kara [5] provided a single level path-based formulation for HTNDP, where a set of alternative paths for each hazmat commodity incorporate carriers' cost concerns in the government's risk-reduction decisions. Amaldi and Bruglieri [6] proved that HTNDP, where a set of arcs can be forbidden, is NP-hard even when a single commodity has to be shipped.

1.2   Robust Shortest Path Problems

Soyster [11] first proposed a linear optimization model with uncertain data to construct a solution which is feasible for all input data. A significant step forward for developing a theory for robust optimization was taken by Ben-Tal et al. [12-14]. Specifically for discrete optimization problems, Kouvelis and Yu[16] proposed a framework for robust discrete optimization and showed that the robust counterparts of a number of polynomial solvable combinatorial problems are NP-hard. Zielinski [18] showed that the problem of minimizing the maximum regret criterion robust shortest path is NP-hard for directed graphs. Averbakh and Lebedev [19] proved that the problem of minimizing the maximum regret criterion robust shortest path is strongly NP-hard for non-directed graphs. Bertsimas and Sim [13] considered the interval model for cost uncertainty and showed that the problem can be solved by solving at most nominal problems. In particular, some models where an interval of possible values is associated with each arc have been studied [22-24].

All the models studied in the field of hazmat network design are based on the deterministic risk on each arc. However, the hazmat transportation accidents are recognized as low probability-high consequence events and the risk is a significant ingredient which separates hazmat transportation problems from other transportation problems. It is not easy to estimate the risk on each arc exactly, since it depends on many unpredictable factors. When the risk changes on the road segments, the prior optimal network might not be an optimal network, then the network is lack of robustness. To design a robust hazmat transportation network is reasonable and necessary.

In this paper, we use the maximum regret criterion called robust deviation solution in HTNDP, and the result is based on the following property, proposed by Karasan et al. [15].

Property 1. Given a path p from s to t, the scenario r which maximizes the robust deviation for p is the one where each arc (i, j) on p has cost u_{ij} and each arc (k, h) not on p has cost l_{kh}, i.e. $c_{ij}^r = u_{ij}, \forall (i, j) \in p$ and $c_{kh}^r = l_{kh}, \forall (k, h) \notin p$.

Particularly, the maximum regret criterion methodology is applied to deal with the risk uncertainty in HTNDP. We combine different risk assessment measure and make a full consideration about that potential influence factors which may cause huge damage to give an interval risk $r_{ijk} \in [\underline{r}_{ijk}, \overline{r}_{ijk}]$ on each arc (i, j) for each commodity k. $\overline{r}_{ijk}$ represents a maximum risk for commodity k on arc (i, j), $\underline{r}_{ijk}$ represents a minimum risk for commodity k on arc (i, j). Each interval represents a range of possible risk values on each arc for each commodity.

The paper is organized as follows. In Section 2, we present a bi-level network design formulation. In Section 3, we discuss the computational complexity of hazardous material transportation network design problem. In Section 4, we describe the robust heuristic solution procedure which guarantees a feasible stable solution. In Section 5, we test our heuristic algorithm on the highway transportation network for the province of Guangdong, China. This network involves 21 nodes and 30 edges. We have four different classified hazmats origin-destination data on this network, explosive solid product flammable gas toxic gas and corrosive substances. In Section 6, we provide some conclusions and remarks.

2   A Bi-level Network Design Formulation

Different from the relative deviation criterion [17], we apply the absolute deviation criterion (maximum regret criterion) to deal with the risk uncertainty in HTNDP where the government can decide which road segments have to be closed to hazmats so as to minimize the overall risk of the shipments; while the carriers choose the routes on the designated network to minimize their route costs. The hazmat transport network design problem consists of finding a sub-graph of the existing road network so that the total risk resulting from the carriers' route choices is minimized. Each hazmat commodity has its own risk on each arc in the transportation network. Suppose a hazmat transportation network is defined on an undirected graph $G = (V, A)$, where V is the set of vertices corresponding to road intersections, and A is the set of arcs corresponding to road segments. There are K hazmat commodities which need to be transported from their origins $s(k)$ to destinations $t(k)$. Let c_{ijk} be a cost associated with a unit flow of commodity k transporting on arc $(i, j) \in A$ and d_k be the corresponding transporting amount. Let $\underline{r}_{ijk}$ and $\overline{r}_{ijk}$ refer to the interval risk associated with a unit flow of commodity on arc $(i, j) \in A$, denoted as $r_{ijk} \in [\underline{r}_{ijk}, \overline{r}_{ijk}]$, this interval risk represents the set of possible values for each commodity $k(k = 1, 2, 3, \ldots, K)$ on arc $(i, j) \in A$.

We define the decision variables on the network below

$$x_{ijk} = \begin{cases} 1 & \text{if arc (i,j) is chosen by the commodity k} \\ 0 & \textit{otherwise} \end{cases}$$

$$y_{ij} = \begin{cases} 1 & \text{if arc (i,j) is available for hazmat transport} \\ 0 & \textit{otherwise} \end{cases}$$

The bi-level multi-commodity hazmat network design integer formulation is

$$\min_{y_{ij} \in \{0,1\}} \sum_{k \in \{1,\ldots,K\}} \sum_{(i,j) \in A} d_r r_{ijk} x_{ijk} \qquad r_{ijk} \in [\underline{r}_{ijk}, \overline{r}_{ijk}] \qquad (1)$$

$$s.t. \quad y_{ij} = y_{ji}, \qquad (i,j), (j,i) \in A \qquad (2)$$

$$x_{ijk} \in arg \min \sum_{k \in \{1,\ldots,K\}} \sum_{(i,j) \in A} d_k c_{ijk} x_{ijk} \qquad (3)$$

$$s.t. \quad \sum_{i \in V} x_{ijk} - \sum_{i \in V} x_{jik} = \begin{cases} -1 & \text{if j=s(k)} \\ 1 & \text{if j=t(k)} \\ 0 & \textit{otherwise} \end{cases} \quad j \in V, k \in 1, \ldots, K \qquad (4)$$

$$x_{ijk} \leq y_{ij}, \qquad (i,j) \in A, k \in 1, \ldots, K \qquad (5)$$

$$x_{ijk} \in \{0,1\}, \qquad (i,j) \in A, k \in 1, \ldots, K \qquad (6)$$

The objective (1) represents the government minimizing the total risk chosen by the carriers, each commodity of risk can vary in the interval on the arc of the network, while that of the carriers (3) is to minimize the cost. Constraints (2) state that both arcs (i, j) and (j, i) can be traversed in both directions used by any of the shipments. Constraints (4) ensure the flow of commodity k from its origin to the destination. Constraints (5) ensure that only edges selected by the government can be used by the carriers. Constraints (6) are binary requirements on the variables.

3 Computational Complexity

In Lemma 1 [6], it was shown that $HTNDP$ is NP-hard.

Lemma 1. *HTNDP is strongly NP-hard even for a single commodity.*

A natural way to handle NP-hard problems is approximation solutions or FPT algorithms. Let (I, k) be an instance of parameterized problem. An FPT algorithm decides (I, k) in time $O(f(k) \cdot n^c)$, where f is an arbitrary computable function that only depends on k and c is a constant. We often use the notation $O^*(f(k))$ to suppress the polynomial term. The class of fixed-parameter tractable parameterized problems is denoted FPT [25].

However, in this section, we prove that HTNDP does not admit any approximation, neither any FPT algorithm, unless $P = NP$.

Theorem 1. *HTNDP does not admit any polynomial time approximation (regardless of its approximation factor), unless $P = NP$.*

Proof. HTNDP is a minimization problem, then the result in [6] implies that deciding whether $OPT = 0$ is NP-hard. Let A be any approximation algorithm for HTNDP with factor α. By definition A returns an approximation solution value APP, with $APP \leq \alpha \times OPT$

When $OPT = 0$, clearly APP must also satisfy $APP = 0$. In other words, A would be able to solve the instance in[6] in polynomial time. This, however, contradicts with the corresponding NP-hard result (unless $P = NP$). □

Theorem 2. *The HTNDP does not admit any FPT algorithm, unless $P = NP$.*

Proof. HTNDP is a minimization problem, then the result in [6] implies that deciding whether $OPT = 0$ is NP-hard. Let B be any algorithm for FPT which runs in $O(f(k) \cdot n^c)$ time. When $OPT = k = 0$, B solves HTNDP in $O(f(0) \cdot n^c) = O(n^c)$ time. In other words, B would be able to solve the instance in [6] in polynomial time. This, again, contradicts with the corresponding NP-hard result, unless $P = NP$. □

4 A Robust Heuristic Approach

Since HTNDP does not admit any approximation, neither any FPT algorithm, unless $P = NP$, we describe a heuristic algorithm inspired by Erkut and Gzara [4] that always finds a solution with stability and robustness. A feasible solution of HTNDP is called stable if the sub-network does not admit for any commodity multiple minimum cost paths with different risk values. We use the maximum regret criterion to find the robust risk shortest path for each commodity k, and then we obtain the sub-network of G formed by k robust risk shortest routes. Let the resulting network be $\psi(G)$ with an associated risk value of R, which is the sum of min maximum regret value of each commodity k. Then, the carriers choose their own routes with an objective of minimum cost on $\psi(G)$, hence obtain a new sub-network $\psi'(G)$,each route chosen by each commodity k has a corresponding maximum regret value. When there are multiple minimum cost routes with different maximum regret value for each commodity, we always use the maximum sum of total maximum regret risk value of each commodity, with an associated total risk value of R_{max}. If $R = R_{max}$, and then a solution is found. Otherwise, there is at least one commodity k uses a different path designated by the government under the robust minimum risk objective. In order to eliminate the difference between $\psi(G)$ and $\psi'(G)$, we remove the maximum upper bound of interval risk arc not used by the government solution but used in the solution of the carriers for some commodity k. We remove arcs iteratively on the original network with the residual network. When the algorithm stops, we obtain a robust stable sub-network.

The detailed steps of designing a robust network algorithm are given below. Step 1: Firstly, we use maximum regret criterion to find the robust risk shortest path for each commodity k, and then we obtain the sub-network of G formed by k risk robust shortest routes, call the resulting network $\psi(G)$ with an associated total min-max risk value of value of R^t. According to the property 1 [4], we follow the mixed integer programming formulation presented by Karasan et al.[15],to solve K robust risk shortest path problems on G

$$\min_{y_{ij} \in \{0,1\}} \sum_{k \in \{1,...,K\}} \sum_{(i,j) \in A} d_k \overline{r}_{ijk} x_{ijk} - x_{t(k)}$$

$$s.t. \quad \sum_{i \in V} x_{ijk} - \sum_{i \in V} x_{jik} = \begin{cases} -1 & \text{if } j=s(k) \\ 1 & \text{if } j=t(k) \\ 0 & otherwise \end{cases} \quad j \in V, k \in \{1,...,K\}$$

$$x_j \leq x_i + \underline{r}_{ijk} + (\overline{r}_{ijk} - \underline{r}_{ijk}) x_{ijk}, \quad (i,j) \in A, k \in \{1,...,K\}$$

$$x_{s(k)} = 0$$

$$x_{ijk} \in \{0,1\}, \quad (i,j) \in A, k \in \{1,...,K\}$$

Where variables x_{ijk} represent a path μ as follows: $x_{ijk} = \begin{cases} 1 & if (i,j) \in \mu \\ 0 & otherwise \end{cases}$, and x_i represents the length of the shortest path from 1 to i under scenario $s(\mu)$; R_k^t represents the corresponding min-max regret value for each commodity k ; R^t represents the total min-max risk value of K commodities.

Step 2: Secondly, the carriers select the minimum cost routes on $\psi(G)$, and the total risk of $\psi(G)$ depends on the routes chosen by the carriers with a maximum regret risk value of each commodity k. If there exist multiple routes with the same cost but different maximum regret risk value, we sort the total maximum regret value as $< R^t_{min}, R^t_1, \ldots, R^t_m, R^t_{max} >$.

Solve K minimum-cost path problems on $\psi(G)$

$$\min \sum_{k \in \{1,\ldots,K\}} \sum_{(i,j) \in A} d_k c_{ijk} x_{ijk}$$

$$s.t. \quad \sum_{i \in V} x_{ijk} - \sum_{i \in V} x_{jik} = \begin{cases} -1 & \text{if } j=s(k) \\ 1 & \text{if } j=t(k) \\ 0 & \text{otherwise} \end{cases} \quad j \in V, k \in \{1,\ldots,K\}$$

$$x_{ijk} \in \{0,1\}, \quad (i,j) \in A, k \in \{1,\ldots,K\}$$

Step 3: If $\frac{R^t_{max}-R^t}{R^t} \le \Delta$, then a potential robust heuristic network is determined by $\psi'(G)$, else $\psi'(G)$ is not a stable network. Go to Step 4.

Step 4: Let (i,j) be an arc found according to the selection rule, which remove arcs (i,j) and (j,i) from the network, update $t = t+1$, $G^{t+1} = G^t - \{(i,j),(j,i)\}$. We remove the maximum upper bound of interval risk arc from the routes in R^t_{max} that is not used in the government solution but used in the solution of the carriers for some commodity k.

Step 5: Go to step 1.

The deterministic risk scenario network design heuristic algorithm is:

Repeat:

1: Solve K minimum-risk path problems on G.

2: Solve K minimum-cost path problems on $\psi(G)$.

3: If $\frac{R^t_{max}-R^t}{R^t} \le \Delta$, then a potential heuristic network is determined by $\psi'(G)$, else $\psi'(G)$ is not a stable network. Go to Step 4.

4: Let (i,j) be an arc found according to the selection rule, which remove arcs (i,j) and (j,i) from the network, update $t = t+1$, $G^{t+1} = G^t - \{(i,j),(j,i)\}$. We remove the maximum risk arc from the routes in R^t_{max} that is not used in the government solution but used in the solution of the carriers for some commodity k.

5: Go to step 1.

---

5   Application on Guangdong Province

In this section, we present some results on HTNDP applied in Guangdong province, China. We first describe the problem data in detail and then discuss the analyses and our interesting findings.

5.1   The Data

The primary source of our data is Statistics China, State Administration of Work Safety and Ministry of Transportation of the People's Republic of China, which contains actual distance between two connecting nodes, the population exposure around the edges, the locations where potential high risk exist such as buildings, bridges and road intersections and population concentration points, such as schools, factories and commercial centers, etc. Our study is focused on shipments of explosive solid product, flammable gas, toxic gas and corrosive substances, these four materials account for about 67% of all the hazmats transported through Guangdong highways. The Guangdong highway system is composed of China-highway (Gao Su Gong Lu) and national highways (Guo Dao), contains 21 vertices and 31 edges, as depicted in Figure 1. In order to see the original network directly, we simply draw the routes between each two nodes with straight line in right side of Figure 1.

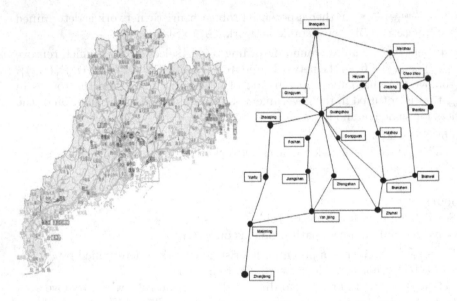

Fig. 1. The highway system of Guangdong province in China

According to the 2010 population census, our model represents the spatial distribution of 93.87 million people, which covers 90% of the total population of Guangdong province. The data set includes the origin and destination of each hazmat shipment, as well as the frequency of four hazmats in the Guangdong highway system, as depicted in Table 1.

We assume that each truck is fully loaded with up to 8 tons and can transport the same kind of hazardous material. Because of the scale of network of Guangdong highway system, we set the number of origin-destination pairs $K = 4, 8, 12, 16$. For each value of K, we generate 4 random instances and fix

Table 1. The frequency of four hazmats in the Guangdong highway system

The probability of four commodities, unit: number of accident times/kilometer	
explosive solid product	0.043×10^{-6}
flammable gas	0.049×10^{-6}
toxic gas	0.028×10^{-6}
corrosive substances	0.025×10^{-6}

the number of distinct origins and distinct destinations, as depicted in Table 2, from which we can see that K different kinds of hazmats transporting from their distinct origins to their distinct destinations in which the origins and destinations are generated randomly. The each edge transportation risk is computed by multiplying the probability of an undesirable event by population figure within 1600 meters of the edges and the actual road distance between two nodes. The transporting cost is given by the actual distance for each kind of commodity. i.e., each kind of commodity has the same cost value on the same arc of network. Based on the deterministic risk on each arc for each kind of commodity, we make a full consideration of detailed factors which may cause high consequence in accidents to give an interval risk for each commodity on each arc.

5.2   Numerical Analysis

The following numbers <0,1,2,3,4,5,6,7,8,9,10,11,12,13,14,15,16,17,18,19,20> represent the 21 cities, respectively <Shaoguan, Qingyuan, Meizhou, Guangzhou, Jieyang, Heyuan, Chaozhou, Shantou, Shanwei, Huizhou, Dongguan, Foshan, Zhuhai, Zhongshan, Jiangshan, Yangjiang, Shenzhen, Zhaoqing, Yunfu, Maoming, Zhenjiang>.

The statistic data provide no information about the origins, destinations and hazmat type of each shipment. So we perform testing on randomly generated pairs. We investigate the significance of robust solution network and compare to the deterministic risk solution network. Table 2 shows the detailed results on the random instances. For example, for 4 commodities originating points $< 6, 3, 16, 4 >$ to destination points $< 4, 16, 17, 9 >$. All tests are performed using the aggregate risk measure and max risk arc selection rules. The results obtained by deterministic risk scenario network and interval risk scenario network respectively for the different commodities with instances are described in Table 3. Firstly, we calculate the total risk of the heuristic solution network on the deterministic bi-level model, and then calculate the total risk when the risk changes on arcs. Secondly, we obtain a robust network, use the same deterministic risk on each arc to calculate the total risk on the robust network, and then use the same changing risk value on the robust network. The deterministic risk of each arc for each commodity belongs to the interval risk of each commodity on each arc.

Table 2.

K	origin vertices	destination vertices
4	<6,3,16,4>	<4,16,17,9>
	<11,1,0,3>	<2,10,16,8>
	<7,2,18,11>	<16,6,2,5>
	<7,2,18,11>	<16,6,2,5>
8	<6,3,16,4,1,0,5,2>	<4,16,17,9,2,1,10,9>
	<6,0,16,8,1,10,5,12>	<1,20,17,3,2,19,11,9>
	<10,1,19,15,5,16,2,11>	<1,2,6,10,17,19,12,3>
	<17,8,2,7,14,12,0,20>	<3,7,11,13,17,20,19,0>
12	<1,7,0,1,6,5,2,1,3,6,2,1>	<6,3,10,3,5,2,6,11,5,2,8,2>
	<1,2,7,8,2,13,2,11,8,7,8,2>	<4,3,2,9,1,14,13,10,12,2,9,3>
	<2,4,3,4,2,1,3,4,6,2,9,1>	<6,2,10,0,11,12,7,5,3,1,8,6>
	<2,3,6,0,2,3,6,8,9,10,5,3>	<7,2,4,2,5,2,1,2,3,11,2,5>
16	<0,1,2,3,4,5,6,7,8,9,10,11,12,13,14,15>	<3,2,5,8,7,9,0,2,3,5,1,9,3,8,2,4>
	<0,1,1,2,3,3,4,5,6,6,11,12,13,13,14,14>	<2,8,2,7,2,1,5,9,2,7,1,3,10,2,11,9>
	<0,0,1,2,2,3,4,5,7,8,10,11,12,12,13,13>	<1,2,5,3,9,5,7,1,6,2,3,4,11,1,7,6>
	<4,5,6,7,7,8,10,11,12,13,13,17,18,2,2,3>	<11,2,9,5,3,12,12,2,11,2,15,18,15,10,6,5>

From the Table 3, we find that the robust heuristic finds good quality solutions, especially when the risk changes on the network compared to the deterministic model.

In Table 3, the first column indicates the number of commodities K, the second column identifies the instance generated for each value of K, the third column is devoted to analyze the behavior of the total risk of network under deterministic risk, and the fourth column presents the total risk under changed arc risk. The fifth column reports the total risk of the robust network using the same arc risk in first column. The sixth column indicates the total risk of robust network using the same changing arc risks in the second column. The seventh column indicates that arcs whose risk value is changed. The eighth column show that the arc risk value before change, while the ninth column show that the arc risk value after change. The tenth column of $[\underline{r}_{ij}, \overline{r}_{ij}]_{(k)}$ show that the commodity k interval risk on arc (i, j), $k = Null$ indicates that no transport on that corresponding arc (i, j).

This section shows the computational results achieved by the deterministic risk hazmat transportation network and the robust hazmat transportation network. We can see from the Table 3, most of the results show that robust network performs relatively well when the risk changes. The robust network always gives robustness and stable solution, which always avoid the arcs have higher upper bound of the interval risk. Because of the total risk of resulting transport network is decided by carriers' route choices, the routes selected by carriers in deterministic network and robust network have four different cases, as depicted in Table 4, where $\sqrt{}$ represents the arc risk change on the network and $\times$ represents the arc risk have no change on the network.

Table 3.

K	Run	R_1	R_2	R_1	R_1	(i,j)	r_{ij}	$\overline{r}_{ij}$	$[\underline{r}_{ij}, \overline{r}_{ij}]_{(k)}$
	1	368	832	467	467	(6,4)	80	554	$[22, 878]_{(4)}$
	2	390	590	465	465	(16,8)	18	218	$[17, 311]_{(3)}$
4	3	414	833	646	646	(2,4)	20	100	$[16, 520]_{(1)}$
						(2,4)	21	130	$[10, 419]_{(2)}$
						(2,4)	10	150	$[8, 316]_{(3)}$
	4	414	833	435	515	(2,4)	20	100	$[16, 220]_{(1)}$
	1	674	974	837	837	(0,2)	14	314	$[12, 222]_{(1)}$
	1	674	974	837	837	(0,3)	19	115	$[19, 146]_{(null)}$
8	2	1138	1438	1158	1158	(7,8)	31	331	$[31, 527]_{(2)}$
	2	1138	1138	1158	1188	(4,6)	30	60	$[27, 68]_{(0)}$
	3	819	833	825	839	(1,3)	30	44	$[18, 54]_{(3)}$
	4	931	1031	938	938	(13,15)	33	235	$[33, 328]_{(7)}$
	1	913	1112	964	964	(12,15)	24	84	$[24, 117]_{(0)}$
						(12,15)	33	100	$[31, 210]_{(6)}$
						(12,15)	45	117	$[43, 310]_{(11)}$
	2	1003	1003	1040	1040	(9,16)	33	97	$[33, 97]_{(null)}$
12	3	925	1032	925	1032	(0,3)	57	80	$[57, 104]_{(5)}$
						(0,6)	64	84	$[64, 96]_{(7)}$
						(4,7)	55	95	$[55, 108]_{(11)}$
						(3,10)	45	69	$[45, 95]_{(12)}$
	4	796	796	836	836	(4,8)	19	84	$[19, 97]_{(null)}$
						(10,16)	37	103	$[21, 142]_{(null)}$
	1	1396	1546	1457	1457	(9,16)	50	200	$[44, 274]_{(13)}$
16	2	1560	1610	1608	1658	(5,9)	33	83	$[33, 87]_{(15)}$
	3	1636	1636	1711	1741	(14,15)	30	60	$[22, 85]_{(14)}$
	4	1679	1739	1679	1739	(3,13)	37	97	$[27, 104]_{(8)}$

Table 4.

scenarios \ cases	I	II	III	IV
Deterministic risk scenario network	√	×	√	×
Interval risk scenario network	×	√	√	×

Case I : If some arcs risk change in the deterministic network but not in the robust network, have the changing increase to the deterministic network and have no influence to the robust network. For example, in 4{1,2,3}, 8{1,2,4}, 12{1},16{1}.Because the robust network does avoid selecting the routes with high potential changing risk, as shown in 4{1,2,3}. When some commoditys upper bound of interval risk is not that high, the robust network may select this link, as shown in 4{4}. So when commodity 1' risk changes on arc(2,4), the same increase to robust network. But the change is not high.

Case II : If some arcs' risk change in the robust network but not in the deterministic network, have the changing increase to the robust network and have no influence to the deterministic network. But the margin of the changing risk is not high.For example, in 8{*2*}, 16{3}.

Case III: If some common arcs risk change, which both selected by carriers in deterministic network and the robust network, the same increase to both deterministic network and the robust network. For example, in 4{*4*},8{3},12{3}, 16{2, 4}.

Case IV : $\Delta > 0$ in step 3 indicates the government allows carriers choose some links which not exactly the links the government designated for some commodities. So some arcs risk change from the external factors which in the network but not selected by carriers have no influence to both deterministic network and the robust network. For example, in 8{*1*},12{2,4}. When $\Delta = 0$, the Case IV does not happen.

We find that if the margin of the interval risk of each commodity on each arc is not that higher, the deterministic network and the robust network almost the same.

Generally speaking, the robust interval scenario network performs not bad comparing to the deterministic scenario network under fixed risk on each arc for each commodity, but performs really better when the risk changes since the robust interval scenario network avoid selecting the potential high risk arcs.

6 Concluding Remarks

We consider the problem of designing a network for hazardous material transportation. In this paper, we have proved that the problem does not admit any approximation, neither any FPT algorithm, unless P=NP. We present a robust optimization-based formulation for HTNDP under edge risk uncertainty and tested a simple heuristic for a robust bi-level network design problem for hazmat transportation that the heuristic algorithm was able to give always robustness and stable solution and always avoid the arcs have higher upper bound of the interval risk. One of the challenges in designing a hazardous network is the use of common road links for different shipments. When some of common links for different shipments removed from original network, we guarantee that after a number of iterations less than the number of network links the algorithms stops with a feasible stable solution for the residual network.

In order to evaluate the effectiveness of the proposed model and algorithm, we concentrated our analysis on a real-world case study. We considered the road network of the Guangdong province in China which contains 21 vertices and 31 edges. Comparing to the solutions of the bi-level model with the results coming from two scenarios, called deterministic risk scenario network and interval risk scenario network respectively, we find that the robust network performed not bad comparing to the deterministic network under deterministic risk on each arc for each commodity, but performed really better when the risk changes since the robust network avoid selecting the potential high risk arcs. The robust optimization for hazmat transport network design is more reasonable and performed good quality in robustness.

The solution to the problem depends on the input data, in particular, the origin-destination flows, the topology of the original road network, the spatial distribution of population centers, the location of the origin-destination pairs, and the type of hazmats being shipped. In general but not absolutely, we find that when the number of hazmats growing, the iteration times and CPU requirement are increasing. Due to the scale of our original road network, the heuristic algorithm performed only several times to gain a stable network, which fully depends on the input data of the original network.

Acknowledgement. The research reported in this paper has been partially supported by the National Natural Science Foundation of China (70971008), and the Ministry of Education, Humanities and Social Sciences Planning Project of China(09YJC630008).

References

[1] China Chemical Safety Association, http://www.chemicalsafety.org.cn
[2] Kara, B.Y., Verter, V.: Designing a Road Network for Hazardous Materials Transportation. Transportation Science 38(2), 188–196 (2004)
[3] Erkut, E., Alp, O.: Designing a road network for hazardous materials shipments. Computers & Operations Research 34(5), 1389–1405 (2007)
[4] Erkut, E., Gzara, F.: Solving the hazmat transport network design problem. Computers & Operations Research 35(7), 2234–2247 (2008)
[5] Verter, V., Kara, B.Y.: A Path-Based Approach for Hazmat Transport Network Design. Management Science 54(1), 29–40 (2008)
[6] Amaldi, E., Bruglieri, M., Fortz, B.: On the Hazmat Transport Network Design Problem. Network Optimization, 327–338 (2011)
[7] Bianco, L., Caramia, M., Giordani, S.: A bilevel flow model for hazmat transportation netwokrk design. Transportation Research Part C: Emerging Technologies 17(2), 175–196 (2009)
[8] Erkut, E., Verter, V.: Modeling of transport risk for hazardous materials. Operations Research 46(5), 625–642 (1998)
[9] Erkut, E., Tjandra, S.A., Verter, V.: Chapter 9 Hazardous Materials Transportation 14, 539–621 (2007)
[10] Zhu, B.: Approximability and Fixed-Parameter Tractability for the Exemplar Genomic Distance Problems. In: Chen, J., Cooper, S.B. (eds.) TAMC 2009. LNCS, vol. 5532, pp. 71–80. Springer, Heidelberg (2009)
[11] Soyster, A.: Convex programming with set-inclusive constraints and applications to inexact linear programming. Oper. Res. 21, 1154–1157 (1973)
[12] Ben-Tal, A., El Ghaoui, L., Nemirovski, A.: Robustness Optimization. Princeton University Press, Princeton (2009)
[13] Bertsimas, D., Sim, M.: Robust discrete optimization and network flows, Math. Program.Ser. B 98, 49–71 (2003)
[14] Bertsimas, D., Sim, M.: The price of robustness. Oper. Res. 52(1), 35–53 (2004)
[15] Karasan, O.E., Pinar, M.C., Yaman, H.: The robust shortest path problem with interval data. Technical report, Bilkent University (2001)
[16] Kouvelis, P., Yu, G.: Robust discrete optimization and its applications. Kluwer Academic Publishers, Boston (1997)

[17] Gabrel, V., Murat, C.: Robust shortest path problems. Annales du LAMSADE (7), 71–93 (2007)

[18] Zielinski, P.: The computational complexity of the relative robust shortest path problem with interval data. European Journal of Operational Research 158, 570–576 (2004)

[19] Averbakh, I., Lebedev, V.: Interval data minmax regret network optimization-problems. Discrete Applied Mathematics 138, 289–301 (2004)

[20] Wen, U., Hsu, S.: Linear Bi-Level Programming Problems – A Review. The Journal of the Operational Research Society 42(2), 125–133 (1991)

[21] Colson, B., Marcotte, P., Savard, G.: An overview of bilevel optimization. Annals of Operations Research 153(1), 235 (2007)

[22] Montemanni, R., Gambardella, L.M.: An exact algorithm for the robust shortest path problem with interval data. Computers and Operations Research 31, 1667–1680 (2004)

[23] Montemanni, R., Gambardella, L.M.: The robust path problem with interval data via benders decomposition. 4OR 3(4), 315–328 (2005)

[24] Montemanni, R., Gambardella, L.M., Donati, A.V.: A branch and bound algorithm for the robust shortest path problem with interval data. Operations Research Letters 32, 225–232 (2004)

[25] Zhang, C., Jiang, H., Zhu, B.: Radiation hybrid map construction problem parameterized. In: Lin, G. (ed.) COCOA 2012. LNCS, vol. 7402, pp. 127–137. Springer, Heidelberg (2012)

Online Bin Packing with (1,1) and (2,R) Bins

Jing Chen[1], Xin Han[2], Kazuo Iwama[1], and Hing-Fung Ting[3]

[1] Graduate School of Informatics, Kyoto University
chen@algo.cce.i.kyoto-u.ac.jp
iwama@kuis.kyoto-u.ac.jp
[2] Software School, Dalian University of Technology
hanxin@dlut.edu.cn
[3] Departent of Computer Science, The University of Hong Kong
hfting@cs.hku.hk

Abstract. We study a variant of online bin packing problem, in which there are two types of bins: $(1,1)$ and $(2,R)$, i.e., unit size bin with cost 1 and size 2 bin with cost $R > 1$, the objective is to minimize the total cost occurred when all the items are packed into the two types of bins. It is not difficult to see that the offline version of the problem is equivalent to the classical bin packing problem when $R > 3$. In this paper, we focus on the case $R \leq 3$, and propose online algorithms and obtain lower bounds for the problem.

1 Introduction

In this paper we consider a variant of online bin packing problem in which there are two types of bins: $(1,1)$ and $(2,R)$, i.e., unit size bin with cost 1 and size 2 bin with cost $R > 1$, items are online given, we are asked to pack all the items into the two types of bins such that the total cost of used bins is minimal. For short, we call the problem as OBP1R. The problem is related to the generalized cost variable sized bin packing problem, which is a generalization of the variable sized bin packing problem([11],[12],[13]) and in special case our problem becomes the classical bin packing problem([2], [3], [4], [5], [6], [7], [16]).

In the Generalized Cost Variable Sized Bin Packing problem(GCVS) [1], there will be infinite supply of r types of bins whose sizes are $0 < b_r < ... < b_1 = 1$, each bin of type i is associated with cost b_i, input items have sizes in $(0,1]$, the goal is to find a feasible solution with a minimal cost. In [1], the authors allow to use a (general) set of bins with different sizes(the cost is the bin size), and give APTAS (Asymptotic Polynomial Time Approximation Scheme). However, the APTAS does not give us too much idea about how the different costs affect packing items. In this paper we consider the most basic case that only two different types of bins are used.

The OBP1R problem is also related to the parametric bin packing problem, in which all the items are with sizes in $(0, 1/r]$, where $r > 0$ is an integer, the goal is to pack all the items in a minimum number of unit size bins. We find that when $R \leq 2$, our problem is almost equivalent to the parametric bin packing problem with $r = 2$.

P. Widmayer, Y. Xu, and B. Zhu (Eds.): COCOA 2013, LNCS 8287, pp. 387–401, 2013.

Formally the offline version of our problem is defined as follows: given an input $L = \{a_1, a_2, ...a_n\}$ of n items with each $a_i \in (0, 1]$, we partition L into subsets $\{L_j\}$, if all the items in L_j are packed in a unit size bin, then define $\pi(L_j) = 1$, else define $\pi(L_j) = 2$. Define vectors $size$ and $cost$ as: $size = \{1, 2\}$ and $cost = \{1, R\}$. The objective is to minimize

$$\sum_j cost[\pi(L_j)] \quad s.t. \quad \sum_{a_i \in L_j} a_i \leq size[\pi(L_j)].$$

In the problem, the value of R has a critical effect on the solution of the problem. It is not difficult to see that the above problem is equivalent to the classical bin packing when $R > 3$, i.e., in an optimal packing all the items have to be packed into unit size bins. Otherwise, if there is a size 2 bin used, then we can use at most three unit size bins to replace it, and the cost decreases since $R > 3$.

In our paper, we study the online case for $R \leq 3$, i.e., items are given one by one, once items are packed, we cannot repack them. To evaluate online algorithms, we use one of standard measures: *competitive ratio*. Formally, let L be an list of input items, let $cost_A(L)$ be the cost used by an online algorithm A on L, let $OPT(L)$ be the cost of an optimum solution of packing L, then the competitive ratio of an online algorithm A is defined as:

$$R_A^\infty = \lim_{n \to \infty} \sup \sup_L \left\{ \frac{cost_A(L)}{OPT(L)} | OPT(L) = n \right\}.$$

Related Work: Bin packing is one of the most well-studied problems, which was first studied by Ullman[2], Lee and Lee[7] provided the first harmonic algorithm based on the idea of interval classification, the method was developed to the most recent version called Harmonic++ by Seiden[9]. It provides the present upper bound 1.58889 for the online bin packing problem for the past decade. The lower bound 1.5401 about the online algorithm for this problem is provide by Vliet[16],[10]. There are many variants of bin packing problems, one of them is the generalized cost variable sized bin packing problem, Epstein[1] provided an APTAS for the problem.

Our Contributions: when $R > 3$, the problem is equivalent to the classical bin packing problem; so we focus on the case $1 < R \leq 3$ and give online algorithms and calculate lower bounds for the problem (refer to Fig. 1). For the large values of R, we tend to use only unit bins to pack all the items, while for the small values of R, we tend to use only size 2 bins to pack all the items, and for the medium values of R, we will use both types of bins to pack the items, therefore we partition R into three intervals, by calculations the intervals are defined as: 1)$2.6915 < R \leq 3$; 2)$2.040 < R \leq 2.6915$; and 3)$1 < R \leq 2.040$.

When $2.6915 < R \leq 3$, we use Harmonic++ [9] to pack all the items into unit size bins and prove the upper bound of the competitive ratio is still 1.58889. When $2.040 < R \leq 2.6915$, we develop an online algorithm called MAIN, in which some items are packed into size two bins and others are packed into unit size bins, and prove the competitive ratio is below 1.58889; when $1 < R \le 2.040$, we pack all the items into size two bins by a Refined Harmonic algorithm and prove the competitive ratio is 1.4078. As to the lower bound of the competitive ratio, we cannot give a general formula of R. However given a value of R, we can calculate the lower bound of the online problem by using linear programs.

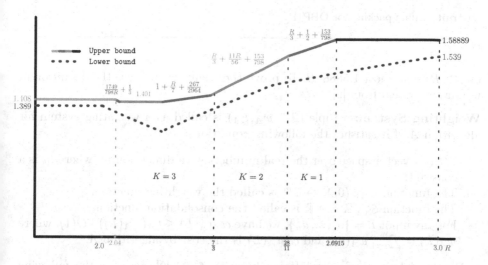

Fig. 1. Upper and Lower bounds

2 Online Algorithms for $R \le 3$

When $R \le 3$, we have three cases. Case 1: $2.6915 < R \le 3$, we use Harmonic++ [9] to pack all the items into unit size bins and prove the upper bound of the competitive ratio is 1.58889. Case 2: $2.040 < R \le 2.6915$, we develop an online algorithm called MAIN, in which some items are packed into size two bins and others are packed into unit size bins, and prove the competitive ratio is below 1.58889. Case 3: $1 < R \le 2.040$, we pack all the items into size two bins by a Refined Harmonic algorithm and prove the upper of the competitive ratio is 1.4078. The general algorithm for OBP1R is summarized as follows.

2.1 Online Algorithm for $R > 2.6915$

In this subsection, we prove that online algorithm H++ [9] is 1.58889 competitive for our problem when $R > 2.6915$, i.e., all the items are packed in unit size bins by H++. Observe if an optimal packing uses only unit size bins, then we are done since the competitive ratio 1.58889 follows directly from [9]. Otherwise instead of calculating how many bins $(2, R)$ used, we prove that the total weight in one

---

Algorithm 1. Online algorithm for OBP1R

Input: cost $R > 1$, items $(a_i)_{1 \leq i \leq n}$, where $a_i \in (0, 1]$.

1. **if** $R > 2.6915$, use algorithm H++ [9] to pack all items in unit size bins.
2. **if** $R \in (2.040, 2.6915]$, first calculate a parameter $K \geq 1$, then call MAIN(K) to pack items.
3. **if** $R \leq 2.040$, use algorithm PAR to pack items to size two bins.

Output: online packing for OBP1R.

---

bin $(2, R)$ is at most $1.58889R$. To prove the result, we will use the definition of weighting system from [9].

Weighting System: a tuple $(\mathbb{R}^m, w_A, \xi_A)$ is called as a weighting system for algorithm A, if it satisfies the following conditions:

- $\mathbb{R}^m$ is a vector space over the real numbers with dimension m, where m is a constant.
- The function $w_A : (0, 1] \to \mathbb{R}^m$ is called the weighting function.
- The function $\xi_A : \mathbb{R}^m \to \mathbb{R}$ is called the consolidation function.
- For any input $L = \{a_1, ..., a_n\}$, we have $cost_A(L) \leq \xi_A(w_A(L)) + O(1)$, where $w_A(L) = \sum_{i=1}^{n} w_A(a_i)$ and $cost_A(L)$ is the cost by algorithm A.

By the definition of consolidation function $\xi_A(\cdot)$ [9], we have the following lemmas.

Lemma 1. *Given any set $L = L_1 \cup L_2$, we have $\xi_A(w_A(L)) \leq \xi_A(w_A(L_1)) + \xi_A(w_A(L_2))$.*

Lemma 2. *[9] For any set $S = \{x_1, x_2, ..., x_s | \sum_{i=1}^{s} x_i \leq 1\}$, we have $\xi_A(w_A(S)) \leq 1.58889$ and $\xi_A(w_A(x_i)) \leq 1$ for all i, if algorithm A is H++.*

Theorem 1. *Online algorithm H++ is still 1.58889 for our problem when $R > 2.63$.*

Proof. In our proof, the weighting function is exactly the same as the one in [9]. Consider an optimal solution, assume the number of $(1, 1)$ bins used is X and the number of $(2, R)$ bins used is Y. So the optimal cost is $X + R \cdot Y$. Assume all the items packed in the unit size bins is in set L_1, all the others is in set L_2. By Lemma 2, we have $\xi_A(w_A(L_1)) \leq 1.58889 \cdot X$. If we have $\xi_A(w_A(L_2)) \leq 1.58889 \cdot R \cdot Y$, then by Lemma 1, we have this theorem. To get the above result, we are going to prove that

$$\xi_A(w_A(T)) \leq 1.58889 \cdot R,$$

where $T = \{x_1, x_2, ..., x_t | \sum_{i=1}^{t} x_i \leq 2\}$ is any set fitted for a size two bin. It is not difficult to see that set T can be partitioned into three sets T_i, where

$1 \leq i \leq 3$, such that all the items in T_i fit in a unit size bin, and there is only one item in T_3. By Lemma 2, we have

$$\xi_A(w_A(T_i)) \leq 1.58889, \qquad \xi_A(w_A(T_3)) \leq 1,$$

then by Lemma 1, we have

$$\xi_A(w_A(T)) \leq 1.58889 \times 2 + 1 \leq 1.58889 \cdot R,$$

where the last inequality holds from $R \geq 2.63$. $\qquad\qquad$ □

2.2 Online Algorithm MAIN for $2.040 < R \leq 2.6915$

In our algorithm, the ideas of Refined Harmonic and interval classification [7] are used. However comparing with Refined Harmonic, we have a different partition for the interval $(0, 1]$ for different values of R. Based on the partition, we group all the items into two classes, F and J. For all the items in F, we pack them into unit size bins by Harmonic algorithm; for all the items in J, we pack them into size two bins by Refined Harmonic algorithm.

Grouping: Let $M = 20$. Given a constant K, we classify all the items by partitioning the interval $(0, 1]$ into sets F and J, where $F = \cup_{k=1}^{M} F_k$ such that $F_k = (2/(2k+1), 1/k]$ for $1 \leq k \leq K$, $F_k = (1/(k+1), 1/k]$, for $K+1 \leq k \leq M-1$, and $F_M = (0, 1/M]$; for $K = 1$, $J = J_1 = (1/2, 2/3]$, for $K \in \{2, 3\}$, $J = J_a \cup J_b \cup_{i=1}^{K} J_i$, where

$$J_a = (\frac{1}{2}, \frac{13}{24}], \quad J_1 = (\frac{13}{24}, \frac{2}{3}], \quad J_b = (\frac{1}{3}, \frac{3}{8}], \quad J_2 = (\frac{3}{8}, \frac{2}{5}], \quad J_3 = (\frac{1}{4}, \frac{2}{7}].$$

For $K = 2$, the definitions of F_* and J_* are showed in Fig. 2.

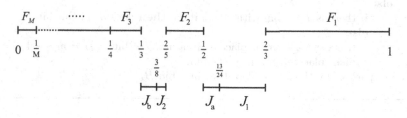

Fig. 2. Partition of Interval $(0, 1]$ for $K = 2$

The value of K is dependent on R, and its value is given in the following table 1.

Naming: An item a_i is called as an F_k-piece if $a_i \in F_k$, and a J_k-pieces if $a_i \in J_k$. Bin B_{F_i} is a unit bin designated to pack F_i-pieces. Except for items J_a and J_b, a B_{J_i}-bin $(1 \leq i \leq K)$ is $(2, R)$ bin designated to pack J_i-pieces exclusively.

Table 1. Setting of K for MAIN

Interval of R	$(2.040, \frac{7}{3}]$	$(\frac{7}{3}, \frac{28}{11}]$	$(\frac{28}{11}, 2.6915]$
Value of K	3	2	1

Packing Items J_a and J_b: We pack items J_a and J_b into size two bins. It is not difficult to verify that a size 2 bin can accept exactly three pieces of J_a-pieces, and the remaining space in the bin is larger than $\frac{3}{8}$, which is enough for one item of J_b-piece.

First we define two types of bins with size 2: i) B_A bin is designed to contain three J_a-pieces and one J_b-piece; ii) B_B bin is designed to contain five J_b-pieces. The packing of J_a and J_b pieces is below: given a J_a piece, if there is a type B_A bin with less than three J_a pieces, then we pack it into this bin, else we open a new B_A bin for the item. Given a J_b piece, if there is a B_A bin with no J_b piece in it, then put the item into this bin, else we open a new $(2, R)$ bin and pack the item there. Let the total number of J_b items be n_b, and the total number of B_B bins be m_{5b}. If $\frac{m_{5b}}{n_b} < \frac{11}{56}$, then name it as B_B bin, else as B_A bin.

Given an item a_i with type $x \in \{F_1, ..., F_M, J_1, ..., J_K\}$, algorithm MAIN is described as below.

Algorithm 2. MAIN

---

1.   **if** x is J_a **then**
        if there is a B_A bin and the number of J_a items is less than three, **then** pack a_i in this bin.
        else pack a_i into a new B_A bin.
2.   **if** a_i is J_b **then** $n_b = n_b + 1$,
        if there is a B_B bin with J_b items less than five, **then** pack a_i in this bin.
        else
                if there is a B_A bin without J_b item; **then** pack a_i in this bin.
                else
                        if $m_{5b} \leq \frac{11 n_b}{56}$ **then** place a_i in a new B_B bin, $m_{5b} = m_{5b} + 1$.
                        else place a_i in a new B_A bin.
3.   **else** pack a_i by Harmonic algorithm into bin B_x.

Observation: there is at most one bin of type B_A with content from the set $\{a, aa, ab, aab\}$. In bins B_A, a bin with content $\{aaa\}$ and a bin with content $\{b\}$ do not coexist.

2.3   Performance Analysis of MAIN

Our analysis is based on weighting functions, which is in the framework of the weighting system. Given an input L, we construct weighting functions $w_i()$ such that the cost by our algorithm $cost(L) \leq \max_i w_i(L) + O(1)$. Then we prove for each i,

$$w_i(S) \leq \rho, \qquad w_i(T) \leq \rho \cdot R,$$

where $S = \{x_1, x_2, ..., x_s | \sum_{i=1}^{s} x_i \leq 1\}$ and $T = \{x_1, x_2, ..., x_t | \sum_{i=1}^{t} x_i \leq 2\}$. Finally our work is to calculate the value of ρ, which is much more complicated than the calculation used in the classical bin packing, since we have many cases to be considered.

Weighting Functions: There are two cases.

Case 1: $K = 1$. Given an item with size x, the weighting functions are defined as follows:

$$w_1(x) = w_2(x) = \begin{cases} \frac{1}{k}, & \text{if } x \in F_k, 1 \leq k \leq M-1, \\ \frac{M \cdot x}{M-1}, & \text{if } x \in F_M, \\ \frac{R}{3}, & \text{if } x \in J_1. \end{cases}$$

Case 2 $K \geq 2$. The weighting functions are defined as follows:

$$w_1(x) = \begin{cases} \frac{1}{k}, & \text{if } x \in F_k \text{ and } 1 \leq k \leq M-1, \\ \frac{Mx}{M-1}, & \text{if } x \in F_M, \\ \frac{12R}{56}, & \text{if } x \in J_b, \\ 0, & \text{if } x \in J_a, \\ \frac{R}{2k+1}, & \text{else } x \in J_k \text{ and } 1 \leq k \leq K. \end{cases} \qquad w_2(x) = \begin{cases} \frac{1}{k}, & \text{if } x \in F_k \text{ and } 1 \leq k \leq M-1, \\ \frac{Mx}{M-1}, & \text{if } x \in F_M, \\ \frac{11R}{56}, & \text{if } x \in J_b, \\ \frac{R}{3}, & \text{if } x \in J_a, \\ \frac{R}{2k+1}, & \text{else } x \in J_k \text{ and } 1 \leq k \leq K. \end{cases}$$

Lemma 3. *Given an input L, the cost of our algorithm is at most $\max_i w_i(L) + O(1)$.*

Proof. Let f_k be the number of F_k-items, and j_k be the number of J_k-items in input list L respectively. Also, let s_M be the total size of all F_M-item in L.

When $K = 1$, the cost of algorithm $MAIN$

$$cost(L) \leq \frac{f_1}{1} + \frac{f_2}{2} + ... + \frac{f_{M-1}}{M-1} + \frac{M \cdot s_M}{M-1} + \frac{j_1}{3} \cdot R + O(1), \tag{1}$$

$$\leq \sum_{x \in L} w_1(x) + O(1). \tag{2}$$

Consider $K \geq 2$. Let n_a denote the number of J_a-items. Let n_b denote the number of J_b-items. Let m_{5b} be the number of bins B_B. Let m_b be the number of bins B_A with only one item J_b and without any J_a. Let m_{ab} be the number of bins B_A with one item J_b and at least one J_a. Let m_{3a} be the number of bins B_A with three J_a items and without J_b.

$$n_b = 5m_{5b} + m_b + m_{ab}, \tag{3}$$

$$3(m_{3a} - 1) + 3(m_{ab} - 1) < n_a \leq 3m_{3a} + 3m_{ab}. \tag{4}$$

The algorithm maintains $\frac{11}{56} \cdot n_b - 1 < m_{5b} < \frac{11}{56} \cdot n_b + 1$. Thus we have

$$\frac{11n_b}{56} - 1 \leq m_{5b} \leq \frac{11n_b}{56} + 1, \tag{5}$$

$$\frac{n_b}{56} - \frac{55}{56} \leq m_b + m_{ab} \leq \frac{n_b}{56} + \frac{55}{56}, \tag{6}$$

The algorithm ensures that a B_A-bin with only J_a-items and a B_A-bin with one J_b-item never coexist. That is $m_{3a} \cdot m_b = 0$, throughout the running of algorithm.

Let m_k be the number of B_{J_k}-bins. Then $m_k = \lceil \frac{j_k}{2k+1} \rceil$, where $1 \le k \le K$. The cost of algorithm $MAIN$

$$cost(L) \le \sum_{k=1}^{M-1} \frac{f_k}{k} + \frac{M \cdot s_M}{M-1} + R(\sum_{k=1}^{K} m_k + m_{3a} + m_b + m_{ab} + m_{5b}) + O(1).$$

We distinguish two cases based on $m_{3a} \cdot m_b = 0$.

Case 1. $m_{3a} = 0$, we will prove that $cost(L) \le w_1(L) + O(1)$.
By (5), (6), we have

$$cost(L) \le \sum_{k=1}^{M-1} \frac{f_k}{k} + \frac{M \cdot s_M}{M-1} + R \cdot (\frac{j_1}{3} + ... + \frac{j_K}{2K+1} + m_{ab} + m_b + m_{5b}) + O(1) \tag{7}$$

$$\le \sum_{k=1}^{M-1} \frac{f_k}{k} + \frac{M \cdot s_M}{M-1} + \sum_{k=1}^{K} \frac{R}{2k+1} j_k + \frac{12R}{56} n_b + O(1), \tag{8}$$

$$\le \sum_{x \in L} w_1(x) + O(1) = w_1(L) + O(1). \tag{9}$$

Case 2. $m_b = 0$, we will prove that $cost(L) \le w_2(L) + O(1)$.
In this case, by (5), (6), we have

$$cost(L) \le \sum_{k=1}^{M-1} \frac{f_k}{k} + \frac{M \cdot s_M}{M-1} + R(\frac{j_1}{3} + ... + \frac{j_K}{2K+1} + m_{3a} + m_{ab} + m_{5b}) + O(1), \tag{10}$$

$$= \sum_{k=1}^{M-1} \frac{f_k}{k} + \frac{M \cdot s_M}{M-1} + \sum_{k=1}^{K} \frac{R}{2k+1} j_k + \frac{R}{3} n_a + \frac{11R}{56} n_b + O(1), \tag{11}$$

$$= \sum_{x \in L} w_2(x) + O(1) = w_2(L) + O(1). \tag{12}$$

$\square$

Definition of W: given $S = \{x_1, x_2, ..., x_s | \sum_{j=1}^{s} x_j \le 1\}$ and $T = \{x_1, x_2, ..., x_t | \sum_{j=1}^{t} x_j \le 2\}$, for $1 \le i \le 2$, we define

$$W_{1,i} = \max_{S} w_i(S), \quad W_{2,i} = \max_{T} w_i(T), \quad W = \max_{i} \max\{W_{1,i}, W_{2,i}/R\}.$$

Lemma 4. For an input L, we have $\max_i w_i(L) \le W \cdot OPT$, where OPT is the optimal cost.

Proof. Consider an optimal solution, assume the number of $(1,1)$ bins used is X and the number of $(2,R)$ bins used is Y. So the optimal cost is $X + R \cdot Y$. Assume all the items packed in the unit size bins is in set L_1, all the others is in L_2. For $1 \le i \le 2$, we have

$$w_i(L) = w_i(L_1) + w_i(L_2).$$

By the definitions of $W_{1,i}$ and $W_{2,i}$, we have

$$w_i(L_1) \le W_{1,i} \cdot X, \quad w_i(L_2) \le W_{2,i} \cdot Y = \frac{W_{2,i}}{R} \cdot R \cdot Y.$$

Then we have

$$w_i(L) \le \max\{W_{1,i}, \frac{W_{2,i}}{R}\}(X + R \cdot Y) \le W \cdot (X + R \cdot Y). \tag{13}$$

$\square$

Calculating W:

Lemma 5. *When* $K = 1$, *we have* $W_{1,j} \le \frac{R}{3} + \frac{552}{798}$, *where* $R < 2.6915$ *and* $1 \le j \le 2$.

Lemma 6. *When* $K = 2$, *i.e.*, $\frac{7}{3} < R \le \frac{28}{11}$, *we have* $W_{1,j} \le \frac{R}{3} + \frac{11R}{56} + \frac{153}{798}$, *for* $j = 1, 2$.

Lemma 7. *When* $R \ge 2.25$, $W_{2,j} \le \frac{4R}{3}$.

Lemma 8. *When* $K = 3$, *i.e.*, $2.040 \le R \le \frac{7}{3}$, $W_{1,j} \le \max\{\frac{4474}{3192}, 1 + \frac{R}{7} + \frac{267}{2964}, \frac{R}{3} + \frac{11R}{56} + \frac{153}{798}\}$.

Lemma 9. *When* $K = 3$, *i.e.*, $2.040 < R \le \frac{7}{3}$, *we have* $W_{2,j} \le \max\{1.381R, \frac{1749}{798} + \frac{R}{3}\}$.

Proofs of the above Lemmas can be referred to the full version.

Theorem 2. *Algorithm 1 applied to OBP1R problem has competitive ratios as Table 2.*

Table 2. Upper bounds for $2.040 < R \le 2.6915$

R	Upper bound
$\frac{28}{11} < R \le 2.6915$	$\frac{R}{3} + \frac{1}{2} + \frac{153}{798}$
$2.3219 < R \le \frac{28}{11}$	$\frac{R}{3} + \frac{11R}{56} + \frac{153}{798}$
$2.1808 < R \le 2.3219$	$1 + \frac{R}{7} + \frac{267}{2964}$
$2.0513 < R \le 2.1808$	$\frac{4474}{3192} = 1.40163$
$2.040 < R \le 2.0513$	$\frac{1749}{798R} + \frac{1}{3}$

Proof. By inequality (13), the competitive ratio of the algorithm is the maximum of $W_{1,j}$ and $\frac{W_{2,j}}{R}$. By Lemmas 5 and 7, for $\frac{28}{11} < R \le 2.6915$, we have the competitive ratio is $\frac{R}{3} + \frac{1}{2} + \frac{153}{798}$. By Lemmas 6 and 7, for $\frac{7}{3} < R \le \frac{28}{11}$, the competitive ratio is $\frac{R}{3} + \frac{11R}{56} + \frac{153}{798}$. By Lemmas 8 and 9, for $2.3219 < R \le \frac{7}{3}$, the competitive ratio is $\frac{R}{3} + \frac{11R}{56} + \frac{153}{798}$; for $2.1808 < R \le 2.3219$, the competitive ratio is $1 + \frac{R}{7} + \frac{267}{2964}$; for $2.0513 < R \le 2.1808$, the competitive ratio is 1.40163; for $2.040 < R \le 2.0513$, the competitive ratio is $\frac{1749}{798R} + \frac{1}{3}$. We summarized these results in Table 2. $\square$

2.4 Algorithm PAR for $R \leq 2.040$

It is not difficult to see that when $1 < R \leq 2$, there will be at most one unit bin in an offline optimum packing, i.e., almost all items are packed into size two bins. For online version, we pack all the items into size two bins by a refined harmonic algorithm, our algorithm is called as PAR, and also useful for the parametric bin packing problem(with parameter $r = 2$).

First we give useful definitions for PAR algorithm. In PAR we set $J_k = (\frac{2}{k+1}, \frac{2}{k}], k \geq 4$, $J_M = (0, 2/M] = (0, 1/20]$, $M = 40$, $J_b = (\frac{1}{2}, \frac{11}{20}]$, $J_3 = (\frac{11}{20}, \frac{2}{3}]$, $J_a = (\frac{2}{3}, \frac{29}{40}]$, and $J_2 = (\frac{29}{40}, 1]$. Bin B_{J_k} of size 2 is desiegned to pack J_k items.

First we define two types of bins: i) B_A bin is designed to contain two J_a-pieces and one J_b-piece; ii) B_B bin is designed to contain three J_b-pieces. The packing of J_a and J_b pieces is below: given a J_a piece, if there is a type B_A with less than two J_a pieces, then we pack it into this bin, else we open a new B_A bin for the item. Given a J_b piece, if there is a B_A bin with no J_b piece in it, then put the item into this bin, else we open a new bin with size two and pack the item there. Let the total number of J_b items be n_b, and the total number B_B bins be m_{53b}. If $\frac{m_{3b}}{n_b} < \frac{6}{19}$, then open a new B_B bin, else B_A bin.

Given an item a_i with type $x \in \{J_1, ..., J_M, J_a, J_b\}$, PAR is described as Algorithm 3.

Algorithm 3. PAR

---

1.   **if** x is J_a **then**
        if there is a B_A bin and the number of J_a items is less than two, **then** pack a_i in this bin.
        else pack a_i into a new B_A bin.
2.   **if** a_i is J_b **then** $n_b = n_b + 1$,
        if there is a B_B bin with J_b items less than three, **then** pack a_i in this bin.
        else
            if there is a B_A bin without J_b items; **then** pack a_i in this bin.
            else
                if $m_{3b} \leq \frac{6n_b}{19}$ **then** place a_i in a new B_B bin, $m_{3b} = m_{3b} + 1$.
                else place a_i in a new B_A bin.
3.   **else** pack a_i by Harmonic algorithm into bin B_x.

We then analyze he performance of PAR. Similarly we have weighting functions for PAR:

$$h_1(x) = \begin{cases} \frac{1}{k}, x \in J_k, 2 \leq k \leq M - 1, \\ \frac{Mx}{2M-2}, x \in J_M, \\ \frac{7}{19}, x \in J_b, \\ 0, x \in J_a. \end{cases} \qquad h_2(x) = \begin{cases} \frac{1}{k}, x \in J_k, 2 \leq k \leq M - 1, \\ \frac{Mx}{2M-2}, x \in J_M, \\ \frac{6}{19}, x \in J_b, \\ \frac{1}{2}, x \in J_a. \end{cases}$$

Let $w_1(x) = h_1(x) \cdot R$, $w_2(x) = h_2(x) \cdot R$, the way of calculating performance of PAR is similar as showed one in MAIN.

Lemma 10. *For $R \leq 2.040$, we have $W_{1,j} \leq \frac{556R}{819}$.*

Lemma 11. *When $R \leq 2.040$, we have $W_{2,j} \leq \frac{219053R}{155610} \leq 1.4078 \cdot R$ for all j.*

Proofs of the Lemmas can be referred to the full version.

By Lemmas 4, 10, 11, and equations (13), We have the following theorem.

Theorem 3. *The competitive ratio c_r of PAR is:*

$$c_r \leq \max\{W_{1,j}, W_{2,j}\} \leq \max\{\frac{556R}{819}, \frac{219053}{155610}\} \leq 1.4078. \tag{14}$$

3   Lower Bounds

We first introduce the framework for calculating the lower bound of our problem, which follows from the previous work for the classical bin packing [14],[15],[16],[10]. Then we give two input sequences based on concrete values of R. We find the previous approach does not go through for a variable R. So, we fix the parameter R as some constants, then apply the previous framework to get the lower bound of our problem. And more our problem is more complicated than the classical bin packing, since the number of all the possible patterns in our problem is much more than the one used in the classical bin packing.

3.1   Framework for Calculating the Lower Bound

Let $\rho = \{s_1, s_2, ..., s_k\}$ be a sequence of item sizes such that $0 < s_1 < s_2 < ... < s_k \leq 1$, where constant k is a positive integer and $s_i = \delta_i + \epsilon$, and ϵ is a sufficiently small positive number. With respect to ρ, a pattern p is defined as below:

$$p = \langle size(p), cost(p), p_1, ..., p_k \rangle, \quad \text{where } \sum_{i=1}^{k} p_i s_i \leq size(p)$$

p_i is the number of items of size s_i, if $size(p) = 1$ then $cost(p) = 1$, else if $size(p) = 2$ then $cost(p) = R$. Set $\mathcal{P}(\rho)$ denotes the set of all patterns p

$$\mathcal{P} = \{p : pattern | size(p) = 2 \text{ or } size(p) = 1\}.$$

Since k is a constant, then $\mathcal{P}(\rho)$ is finite.

Online Input: Define L_0 to be the empty input. Input L_i consists of input L_{i-1} and follows with items of size $\delta_i + \epsilon$ for $1 \leq i \leq k$. The online input L_i is constructed as below:

$$\overbrace{\underbrace{\delta_1 + \epsilon, ..., \delta_1 + \epsilon,}_{L_1} \quad \overbrace{\delta_2 + \epsilon, ..., \delta_2 + \epsilon,}^{L_2} \quad ... \quad \delta_k + \epsilon, ..., \delta_k + \epsilon.}^{}$$

$$\underbrace{}_{L_k}$$

Given an input with sizes from ρ, any algorithm A is defined as a function $\Phi : \mathcal{P}(\rho) \to \mathbb{N}$. The algorithm uses $\Phi(p)$ bins containing items as content of pattern p.

Consider the function Φ that determines by online algorithm A for list L_k. Each pattern is assigned to a class, defined as

$$class(p) = min\{i|p_i \neq 0\}.$$

Define

$$\mathcal{P}_i = \{p \in \mathcal{P}(\rho)|class(p) \leq i\}.$$

Then, the cost of algorithm A for L_i can be calculated by Φ

$$\sum_{p \in \mathcal{P}_i} \Phi(p)\, cost(p).$$

Since the algorithm must pack every item, it comes the following constraints:

$$\sum_{p \in \mathcal{P}(\rho)} \Phi(p)p_i \geq |L_i - L_{i-1}|, \quad \text{for } 1 \leq i \leq k \text{ and } |L_i| \text{ is the length of } L_i.$$

Define $X(L_i)$ to be the optimal offline cost for packing the items in L_i, which can be calculated by the following linear program.

$$Minimize \quad \sum_{p \in \mathcal{P}_i} cost(p)\Phi(p) \tag{15}$$

$$subject\ to \quad |L_j - L_{j-1}| \leq \sum_{p \in \mathcal{P}(\rho)} \Phi(p)p_j, \quad \text{for } 1 \leq j \leq k, \tag{16}$$

where constraints (16) assure that Φ is a feasible solution for L_i.

Let $cost_A(L_i)$ be the cost by online algorithm A for packing L_i. Then the lower bound of the online problem can be expressed as below:

$$c = \min_{A \in \mathcal{A}} \max \{\frac{cost_A(L_1)}{X(L_1)}, \frac{cost_A(L_2)}{X(L_2)}, \cdots, \frac{cost_A(L_k)}{X(L_k)}\}$$

We use the following linear program to estimate c. Note that the optimal value of the following linear programming is not larger than c.

$$Minimize\ c$$
$$subject\ to\ \ c \geq \frac{\sum_{p \in \mathcal{P}_i} cost(p)\Phi(p)}{X(L_i)}, \quad \text{for } 1 \leq i \leq k$$

$$|L_j - L_{j-1}| \leq \sum_{p \in \mathcal{P}(\rho)} \Phi(p)p_j, \quad \text{for } 1 \leq j \leq k.$$

Dominant Patterns: A pattern p of class i is dominant if

$$s_i + \sum_{j=1}^{k} p_j s_j > size(p),$$

which helps us to reduce the cases when we estimate the lower bound c.

3.2   Two Sequences and Lower Bounds

We calculate the lower bound of the OBP1R problem by the method described above. When $1 < R \leq 2.178$, we use *Parametric sequence*, when $2.178 < R \leq 3$, we use *Greedy sequence*.

Given a sequence, we have enumerate its dominant patterns first, then calculate the optimal solution for L_i, finally we use LP to estimate the value of c.

Greedy Sequence: we have $k = 4$ and $s_1 = \frac{1}{43}+\epsilon$, $s_2 = \frac{1}{7}+\epsilon$, $s_3 = \frac{1}{3}+\epsilon$, $s_4 = \frac{1}{2} + \epsilon$ and $|L_i - L_{i-1}| = n$ for all i. The dominant patterns of the sequence is given in the full version.

Consider integer programming (15) and (16), there are at most k non-zero variables in an optimal solution. Since $cost(p)$ is at most R, where R is the cost of a bin with size two, suppose the objective value will decrease μ, then μ will be at most $k \cdot R$ if we solve (15) and (16) as all variables are real numbers rather than integers. Let

$$X_i^* = \lim_{|L_i| \to \infty} X(L_i)/|L_i|.$$

By the definition of X_i^*, we have X_i^* is also equal to $\lim_{|L_i| \to \infty} \frac{X(L_i)-\mu}{|L_i|}$, $\mu \leq k \cdot R$. Define $\phi(p) = \Phi(p)/n$. Therefore we can use the following linear programming to calculate X_i^*.

LP 31:

$$Minimize \sum_{p \in \mathcal{P}_i} cost(p)\phi(p)$$

$$subject\ to \quad 1 \leq \sum_{p \in \mathcal{P}(\rho)} \phi(p)p_j, \quad for\ 1 \leq j \leq i.$$

we use the X_i^*, $1 \leq i \leq k$, and the following linear programming to calculate lower bound.

LP 32:

$$Minimize\ c$$

$$subject\ to\ \ c \geq \frac{1}{X_i^*} \sum_{p \in \mathcal{P}_i} cost(p)\phi(p), \quad for\ 1 \leq i \leq k$$

$$1 \leq \sum_{p \in \mathcal{P}(\rho)} \phi(p)p_i, \quad for\ 1 \leq i \leq k.$$

Lemma 12. *We get results as following:*

Table 3. Lower bounds of OBP1R for $2.178 < R \leq 3$

R	2.178	2.25	2.4	2.5	2.545	2.6915	2.832	3
Lower bound	1.314	1.346	1.410	1.452	1.461	1.488	1.512	1.539

Parametric Sequence: we have $k = 4$ and $s_1 = \frac{2}{157} + \epsilon$, $s_2 = \frac{2}{13} + \epsilon$, $s_3 = \frac{1}{2} + \epsilon$, $s_4 = \frac{2}{3} + \epsilon$ and $|L_i - L_{i-1}| = n$ for all $i \leq 3$ and $|L_4 - L_3| = 2n$. The dominant patterns of the sequence is left in the full version.

Define $\phi(p) = \Phi(p)/n$. Let $X_i^* = \lim_{|L_i| \to \infty} X(L_i)/|L_i|$. We use the following linear programming to calculate X_i^* and the lower bound.

LP 33:

$$Minimize \ \sum_{p \in \mathcal{P}_i} cost(p)\phi(p)$$

$$subject \ to \quad 1 \leq \sum_{p \in \mathcal{P}(\rho)} \phi(p)p_j, \ for \ 1 \leq j \leq min\{i, k-1\},$$

$$2 \leq \sum_{p \in \mathcal{P}(\rho)} \phi(p)p_k.$$

LP 34:

$$Minimize \ c$$

$$subject \ to \quad c \geq \frac{1}{X_i^*} \sum_{p \in \mathcal{P}_i} cost(p)\phi(p), \ for \ 1 \leq i \leq k,$$

$$1 \leq \sum_{p \in \mathcal{P}(\rho)} \phi(p)p_i, \ for \ 1 \leq i \leq k-1,$$

$$2 \leq \sum_{p \in \mathcal{P}(\rho)} \phi(p)p_k.$$

Lemma 13. *We get results as following:*

Table 4. Lower bounds of OBP1R for $1 \leq R \leq 2.178$

R	1	2	2.010	2.030	2.040	2.071	2.074	2.15	2.178
Lower bound	1.3896	1.3896	1.3851	1.375	1.372	1.356	1.356	1.325	1.314

Accuracy in Processing: As mentioned we enumerated all the dominant patterns using a recursion program, the recursion is not difficult to construct, and the point is that we should avoid float operations in the processing, since float operations will drop small figures, but the small operator like ϵ are important in our lower bound calculating, and we defined integer operation for our calculation, every operator are presented in fraction of integers.

We define an extra-long integer data type XLong, which can handle integers with more than hundreds or thousands figures. This type of data is defined to avoid the inaccurateness of calculation in enumerating the dominant patterns when using float arithmetic operations. In our calculation, any data is present in fraction, and arithmetic operations are redefined for XLong data in the forms of fraction operations. For eaxmple, subtraction $M - S$ can be described as :$M - S = \frac{M_N}{M_D} - \frac{S_N}{S_D}$, where $M = \frac{M_N}{M_D}$, $S = \frac{S_N}{S_D}$, M_N, M_D, S_N, S_D, are integers, then the subtraction becomes $\frac{M_N * S_D - S_N * M_D}{M_D * S_D}$. The fraction operation we defined can help us enumerating the dominant patterns exactly, but the disadavantage of these integer operators is that the figures of the Numerator and Denominator may becomes increasing as the recursion progress, and typical integer operations with the figures of argument integer under hundred is not sufficient to handle these, we defined XLong data types which can handle integers with the number of figures we assigned(thousand figures or more).

Acknowledgements. This research has been partially supported by "the Fundamental Research Funds for the Central Universities(DUT12LK09)" and NSFC(11101065), RGC(HKU716412E).

References

1. Epstein, L., Levin, A.: An APTAS for Generalized Cost Variable-Sized Bin Packing. SIAM J. Comput. 38(1), 411–428 (2008)
2. Ullman, J.D.: The performance of a memory allocation algorithm. Technical Report 100, Princeton University, Princeton, NJ (1971)
3. Johnson, D.S.: Fast algorithm for bin packing. Journal of Computer and System Sciences 8, 272–314 (1974)
4. Johnson, D.S., Demers, A., Ullman, J.D., Garey, M.R., Graham, R.L.: Worst-case performance bounds for simple one-dimensional packing algorithms. SIAM J. Comput. 3, 256–278 (1974)
5. Yao, A.C.C.: New algorithms for bin packing. J. ACM 27, 207–227 (1980)
6. Coffman, E.G., Garey, M.R., Johnson, D.S.: Approximation algorithms for bin packing: A survey. In: Hochbaum, D. (ed.) Approximation Algorithms. PWS Publishing Company (1997)
7. Lee, C.C., Lee, D.T.: A simple on-line bin packing algorithm. J. ACM 32(3), 256–278 (1985)
8. Van Vliet, A.: An improved lower bound for on-line bin packing algorithms. Information Processing Letters 43(5), 277–284 (1992)
9. Seiden, S.S.: On the online bin packing problem. J. ACM 49, 640–671 (2002)
10. Seiden, S.S., Van Stee, R., Epstein, L.: New bounds for variable-sized online bin packing. SIAM J. Comput. 32(2), 455–469 (2002)
11. Friesten, D.K., Langston, M.A.: Variable sized bin packing. SIAM J. Comput. 15, 222–230 (1986)
12. Kinnerseley, N.G., Langston, M.A.: Online variable-sized bin packing. Discrete Applied Mathematics 22(2), 143–148 (1988)
13. Csirik, J.: An on-line algorithm for variable-sized bin packing. Acta Informatica 26(8), 697–709 (1989)
14. Brown, D.J.: A lower bound for on-line one-dimensional bin packing algorithms. Tech. report -864. Coordinated Science Laboratory Urbana IL (1979)
15. Liang, F.M.: A lower bound for on-line bin packing. Information Processing Letters 10, 76–79 (1980)
16. Van Vliet, A.: An improved lower bound for online bin packing algorithm. Inform. Process. Lett. 43, 277–284 (1992)
17. Van Vliet, A.: Lower and upper bounds for online bin packing and scheduling heuristics. Thesis Publishers, Amsterdam (1995)
18. Blitz, D., Van Vliet, A., Woeginger, G.J.: Lower bounds on the asymptotic worst-case ratio of online bin packing algorithms (1996) (unpublished manuscript)
19. Valerio de Carvalho, J.M.: LP models for bin packing and cutting stock problem. European Journal of Operational Research 141, 253–273 (2002)

Disclosing Barriers: A Generalization of the Canonical Partition Based on Lovász's Formulation

Nanao Kita

Keio University, Yokohama, Japan
kita@a2.keio.jp

Abstract. Given a graph, a barrier is a set of vertices determined by the Berge formula—the min-max theorem characterizing the size of maximum matchings. The notion of barriers plays important roles in numerous contexts of matching theory, since barriers essentially coincides with dual optimal solutions of the maximum matching problem. In a special class of graphs called the elementary graphs, the family of maximal barriers forms a partition of the vertices; this partition was found by Lovász and is called the canonical partition. The canonical partition has produced many fundamental results in matching theory, such as the two ear theorem. However, in non-elementary graphs, the family of maximal barriers never forms a partition, and there has not been the canonical partition for general graphs. In this paper, using our previous work, we give a canonical description of structures of the odd-maximal barriers—a class of barriers including the maximal barriers—for general graphs; we also reveal structures of odd components associated with odd-maximal barriers. This result of us can be regarded as a generalization of Lovász's canonical partition.

1 Introduction

A *matching* of a graph G is a set of edges no two of which have common vertices. A matching of cardinality $|V(G)|/2$ (resp. $|V(G)|/2 - 1$) is called a *perfect matching* (resp. a *near-perfect matching*). We call a graph *factorizable* if it has at least one perfect matching. Now let G be a factorizable graph. An edge $e \in E(G)$ is called *allowed* if there is a perfect matching containing e. Let $\widehat{M}$ be the union of all the allowed edges of G. For each connected component C of the subgraph of G determined by $\widehat{M}$, we call the subgraph of G induced by $V(C)$ as *factor-connected component* or *factor-component* for short. The set of all the factor-components of G is denoted by $\mathcal{G}(G)$. Therefore, a factorizable graph is composed of factor-components and some edges joining between different factor-components. A factorizable graph with exactly one factor-component is called *elementary*.

Matching theory is of central importance in graph theory and combinatorial optimization, with numerous practical applications [1]. In matching theory, the

P. Widmayer, Y. Xu, and B. Zhu (Eds.): COCOA 2013, LNCS 8287, pp. 402–413, 2013.

notion of barriers plays significant roles. Given a graph, we call a connected component of it with an odd (resp. even) number of vertices *odd component* (resp. *even component*). Given $X \subseteq V(G)$ of a graph G, we denote as $q_G(X)$ the number of odd components that the graph resulting from deleting X from G has; we denote the cardinality of a maximum matching of G as $\nu(G)$. There is a min-max theorem called the *Berge formula* [2] that for any graph G, $|V(G)| - 2\nu(G) = \max\{q_G(X) - |X| : X \subseteq V(G)\}$. A set of vertices that attains the maximum in the right side of the equation is called a *barrier*. Roughly speaking, barriers essentially coincide with dual optimal solutions of the maximum matching problem, and decompose graphs so that one can see the structures of maximum matchings. However, compared to numerous results on maximum matchings, "much less is known about barriers [2]".

There is a structure of elementary graphs called the *canonical partition*; Kotzig first introduced it as the equivalence classes of a certain equivalence relation, and later Lovász reformulated it from the point of view of barriers, stating that the family of maximal barriers forms a partition of the vertices in elementary graphs. This reformulation by Lovász has produced many fundamental properties in matching theory such as the *two ear theorem* [1,2], and the *brick decomposition* or the *tight cut decomposition*, and underlies polyhedral studies of matching theory; see the survey article [3].

However, in non-elementary graphs, the family of maximal barriers never forms a partition of the vertices, and there has not been known the counterpart structure of Lovász's canonical partition for general graphs. In this paper, therefore, we reveal canonical structures of maximal barriers and obtain a generalization of Lovász's canonical partition for general graphs; here, our previous work on canonical structures of general factorizable graphs [4,5], the *generalized cathedral structure* (see Section 2.3), serves as a language to describe barriers. (Actually, we work on a wider notion called *odd-maximal barriers*; see Section 2.2.) In [4,5], we defined an equivalence relation and introduced a generalization of the canonical partition based on Kotzig's formulation: the *generalized canonical partition*. In this paper, we show that it can be also regarded as a generalization based on Lovász's formulation, stating that the family of equivalence classes of the generalized canonical partition are "atoms" that constitute (odd-)maximal barriers in general graphs (which shall be introduced in Section 3). We also reveal the structure of odd components associated with (odd-)maximal barriers.

Because the canonical partition and the notion of barriers are important, we are sure that our result will produce many applications in matching theory. There has been known a close relationship between algorithms in matching theory, barriers, and canonical structure theorems [1,2]; therefore, our result will have algorithmic applications. Lovász's canonical partition has been the foundation in the study of polyhedral aspects of matchings; therefore, our results will make a contribution to this field. So far we have already obtained some consequences [6] on the *optimal ear-decomposition* [7].

2   Preliminaries

2.1   Definitions and Some Preliminary Facts

In this paper we mostly observe those given by Schrijver [8] for standard definitions and notations. We list here those additional or non-standard.

Hereafter for a while let G be a graph. For $X \subseteq V(G)$, we define the *contraction* of G by X as the graph obtained by contracting X into one vertex, and denote it as G/X. For simplicity, we identify vertices, edges, subgraphs of G/X with those of G naturally corresponding to them.

In many contexts, we often regard a subgraph H of G as a vertex set $V(H)$. For example, G/H means $G/V(H)$. We treat paths and circuits as graphs. For a path P and $x, y \in V(P)$, xPy means the subpath of P whose end vertices are x and y.

We say a matching M of G *exposes* $v \in V(G)$ if $\delta(v) \cap M = \emptyset$, otherwise say it *covers* v. For a matching M of G and $u \in V(G)$ covered by M, u' denotes the vertex to which u is matched by M. For $X \subseteq V(G)$, M_X denotes $M \cap E(G[X])$.

Hereafter for a while let M be a matching of G. For a subgraph Q of G, which is a path or circuit, we call Q M-*alternating* if $E(Q) \setminus M$ is a matching of Q. Let P be an M-alternating path of G with end vertices u and v. If P has an even number of edges and $M \cap E(P)$ is a near-perfect matching of P exposing only v, we call it an M-*balanced path* from u to v. We regard a trivial path, that is, a path composed of one vertex and no edges as an M-balanced path. If P has an odd number of edges and $M \cap E(P)$ (resp. $E(P) \setminus M$) is a perfect matching of P, we call it M-*saturated* (resp. M-*exposed*).

Let $X \subseteq V(G)$. We say a path P of G is an *ear relative to* X if both end vertices of P are in X while internal vertices are not. So do we to a circuit if exactly one vertex of it is in X. For simplicity, we call the vertices of $V(P) \cap X$ *end vertices* of P, even if P is a circuit. For an ear P of G relative to X, we call it an M-*ear* if $P - X$ is an M-saturated path. Given an ear P and $Y \subseteq V(G)$, we say P is *through* Y if P has some internal vertices in Y.

Factor-components of a bipartite factorizable graph are known to have the following partially ordered structure[1]:

Theorem 1 (The Dulmage-Mendelsohn Decomposition [2, 9–12]). *Let* $G = (A, B; E)$ *be a bipartite factorizable graph, and let* $\mathcal{G}(G) =: \{G_i\}_{i \in I}$. *Let* $A_i := A \cap V(G_i)$ *and* $B_i := B \cap V(G_i)$ *for each* $i \in I$. *Then, there exists a partial order* $\preceq_A$ *on* $\mathcal{G}(G)$ *such that for any* $i, j \in I$,

(i) $E[B_j, A_i] \neq \emptyset$ *yields* $G_j \preceq_A G_i$, *and*
(ii) *if* $G_j \preceq_A H \preceq_A G_i$ *yields* $G_i = H$ *or* $G_j = H$ *for any* $H \in \mathcal{G}(G)$, *then* $E[B_j, A_i] \neq \emptyset$.

We call this decomposition of G into a poset *the Dulmage-Mendelsohn decomposition* (in short, *the DM-decomposition*), and each element of $\mathcal{G}(G)$, in this context, a *DM-component*. The DM-decomposition is uniquely determined by a graph, up

---

[1] This is different from the one in [4,5]. Though it is sometimes presented as a theorem for general bipartite graphs, we introduce it as one for bipartite factorizable graphs.

to the choice of roles of color classes. In this paper, we call the DM-decomposition of $G = (A, B; E)$ as in Theorem 1 the DM-decomposition *with respect to* A.

Proposition 1 (Dulmage and Mendelsohn [9–12]). *Let $G = (A, B; E)$ be a bipartite factorizable graph, and M be a perfect matching of G. Let $G_1, G_2 \in \mathcal{G}(G)$, and let $u \in A \cap V(G_1)$, $v \in A \cap V(G_2)$, and $w \in B \cap V(G_2)$. Then there is an M-balanced path from u to v if and only if $G_1 \preceq_A G_2$; additionally, there is an M-saturated path between u to w if and only if $G_1 \preceq_A G_2$.*

Hereafter in this section we present some basic properties used explicitly or implicitly throughout this paper. These are easy to see and the succeeding two propositions are well-known and might be folklores. A graph is called *factor-critical* if any deletion of an arbitrary vertex leaves an empty graph or a factorizable graph.

Proposition 2 (folklore). *Let M be a near-perfect matching of a graph G that exposes $v \in V(G)$. Then, G is factor-critical if and only if for any $u \in V(G)$ there exists an M-balanced path from u to v.*

Given a graph G and $X \subseteq V(G)$, we denote the vertices contained in the odd components of $G - X$ as D_X, and $V(G) \setminus X \setminus D_X$ as C_X. The next proposition can be easily observed by the Berge formula.

Proposition 3 (folklore). *Let G be a factorizable graph, and $X \subseteq V(G)$ be a barrier of G. Then for any perfect matching M of G,*

(i) *each vertex of X is matched to a vertex of D_X,*
(ii) *for each component K of $G[D_X]$, M_K is a near-perfect matching of K, accordingly $|\delta(K) \cap M| = 1$,*
(iii) *M contains a perfect matching of $G[C_X]$, and*
(iv) *no edge in $E[X, C_X]$ nor $E(G[X])$ is allowed.*

Now let G be a factorizable graph. We say $X \subseteq V(G)$ is *separating* if any $H \in \mathcal{G}(G)$ satisfies $V(H) \subseteq X$ or $V(H) \cap X = \emptyset$. The next one is easy to see by the definitions.

Proposition 4. *Let G be a factorizable graph, and let $X \subseteq V(G)$. Then, the following four properties are equivalent:*

(i) *X is separating.*
(ii) *X is an empty set, or there exists $H_1, \ldots, H_k \in \mathcal{G}(G)$ such that $X = V(H_1) \dot\cup \cdots \dot\cup V(H_k)$.*
(iii) *For any perfect matching M of G, $\delta(X) \cap M = \emptyset$.*
(iv) *For any perfect matching M of G, M_X forms a perfect matching of $G[X]$.*

2.2 Our Aim

Given an elementary graph G, we say $u \sim v$ for $u, v \in V(G)$ if $u = v$ holds or $G - u - v$ is not factorizable. Kotzig [13–15] found that $\sim$ is an equivalence relation. Later Lovász redefined it:

Theorem 2 (Lovász [2]). *Let G be an elementary graph. Then, the family of maximal barriers forms a partition of $V(G)$. Additionally, this partition coincides with the equivalence classes by $\sim$.*

This partition by the maximal barriers is called the *canonical partition*. As we mention in Section 1, it plays fundamental and significant roles in matching theory. On the other hand, as for non-elementary graphs, the family of maximal barriers never forms a partition of the vertices (see [2]). The question remains: how all the maximal barriers exist and what is the counterpart in general graphs? Therefore, we are going to investigate it. Actually, we work on a wider notion: *odd-maximal barriers*.[2]

Definition 1. *Let G be a graph. A barrier $X \subseteq V(G)$ is called an odd-maximal barrier if it is a barrier which is maximal with respect to $X \cup D_X$, i.e., no $Y \subseteq D_X$ with $Y \neq \emptyset$ satisfies that $X \cup Y$ is a barrier of G.*

Odd-maximal barriers have some nice properties (see [16, 17]): First, A maximal barrier is an odd-maximal barrier. Second, for elementary graphs, the notion of maximal barriers and the notion of odd-maximal barriers coincide. Hence, it seems reasonable to work on the odd-maximal barriers. Actually, with the Gallai-Edmonds structure theorem and the theorem by Király [16], we can see that it suffices to work on factorizable graphs. Given the above facts, in this paper we give canonical structures of odd-maximal barriers in general factorizable graphs that can be regarded as a generalization of Lovász's canonical partition, aiming to contribute to the foundation of matching theory.

2.3   The Generalized Cathedral Structure

In this section we are going to introduce the canonical structure theorems of factorizable graphs, which shall serve as a language to describe odd-maximal barriers. They are composed of three parts: a partially ordered structure on the factor-components (Theorem 3), a generalization of the canonical partition (Theorem 4), and a relationship between these two (Theorem 5).[3]

Definition 2. *Let G be a factorizable graph, and let $G_1, G_2 \in \mathcal{G}(G)$. We say $X \subseteq V(G)$ is a critical-inducing set for G_1 to G_2 if X is separating, $V(G_1) \cup V(G_2) \subseteq X$ holds, and $G[X]/G_1$ is factor-critical. Additionally, we say $G_1 \lhd G_2$ if there is a critical-inducing set for G_1 to G_2.*

Theorem 3 (Kita [4, 5]). *For any factorizable graph G, $\lhd$ is a partial order on $\mathcal{G}(G)$.*

Definition 3. *Let G be a factorizable graph. For $u, v \in V(G)$ we say $u \sim_G v$ if u and v are contained in the same factor-component of G, and $G - u - v$ is NOT factorizable.*

---

[2] This is identical to those Király calls strong barriers [16], however we call it in the different way so as to avoid the confusion with the notion of strong end by Frank [7].

[3] All the statements in [5] can be also found in [4].

Theorem 4 (Kita [4, 5]). *For any factorizable graph G, $\sim_G$ is an equivalence relation on $V(G)$.*

As you can see by the definition, if G is an elementary graph then $\sim$ and $\sim_G$ coincide. Therefore, we call the equivalence classes by $\sim_G$, i.e. $V(G)/\sim_G$, the *generalized canonical partition* or just the *canonical partition*, and denote by $\mathcal{P}(G)$. For each $H \in \mathcal{G}(G)$, we define $\mathcal{P}_G(H) := \{S \in \mathcal{P}(G) : S \subseteq V(H)\}$; then, $\mathcal{P}_G(H)$ forms a partition of $V(H)$, since by the definition each equivalence class is respectively contained in one of the factor-components. Note that $\mathcal{P}_G(H)$ is always a refinement of $\mathcal{P}(H)$, which equals to $\mathcal{P}_H(H)$.

For each $H \in \mathcal{G}(G)$, we denote the family of the upper bounds of H in the poset $(\mathcal{G}(G), \lhd)$ as $\mathcal{U}_G^*(H)$, and $\mathcal{U}_G^*(H) \setminus \{H\}$ as $\mathcal{U}_G(H)$. Moreover, we denote the vertices contained in $\mathcal{U}_G^*(H)$ as $U_G^*(H)$; i.e., $U_G^*(H) := \bigcup_{H' \in \mathcal{U}_G^*(H)} V(H')$. We also denote $U_G^*(H) \setminus V(H)$ as $U_G(H)$. Actually, the next theorem states that each strict upper bound of $H \in \mathcal{G}(G)$ in $(\mathcal{G}(G), \lhd)$ is respectively "assigned" to some $S \in \mathcal{P}_G(H)$:

Theorem 5 (Kita [4, 5]). *Let G be a factorizable graph, and let $H \in \mathcal{G}(G)$. For each connected component K of $G[U_G(H)]$, there exists $S_K \in \mathcal{P}_G(H)$ such that $N(K) \cap V(H) \subseteq S_K$.*

Based on Theorem 5, we define $\mathcal{U}_G(S)$ as follows: $H' \in \mathcal{U}_G(S)$ if and only if $H \lhd H'$ and $H \neq H'$ holds and there exists a connected component K of $G[U(H)]$ with $N(K) \cap V(H) \subseteq S$ such that $V(H') \subseteq V(K)$. Additionally, we denote the vertices contained in $\mathcal{U}_G(S)$ as $U_G(S)$; i.e., $U_G(S) := \bigcup_{H' \in \mathcal{U}_G(S)} V(H')$. We also define $U_G^*(S) := U_G(S) \cup S$. Regarding these eight notations we sometimes omit the subscripts "G" if they are apparent from the contexts. Note that $\bigcup_{T \in \mathcal{P}_G(H)} \mathcal{U}(T) = \mathcal{U}(H)$.

We call the canonical structures of factorizable graphs given by Theorems 3, 4, and 5 the *generalized cathedral structures* or just the *cathedral structures*. Now let us add some propositions used later in this paper:

Proposition 5 (Kita [4, 5]). *Let G be a factorizable graph, and let $H \in \mathcal{G}(G)$. Then, $G[U^*(H)]/H$ is factor-critical, so is each block of it.*

Proposition 6 (Kita [4, 5]). *Let G be a factorizable graph and M be a perfect matching of G, and let $H \in \mathcal{G}(G)$. Let P be an M-ear relative to H.*

(i) *Let $H' \in \mathcal{G}(G)$. If P is through H', then $H \lhd H'$.*
(ii) *The end vertices $u, v \in V(H)$ of P satisfies $u \sim_G v$.*

3   A Generalization of Lovász's Canonical Partition

3.1   Our Main Result

Our main result is the following:

Main Theorem. *Let G be a factorizable graph, and $X \subseteq V(G)$ be an odd-maximal barrier of G. Then, X is a disjoint union of some members of $\mathcal{P}(G)$;*

namely, there exists $S_1, \ldots, S_k \in \mathcal{P}(G)$ such that $X = S_1 \dot{\cup} \cdots \dot{\cup} S_k$. Additionally, odd components of $G - X$ have structures as follows: $D_X = (U^*(G_1) \setminus U^*(S_1)) \dot{\cup} \cdots \dot{\cup} (U^*(G_k) \setminus U^*(S_k))$, where $G_i \in \mathcal{G}(G)$ is such that $S_i \in \mathcal{P}_G(G_i)$ for each $i \in \{1, \ldots, k\}$.

This theorem states that in general graphs the equivalence classes of the generalized canonical partition are the "atoms" that constitute odd-maximal barriers, and that odd components associated to odd-maximal barriers are also described canonically by the generalized cathedral structure. As we see in previous sections, among two formulations of the canonical partition of elementary graphs, the generalization of the canonical partition introduced in [4, 5] is attained based on Kotzig's formulation; here we show it is as well a generalization based on Lovász's formulation.

This theorem is an immediate corollary of Theorem 8, and the rest of this paper is to prove Theorem 8. We shall prove it by examining the reachability of alternating paths from two viewpoints— regarding odd-maximal barriers and regarding the generalized cathedral structure—and showing their equivalence. Let us mention an additional property used later in this paper.

Proposition 7 (Király [16]). *A barrier $X \subseteq V(G)$ of a graph G is odd-maximal if and only if all the odd components of $G - X$ are factor-critical.*

3.2   Barriers vs. Alternating Paths

In this subsection we introduce some lemmas on the reachability of alternating paths regarding odd-maximal barriers. Given an odd-maximal barrier X of a factorizable graph G, we generate a bipartite graph, thus canonically decompose $X \cup D_X$ and state the reachability using the DM-decomposition as a language. This technique of generating a bipartite graph has been known [2, 7] and essences of ideas are found there. However, we first reveal it thoroughly to obtain Proposition 10 and Theorem 6.

Proposition 8 (might be a folklore). *Let G be a factorizable graph, M be a perfect matching, and $X \subseteq V(G)$ be an odd-maximal barrier. Then, for any $u \in X$ and $v \in X \cup C_X$ there is no M-saturated path between u and v.*

Definition 4. *Let G be a graph, $X \subseteq V(G)$, and $K_1, \ldots, K_l$ be the odd components of $G - X$. We denote the bipartite graph resulting from deleting the even components of $G - X$, removing the edges whose vertices are all contained in X, and contracting each K_i, where $i = 1, \ldots, l$, respectively into one vertex, as $H_G(X)$. Namely, $H_G(X) := (G - C_X - E(G[X]))/K_1/\cdots/K_l$.*

The next proposition is easily seen by Propositions 3 and 7 and enables us to discuss Proposition 10 and so on.

Proposition 9 (might be a folklore). *Let G be a factorizable graph and X be an odd-maximal barrier of G. If $M \subseteq E(G)$ is a perfect matching of G, then $M \cap \delta(X)$ forms a perfect matching of $H_G(X)$. Conversely, if M' is a perfect matching of $H_G(X)$, there is a perfect matching M of G such that $M' = M \cap \delta(X)$.*

The next proposition shows that the reachabilities of alternating paths are equivalent between G and $H_G(X)$, which, with Proposition 1, derives Theorem 6 immediately.

Proposition 10. *Let G be a factorizable graph, $X \subseteq V(G)$ be an odd-maximal barrier of G, and $\mathcal{K} := \{K_i\}_{i=1}^l$ be the family of odd components of $G - X$, where $l = |X|$. Let M be a perfect matching of G, and M' be the perfect matching of $H_G(X)$ such that $M' = M \cap \delta(X)$. Let $u, v \in X$, and $w \in V(K)$, where $K \in \mathcal{K}$, and let w_K be the contracted vertex of $H_G(X)$ corresponding to K.*

(i) *Then, for any M-balanced path (resp. M-saturated path) P of G from u to v (resp. between u and w), $P' = P/K_1/ \cdots /K_l$ is an M'-balanced path (resp. M'-saturated path) of $H_G(X)$ from u to v (resp. between u and w_K).*

(ii) *Conversely, for any M'-balanced path (resp. M'-saturated path) P' from u to v in $H_G(X)$ (resp. between u and w_K), there is an M-balanced path (resp. M-saturated path) P from u to v in G (resp. between u and w) such that $P' = P/K_1/ \cdots /K_l$.*

Given a factorizable graph G and an odd-maximal barrier X, we denote the DM-decomposition of $H_G(X)$ with respect to X as just the DM-decomposition of $H_G(X)$. In this case, we sometimes denote $\preceq_X$ as just $\preceq$, omitting the subscript "X".

Definition 5. *Let G be a factorizable graph, and X be an odd-maximal barrier of G. Let D be a DM-component of $H_G(X)$, whose vertices in $V(D) \setminus X$ are the contracted vertices resulting from some odd components of $G - X$, say $K_1, \ldots, K_l$, where $l \leq |X|$. We say $\widehat{D}$ is the expansion of D if it is the subgraph of G induced by $(V(D) \cap X) \cup \bigcup_{i=1}^l V(K_i)$.*

The next proposition is a basic observation on expansions.

Proposition 11. *Let G be a factorizable graph, and X be an odd-maximal barrier of G. Let $D_1, \ldots, D_k$ be the DM-components of $H_G(X)$. For each $i = 1, \ldots, k$, let $\widehat{D}_i$ be the expansion of D_i. Then,*

(i) *$\{V(\widehat{D}_i)\}_{i=1}^k$ forms a partition of $X \cup D_X$,*

(ii) *$V(\widehat{D}_i)$ is separating, accordingly $\widehat{D}_i$ is factorizable,*

(iii) *$X \cap V(\widehat{D}_i)$ is an odd-maximal barrier of $\widehat{D}_i$, and*

(iv) *$H_{\widehat{D}_i}(X \cap V(\widehat{D}_i))$ is isomorphic to D_i, for each $i = 1, \ldots, k$.*

Theorem 6. *Let G be a factorizable graph, X be an odd-maximal barrier, and M be a perfect matching of G. Let $u, v \in X$, and $w \in D_X$, and for each $\alpha = u, v, w$ let D_α be the DM-component of $H_G(X)$ whose expansion $\widehat{D}_\alpha$ contains α. Then, there is an M-balanced path from u to v (resp. an M-saturated path from u to w) in G if and only if $D_u \preceq D_v$ (resp. $D_u \preceq D_w$).*

The following lemma is obtained by Propositions 10 and 11, and Theorem 6.

Lemma 1. *Let G be a factorizable graph, X be an odd-maximal barrier, and M be a perfect matching of G. Let $\widehat{D}_1$ and $\widehat{D}_2$ be the subgraphs of G which are respectively the expansions of DM-components D_1 and D_2 such that $D_1 \preceq D_2$. Then, for any $u \in X \cap V(\widehat{D}_1)$ and $w \in V(\widehat{D}_2) \setminus X$, any M-saturated path P between u and w traverses $X \cap V(\widehat{D}_2)$.*

3.3   Canonical Structures of Odd-Maximal Barriers

In this subsection we examine the reachability of alternating paths regarding the cathedral structure and derive the main theorem. The next lemma is obtained by Proposition 5 and Proposition 2.

Lemma 2. *Let G be a factorizable graph and M be a perfect matching of G, and let $H \in \mathcal{G}(G)$ and $S \in \mathcal{P}_G(H)$. Then, for any $x \in U^*(S)$, there is an M-balanced path from x to some vertex $y \in S$, whose vertices except y are contained in $U(S)$.*

Immediately by Theorem 4, we can see the next proposition:

Proposition 12. *Let G be a factorizable graph and M be a perfect matching of G, and let $H \in \mathcal{G}(G)$. A set of vertices $S \subseteq V(H)$ is a member of $\mathcal{P}_G(H)$ if and only if it is a maximal subset of $V(H)$ satisfying that there is no M-saturated path between any two vertices of it.*

The next one is by Proposition 6 and Lemma 2.

Lemma 3. *Let G be a factorizable graph and M be a perfect matching of G, and let $H \in \mathcal{G}(G)$ and $S \in \mathcal{P}_G(H)$. Then, for any $s \in S$ and $x \in U(S)$, there is no M-saturated path between s and x nor M-balanced path from s to x.*

The next one, Lemma 4, is rather easy to see by Proposition 6, and combining it with Lemma 2 we can obtain Lemma 5.

Lemma 4. *Let G be a factorizable graph and M be a perfect matching of G. Let $H \in \mathcal{G}(G)$, and let $u, v \in V(H)$ be such that $u \not\sim_G v$. Let P be an M-saturated path between u and v such that $E(P) \setminus E(H) \neq \emptyset$, and let $P_1, \ldots, P_l$ be the components of $P - E(H)$. Let $S_0, S_{l+1} \in \mathcal{P}_G(H)$ be such that $u \in S_0$ and $v \in S_{l+1}$. Then,*

  (i)   *two end vertices of P_i belong to the same member of $\mathcal{P}_G(H)$, say S_i,*
  (ii)   *P_i is, except its end vertices, contained in $U(S_i)$ for each $i = 1, \ldots, l$, and*
  (iii)   *for any $i, j \in \{0, \ldots, l+1\}$ with $i \neq j$, $S_i \neq S_j$.*

Lemma 5. *Let G be a factorizable graph and M be a perfect matching of G. Let $H \in \mathcal{G}(G)$, and let $S, T \in \mathcal{P}_G(H)$ be such that $S \neq T$. Then, for any $s \in S$ and $t \in U^*(T)$, there is an M-saturated path P between s and t, which is contained in $U^*(H) \setminus U(S)$.*

Lemma 5 immediately yields the following: Lemma 6.

Lemma 6. *Let G be a factorizable graph and M be a perfect matching of G. Let $H \in \mathcal{G}(G)$, and let $S, T \in \mathcal{P}_G(H)$ be such that $S \neq T$. Then, for any $s \in S$ and $t \in U^*(T)$, there is an M-saturated path P between s and t such that for any $u \in S$ and $v \in V(P) \setminus S$ there is an M-saturated path between u and v.*

Theorem 7. *Let G be a factorizable graph, M be a perfect matching of G, and $u, v \in V(G)$ be such that $G - u - v$ is not factorizable. If there are M-balanced paths respectively from u to v and from v to u, then u and v are in the same factor-component of G.*

Now we are ready to prove the main theorem, combining up the results in this section.

Theorem 8. *Let G be a factorizable graph, and X be an odd-maximal barrier of G. Let $D_1, \ldots, D_k$ be the DM-components of $H_G(X)$. Let $\widehat{V}_1, \ldots, \widehat{V}_k$ be the partition of $X \cup D_X$ such that for each $i = 1, \ldots, k$, $\widehat{D}_i := G[\widehat{V}_i]$ is the expansion of D_i. Then, for each $i = 1, \ldots, k$, $S_i := X \cap \widehat{V}_i$ coincides with a member of $\mathcal{P}_G(H_i)$ for some $H_i \in \mathcal{G}(G)$, and $\widehat{V}_i$ coincides with $U^*(H_i) \setminus U(S_i)$.*

Proof. Note that such a partition of $X \cup D_X$ surely exists by Proposition 11. Let M be a perfect matching of G. Let $i \in \{1, \ldots, k\}$.

Claim 1. There is no M-saturated path between any two vertices of S_i.

Proof. This is immediate from Proposition 8. □

Claim 2. S_i is contained in the same factor-component of G, say H_i.

Proof. Take $u, v \in S_i$ arbitrarily. Note first that there is no M-saturated path between u and v, by Claim 1. Additionally, there are M-balanced paths from u to v and from v to u respectively, which is immediate from Theorem 6 and Proposition 1. Therefore by Theorem 7, u and v are contained in the same factor-component. Thus, we have the claim. □

Since $\widehat{V}_i$ is separating by Proposition 11,

Claim 3. $V(H_i) \subseteq \widehat{V}_i$.

Claim 4. For any $u \in S_i$ and any $v \in \widehat{V}_i \setminus S_i$, there is an M-saturated path between u and v whose vertices are contained in $\widehat{V}_i$.

Proof. Note that $M_{\widehat{V}_i}$ is a perfect matching of $\widehat{D}_i$, S_i is an odd-maximal barrier of $\widehat{D}_i$, and $H_{\widehat{D}_i}(S_i)$ is a factorizable bipartite graph with exactly one DM-component by Proposition 11. Thus, by applying Theorem 6 to $\widehat{D}_i$, $M_{\widehat{V}_i}$ and S_i, there is an M-saturated path between any $u \in S_i$ and any $v \in \widehat{V}_i \setminus S_i$, which is contained in $\widehat{V}_i$. □

By combining Claims 1, 2, 3, and 4, we obtain that S_i is a maximal subset of $V(H_i)$ such that there is no M-saturated path between any two vertices of it. Hence, by Proposition 12, $S_i \in \mathcal{P}_G(H_i)$ holds.

Claim 5. $\widehat{V}_i \supseteq U^*(H_i) \setminus U(S_i)$.

Proof. Take $y \in U^*(H_i) \setminus U(S_i)$ arbitrarily. If $y \in S_i$, then of course $y \in \widehat{V}_i$. Hence hereafter let $y \in U^*(H_i) \setminus U^*(S_i)$, and let $T \in \mathcal{P}_G(H_i) \setminus \{S_i\}$ be such that $y \in U^*(T)$.

Let $u \in S_i$. There is an M-saturated path P between u and y by Lemma 5. Hence, by Proposition 8, $y \in D_X$. Therefore, there exists $j \in \{1, \ldots, k\}$ such that $y \in \widehat{V}_j$. By Theorem 6 and Proposition 1, $D_i \preceq D_j$.

If $i \neq j$, then by Lemma 1, P has some internal vertices which belong to S_j. However, by Proposition 8, there is no M-saturated path between any two vertices respectively in S_i and S_j, and of course $V(P) \cap S_j$ is disjoint from S_i. This contradicts Lemma 6. Hence, we obtain $i = j$; accordingly, $U^*(H_i) \setminus U(S_i)$ is contained in $\widehat{V}_i$. □

Claim 6. $\widehat{V}_i \subseteq U^*(H_i) \setminus U(S_i)$.

Proof. Let $z \in \widehat{V}_i \setminus V(H_i)$. By Claim 4, there is an M-saturated path P between z and some vertex of S_i which is contained in $\widehat{V}_i$. Trace P from z and let w be the first vertex we encounter that is in $V(H_i)$. Since $V(H_i)$ is separating, zPw is an M-balanced path from z to w by Proposition 4. In $\widehat{D}_i/H_i$, zPw corresponds to an M-balanced path from z to the contracted vertex h, corresponding to H_i. Obviously, M contains a near-perfect matching of $\widehat{D}_i/H_i$ exposing only h.

Therefore, $\widehat{D}_i/H_i$ is factor-critical by Proposition 2; accordingly, $\widehat{V}_i$ is contained in $U^*(H_i)$. Additionally, by Claim 4 again and Lemma 3, we can see that $\widehat{V}_i$ is disjoint from $U(S_i)$ and that $\widehat{V}_i$ is contained in $U^*(H_i) \setminus U(S_i)$. □

Thus, by Claims 5 and 6, we have $\widehat{V}_i = U^*(H_i) \setminus U(S_i)$. □

Remark 1. If G in Theorem 8 is elementary, then Theorem 8 claims that $\mathcal{P}(G)$ is the family of (odd-)maximal barriers; namely, Theorem 8 coincides with Theorem 2. Therefore, Theorem 8 can be regarded as a generalization of Theorem 2.

Remark 2. Let G be a factorizable graph. For an arbitrary vertex $x \in V(G)$, take a maximal barrier of $G - x$, say X. Then, $X \cup \{x\}$ is a maximal barrier of G; namely, for any vertex x there is an odd-maximal barrier that contains x. Therefore, for any $S \in \mathcal{P}(G)$, there exists an odd-maximal barrier that contains S.

Remark 3. Let $A(G), D(G), C(G) \subseteq V(G)$ be those of so-called the Gallai-Edmonds structure theorem [2]. With Király [16], if G is a non-factorizable graph, then $\{A(G)\} \cup \mathcal{P}(G[C(G)])$ are the "atoms" that constitute odd-maximal barriers. For each odd-maximal barrier X, the odd components of $G - X$ are the components of $G[D(G)]$ and the odd components of $G[C(G)] - (X \setminus A(G))$; here $G[C(G)]$ forms a factorizable graph and $X \setminus A(G)$ is an odd-maximal barrier.

Acknowledgment. The author is grateful to Prof. Y. Oda and Prof. T. Sei for carefully reading the paper and giving useful comments.

References

1. Carvalho, M.H., Cheriyan, J.: An $O(VE)$ algorithm for ear decompositions of matching-covered graphs. ACM Transactions on Algorithms 1(2), 324–337 (2005)
2. Lovász, L., Plummer, M.D.: Matching Theory. AMS Chelsea Publishing (2009)
3. Carvalho, M.H., Lucchesi, C.L., Murty, U.S.R.: The matching lattice. In: Reed, B., Sales, C.L. (eds.) Recent Advances in Algorithms and Combinatorics. Springer (2003)
4. Kita, N.: A partially ordered structure and a generalization of the canonical partition for general graphs with perfect matchings. CoRR abs/1205.3816 (2012)
5. Kita, N.: A partially ordered structure and a generalization of the canonical partition for general graphs with perfect matchings. In: Chao, K.-M., Hsu, T.-S., Lee, D.-T. (eds.) ISAAC 2012. LNCS, vol. 7676, pp. 85–94. Springer, Heidelberg (2012)
6. Kita, N.: A generalization of the Dulmage-Mendelsohn decomposition for general graphs (preprint)
7. Frank, A.: Conservative weightings and ear-decompositions of graphs. Combinatorica 13(1), 65–81 (1993)
8. Schrijver, A.: Combinatorial Optimization: Polyhedra and Efficiency. Springer (2003)
9. Dulmage, A.L., Mendelsohn, N.S.: Coverings of bipartite graphs. Canadian Journal of Mathematics 10, 517–534 (1958)
10. Dulmage, A.L., Mendelsohn, N.S.: A structure theory of bipartite graphs of finite exterior dimension. Transactions of the Royal Society of Canada, Section III 53, 1–13 (1959)
11. Dulmage, A.L., Mendelsohn, N.S.: Two algorithms for bipartite graphs. Journal of the Society for Industrial and Applied Mathematics 11(1), 183–194 (1963)
12. Murota, K.: Matrices and matroids for systems analysis. Springer (2000)
13. Kotzig, A.: Z teórie konečných grafov s lineárnym faktorom. I. Mathematica Slovaca 9(2), 73–91 (1959) (in slovak)
14. Kotzig, A.: Z teórie konečných grafov s lineárnym faktorom. II. Mathematica Slovaca 9(3), 136–159 (1959) (in slovak)
15. Kotzig, A.: Z teórie konečných grafov s lineárnym faktorom. III. Mathematica Slovaca 10(4), 205–215 (1960) (in slovak)
16. Király, Z.: The calculus of barriers. Technical Report TR-9801-2, ELTE (1998)
17. Kita, N.: A canonical characterization of the family of barriers in general graphs. CoRR abs/1212.5960 (2012)

A Portable Parallel Implementation of the *lrs* Vertex Enumeration Code

David Avis[1] and Gary Roumanis[2]

[1] School of Informatics, Kyoto University, Kyoto, Japan and
School of Computer Science, McGill University, Montréal, Québec, Canada
avis@cs.mcgill.ca
[2] Microsoft, Seattle, USA
me@garyroumanis.com

Abstract. We describe a parallel implementation of the vertex enumeration code *lrs* that automatically exploits available hardware on multi-core computers and runs on a wide range of platforms. The implementation makes use of a C++ wrapper that essentially uses the existing *lrs* code with only minor modifications. This allows the simultaneous development of the existing single processor code with the speedups available from multi-core systems. It makes use of the restart feature of reverse search that allows for independent subtree search and the fact that no communication is required between these searches. As such it can be readily adapted for use in other reverse search enumeration codes.

Keywords: vertex enumeration, reverse search, parallel processing.

1 Introduction

Since its discovery in the 1990s the reverse search technique[3] [4] has been used to solve a large number of unstructured enumeration problems of which perhaps the most widely used is vertex enumeration using the *lrs* program [2]. From the outset it was realized that reverse search was imminently suitable for parallelization. The first such code, *prs* was developed by Ambros Marzetta using his ZRAM parallelization platform, as described in [6] and available online at [10]. In this case the parallelization was built into the *lrs* code itself leading to problems of maintenance and upgrading as newer parallel libraries developed. Another parallelization of reverse search was developed by Christophe Weibel for computing Minkowski sums [11]. This used a recursive version of reverse search where a backtrack stack is employed and some message passing is allowed during parallel execution. We discuss it further in Section 2.3.

The *lrs* code is rather complex and has been under development for over twenty years incorporating a multitude of different functions. It has been used extensively and basic functionality is very stable. Directly adding parallelization code to such legacy software is extremely delicate and can easily produce bugs that are difficult to find. The approach we use avoids this completely as the parallelization occurs in a separate layer. This allows independent development

P. Widmayer, Y. Xu, and B. Zhu (Eds.): COCOA 2013, LNCS 8287, pp. 414–429, 2013.

of both parallelization ideas and basic improvements in the underlying code. Parallelization is obtained by using the built in restart features of *lrs* with a completely separate multi-thread scheduler. The concept was tested by a shell script, *tlrs* developed by John White in 2009. Here the parallelization is achieved by scheduling independent processes for subtrees via the shell. Although good speedups were obtained several limitations of this approach materialized as the number of processors available increased. In particular job control becomes a major issue: there is no single controlling process. A strong point of the approach used in *tlrs* was that no modification of the underlying *lrs* code was required.

The approach we describe here lies somewhere between the approaches of Marzetta and White. We built a C++ wrapper that compiles in the original *lrslib* library essentially maintaining the integrity of the underlying *lrs* code. The parallelization is achieved by multithreading using an initial bounded depth run of *lrs* and and an additional process is used to concatenate the output streams. Job control is easily available since one process is in charge of all threads. Furthermore the development of the parallelization techniques can proceed independently of the original *lrs* code itself.

The paper is organized as follows. In the next section we begin with background on reverse search and explain the simple modifications necessary to prepare for parallelization. We use the example of generating permutations as an illustration. We then give a high level description of the parallelization technique illustrating on the permutation example. In Section 3 we describe the vertex enumeration problem and some of the properties that may potentially limit parallel speedups. In Section 4 we describe the wrapper constructed to schedule the parallel *lrs* executions, detailing various design decisions taken. Section 5 gives numerical experiments on a wide variety of polyhedra with bench marks against the standard solvers *cddr+* [8] and *lrs* . We conclude with some observations and directions for improving the parallelization performance.

2   Background

2.1   Reverse Search

Reverse search is a technique for generating large, relatively unstructured, sets of discrete objects. We give an outline of the method here referring the reader to [3] [4] for further details.

In its most basic form, it can be viewed as the traversal of a spanning tree, called the reverse search tree T, of a graph $G = (V, E)$ whose nodes are the objects to be generated. Edges in the graph are specified by an adjacency oracle, and the subset of edges of the reverse search tree are determined by an auxiliary function, which can be thought of as a local search function f for an optimization problem defined on the set of objects to be generated. One vertex, v^*, is designated as the *target* vertex. For every other vertex $v \in V$ repeated application of f must generate a path in G from v to v^*. The set of these paths defines the reverse search tree T, which has root v^*.

A reverse search is initiated at v^*, and only edges of the reverse search tree are traversed. When a node is visited the corresponding object is output. Since there is no possibility of visiting a node by different paths, the nodes are not stored. Backtracking can be performed in the standard way using a stack, but this is not required as the local search function can be used for this purpose. This means that it is not necessary to keep more than one node of the tree at any given time, and this memoryless property is the main feature of reverse search. For a given problem, there may be many choices of adjacency oracle and local search function.

However, in the basic setting described here, a few properties are required. Firstly, the underlying graph G must be connected and an upper bound on the maximum vertex degree, Δ, must be known. The performance of the method depends on G having Δ as low as possible. The adjacency oracle must be capable of generating the adjacent vertices of some given vertex v sequentially and without repetition. This is done by specifying a function $Adj(v, j)$, where v is a vertex of G and $j = 1, 2, ..., \Delta$. Each value of $Adj(v, j)$ is either a vertex adjacent to v or null. Each vertex adjacent to v appears precisely once as j ranges over its possible values. For each vertex $v \neq v^*$ the local search function $f(v)$ returns the tuple (u, j) where $v = Adj(u, j)$ such that u is v's parent in T. The algorithm is shown in Algorithm 1. The order that the vertices are output is called the *reverse search order*. For convenience later, we do not output the root vertex v^*.

Algorithm 1. Generic Reverse Search

```
 1: procedure RS(v*, Δ, Adj, f)
 2:     v ← v*  j ← 0
 3:     repeat
 4:         while j < Δ do
 5:             j ← j + 1
 6:             if f(Adj(v, j)) = v then                    ▷ forward step
 7:                 v ← Adj(v, j)
 8:                 output v
 9:                 j ← 0
10:             end if
11:         end while
12:         if v ≠ v* then                                 ▷ backtrack step
13:             (v, j) ← f(v)
14:         end if
15:     until v = v* and j = Δ
16: end procedure
```

These ideas can be illustrated on a simple example: generating all permutations of a set of integers. Here the goal is to generate all permutations of the integers $\{1, 2, ..., n\}$. The underlying graph $G_n = (V, E)$ is defined as follows. The vertices are n-tuples, $v = v_1 v_2 ... v_n$, representing the $n!$ permutations of the n integers. We set the target $v^* = (12...n)$. The adjacency oracle simply interchanges two consecutive integers in a permutation, and is given by

$$Adj(v, i) = (v_1 v_2 ... v_{i-1} v_{i+1} v_i ... v_n) \qquad i = 1, 2, ..., n - 1.$$

So G_n is regular of degree $\Delta = n-1$. Finally we set the local search function f to interchange the first two consecutive integers that are out of order numerically:

$$f(v) = (v_1v_2...v_{i-1}v_{i+1}v_i...v_n) \quad \text{for the smallest } i \text{ s.t. } v_i > v_{i+1}.$$

Figure 1 shows G_4 which has 24=4! vertices and is regular of degree 3. The edges chosen by the local search function f are shown with arrows directed towards the root 1234. So for example starting at vertex 4231 f generates the path

$$4231 \mapsto 2431 \mapsto 2314 \mapsto 2134 \mapsto 1234.$$

The set of all arcs with arrows defines the reverse search tree T.

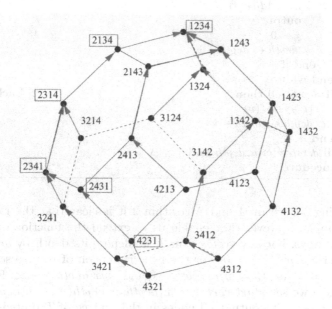

Fig. 1. Permutahedron: $n = 4$

2.2   Extended Reverse Search

To achieve parallelization of Algorithm 1 we make use of the lack of memory property that allows it to be restarted from any node in the reverse search tree T. After a restart, all remaining nodes of T will be generated. We adapt this to allow for a subtree to be enumerated from its given root.

When calling the reverse search procedure we now supply four additional parameters:

- *start_vertex* is vertex from which the reverse search should be initiated and replaces v^*
- *depth* is initially the depth in T of *start_vertex* and will be updated to be the depth in T of the vertex v currently being considered in the search

- *max_depth* is the depth at which forward steps are terminated
- *min_depth* is the depth at which backtrack steps are terminated

The modified algorithm is shown in Algorithm 2.

Algorithm 2. Extended Reverse Search

1: **procedure** RS2($start\_vertex$, Δ, Adj, f, $depth$, max_depth, min_depth)
2:      $j \leftarrow 0$ $v \leftarrow start\_vertex$
3:      **repeat**
4:         **while** $j < \Delta$ and $depth < max\_depth$ **do**
5:            $j \leftarrow j + 1$
6:            **if** $f(Adj(v,j)) = v$ **then**   ▷ forward step
7:               $v \leftarrow Adj(v,j)$
8:               **output** v
9:               $j \leftarrow 0$
10:               $depth \leftarrow depth + 1$
11:            **end if**
12:         **end while**
13:         **if** $depth > 0$ **then**   ▷ backtrack step
14:            $(v,j) \leftarrow f(v)$
15:            $depth \leftarrow depth - 1$
16:         **end if**
17:      **until** $depth = min\_depth$ and $j = \Delta$
18: **end procedure**

Comparing Algorithm 1 and Algorithm 2 it is clear that the modifications are very simple. However they enable us to extend the function of Algorithm 1 in several ways. For any vertex v in T we denote its depth by $depth(v)$. Initially we have $depth(v^*) = 0$. For the generic version of reverse search we set $start\_vertex = v^*, depth = min\_depth = 0$ and $max\_depth = +\infty$. For a restart from vertex v we set $start\_vertex = v, depth = depth(v), min\_depth = 0$ and $max\_depth = +\infty$. To output all nodes in the subtree of T rooted at v we set $start\_vertex = v, depth = depth(v), min\_depth = depth(v)$ and $max\_depth = +\infty$. To initialize the parallelization process we will generate the tree T down to a fixed depth k by setting $start\_vertex = v^*, depth = min\_depth = 0$ and $max\_depth = k$.

Returning to the example in Figure 1 we could do a restart from $v = 2143$ with $depth = 2$ obtaining the output 2413 4213 1423 4123. To list all nodes in the subtree rooted at v we would in addition set $min\_depth = 2$ producing the output 2413 4213. To do a partial enumeration down to $depth = 2$ we would set $start\_vertex = 1234, depth = min\_depth = 0, max\_depth = 2$ generating the output 2134 2314 1324 3124 1342 1243 2143 1423.

2.3 Parallelization

In this subsection we describe how the extended reverse search algorithm can be parallelized without requiring further modification. We give a rather generic description of the parallelization which is by nature somewhat oversimplified. The details of the actual implementation with the *lrs* program will be given in Section 4.

We proceed in three phases. In the first phase we generate the reverse search tree T down to a fixed depth *init_depth*. Rather than output the nodes of the tree, we store them in a list L. In the second phase we schedule threads in parallel from L using the subtree enumeration feature. For this we require the parameter *max_threads* giving the maximum number of parallel threads to that can run at the same time. We will also control where the output stream is sent. In Phase 1 it will be directed to the list L. From L all vertices that have depth less than *init_depth* are removed and output. In Phase 2 we schedule threads from the nodes in L up to the number specified by *max_threads*. Each thread uses *lrs* to enumerate the subtree rooted at the node removed for it from L. When the list L becomes empty we move to Phase 3 in which the threads terminate one by one until there are no more running and the procedure terminates. We make use of a *collection process* which concatenates the output from the threads into a single output stream. The procedure is outlined in Algorithm 3. It is clear from the pseudocode the only interaction between the parallel threads is the common output collection process. The only signalling required is when a thread terminates. Let us return to the example in Figure 1. Suppose we set the *init_depth* = 2 and *max_threads* = 3. We initiate the computation with the call

$$PRS(1234, 3, Adj, f, 2, 3)$$

Algorithm 3. Parallel Reverse Search

1: **procedure** PRS(*start_vertex*, Δ, *Adj*, *f*, *init_depth*, *max_threads*)
2:     *num_threads* $\leftarrow$ 0
3:     **redirect output** to a list L ▷ Phase 1
4:     RS2(*start_vertex*, Δ, *Adj*, *f*, 0, *init_depth*, 0)
5:     **redirect output** to **collection process**
6:     **remove all** $v \in L$ with *depth*(v) < *init_depth* and **output**(v)
7:     **while** *num_threads* < *max_threads* and $L \neq \emptyset$ **do**   ▷ Phase 2
8:         **remove any** $v \in L$
9:         RS2(v, Δ, *Adj*, *f*, *depth*(v), ∞, *depth*(v))
10:         *num_threads* $\leftarrow$ *num_threads* + 1
11:     **end while**
12:     **while** *num_threads* > 0 **do**
13:         **wait** until a termination signal is received
14:         **if** $L \neq \emptyset$ **then**
15:             **remove any** $v \in L$
16:             RS2(v, Δ, *Adj*, *f*, *depth*(v), ∞, *depth*(v))
17:         **else**   ▷ Phase 3
18:             *num_threads* $\leftarrow$ *num_threads* - 1
19:         **end if**
20:     **end while**
21: **end procedure**

This will generate the output list

$$L = \{2134\ 2314\ 1324\ 3124\ 1342\ 1243\ 2143\ 1423\}$$

in line 3. In line 6 we remove and output 2134 1324 1243 which have $depth < 2$ leaving
$$L = \{2314\ 3124\ 1342\ 2143\ 1423\}$$

which are at $depth = 2$. We assume L is processed in left to right order. In lines 8-10 we initiate three calls to RS2:

$$RS2(2314, 3, Adj, f, 2, \infty, 2),\ RS2(3124, 3, Adj, f, 2, \infty, 2),\ \text{and}\ RS2(1342, 3, Adj, f, 2, \infty, 2).$$

After each of the first two threads terminate, in lines 15-16, two further calls are made:

$$RS2(2143, 3, Adj, f, 2, \infty, 2)\ \text{and}\ RS2(1423, 3, Adj, f, 2, \infty, 2).$$

Then $L = \emptyset$ and each subsequent termination decrements $num\_threads$ until all threads have completed.

In analyzing Algorithm 3 we observe that in Phase 1 there is no parallelization, in Phase 2 all available cores are used, and in Phase 3 the level of parallelization drops monotonically as threads terminate. Looking at the overhead compared with Algorithm 1 we see that this almost entirely consists of the amount of time required to restart the reverse search process. This leads to conflicting issues in setting the critical $init\_depth$ parameter. A larger value implies that:

- only a single thread is working for a longer time
- the list L will be typically be larger requiring more overhead in restarts, but
- the time spent in Phase 3 will typically be reduced.

The success in parallelization clearly depends on the structure of the tree T. In the worst case it is a path and no parallelization occurs in Phase 2. Therefore in the worst case we have no improvement in complexity over that reported by Avis and Fukuda [3] for basic reverse search. In the best case the tree is balanced so that the list L can be short reducing overhead and all threads terminate at more or less the same time. Success therefore heavily depends on the structure of the underlying enumeration problem.

For the vertex enumeration problem, discussed in the next section, both of these extremes and everything in between is possible. We will see experimental results to illustrate this in Section 5.

We conclude by comparing the method of Algorithm 3 with that used by Weibel [11] for computing Minkowski sums by reverse search. The latter method uses a more sophisticated approach. Firstly the search is recursive so that all nodes are stored in the backtrack path. As we noted, for vertex enumeration it is not possible in general to keep a full backtrack stack since it may contain all of the LP dictionaries and exhaust memory. An approximation to this included in the original *lrs* code, which employs a user specified parameter k and caches the last k nodes of the backtrack stack. In this way memory is not exhausted and the number of cache misses is usually rather low.

Secondly the rather than executing a distinct Phase 1, in Weibel's method a given process is designated the *boss* and can either execute normally or spin off nodes to other threads to be executed in parallel. When the boss runs out of work another node is designated to be the boss and messages are sent to inform all other nodes.

Computational experience is given for up to 8 parallel processors with reported speedups of 5.5 to 8 times.

3 Vertex Enumeration

3.1 Reverse Search Vertex Enumeration Method

The initial application of reverse search was to the vertex enumeration problem [3]. From this paper the *lrs* program was derived and a full description of its implementation is given in [1]. We give a simplified description here.

Given an $m \times n$ matrix $A = (a_{ij})$ and an m dimensional vector b, a *convex polyhedron*, or simply *polyhedron*, P is defined as:

$$P = \{x \in R^n : b + Ax \geq 0\}.$$

A *polytope* is a bounded polyhedron. For simplicity in this description we will assume that we are dealing input data A, b that define full dimensional polytopes. A point $x \in P$ is a *vertex* of P iff it is the unique solution to a subset of n inequalities solved as equations. The *vertex enumeration problem* is to output all vertices of a polytope P. Figure 2 shows a typical input which defines the polytope P sketched in Figure 3 with 5 vertices.

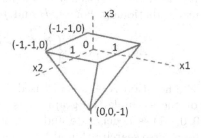

$$A = \begin{bmatrix} -1 & 0 & 1 \\ 0 & -1 & 1 \\ 1 & 0 & 1 \\ 0 & 1 & 1 \\ 0 & 0 & -1 \end{bmatrix} \quad b = \begin{bmatrix} 1 \\ 1 \\ 1 \\ 1 \\ 0 \end{bmatrix} \quad \begin{array}{l} 1 - x_1 + x_3 \geq 0 \\ 1 - x_2 + x_3 \geq 0 \\ 1 + x_1 + x_3 \geq 0 \\ 1 + x_2 + x_3 \geq 0 \\ -x_3 \geq 0 \end{array}$$

Fig. 2. A, b and its polyhedron P **Fig. 3.** P has 5 vertices

The computations are based on *dictionaries*, as is done for the simplex method of linear programming. To get a dictionary for $P = \{x \in R^n : b + Ax \geq 0\}$ we add one new nonnegative variable for each inequality:

$$x_{n+i} = b_i + \sum_{j=1}^{n} a_{ij} x_j, \quad x_{n+i} \geq 0 \quad i = 1, 2, ..., m.$$

These new variables are called *slack variables* and the original variables are called *decision variables*.

In order to have any vertex at all we must have $m \geq n$, and normally m is significantly larger than n, allowing us to solve the equations for various sets of variables on the left hand side. The variables on the left hand side of a dictionary are called *basic*, and those on the right hand side are called *non-basic* or, equivalently, *co-basic*. We use the notation $B = \{i : x_i \text{ is basic}\}$ and $N = \{j : x_j \text{ is co-basic}\}$.

A *pivot* interchanges one index from B and N and solves the equations for the new basic variables. A *basic solution* from a dictionary is obtained by setting $x_j = 0$ for all $j \in N$. It is a *basic feasible solution(BFS)* if $x_j \geq 0$ for every slack variable x_j. A dictionary is called *degenerate* if it has a slack basic variable $x_j = 0$. As is well known, each BFS defines a vertex of P and each vertex of P can be represented as one or more (in the case of degeneracy) BFSs. For the example a typical BFS and dictionary are shown in Figure 4.

$$\begin{aligned}
x_1 &= -1 + x_6 + x_8 \geq 0 \\
x_2 &= -1 + x_7 + x_8 \geq 0 \\
x_3 &= -x_8 \geq 0 \\
x_4 &= 2 - x_6 + x_8 \geq 0 \\
x_5 &= -2 - x_7 - 2x_8 \geq 0
\end{aligned}$$

Fig. 4. Decision variables are all basic, $N = \{6, 7, 8\}$

To apply reverse search to this problem we first define the relevant graph $G = (V, E)$. Each node in V corresponds to a BFS and is labelled with the cobasic set N. Each edge in E corresponds to a pivot between two BFSs. Formally we may define the adjacency oracle as follows. Let B and N be index sets for the current dictionary. For $i \in B$ and $j \in N$

$$Adj(N, i, j) = \begin{cases} N - j + i & \text{if this gives a feasible dictionary} \\ \emptyset & \text{otherwise} \end{cases}$$

(The notation $N - j + i$ is used as a convenient shorthand for $N \setminus \{j\} \cup \{i\}$.) For the example the graph G is shown in Figure 5. Observe that the vertex $(0, 0, -1)$ is degenerate and is represented by four cobases. The target v^* for the reverse search is found by solving a linear program over this dictionary with any objective function $z = c^T x$ that defines a unique optimum vertex. We use the objective function z and a non-cycling pivot selection rule to define the local search function f. In the case of *lrs* we use Bland's least subscript rule for selecting the variable which enters the basis and a lexicographic ratio test to select the leaving variable. This lexicographic rule simulates a simple polytope which greatly reduces degeneracy. In the example only two of the four bases defining vertex $(0, 0, -1)$ would be generated. For details see [1].

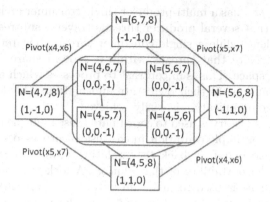

Fig. 5. The graph of feasible dictionaries for Figure 2

3.2 Parallelization Issues

In this subsection we discuss issues affecting how successful we can expect Algorithm 3 to be when applied to vertex enumeration. Referring back to the analysis at the end of Section 2.3 we recall the worst case is when the reverse search tree T is a path, as no parallelization is achieved. This in fact can happen!

The Klee-Minty examples [9], and their relatives, are specially constructed polytopes so that the simplex method with a given pivot rule will follow a Hamiltonian path on the polytope's skeleton. This is precisely the case when no parallelization occurs. This creates no problem for single processor codes, as the tree shape is largely irrelevant. Examples of various types of polytopes are given in Section 5.

4 Implementation Description

As discussed earlier, the first attempt at parallelization was unfortunately ineffective, unportable and most importantly unmanageable. The approach essentially used POSIX threads to initiate a system call of lrs on subtrees. Concatenating the output to a standard location proved difficult as interprocess communication is not easily achieved. To circumvent this issue, temporary files were used to store output data. This was inefficient with regards to memory requirements for a given problem. These short comings were carefully reviewed when the parallelization problem was attacked for a second time.

In the second approach, several open source, multi-threading libraries were considered. It was decided that the C++ Boost library offered the greatest performance, adaptability and maintainability. Moreover, Boost works on almost any modern operating system, including UNIX and Windows variants; ensuring the portability of the final solution. Although C++ is not a strict superset of C, the language provides mechanisms for mixing code that is compiled by compatible C and C++ compilers. This allowed us to create a lightweight C++ wrapper around the lrs codebase using the g++ compiler.

On a high level, *plrs* has a multi-producer single consumer architecture. What this equates to is that several producer threads traverse subtrees of the vertex enumeration problem, while a single consumer thread concatenates output to a unified location. Note that threads within a process share the same state, and same memory space. This is in contrast to processes which are independent execution units that contain their own state information. This leads to the fact that inter-thread communication is easily achieved.

The Boost.Atomic library is used to coordinate these multiple threads through atomic variables. The implementation makes use of processor-specific instructions where possible and falls back to emulating atomic operations through locking; ensuring the portability of the solution. A lock-free multiple producer single consumer queue is used to maintain output. The specific function compare_exchange_weak is used to post output from the producer threads to the single consumer thread. For more details, please visit the Boost.org web site [5].

The *boost::thread* class is responsible for launching the consumer thread while the *boost::thread_group* class is used for launching and managing all producer threads. In order to wait for the execution of all producer threads to finish, the *join_all()* member function is used. Essentially, this blocks the main process thread from completing until vertex enumeration has exhausted. A similar function, *join()*, is used on the consumer thread to ensure all output is captured before the completion of the entire process.

5   Numerical Experiments

We describe here some experimental results using the plrs code on two computers, *mai12* [1] and *mai64* [2] with respectively 12 and 64 cores and similar processor speeds. When we performed the experiments, the machines were idle except for systems functions.

Using the *top* command we can measure the amount of work each machine is doing. With no work, the load average is zero and with k threads running *lrs* the load average is very close to k. We initially benchmarked the two computers by running an *lrs* job (single thread) when the computers were idle and then with increasing load averages. On *mai12* , with a load average of 12, the time for the *lrs* run was essentially the same as with a load average of one, as one would wish for in an ideally constructed parallel computer. In a similar test on *mai64* the performance deteriorated noticeably with high load averages. At load averages of 32, 48, and 64 on *mai64* the *lrs* times were respectively 1.15, 1.41 and 1.46 times longer then with a load average of one. We therefore restricted tests to a maximum of 32 threads on this machine and, even so, these results probably underestimate the speedup by about 15% compared to a machine such as *mai12* which performs close to the ideal.

We chose a few representative polyhedra that are shown in Table 1. For each example we first give the input file name, type (H or V-representation) and input

---

[1] Xeon X5640, 2.66GHz, 12 core, 24GB memory, 60GB hard drive.

[2] Opteron 16core 6272 X 4, 2.1GHz, 64 core, 64GB memory, 500GB hard drive.

Table 1. Polyhedra tested: *lrs* ,*cddr+* times on *mai12*

Name	Input			Output		*lrs*			cddr+
	H/V	m	n	V/H	size	bases	depth	secs	secs
mit	H	729	9	4861	196K	1375608	101	809	505
bv7	H	69	57	5040	867K	84707280	17	11851	
perm7	H	127	8	5040	127K	5040	21	0.6	15.0
c30-15	V	30	16	341088	73.8M	319770	14	80	4652
perm10	H	1023	11	3628800	127M	3628800	45	3193	
c40-20	V	40	21	40060020	15.6G	20030010	19	22458	

dimensions (m rows and n columns). We then give the output size (number of vertices or facets, respectively) and space, which ranges from 127K to an enormous 15.6G. For *lrs* we give the number of bases generated, the maximum tree depth and running time in seconds. The *cddr+* times were obtained using the default settings and are only given to emphasize the difference between pivoting and double description methods. No attempt was made to optimize the settings. No value implies that the *cddr+* run did not terminate with 48 hours. Input files are available from the web site [2].

The polytope *mit* is a configuration polytope which required about a month of computer time for its vertex enumeration by *cdd* and *lrs* when first run in 1993 [7]. It is a rather degenerate polytope. *c40-20* is a cyclic polytope and is simple, ie, non-degenerate. *perm7* and *perm10* are permutation polytopes written in their standard formulations, and are also simple polytopes. The vertices of *perm_n* are the $n!$ permutations of $1, 2, ..., n$. The standard formulation using n variables has $2^n - 2$ inequalities and one linearity. *bv7* is an alternative formulation that has polynomial size in n as it is based on the Birkhoff-Von Neumann polytope. It has n^2 inequalities and $3n - 1$ linearities in $n^2 + n$ variables. We included *perm7* only for comparison purposes with *bv7* and do not use it in parallelization experiments.

Table 2. Times and speedups (su): no. of threads =mt, initial depth=id (*mai12*)

Name	*lrs*	mt = 4		mt=8		mt=12	
	secs	secs	su	secs	su	secs	su
		L	id	L	id	L	id
mit	809	232	3.5	142	5.7	104	7.8
		284	4	613	5	1213	6
bv7	11851	3117	3.8	1580	7.5	1104	10.7
		645	2	645	2	7554	3
c30-15	80	27	3.0	15	5.3	12	6.7
		1716	6	1716	6	1716	6
perm10	3193	983	3.2	517	6.2	421	7.6
		4489	7	4489	7	4489	7
c40-20	22458	9633	2.3	5600	4.0	3697	6.1
		220	3	715	4	2002	5

Table 3. Times and speedups (su): no. of threads =mt, initial depth=id (*mai64*)

Name	lrs secs	mt = 4 secs	su	mt=8 secs	su	mt=16 secs	su	mt=32 secs	su
		L	id	L	id	L	id	L	id
mit	1125	339	3.3	190	5.9	123	9.1	110	10.2
		284	4	613	5	1213	6	2121	7
bv7	17381	4513	3.85	2345	7.4	1215	14.3	707	24.5
		645	2	645	2	7554	3	7554	3
c30-15	75	34	2.2	22	3.4	20	3.8	21	3.6
		1716	6	1716	6	1716	6	1716	6
perm10	4295	1317	3.3	683	6.3	566	7.6	570	7.6
		4489	7	4489	7	4489	7	4489	7
c40-20	17538	9802	1.8	6707	2.6	4902	3.6	4106	4.3
		220	3	715	4	2002	5	5005	6

In Tables 2 and 3 we present speedup results for *plrs* runs on *mai12* and *mai64* respectively for the problems presented in Table 1. The initial depth parameter was chosen fairly arbitrarily to give a reasonable size list L of problems to solve in parallel. On *mai12* we observe that the speedups are roughly comparable except for the last problem, *c40-20*, which are considerably smaller. As remarked, it has huge output size. On *mai64* in addition *c30-15* shows very small speedups as the number of threads increases. Note that it has short running time relative to its output size. On both machines the speedups are largest for the highly degenerate *bv7* which generates very little output. Together this is evidence that the collection process may be the bottleneck in these cases.

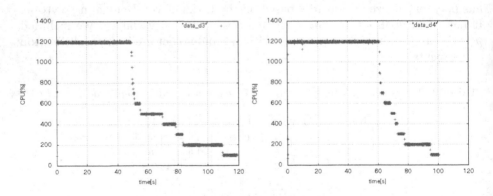

Fig. 6. *mit*: mt=12, id=3, *mai12*   **Fig. 7.** *mit*: mt=12, id=4, *mai12*

The only user parameter for *plrs* is the initial depth parameter. If this parameter is too low the list of jobs L may be too short to provide adequate parallelism. On the other hand if it is too large a relatively large amount of time will be spent in phase 1 using only one thread, and also in restarting each job in L during

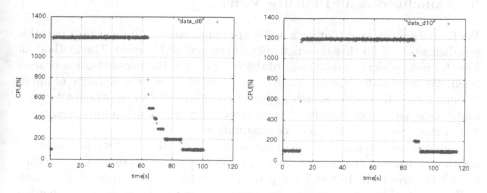

Fig. 8. *mit*: mt=12, id=6, *mai12*   **Fig. 9.** *mit*: mt=12, id=10, *mai12*

phase 2. This is illustrated in Figures 6 to 9. These show the load average during runs of *plrs* on *mit* using 12 cores and depths 3,4,6, 10 respectively. One can clearly see the three phases of execution: a short start up with one core, a period with all 12 cores active while L is depleted, then the final phase as each process terminates. The area under the graph is the total execution time. Depths of 4 and 6 achieve the fastest elapsed time, but total execution time is higher at depth 6. With depth 10 the first two phases are longer, the third phase shorter, the total execution time longer. However the elapsed time to complete the run is about 20% longer than at depth 4.

In Table 4 we show the dependence on the initial depth of speedup results for *plrs* on four of the polytopes. Although there are differences in performance they are less important than we expected. Given the previous discussion, one would expect increasing speedups as the depth increases from a small value to a minimum then decreasing speedups as the depth increases. Although this is somewhat observed in the data there are obviously other competing factors at play and the situation is more complicated.

Table 4. Comparison of various init_depth(id) values with max_threads(mt)=12 (*mai12*)

Name	*lrs*	id = 2		id = 3		id=4		id=5		id=6	
	secs	L	secs	L	secs	L	secs	L	secs	L	secs
mit	809	35	183	115	127	284	102	613	104	1213	108
bv7	11851	645	1144	7554	1104	60966	1126	349984	1499	-	-
c30-15	80	36	34	120	23	330	17	792	13	1716	12
perm10	3193	44	405	155	457	440	507	1068	530	2298	551

6   Conclusions and Future Work

We have demonstrated that very useful speedups can be obtained by a portable parallelization of *lrs* that does not disturb the underlying code. The method allows for independent development of the *lrs* code and the parallelization process itself. We expect that similar results can be obtained by a wide range of applications using the reverse search approach. The installation is straight forward and no special purpose hardware is required. Very noticeable improvements are found using just quad-core personal computers.

Figure 7 shows the limitations of our approach. For the polytope *mit* an initial depth of 4 achieves the shortest elapsed time. However for only 60% of the time are all 12 cores busy. In fact for a quarter of the time only two cores are active.

To remedy this one can imagine interrupting long running tasks and then using plrs recursively to split them into subproblems, repopulating *L*. We performed some preliminary experiments along these lines, but the results were mixed. As this increases overhead, the final result may sometimes be worse, depending on the search tree shape. It is a fruitful area for future research. Another possibility is to use the built in estimator function of *lrs* . For each leaf obtained in phase 1 it is possible to get an unbiased estimate of the size of the subtree that it roots by using a random probe. One could then schedule jobs from L using a list decreasing heuristic, so that longer runs are done first. The trade off is again overhead: the random probes may require a lot of processing time if the tree is unbalanced.

Acknowledgments. The authors would like to thank Kenji Okuda for preparing Figures 6 to 9. The research was supported by grants from NSERC and JSPS.

References

1. Avis, D.: lrs: A Revised Implementation of the Reverse Search Vertex Enumeration Algorithm. In: Kalai, G., Ziegler, G. (eds.) Polytopes - Combinatorics and Computation, pp. 177–198. Springer (2000)
2. Avis, D.: (2013), http://cgm.cs.mcgill.ca/~avis/C/lrs.html
3. Avis, D., Fukuda, K.: A pivoting algorithm for convex hulls and vertex enumeration of arrangements and polyhedra. Discrete & Computational Geometry 8, 295–313 (1992)
4. Avis, D., Fukuda, K.: Reverse search for enumeration. Discrete Applied Mathematics 65, 21–46 (1993)
5. Boost.org (2013), http://www.boost.org/doc/libs/1_53_0/doc/html/lockfree.html
6. Brungger, A., Marzetta, A., Fukuda, K., Nievergelt, J.: The parallel search bench ZRAM and its applications. Ann. Oper. Res. 90, 45–63 (1999)
7. Ceder, G., Garbulsky, G., Avis, D., Fukuda, K.: Ground states of a ternary fcc lattice model with nearest- and next-nearest-neighbor interactions. Phys. Rev. B Condens. Matter 49(1), 1–7 (1994)

8. Fukuda, K.: (2012), http://www.inf.ethz.ch/personal/fukudak/cdd_home
9. Klee, V., Minty, G.J.: How Good is the Simplex Algorithm? In: Shisha, O. (ed.) Inequalities III, pp. 159–175. Academic Press Inc., New York (1972)
10. Marzetta, A.: (2008) maintained by D. Bremner: http://www.cs.unb.ca/~bremner/software/zram/
11. Weibel, C.: Implementation and parallelization of a reverse-search algorithm for minkowski sums. In: ALENEX, pp. 34–42 (2010)

Author Index